ADVANCES IN EXPERIMENTAL MEDICINE AND BIOLOGY
VOLUME 520

Cytokines and Chemokines in Autoimmune Disease

Pere Santamaria, M.D. Ph.D.
University of Calgary
Microbiology and Infectious Diseases
Associate Professor
Health Sciences Centre
Calgary, Alberta, Canada

SPRINGER SCIENCE+BUSINESS MEDIA, LLC

Cytokines and Chemokines in Autoimmune Disease

Advances in Experimental Medicine and Biology Volume 520

Originally published by Kluwer Academic/Plenum Publishers in 2003
Softcover reprint of the hardcover 1st edition 2003

Please address all inquiries to Landes Bioscience / Eurekah.com:
Landes Bioscience / Eurekah.com, 810 South Church Street, Georgetown, Texas, U.S.A. 78626
Phone: 512/ 863 7762; FAX: 512/ 863 0081; www.Eurekah.com; www.landesbioscience.com
Landes tracking number: 1-58706-088-4

Cytokines and Chemokines in Autoimmune Disease, edited by Pere Santamaria, Landes / Kluwer dual imprint/ Advances in Experimental Medicine and Biology Volume 520,
ISBN 978-0-306-47693-8 ISBN 978-1-4615-0171-8 (eBook)
DOI 10.1007/978-1-4615-0171-8

Library of Congress Cataloging-in-Publication Data

CIP applied for but not received at time of publication.

ADVANCES IN EXPERIMENTAL MEDICINE AND BIOLOGY

Recent Volumes in this Series

Volume 506
LACRIMAL GLAND, TEAR FILM, AND DRY EYE SYNDROME 3: BASIC SCIENCE AND CLINICAL RELEVANCE
Edited by David A. Sullivan, Michael E. Stern, Kazuo Tsubota, Darlene A. Dartt, Rose M. Sullivan, and B. Britt Bromberg

Volume 507
EICOSANOIDS AND OTHER BIOACTIVE LIPIDS IN CANCER, INFLAMMATION, AND RADIATION INJURY 5
Edited by Kenneth V. Honn, Lawrence J. Marnett, Santosh Nigam, and Charles Serhan

Volume 508
SENSORIMOTOR CONTROL OF MOVEMENT AND POSTURE
Edited by Simon C. Gandevia, Uwe Proske and Douglas G. Stuart

Volume 509
IRON CHELATION THERAPY
Edited by Chiam Hershko

Volume 510
OXYGEN TRANSPORT TO TISSUE XXIII: OXYGEN MEASUREMENTS IN THE 21ST CENTURY: BASIC TECHNIQUES AND CLINICAL RELEVANCE
Edited by David F. Wilson, John Biaglow and Anna Pastuszko

Volume 511
PEDIATRIC GENDER ASSIGNMENT: A CRITICAL REAPPRAISAL
Edited by Stephen A. Zderic, Douglas A. Canning, Michael C. Carr and Howard McC. Snyder III

Volume 512
LYMPHOCYTE ACTIVATION AND IMMUNE REGULATION IX: LYMPHOCYTE TRAFFIC AND HOMEOSTASIS
Edited by Sudhir Gupta, Eugene Butcher and William Paul

Volume 513
MOLECULAR AND CELLULAR BIOLOGY OF NEUROPROTECTION IN THE CNS
Edited by Christian Alzheimer

Volume 514
PHOTORECEPTORS AND CALCIUM
Edited by Wolfgang Baehr and Krzysztof Palczewski

Volume 515
NEUROPILIN: FROM NERVOUS SYSTEM TO VASCULAR AND TUMOR BIOLOGY
Edited by Dominique Bagnard

Volume 516
TRIPLE REPEAT DISORDERS OF THE NERVOUS SYSTEM
Edited by Lubov T. Timchenko

Volume 517
DOPAMINERGIC NEURON TRANSPLANTATION IN THE WEAVER MOUSE MODEL OF PARKINSON'S DISEASE
Edited by Lazaros C. Triarhou

Volume 518
ADVANCES IN MALE MEDIATED DEVELOPMENTAL TOXICITY
Edited by Bernard Robaire and Barbara F. Hales

Volume 519
POLYMER DRUGS IN THE CLINICAL STAGE
Edited by Maeda, et al.

Volume 520
CYTOKINES AND CHEMOKINES IN AUTOIMMUNE DISEASE
Edited by Pere Santamaria

Volume 521
IMMUNE MECHANISMS OF PAIN AND ANALGESIA
Edited by Halina Machelska and Christoph Stein

A Continuation Order Plan is available for this series. A continuation order will bring delivery of each new volume immediately upon publication. Volumes are billed only upon actual shipment. For further information please contact the publisher.

DEDICATION

To Joan and Josefa, my parents, for their boundless love and devotion, and to Chus, my wife, for her unwavering support.

CONTENTS

EDITOR

Pere Santamaria, M.D., Ph.D.
University of Calgary
Microbiology and Infectious Diseases
Associate Professor
Health Sciences Centre
Calgary, Alberta, Canada
Chapter 1

CONTRIBUTORS

Guillermo A. Arreaza
The Robarts Research Institute
and University of Western Ontario
Department of Microbiology
and Immunology, and Medicine
London, Ontario, Canada
Chapter 9

Alicia Babcock
McGill University
Montreal Neurology Institute
Montreal, Quebec, Canada
Chapter 8

Mark J. Cameron
The Robarts Research Institute
and University of Western Ontario
Department of Microbiology
and Immunology, and Medicine
London, Ontario, Canada
Chapter 2, 9

Yiguang Chen
University of Pennsylvania
Philadelphia, Pennsylvania, U.S.A.
Chapter 4

Youhai Chen
University of Pennsylvania
Philadelphia, Pennsylvania, U.S.A.
Chapter 4

Mark M.W. Chong
The Walter and Eliza Hall Institute
Burnet Clinical Research Unit
Parkville, Victoria, Australia
Chapter 5

Urs Christen
The Scripts Research Institute
Division of Virology
La Jolla, California, U.S.A.
Chapter 12

Helene Coppin
Laboratorie d'immunologie Cellulaire
INSERM CJF 97-11
Hospital Pitie-Salpetriere
Paris, France
Chapter 3

Rima Darwiche
The Walter and Eliza Hall Institute
Burnet Clinical Research Unit
Parkville, Victoria, Australia
Chapter 5

Terry L. Delovitch
The Robarts Research Institute
and University of Western Ontario
Department of Microbiology
and Immunology, and Medicine
London, Ontario, Canada
Chapter 9

Sylvie Ferrari-Lacraz
Beth Israel Deaconess Medical Centre
Boston, Massachusetts, U.S.A.
Chapter 6

Frederick A. Houssiau
Rheumatology Unit, Christian de Duve
Institute of Cellular Pathology
Universite Catholique de Louvain
Bruxelles, Belgium
Chapter 14

S. Hussain
The Robarts Research Institute
and University of Western Ontario
Department of Microbiology
and Immunology, and Medicine
London, Ontario, Canada
Chapter 9

Windy Irawaty
The Walter and Eliza Hall Institute
Burnet Clinical Research Unit
Parkville, Victoria, Australia
Chapter 5

Thomas W.H. Kay
The Walter and Eliza Hall Institute
Burnet Clinical Research Unit
Parkville, Victoria, Australia
Chapter 5

David J. Kelvin
Laboratory of Molecular Immunology
and Inflammation
Robarts Institute
London, Ontario, Canada
Chapter 2

B.R. Lauwerys
Rheumatology Unit, Christian de Duve
Institute of Cellular Pathology
Universite Catholique de Louvain
Bruxelles, Belgium
Chapter 14

Roland S. Liblau
Laboratorie d'immunologie Cellulaire
INSERM CJF 97-11
Hospital Pitie-Salpetriere
Chapter 3

Erik Lubberts
Rheumatology Research Laboratory
Department of Rheumatology
University Hospital Nijmegen
Nijmegen, The Netherlands
Chapter 11

Wlodzmierz Maslinski
Beth Israel Deaconess Medical Centre
Boston, Massachusetts, U.S.A.
Chapter 6

C. Meagher
The Robarts Research Institute
and University of Western Ontario
Department of Microbiology
and Immunology, and Medicine
London, Ontario, Canada
Chapter 9

Dorothée Nashan
Ludwig Boltzmann Institute for Cell
Biology and Inmmunobiology
of the Skin
Department of Dermatology
University of Munster
Munster, Germany
Chapter 13

Deborah O'Neil
Department of Molecular Biology
Ghent University and Flanders
Interuniversity Institute
for Biotechnology
Gent, Belgium
Chapter 15

Trevor Owens
McGill University, Montreal Neurology
Institute
Montreal, Quebec, Canada
Chapter 8

Helen L. Pennington
The Walter and Eliza Hall Institute
Burnet Clinical Research Unit
Parkville, Victoria, Australia
Chapter 5

Alex Rabinovitch
University of Alberta,
Department of Medicine
Edmonton, Alberta, Canada
Chapter 10

Marie-Paule Roth
Laboratorie d'immunologie Cellulaire
INSERM CJF 97-11
Hospital Pitie-Salpetriere
Paris, France
Chapter 3

Pere Santamaria
Department of Microbiology
and Infectious Diseases
Faculty of Medicine, University
of Calgary
Calgary, Alberta, Canada
Chapter 1

Thomas Schwarz
Ludwig Boltzmann Institute for Cell
Biology and Inmmunobiology
of the Skin, Department of
Dermatology
University of Munster
Munster, Germany
Chapter 13

S. Sharif
The Robarts Research Institute
and University of Western Ontario
Department of Microbiology
and Immunology, and Medicine
London, Ontario, Canada
Chapter 9

Maria A. Staykova
Neurosciences Research Unit, Canberra
Hospital
Woden, Australia
Chapter 7

Lothar Steidler
Department of Molecular Biology
Ghent University and Flanders
Interuniversity Institute
for Biotechnology
Gent, Belgium
Chapter 15

Terry B. Strom
Beth Israel Deaconess Medical Centre
Boston, Massachusetts, U.S.A.
Chapter 6

Helen E. Thomas
The Walter and Eliza Hall Institute
Burnet Clinical Research Unit
Parkville, Victoria, Australia
Chapter 5

Wim B. van den Berg.
Rheumatology Research Laboratory
Department of Rheumatology
University Hospital Nijmegen
Nijmegen, The Netherlands
Chapter 11

Matthias G. von Herrath
The Scripts Research Institute
Division of Virology
La Jolla, California, U.S.A.
Chapter 12

David O. Willenborg
Neurosciences Research Unit, Canberra
Hospital
Woden, Australia
Chapter 7

Xin X. Zheng
Beth Israel Deaconess Medical Centre
Boston, Massachusetts, U.S.A.
Chapter 6

PREFACE

The field of immunoregulation by cytokines and chemokines has witnessed a remarkable progress over the last decade. The number of cytokines, chemokines and cytokine/chemokine receptors has dramatically increased and their physiological functions explored to an extent that was unforeseable a few years ago. Technological advances in genomics and genetic engineering in rodents have provided a wealth of information on cytokines and chemokines that spills over into different fields of biology and pathology. This book is an attempt to capture current knowledge on the role of cytokines and chemokines in autoimmunity by focusing on some of the most prevalent organ-specific or systemic autoimmune disorders that affect humankind. The lessons taught by research in the disorders dealt with in this work are likely applicable to other, less prevalent (albeit arguably as equally important) autoimmune disorders. Diseases not touched upon here include, for example, mysasthenia gravis, autoimmune thyroid diseases, autoimmune disorders resulting from immune complex deposits and other, where cytokines and chemokines undoubtedly also play a role.

This book is divided into 3 parts and contains 15 chapters written by world-class experts in their respective fields. Part I has two chapters. Chapter 1 provides an overview on the role of different cytokines and chemokines in autoimmunity, as a summary of what is discussed in depth in other chapters of the book (for specific autoimmune disorder or groups of disorders). Chapter 2 is a detailed synopsis of the function, genomics and structure of the known cytokines, chemokines and respective receptors. Part II is divided into four chapters that deal with the "genetics and mechanisms of action" of cytokines and their receptors in the context of autoimmunity. Part III groups nine chapters exploring the role of different cytokines and chemokines in various autoimmune disorders, including discussions on the proven and potential use of cytokines, chemokines or inhibitory reagents (i.e. antibodies or soluble receptors) in the clinic. The reader will realize that, as key communicators in immunobiology, cytokines and chemokines are promising targets for the prevention and/or therapy of autoimmune disorders. It will also become obvious to the reader that despite the enormous progress made to date, our knowledge in this area remains limited. It is my hope that this book will appeal to basic immunologists interested in clinical implications of cytokine and chemokine biology, as well as to clinicians interested in gaining an in depth understanding of the role of cytokines and chemokines in the pathogenesis and/or treatment of the autoimmune disorders that affect their patients.

Lastly, I would like to take this opportunity to thank the authors of this book for contributing their valuable time and expertise in preparing the different chapters. I am grateful to each of them for writing outstanding, thorough reviews of their respective areas of expertise.

Pere Santamaria

CHAPTER 1

Cytokines and Chemokines in Autoimmune Disease: An Overview

Pere Santamaria

Introduction

Autoimmune diseases result from complex interactions among different immune cell types, including both T and B lymphocytes and professional antigen-presenting cells, such as macrophages and dendritic cells. These cellular interactions result in auto-aggressive responses that target a number of different cell types in different tissues and organs in a relatively large number of autoimmune disorders. Although the etiology of most autoimmune diseases is unknown, recent years have witnesed important advances in our understanding of how the different immune cell types involved in autoimmunity communicate with one another, how they trigger autoimmune inflammation and how they cause tissue damage. As key elements of this communication network, cytokines and chemokines orchestrate the recruitment, survival, expansion, effector function and contraction of autoreactive lymphocytes in autoimmunity. The different Chapters of this book detail the role of different cytokines and chemokines in specific autoimmune disorders. In this Chapter, I highlight the contributions of individual cytokines and chemokines to multiple autoimmune diseases as discussed in detail throughout the book.[1-14] The reader is referred to specific Chapters for details.

Pro-Inflammatory Cytokines

Interleukin-1 (IL-1)

Interleukin1-α (IL-1α) and IL-1β , along with TNF-α are key inflammatory cytokines in rheumatoid arthritis (RA),[9] dermatomyositis and pemphigus.[14] In vitro data suggest that IL-1 is also an important effector cytokine in type 1 diabetes (T1D), through a number of direct (i.e., beta cell toxicity) and indirect means (i.e., by marking beta cells for Fas-dependent destruction by autoreactive cytotoxic T-lymphocytes).[7] IL-1 is also expressed in the central nervous system (CNS) of animals with experimental autoimmune encephalomyelitis (EAE) and IL-1R antagonsists have been shown to have a moderate therapeutic effect in EAE. IL-1 may contribute to disease severity, rather than to susceptibility in this animal model.[5]

Tumor Necrosis Factor-alpha (TNFα)

TNFα has direct cytotoxic effects on the intestinal mucosa in Crohn's disease and ulcerative colitis but also contributes to the systemic manifestations seen in these diseases. Anti-TNFα antibodies have shown a clear anti-inflammatory effect in patients with Crohn's disease, but the authors raise a note of caution about the long-term effects of TNFα blockade in

Cytokines and Chemokines in Autoimmune Disease, edited by Pere Santamaria.

vivo, particularly in children.[14] There is evidence that some animal models of systemic lupus erythematosus (SLE) produce reduced levels of TNFα. Although the pathogenic role of TNFα in SLE remains unclear, both Lawerys and Houssiau and Nashan and Schwarz point to the observation that RA patients treated with anti-TNFα mAb tend to develop anti-DNA antibodies, and that low TNFα producers have increased susceptibility to develop SLE.[12,13] TNFα appears to have a pathogenic role in the blister lesions of bullous phemphigoid.[12] TNFα plays a critical role in the pathogenesis of RA, and treatment with TNFα and IL-1 blockers offers the highest degree of protection in animal models.[9,11] Furthermore, there is evidence indicating that some TNFα gene variants are markers of RA severity.[2]

TNFα is also a key cytokine in the development of T1D, contributing to beta cell dysfunction and death, as well as orchestrating antigen-presentation and T-cell activation in situ. The effects of TNFα in vivo, however, are age-dependent and there is evidence that TNFα can also have anti-diabetogenic effects.[7,8] TNFα may be key to the breakdown of tolerance to self antigens in virus-induced diabetes. Interestingly, late expression of TNFα in this model could restore normal beta cell function, possibly by inducing T-cell apoptosis.[10] This dichotomy is a recurrent issue with other cytokines as well.

TNFα has been suggested to play a divergent role in the development of EAE and MS, by causing demyelination and fostering the chronicity of the disease (EAE) or by downregulating the disease process (MS).[5] Willenborg et al, however, point out the existence of diametrically opposed views on the effects of TNFα in EAE in the literature, ranging from pro-EAE to anti-EAE. Some studies have indicated that EAE can be inhibited by TNFα blockade, whereas studies in humans have suggested that it may increase the number of clinical exacerbations.[5]

TNF-Related Apoptosis-Inducing Ligand (TRAIL)

Chen and Chen discuss the role of TRAIL in autoimmune responses. TRAIL may contribute to suppression of autoimmune inflammation, such as autoimmune arthritis and thus may have therapeutic value in autoimmune diseases.[3]

RANK-Ligand (RANKL)

RANKL appears to play a critical role in the bone erosion process that occurs in RA and RANKL blockade in vivo may have therapeutic value.[9]

TALL-1/BAFF

Lauwerys and Houssiau discuss a role for this tumor necrosis factor family member in SLE. BAFF is a TNF-family member that induces B-cell proliferation by engaging BCMA or TACI receptors on B-cells. Autoantibody production and lupus-like syndromes have been noted in BAFF-transgenic mice, and the levels of TALL/BAFF-1 are elevated in animal models of SLE and human SLE patients.[13]

Interleukin-2 (IL-2)

IL-2- and IL-2R-deficient mice develop an autoimmune syndrome characterized by haemolytic anemia and ulcerative bowel disease. The contribution of IL-2 to autoimmune phenomena may be indirect, i.e., by virtue of the role it plays in T-cell homeostasis.[3]

Interferon-gamma (IFN-γ)

Peripheral blood mononuclear cells (PBMCs) from SLE patients tend to produce lower levels of IFN-γ than control PBMCs ex vivo. Furthermore, exogenous IFN-γ increases disease severity in some animal models of lupus, and IFN-γ or IFN-αR blockade have beneficial effects.[12,13] Extensive evidence indicates that IFN-γ contributes to the pathogenesis of T1D, but neither IFN-γ nor IFN-γ Rβ-deficient NOD mice are resistant to the disease.[7,8] IFN-γ, however, appears to play a critical role in virus-induced diabetes.[10] There is also some evidence suggesting that INF-γ may be necessary for the development of regulatory T-cells and, when administered systemically for example, inhibits insulitis development.[8] IFN-γ appears to play a downregulating role in EAE (possibly by inducing the production of nitric oxide). On the other hand, IFN-γ blockade appears to alleviate recurrent-relapsing MS (RR-MS).[5]

Interferon-alpha (IFN-α)

IFN-α appears to have a pro-inflammatory effect when expressed as a transgene in beta cells, but is anti-diabetogenic when administered systemically.[7]

Interleukin-6 (IL-6)

IL-6 is elevated in ex vivo organ cultures of inflammed colonic mucosa from both ulcerative colitis and Crohn's disease affected patients, and likely contributes to disease pathogenesis by inhibiting T-cell apoptosis, thereby perpetuating inflammation. Its contribution to pathology is reflected on the observation that IL-6R blockade suppresses colitis in animal models of inflammatory bowel disease.[14] There is also consensus in the literature implicating IL-6 in the development of EAE and possibly MS, but it may have an insignificant effect in disease pathology.[5]

Several lines of experimentation in mice have suggested an important role for IL-6 in the development of islet inflammation, as well as an inhibitory effect on its progression to overt diabetes, perhaps by inducing regulatory Th2 cells.[7]

IL-6 may also play a role in the pathogenesis of SLE and systemic sclerosis (SSc). IL-6 is elevated in sera of SLE and SSc patients and in the cerebrospinal fluid and urine of patients with cerebral lupus and lupus nephritis, respectively. IL-6 may also have a pathogenic role in skin lesions of SLE patients, as discussed extensively by Nashan and Schwarz.[12] IL-6 blockade improves disease outcome in (NZB x NZW) F1 mice, and IL-6 administration exacerbates disease progression.[12] IL-6 may also have a pathogenic role in the blister lesions of bullous phemphigoid.[12]

Interleukin-12 (IL-12)

O'Neil and Steidler note that the small intestine of patients with Crohn's disease contains elevated numbers of IL-12-producing macrophages. These cells are rare in ulcerative colitis lesions and thus may be key to immunopathological differences between these two disorders. IL-12 may contribute to damage of the gut wall by inducing the activation of matrix metalloproteinases. Anti-IL-12 therapy has shown promising results in reversing inflammation in the TNBS-induced model of colitis.[14]

IL-12 is indispensable for the induction of EAE, as indicated by studies of IL-12-deficient mice as well as anti-IL-12 mAb-treated animals.[5]

In contrast, there is impaired production of IL-12 in human SLE and murine models of the disease. As discussed by Lauwerys and Houssiau, IL-12 regulates immunoglobulin and autoantibody production and impaired IL-12 secretion may contribute to the pathogenesis of SLE.[13]

IL-12, as a Th1-driving cytokine, appears to play a key role in the initial phases of RA. IL-12 blockade or IL-12 administration inhibit or accelerate the development of RA, respectively.[9] The role of IL-12 in diabetogenesis is less clear. On the one hand IL-12 administration accelerates diabetes development in NOD mice, and anti-IL-12 treatment is anti-diabetogenic if initiated early, i.e., before development of insulitis. On the other hand, IL-12-deficient NOD mice develop diabetes, implying that Il-12 is dispensable in diabetogenesis.[7,8]

Interleukin 15 (IL-15)

IL-15 is increased in ex vivo cultures of Crohn's biopsy samples, but is absent in ulcerative colitis tissue and may play a role in driving local Th1 responses in Crohn's disease.[14] This cytokine is elevated in the synovial fluid of RA patients and may perpetuate the survival of autoreactive T-cells and promote the secretion of arthritogenic cytokines such as TNFα and IL-17. Zheng et al extensively discuss the therapeutic value of IL-15 blockade strategies in autoimmunity, particularly in RA.[11]

Interleukins-16, -17 and -18 (IL-16, IL-17 and IL-18)

Crohn's disease (but not ulcerative colitis)-affected tissue also contains elevated levels of IL-16. Anti-IL-16 blockade downregulates intestinal mucosal inflammation and damage in the TNBS-induced model of colitis. Crohn's lesions show elevated levels of IL-18 mRNA and active form of the IL-18 protein, which contributes to the Th1-bias seen in Crohn's versus ulcerative colitis disease.[14] IL-16, IL-17 and IL-18 are elevated in serum from SLE and/or SSc patients and may contribute to disease pathogenesis or to some clinical manifestations of the disease process.[12,13] Synovial fluid from RA patients also contains high levels of IL-17 and there is evidence to indicate that it has a direct role in arthritogenesis, possibly by inducing the expression of RANKL.[9] IL-18, which costimulates induction of IFN-γ by IL-12, may also be involved in RA.[9] IL-18 appears to prevent the progression of non-destructive to destructive insulitis, but its role in diabetogenesis remains unclear.[7] Some studies have suggested a role for IL-18 in EAE pathogenesis.[5]

Chemokines

Interleukin-8 (IL-8)/Macrophage Inflammatory Proteins (MIP-1 and MIP-2)

IL-8 is key to recruitment of polymorphonuclear leukocytes to the intestinal mucosa of patients affected with Crohn's disease or ulcerative colitis. Its levels are elevated in lesions from both types of inflammatory bowel disease (IBD).[14] IL-8, along with TNFα, IL-4, IL-5, and IL-13, is also elevated in skin lesions of dermatitis herpetiformis.[12] MIP-1α and MIP-1β have been implicated early in the development of EAE, although there appear to be some differences depending on the inducing antigen (i.e., PLP vs. MBP). MIP-1α blockade with antibodies has been shown to prevent EAE induction, by reducing the recruitment of macrophages. Furthermore, CCR1-deficient mice develop a less severe form of MOG-induced EAE. However, MIP-1α-deficient mice are susceptible to MOG-induced EAE, suggesting a role for MIP-1β.[6] MIP-2, which is considered to be the functional counterpart of human IL-8, has been detected in the CNS of EAE-affected IFN-γ-deficient Balb/c mice and may be involved in the recruitment of neutrophils to the CNS.[6]

The role of chemokines in diabetes is poorly understood. MIP-1α and MCP-1 appear to play a role in the development of insulitis, and both chemokines can be secreted by autoreactive Th1 cells.[7] Interestingly, the *Idd4* locus, which is associated with diabetes susceptibility in nonobese diabetic (NOD) mice, is linked to the CC chemokine gene cluster. MIP-1α is also expressed in pancreatic islets of a virus-induced diabetes model, at a time when most of the viral particles have already been cleared from the body.[10]

Monocyte Chemotactic Proteins (MCPs)

MCP-1 and MCP-3 expression is significantly increased in Chron's and ulcerative colitis lesions and this probably contributes to the recruitment of mononuclear leukocytes to inflammed areas of the intestinal mucosa.[14] MCP-1 may amplify CNS inflammation in PLP-induced EAE, but its role in EAE is unclear, as antibody blocking did not interefere with its induction.[5] Nevertheless, the *eae7* locus, containing genes affording EAE susceptibility, is linked to polymorphisms in TCA-3, MCP-1 and MCP-5.[2] Some studies have suggested a role for MCP-1 in disease relapses. CCR2 deficiency affords resistance to MOG-induced EAE, possibly by interfering with MCP-1-driven recruitment of macrophages. MCP-1, along with RANTES, MCP-2 and MCP-3, have been detected in MS lesions.[6]

Interferon-γ-Inducing Protein-10 (Crg-2, IP-10), Mig

Christen and von Herrath provide a detailed analysis of chemokine gene expression in the pancreas in a model of virus-induced diabetes. They show that Crg-2 and Mig are expressed soon after virus infection. They are followed, a few days later, by other chemokines, including

RANTES, MIP-1α and Eotaxin. Whether any of these chemokines is necessary for diabetes development in this model, however, remains to be determined.[10] IP-10 may be involved in PLP-induced EAE and it has been detected in macrophages/microglia of MS patients, along with other chemokines, including Mig.[6]

Regulated Upon Activation Normally T Expressed and Secreted (RANTES)

RANTES is a pro-inflammatory cytokine with chemotactic properties for T-cells, macrophages, monocytes, eosinophils and NK cells and is overexpressed in the mucosa of Chron's disease and ulcerative colitis patients.[14] RANTES is expressed in pancreatic islets in a virus-induced diabetes model, at a time when most of the viral particles have already been cleared from the body, and can be secreted by autoreactive Th1 cells.[10] RANTES is also expressed in the CNS of EAE-affected animals, but its role in the disease process is less clear than that for other chemokines. However, as pointed out by Babcock and Owens, RANTES blockade does not prevent EAE.[6]

C10, KC/Gro-α

The C10 chemokine has been implicated in the recruitment of macrophages in MOG-induced EAE.[6] KC/Gro-α is expressed early in the course of EAE and may be responsible for recruiting neutrophils into the CNS.[6]

Regulatory Cytokines

Interleukin-3 (IL-3)

Meagher et al point out that IL-3 can inhibit diabetogenesis when given to young NOD mice, but whether this cytokine is involved in the disease process is not known.[7]

Interleukin-10 (IL-10)

IL-10-deficient mice develop a form of inflammatory bowel disease that is histologically similar to human IBD. Anti-IL-10 treatment exacerbates mucosal inflammation in the dextran solium sulphate (DSS) model of colitis, and IL-10 has shown promising results in human clinical trials. IL-10 is increased in the mucosa of IBD patients. Of special note is the efficacy of local delivery of IL-10 using recombinant *Lactobacillus lactis* in IL-10-deficient mice and in the DSS-induced model of colitis, a strategy pioneered by Steidler and co-workers.[14]

Conversely, PBMC from SLE patients produce elevated levels of IL-10 and Lauwerys and Houssiau and Nashan and Schwarz argue that this may play a role in driving auto-antibody production in SLE patients, a point supported by the apparent success of a small clinical trial involving anti-IL-10 administration.[12,13] IL-10 may have an opposite effect in pemphigus vulgaris, as IL-10-deficient mice display increased susceptibility to this disease.[12]

Although the levels of IL-10 are increased in the synovial fluid of RA patients it does not appear to have a pathogenic role. IL-10, however, can prevent collagen-induced arthritis.[9]

The role of IL-10 in diabetogenesis is paradoxical. On the one hand, there is ample evidence for an anti-diabetogenic role of IL-10 in vivo, but on the other hand there is equally convincing evidence in support of just the opposite.[7,8]

IL-10 also has a protective effect against EAE, when administered in vivo, and IL-10 blockade increases the incidence and severity of relapses. Most studies, including those involving IL-10-deficient mice, indicate that IL-10 contributes to disease recovery in EAE. The contribution of IL-10 in MS is less clear.[5]

Transforming Growth Factorβ (TGFβ)

TGFβ1 has inhibitory effects on many immune functions and antagonizes the action of a number of pro-inflammatory cytokines, including IFN-γ, IL-1, IL-6, IL-12 and TNFα. As a result, TGFβ1-deficient mice develop systemic inflammatory processes that result in death.[3]

Local TGFβ expression is increased in IBD. Furthermore, blockade of TGFβ with antibodies exacerbates intestinal inflammation in animal models, and local administration of TGFβ ameliorates the disease process. TGFβ, however, does not appear to play a critical role in IBD pathogenesis.[14] The usefulness of TGFβ therapy in IBD is questionable, as chronic adminstration of TGFβ may result in fibrosis and stenosis and may impair kidney function.[5] SLE PBMCs appear to secrete lower levels of TGFβ than control PBMCs, and this cytokine may have a pathogenetic role in this disease.[12] TGFβ may play a key role in the fibrosis seen in SSc[12] and in the resolution of inflammatory responses, in general.[3] TGFβ has been shown to be a key cytokine in the induction of oral tolerance to autoantigens in a number of models, including experimental colitis, adjuvant arthritis, and EAE,[3] and downregulates EAE.[5] TGFβ has also been shown to have a protective role against type 1 diabetes, by inducing regulatory T-cells, but whether or not this cytokine plays a role in the disease process remains unclear.[7,8]

Interferon-β (IFN-β)

Interferon-β1b has been shown to reduce the number and severity of relapses in EAE and RR-MS patients.[5]

Interleukins-4, -11 and -13 (IL-4, IL-11 and IL-13)

IL-4 and IL-13 can downregulate the production of a number of pro-inflammatory mediators involved in IBD, but they can also lead to Th2-mediated forms of the disease. Their production is reduced in the inflamed mucosa of Crohn's disease and/or ulcerative colitis, yet IL-4-producing Th2 cells appear to have a pathogenic role in oxazolone-induced colitis and in the form of IBD that develops in TCRα-deficient mice. Recombinant IL-11, which promotes Th2 responses, ameliorates colitis in HLA-B27 transgenic rats, and colitis induced in rats by TNBS and acetic acid.[14]

IL-4 may have a pathogenic role in some models of SLE, but its role in the human disease process is unclear.[12] Lauwerys and Houssiau point out that anti-IL-4 antibodies inhibit autoantibody production and delay the onset of glomerulonephritis in (NZB x NZW) F1 mice.[13] Likewise, the affected skin of SSc patients exhibits increased levels of IL-4 and may be responsible for some of its clinical manifestations.[12]

IL-4 is absent in the synovial fluid of RA patients. Although overexpression exacerbates inflammation, it appears to protect against cartilage and bone destruction.[9]

Meagher et al and Rabinovitch discuss the role of IL-4 in T1D. There is some evidence for deficient IL-4 production by NK T cells in this autoimmune disorder, and for an anti-diabetogenic effect of IL-4 in animal models using diverse delivery strategies. Administration of IL-11 and IL-13 to young NOD mice can also inhibit diabetogenesis, possibly by reducing the production of pathogenic cytokines, such as TNFα and IFN-γ, and by increasing the production of regulatory cytokines, such as IL-4.[7,8]

The role of IL-4 in the pathogenesis of EAE and MS is unclear. IL-4-deficient mice display an increased susceptibility to MOG-induced EAE, and IL-4 can delay the onset and progression of EAE when delivered into the CNS; however, the effects varied depending on the genetic background. Willenborg et al raise the possibility that IL-4 may contribute to disease regulation in MS, as suggested by some studies.[5] IL-13 has protective effects against EAE in the rat.[5]

Concluding Remarks

Much has been learned during the last decade about the role of cytokines and chemokines in autoimmune disease. This Chapter is an attempt to capture some of the highlights of what is known about a significant number of cytokines and chemokines in the context of autoimmunity, as detailed in the Chapters that follow. However, the sometimes paradoxical observations made in similar models of autoimmunity, and the apparently contradictory results that have been reported for some cytokines and/or chemokines in the same models underscore the fact that there are lots yet to be learned. The discovery of new cytokines and chemokines, and the development of reductionist models of autoimmunity with relevance to specific autoimmune diseases will undoubtedly foster much needed progress in this field.

Acknowledgments

I thank the members of my laboratory for exciting discussions and feedback. I also thank Daniela Minardi for assistance in the preparation of this manuscript. The work in my laboratory is supported by grants from the Canadian Institutes of Health Research, the Canadian Diabetes Association, The Juvenile Diabetes Research Foundation and the Natural Sciences and Engineering Research Council of Canada.

References

1. Cameron M, Kelvin D. Cytokines, Chemokines and their receptors. In: Santamaria P, ed. Cytokines and Chemokines in Autoimmune Disease. Austin: RG Landes Co., 2001.
2. Coppin H, Roth M-P, Liblau R. Cytokine and cytokine receptor genes in the susceptibility and resistance to organ-specific autoimmune diseases. In: Santamaria P, ed. Cytokines and Chemokines in Autoimmune Disease. Austin: RG Landes Co., 2001.
3. Chen Y, Chen Y. Cytokines, lymphocyte homeostasis and self tolerance. In: Santamaria P, ed. Cytokines and Chemokines in Autoimmune Disease. Austin: RG Landes Co., 2001.
4. Kay T, Darwiche R, Irawaty W, Chong M, Pennington H, Thomas H. The role of cytokines as effectors of tissue destruction in autoimmunity. In: Santamaria P, ed. Cytokines and Chemokines in Autoimmune Disease. Austin: RG Landes Co., 2001.
5. Willenborg D, Staykova M. Cytokines in the pathogenesis and therapy of autoimmune encephalomyelitis and multiple sclerosis. In: Santamaria P, ed. Cytokines and Chemokines in Autoimmune Disease. Austin: RG Landes Co., 2001.
6. Babcock A, Owens T. Chemokines in experimental autoimmune encephalomyelitis and multiple sclerosis. In: Santamaria P, ed. Cytokines and Chemokines in Autoimmune Disease. Austin: RG Landes Co., 2001.
7. Meagher C, Sharif S, Hussain S, Cameron M, Arreaza G, Delovitch T. Cytokines and chemokines in the pathogenesis of murine type 1 diabetes. In: Santamaria P, ed. Cytokines and Chemokines in Autoimmune Disease. Austin: RG Landes Co., 2001.
8. Rabinovitch A. Immunoregulation by cytokines in autoimmune diabetes. In: Santamaria P, ed. Cytokines and Chemokines in Autoimmune Disease. Austin: RG Landes Co., 2001.
9. Lubberts E, Berg W. Cytokines in the pathogenesis of rheumatoid arthritis and collagen-induced arthritis. In: Santamaria P, ed. Cytokines and Chemokines in Autoimmune Disease. Austin: RG Landes Co., 2001.
10. Christen U, Herrath Mv. Cytokines and chemokines in virus-induced autoimmunity. In: Santamaria P, ed. Cytokines and Chemokines in Autoimmune Disease. Austin: RG Landes Co., 2001.
11. Zheng X, Maslinski W, Ferrari-Lacraz S, Strom T. Cytokines in the treatment and prevention of autoimmune response: a role for IL-15. In: Santamaria P, ed. Cytokines and Chemokines in Autoimmune Disease. Austin: RG Landes Co., 2001.
12. Nashan D, Schwarz T. Cytokines and chemokines in human autoimmune skin disorders. In: Sanatamaria P, ed. Cytokines and Chemokines in Autoimmune Disease. Austin: RG Landes Co., 2001.
13. Lauwerys B, Houssiau F. Involvement of cytokines in the pathogenesis of systemic lupus erythematosus. In: Santamaria P, ed. Cytokines and Chemokines in Autoimmune Disease. Austin: RG Landes Co., 2001.
14. O'Neil D, Steidler L. Cytokines and chemokines in the pathogenesis and treatment of inflammatory bowel disease. In: Santamaria P, ed. Cytokines and Chemokines in Autoimmune Disease. Austin: RG Landes Co., 2001.

Chapter 2

Cytokines and Chemokines—Their Receptors and Their Genes: An Overview

Mark J. Cameron and David J. Kelvin

Introduction

The immune system is skilled in communication and designed to respond quickly, specifically and globally to protect an organism against foreign invaders and disease. The cytokine superfamily of proteins is an integral part of the signaling network between cells and is essential in generating and regulating the immune system. Much progress has been made recently in interpreting how the immune system communicates with, or is mediated by, cytokines and chemotactic cytokines (chemokines). These interacting biological signals have remarkable capabilities, such as influencing growth and development, hematopoiesis, lymphocyte recruitment, T cell subset differentiation and inflammation. This chapter provides brief synopses for a comprehensive list of immune-related cytokines and chemokines. Information such as gene cloning and mapping details, protein characteristics and expression, receptor usage, source and target cells, major biological functions and knockout phenotype is described for each cytokine and chemokine. With an approach that organizes cytokines and chemokines into interacting groups with related physical and/or functional properties, this chapter aims to highlight the capability of this system to maintain widespread impact and functional complementation while not sacrificing regulation and specificity of action. A more complete understanding of these properties may lead to more advanced means of correcting improper cytokine- or chemokine-mediated immune responses, such as those causing autoimmune disease.

Detailed and reliable communication must occur through a complex system of network connections to accomplish a task at a modern workstation. In parallel, the immune system is an interdependent biological network charged with developmental tasks and the responsibility of protecting its host against injury and infection. An immune cell within a given microenvironment can respond to signals received through its receptors with its own protein-based language that will influence the cell itself (autocrine effect) or other cells throughout the organism (paracrine effect). The language of cytokines is critical in this communication. Cytokines are small soluble factors with pleiotropic functions that are produced by many cell types as part of a gene expression pattern that can influence and regulate the function of the immune system.

The term cytokine was proposed by Cohen et al in 1974[1] to replace lymphokine, a term coined in the late 1960's to denote lymphocyte-derived soluble proteins that possess immunological effects.[2] Since the latter designation misleadingly suggested that lymphocytes were the only source for these secreted proteins, the term cytokine slowly became preferred. Following the introduction of this general term, the Second International Lymphokine Workshop held in 1979 proposed the interleukin (IL) system of nomenclature to simplify the growing list of identified cytokines. Ironically, this partially adopted system introduced confusion in that the interleukins, presently numbering at least 23, affect many cell types but their name implies that they act only among leukocytes. As a result, modern cytokine nomenclature is a mix of the

Cytokines and Chemokines in Autoimmune Disease, edited by Pere Santamaria.

widely accepted, but slightly misleading, interleukin designations and other proteins still known by their original names. A good example of these potential points of confusion is the chemotactic cytokine (chemokine) IL-8, which is produced by and targets a wide variety of cell types including leukocytes and nonleukocytes.

As this chapter unfolds, repeated mention of a number of cytokines and chemokines will make it clear that these proteins can be part of a bigger immune program, e.g., T cell subset differentiation. Mature $CD4^+$ and $CD8^+$ T cells leave the thymus with a naive phenotype and produce a variety of cytokines. In the periphery, these T cells encounter antigen presenting cells (APCs) displaying either major histocompatibility complex (MHC) class I molecules (present peptides generated in the cytosol to $CD8^+$ T cells) or MHC class II molecules (present peptides degraded in intracellular vesicles to $CD4^+$ T cells). Following activation, characteristic cytokine and chemokine secretion profiles allow the classification of $CD4^+$ T helper (Th) cells into two major subpopulations in mice and humans.[3-7] Th1 cells secrete mainly IL-2, interferon-γ (IFN-γ) and tumor necrosis factor-β (TNF-β), whereas Th2 cells secrete mainly IL-4, IL-5, IL-6, IL-10 and IL-13. Th1 cells support cell-mediated immunity and as a consequence promote inflammation, cytotoxicity and delayed-type hypersensitivity (DTH). Th2 cells support humoral immunity and serve to downregulate the inflammatory actions of Th1 cells. This paradigm is a great example of an integrated biological network and is very useful in simplifying our understanding of typical immune responses and those that turn pathogenic. For example, the failure to communicate "self" can lead to a loss of tolerance to our own antigens and prompt destructive immune responses to self-tissues and autoimmune disease. Autoimmunity, the major focus of this book, is the underlying mechanism of a set of conditions, such as type 1 diabetes mellitus, multiple sclerosis and rheumatoid arthritis. Autoimmune diseases may be caused in part by cytokine- and chemokine-mediated dysregulation of Th cell subset differentiation. The main factors affecting the development of Th subsets, aside from the context in which the antigen and costimulatory signals are presented, are the cytokines and chemokines in the stimulatory milieu. A better understanding of the properties and interactions of the individual cytokines and chemokines that play a role in Th cell activation may lead to more advanced treatments for autoimmune disease.

The proceeding sections will introduce many of the currently identified cytokines and chemokines, along with their receptors. You will find that cytokines and chemokines with related structure and/or function are clustered into groups of interdependent homologues, e.g., the IL-1-like cytokines. A particular group of cytokines or chemokines can exhibit functional redundancy with, and widespread impact on, other groups of cytokines or chemokines, e.g., IL-1-like cytokines and IL-6-like cytokines. Interestingly, this can occur while maintaining several regulatory features, such as internal checkpoints and specificity of action. It is therefore hoped that this chapter may serve as more than a brief catalogue of the field of cytokines, chemokines and their receptors, but may also highlight the remarkable capabilities of this interacting network of biological signals.

Cytokines—Their Receptors and Their Genes

Table 2.1 introduces the human cytokines and lists some of their properties, such as receptor usage and physical characteristics. Each human cytokine described in Table 2.1 has a murine counterpart so the basic list can be used interchangeably in regards to terminology. Hundreds of cytokines have been identified. In the interest of conciseness the table includes only common cytokines with recognized immune function, many of which are discussed in more detail below. Excluded are the 'growth factors', neurobiological proteins and 'trophins', for example. It is also beyond the scope of this chapter to describe how cytokines signal through their receptors in any detail. One popular cytokine signaling mechanism used by cytokines such as IL-2, IL-4, IL-6, IL-7, IL-10, IL-12, IL-13, IL-15 and the interferons, however, begins with dimerization of the appropriate receptor chains upon ligand binding. Following this, different types of receptor-associated Janus family tyrosine kinases (Jak) are activated which

Table 2.1. Common human cytokines and their receptors[1]

Name	Synonym(s)	Amino Acids	Chromosome	Molecular Weight[2]	Cytokine Receptor(s) (Da) and Form	Receptor Location(s)
'Interleukins'						
IL-1-like						
IL-1α	hematopoietin-1	271	2q14	30606	CD121a, CDw121b	2q12, 2q12-q22
IL-1β	catabolin	269	2q14	20747	CD121a, CDw121b	2q12,2q12-q22
IL-1RA	IL-1 receptor antagonist	177	2q14.2	20055	CD121a	2q12
IL-18	interferon-γ inducing factor	193	11q22.2-q22.3	22326	IL-18Rα, β	2θ12
Common γ chain (CD132)						
IL-2	T cell growth factor	153	4q26-q27	17628	CD25, 122,132	10p15-p14, 22q13.1, Xq13.1
IL-4	BSF-1	153	5q31.1	17492	CD124,213a1[3], 132	16p11.2-12.1, X, Xq13.1
IL-7		177	8q12-q13	20186	CD127, 132	5p13, Xq13.1
IL-9	T cell growth factor P40	144	5q31.1	15909	IL-9R, CD132	Xq28 or Yq12, Xq13.1
IL-13	P600	132	5q31.1	14319	CD213a1, 213a2, CD124[3], 132	X, Xq13.1-q28, 16p11.2-12.1, Xq13.1
IL-15		162	4q31	18086	IL-15Rα, CD122, 132	10p14-p14, 22q13.1, Xq13.1
Common β chain (CD131)						
IL-3	multipotential CSF, MCGF	152	5q31.1	17233	CD123, CDw131	Xp22.3 or Yp11.3, 22q13.1
IL-5	BCDF-1	134	5q31.1	15238, homodimer	CDw125, 131	3p26-p24, 22q13.1
Also related						
GM-CSF	CSF-2	144	5q31.1	16295	CD116, CDw131	Xp22.32 or Yp11.2, 22q13.1

continued on next page

Table 2.1. cont.

IL-6-like						
IL-6	IFN-β_2, BSF-2	212	7p21	23718	CD126, 130	1q21, 5q11
IL-11	AGIF	199	19q13.3-13.4	21429	IL-11Ra, CD130	9p13, 5q11
Also related						
G-CSF	CSF-3	207	17q11.2-q12	21781	CD114	1p35-p34.3
IL-12	NK cell stimulatory factor	219/328	3p12-p13.2/ 5q31.1-q33.1	24844/37169 heterodimer	CD212	19p13.1, 1p31.2
LIF	leukemia inhibitory factor	202	22q12.1-q12.2	22008	LIFR, CD130	5p13-p12
OSM	oncostatin M	252	22q12.1-q12.2	28484	OSMR, CD130	5p15.2-5p12
IL-10-like						
IL-10	CSIF	178	1q31-q32	20517, homodimer	CDw210	11q23
IL-20		176	2q32.2	20437	IL-20Rα, β	?
Others						
IL-14	HMW-BCGF	498	1	54759	IL-14R	?
IL-16	LCF	631	15q24	66694, homotetramer	CD4	12pter-p12
IL-17	CTLA-8	155	2q31	17504, homodimer	CDw217	22q11.1
'Interferons'						
IFN-α		189	9p22	21781	CD118	21q22.11
IFN-β		187	9p21	22294	CD118	21q22.11
IFN-γ		166	12q14	19348, homodimer	CDw119	6q23-q24

continued on next page

Table 2.1. Cont.

'TNF'						
CD154	CD40L, TRAP	261	Xq26	29273, homotrimer	CD40	20q12-q13.2
LT-β		244	6p21.3	25390, heterotrimer	LTβR	12p13
TNF-α	cachectin	233	6p21.3	25644, homotrimer	CD120a, b	12p13.2, 1p36.3-p36.2
TNF-β	LT-α	205	6p21.3	22297, heterotrimer	CD120a, b	12p13.2, 1p36.3-p36.2
4-1BBL		254	19p13.3	26624, trimer?	CDw137 (4-1BB)	1p36
APRIL	TALL-2	250	17p13.1	27433, trimer?	BCMA, TACI	16p13.1, 17p11.2
CD70	CD27L	193	19p13	21146, trimer?	CD27	12p13
CD153	CD30L	234	9q33	26017, trimer?	CD30	1p36
CD178	FasL	281	1q23	31485, trimer?	CD95 (Fas)	10q24.1
GITRL		177	1q23	20307, trimer?	GITR	1p36.3
LIGHT		240	16p11.2	26351, trimer?	LTβR, HVEM	12p13, 1p36.3-p36.2
OX40L		183	1q25	21050, trimer?	OX40	1p36
TALL-1		285	13q32-q34	31222, trimer?	BCMA, TACI	16p13.1, 17p11.2
TRAIL	Apo2L	281	3q26	32509, trimer?	TRAILR1-4	8p21
TWEAK	Apo3L	249	17p13.3	27216, trimer?	Apo3	1p36.2
TRANCE	OPGL	317	13q14	35478, trimer?	RANK, OPG	18q22.1, 8q24

continued on next page

Table 2.1. Cont.

'TGF-β'						
TGF-β1	TGF-β	390	19q13.1	44341, homodimer	TGF-βR1	9q22
TGF-β2		414	1q41	47747, homodimer	TGF-βR2	3p22
TGF-β3		412	14q24	47328, homodimer	TGF-βR3	1p33-p32
'Miscellaneous hematopoietins'						
Epo	erythropoietin	193	7q21	21306	EpoR	19p13.3-p13.2
Tpo	MGDF	353	3q26.3-q27	37822	TpoR	1p34
Flt-3L		235	19q13.1	26416	Flt-3	13q12
SCF	stem cell factor, c-kit ligand	273	12q22	30898, homodimer	CD117	4q11-q12
M-CSF	CSF-1	554	1p21-p13	60119, homodimer	CD115	5q33-q35
MSP	Macrophage stimulating factor, MST-1	711	3p21	80379	CDw136	3p21.3

[1] List assembled using data from Gene Cards (World Wide Web URL: http://genome-www.stanford.edu/genecards). Note that some of the cytokines listed are not discussed in this chapter.
[2] Data describes the unprocessed precursor.
[3] Can be found in complexes.

phosphorylate the receptor chains and allow the recruitment and activation of other kinases and transcription factors, such as those of the signal transducer and activator of transcription (STAT) family. This promotes the rapid translocation of these proteins to the nucleus and stimulation of target gene transcription (see references 8 and 9 for more details on cytokine signaling).

IL-1-Like Cytokines

Firstly, the interleukins are comprised mostly of hematopoietic growth factors and can be further divided into groups of proteins as shown in Table 2.1. The IL-1-related group of proinflammatory cytokines consists of IL-1α, IL-1β, IL-1 receptor antagonist (IL-1RA) and IL-18. IL-1α and IL-1β are produced mainly by mononuclear and epithelial cells upon inflammation, injury and infection.[10] These two proteins are of primary importance to the outcome of these challenges to the immune system in that they trigger fever, induce a wide variety of acute phase response (APR) genes and activate lymphocytes.[10] IL-1α and IL-1β arise from two closely linked genes that, along with the IL-1RA gene, lie on human (and mouse) chromosome 2.[10,11] The two forms of IL-1 are quite similar in function since they both signal through the IL-1 type 1 receptor (IL-1-R1/CD121a).[12] Both proteins can also bind to the IL-1 type 2 receptor (IL-1-R2/CDw121b) which does not appear to be involved in signaling, except as a possible decoy.[13] The IL-1 receptor genes are located on human chromosome 2 along with their ligands, albeit at a distance.

Murine knockout studies confirm the importance of IL-1 in fever responses and the APR. While at least three studies involving the IL-1β knockout mouse demonstrate that fever development is suppressed upon turpentine or lipopolysaccharide (LPS) challenge,[14-16] one study demonstrates that the role of IL-1β as a pyrogen is not obligatory and that its absence can in fact exacerbate an induced fever response.[17] The latter conflicting result may stem from differences in experimental protocol or reagents.[14] Knockout studies also show that while both forms of IL-1 can induce fever responses, fever induction is not reduced in IL-1α knockout mice indicating that IL-1β can compensate for IL-1α but not vice-versa.[14] The role for IL-1 in the APR (a series of cellular and cytokine cascades in reaction to trauma or infection that help limit damage) was confirmed in a localized tissue damage model of turpentine injection where challenged IL-1β-deficient mice did not develop an APR.[18] Accordingly, IL-1R1 knockout mice are irresponsive to IL-1 in the induction of IL-6, E-selectin and fever.[18] These mice also have a reduced APR to turpentine.[19]

IL-1RA is produced by virtually any cell that can produce IL-1 and is similar in structure to IL-1β but lacks its agonist activity.[20] The different species of IL-1RA, a secreted form with a signal peptide and at least two intracellular forms, arise from alternative splicing of different first exons on chromosome 2.[20, 21] IL-1RA represents an intriguing example of a naturally occurring cytokine receptor antagonist. IL-1RA may be an acute phase protein that may serve to regulate the agonist effects of IL-1 during chronic inflammatory and infectious disease because its expression is influenced by cytokines, viral and bacterial products, bound antibody and acute phase proteins, such as IL-1, IL-4, IFN-γ and LPS.[20] Consistent with this notion are two studies of IL-1RA-deficient mice which exhibit growth retardation, an exacerbated fever response to turpentine injection, increased lethality following LPS injection and decreased susceptibility to *Listeria monocytogenes*.[14, 22] These observations verify the importance of balance in the IL-1 system in mediating these immune challenges.

IL-18, initially termed interferon-γ inducing factor (IGIF), is a pro-inflammatory cytokine that is encoded on human chromosome 11 and mouse chromosome 9.[23] IL-18 has been placed in the IL-1 group of interleukins because it bears structural homology to IL-1α and β, is converted into a mature form by IL-1β converting enzyme (ICE) along with IL-1β and binds to the IL-18 receptor (IL-18R or IL-1R related protein).[23] The IL-18R resembles the IL-1R and transduces IL-1R signaling.[23] IL-18 shares biological function with IL-12 in that it induces IFN-γ secretion (in synergy with IL-12), enhances natural killer (NK) cell activity and promotes inflammatory Th1 cell responses.[23] Accordingly, when IL-18[24] or its receptor[25] is knocked out, mice exhibit defective NK cell activity and Th1 responses. More recently, however, the role of IL-18 as a pro-inflammatory cytokine has been questioned because IL-18 can also potentiate regulatory Th2 responses, perhaps by inducing IL-4 production by natural killer T (NKT) cells in certain situations.[26-28]

Common γ Chain Cytokines

Cytokines that utilize the common γ chain (γc/CD132) in their receptor comprise the next group of interleukins, namely IL-2, IL-4, IL-7, IL-9, IL-13 and IL-15. These diverse cytokines invoke lymphocyte activation and differentiation (the outcome of which can vary) and possess some redundancy in biological function because of their common receptor subunit.[29] The γc itself cannot bind cytokines, however new evidence suggests that it can be shed as a soluble negative modulator.[30] Indeed, γc-deficient mice are severely immunocompromised, as are humans with γc defects.[31, 32]

IL-2 is expressed from a gene on human chromosome 4 or mouse chromosome 3 and is mainly secreted by activated T cells. IL-2 and the heteromultimeric IL-2 receptor (IL-2R) complex (combinations of IL-2Rα/CD25, IL-2Rβ/CD122 and γc) are upregulated on T cells following antigenic or mitogenic stimulation leading to clonal expansion. As such, IL-2 is commonly regarded as an autocrine or paracrine T cell growth factor but it actually has effects on many cell types, such as B cells, NK cells, macrophages and neutrophils.[29, 33, 34] The IL-2 knockout mouse exhibits immune dysregulation caused by defects in T cell responsiveness in vitro, however only delays in normal T cell functionality were found in vivo.[35, 36] Interestingly, IL-2Rα-[37] and IL-2Rβ-deficient[38] mice exhibit loss of T cell regulation and autoimmunity indicating that proper IL-2 signaling may be required to induce regulatory T cells and/or eliminate abnormally activated T cells via the reversal of T cell anergy or apoptosis (programmed cell death) induction, respectively.[39]

The IL-4 gene is located on human chromosome 5 (along with the IL-3, IL-5, IL-9, IL-13 and granulocyte macrophage colony stimulating factor (GM-CSF) genes) and murine chromosome 11 (along with the IL-3, IL-5, IL-13 and GM-CSF genes). Short or long isoforms of IL-4 can exist arising from alternative splicing.[40] IL-4 is produced by activated T cells, mast cells, basophils and NKT cells and targets many cell types, including B cells, T cells, macrophages and a wide variety of hematopoietic and nonhematopoietic cells.[29, 41] Physiologic signal transduction via IL-4 depends on heterodimerization of the IL-4 receptor α chain (IL-4Rα/CD124), with γc and possibly the IL-13 receptor α chain (IL-13Rα/CD213a1).[42] IL-4 is the principal cytokine required by B cells to switch to the production of immunoglobulin (Ig)E antibodies, which mediate immediate hypersensitivity (allergic) reactions and help defend against helminth infections.[41] IL-4 also inhibits macrophage activation and most of the effects of IFN-γ on macrophages. However, the most important biological effect of IL-4 with respect to immune modulation is the growth and differentiation of Th2 cells. As described earlier, Th2 cells support humoral immunity and serve to downregulate the inflammatory actions of Th1 cells. Moreover, stimuli that favour IL-4 production early after antigen exposure favour the development of Th2 cells.[3] IL-13 is also associated with this subset of T cells.[43] Like IL-4, and along with the fact that it maps closely to IL-4 and shares receptor α subunits with IL-4, IL-13 is expressed by activated T cells, induces IgE production by B cells and inhibits inflammatory cytokine production.[44] These properties of IL-4 and IL-13 have been convincingly demonstrated in mice lacking the IL-4 or IL-13 gene.[45-48] These mice are deficient in the development and maintenance of Th2 cells.

The remaining γc cytokines, IL-7, IL-9 and IL-15, are potent hematopoietic factors expressed from genes on human chromosome 8 and mouse chromosome 3, human chromosome 5 and mouse chromosome 13, and human chromosome 4 and mouse chromosome 8, respectively. IL-7, expressed by stromal and epithelial cells, stimulates immature B cells, thymocytes and mature T cells via its receptor consisting of the IL-7 receptor α chain (IL-7Rα/CD127) and the γc.[49-51] Knocking out IL-7 or IL-7Rα/CD127 causes severe defects in thymic T cell and B cell development consistent with the critical roles that IL-7 and its receptor play in maturation of the immune system.[51-56] IL-9 promotes the growth of mast cells, B cells and other T cells and is mainly expressed by activated T cells, especially Th2 cells.[29, 43, 57] Confirming only the role of IL-9 in enhancing mast cells, the recently generated IL-9 knockout mouse exhibits normal T cell (Th2) responses but not characteristic mast cell expansion upon lung challenge.[58] IL-15, produced by activated monocytes, epithelial cells, and a variety of tissues, shares biological activities with IL-2 in that it stimulates NK cells, B cells and activated T cells.[29, 59-61] The IL-15 receptor (IL-15R) consists of combinations of IL-15Rα, IL-2Rβ/CD122 and γc. Similarities in function between IL-2 and IL-15 are partially due to receptor subunit sharing. A recent study, however, provides evidence that IL-2 and IL-15 control different aspects of primary T-cell expansion in vivo. IL-15 is critical for initiating T cell divisions, whereas IL-2 can limit T cell expansion by decreasing γc expression and rendering cells susceptible to apoptosis.[62] The α chain ligand specificity and broad cellular expression range of IL-15 allows for differential activity even outside of the immune system.[29] IL-15- and IL15Rα-deficient mice were recently generated. Initial studies confirm the role of IL-15 in NK cell stimulation and indicate a role for IL-15 in peripheral $CD8^+$ T cell maintenance upon immune challenge.[63,64]

Common β Chain Cytokines

Cytokines that utilize the common β chain (βc/CDw131) in their receptor comprise the next group of interleukins, namely IL-3, IL-5 and GM-CSF. The genes for IL-3, IL-5 and GM-CSF are closely linked and lie on human chromosome 5 and mouse chromosome 11.[65] Like the γc cytokines, these associated (but not particularly homologous at the amino acid sequence level) βc cytokines overlap in biological function because of their common receptor subunit.[65] When the βc is mutated, normal hematopoiesis is noted but impaired immune responses can be observed that are most likely due to a loss of responsiveness to IL-5 and GM-CSF, rather than IL-3.[66, 67]

IL-3, originally termed multicolony stimulating factor (multi-CSF), is produced by activated T cells and stimulates both multipotential hematopoietic cells (stem cells) and developmentally committed cells such as granulocytes, macrophages, mast cells, erythroid cells, eosinophils, basophils and megakaryocytes.[68-70] The human IL-3 receptor consists of CD123 and βc/CDw131. The mouse IL-3 receptor has an additional β chain called β_{IL-3}, the function of which can be compensated for by CD123 if knocked out.[67] Knocking out CD123 itself also has little effect on hematopoiesis.[71] On the other hand, if IL-3 is knocked out, mast cell and basophil development upon challenge is affected,[66] as well as some forms of DTH,[72] confirming a role for IL-3 in host defense and expanding hematopoietic effector cells.

IL-5, originally identified as a B cell differentiation factor, is produced mainly by activated T cells (especially Th2 cells) and aids in the growth and differentiation of eosinophils and late-developing B cells.[73-75] When IL-5 or CDw125 is absent, mice exhibit developmental defects in certain B cells ($CD5^+$/B-1 B cells) and a lack of eosinophilia upon parasite challenge.[76, 77]

Lastly, GM-CSF, as its name suggests, was originally found to stimulate granulocytes and macrophages. GM-CSF has since been found to be expressed by many cell types, including macrophages and T cells, and shares many of the functions of IL-3 in stimulating a variety of precursor cells, including macrophages, neutrophils and eosinophils.[78-80] Interestingly, GM-CSF-deficient mice have normal hematopoietic development but suffer from pulmonary disease perhaps caused by a lack of lung surfactant clearance by alveolar epithelial cells or macrophages.[81]

IL-6-Like Cytokines

IL-6 is the prototype cytokine representing the next group of interleukins. Most of the members of this group utilize the glycoprotein 130 (gp130) or CD130 receptor. IL-6, IL-11, leukemia inhibitory factor (LIF), oncostatin M (OSM), granulocyte colony-stimulating factor (G-CSF) and IL-12 have partially overlapping functions and are key mediators in various immune processes including hematopoiesis and the APR. CD130-deficient mice exhibit embryonic lethality, a finding that appears to be linked to a significant role for CD130-dependent signaling in homeostasis.[82]

IL-6, with its gene situated on human chromosome 7 and mouse chromosome 5, utilizes the CD130 receptor and the IL-6 receptor α chain (IL-6Rα/CD126). The IL-6Rα/CD126 can exist in a soluble form and serves as an important cofactor by extending the cytokine's half-life.[83] IL-6 was originally characterized as a differentiation factor of B cell hybridomas.[84, 85] Producers of IL-6 include fibroblasts, endothelial cells, macrophages, T cells (Th1) and B cells. IL-6 is a primary inducer of fever, hormones, acute phase proteins and T and B cell expansion upon injury and infection.[86] It can also act as a cofactor in hematopoiesis by increasing GM-CSF and macrophage colony stimulating factor (M-CSF) expression.[87] IL-6-deficent mice exhibit a severely blunted APR following infection or injury,[88, 89] problems in early hematopoiesis and T and B cell function and Th1 development.[90] Interestingly, IL-6 can nonetheless act as an anti-inflammatory agent in some instances.[91]

IL-11, originally identified as a pleiotropic stromal cell-derived cytokine, is encoded on chromosome 19 in humans and chromosome 7 in mice.[92, 93] IL-11 also utilizes the CD130 receptor along with the IL-11 receptor α chain (IL-11Rα). IL-11 is produced by, and has effects on, many hematopoietic and nonhematopoietic cell types.[94, 95] IL-11, like IL-6, is known to stimulate acute phase protein synthesis in the liver.[94, 95] IL-11 also collaborates with other cytokines we have already discussed, such as IL-3, IL-4, IL-7, IL-13 and GM-CSF, to stimulate (by shortening cell-cycle time) the proliferation of hematopoietic stem cells and progenitor cells and induce the differentiation of megakaryocytes.[94, 95] The collaborative nature of IL-11 in vivo may explain why knockout studies have yet to identify a defective phenotype (at least in the hematopoietic compartment) associated with a lack of IL-11 signaling.[96] Interestingly, IL-11 could also be an anti-inflammatory mediator as it inhibits macrophage pro-inflammatory cytokine production and can exert protective effects in several disease models.[91]

LIF is a ligand for CD130 and the LIF receptor (LIFR). LIF is associated with the differentiation of many cell types.[91, 97, 98] In this regard, LIF can both inhibit the differentiation of embryonic stem cells and promote the survival of hematopoietic precursors. LIF can stimulate inflammatory cytokine production. Its expression can be upregulated or downregulated in response to inflammatory cytokines such as IL-1 and TNF or regulatory cytokines such as IL-4, respectively. LIF is therefore often classified as a pro-inflammatory cytokine, however recent evidence may suggest otherwise in some situations.[91] LIF knockout mice display several phenotypes depending on the disease model.[91] This may be due to the observation that loss of LIF expression perturbs the establishment of a normal pool of stem cells, but not the terminal differentiation of these cells.[99] Unlike IL-6, LIF can also stimulate the hypothalamic-pituitary-adrenal axis in response to stress and disease. This property has been elegantly demonstrated in a recent study of the LIF knockout mouse where mice did not respond to immobilization-induced stress with the normal indicators.[100] It is also interesting to note that the genes for LIF and OSM lie in tandem on human chromosome 22 and mouse chromosome 11 and are transcribed in the same orientation.[101, 102] OSM is a very similar cytokine produced mainly by activated macrophages and T cells with inflammatory and growth factor properties.[101, 102]

G-CSF (or colony stimulating factor-3) is produced by fibroblasts and monocytes and stimulates granulocyte progenitor cells and neutrophils.[103-105] The G-CSF gene is located on human chromosome 17 and mouse chromosome 11 and creates two active polypeptides (differing by only three amino acids) by differential mRNA splicing.[103] The G-CSF receptor (G-CSFR) is expressed on multipotential hematopoietic progenitor cells and in cells of the myeloid lineage.[104] The importance of G-CSF in granulocyte differentiation and neutrophil development has been verified in G-CSF- and G-CSFR-deficient mice. These mice have lower numbers of circulating neutrophils, a decrease in granulocytic precursors and impaired terminal differentation of granulocytes.[106,107]

In discussing the IL-6-like cytokines, it bears to mention the heterodimeric cytokine IL-12. IL-12 was originally called NK cell stimulatory factor and can be regarded as a cytokine and soluble receptor complex.[108-110] The "cytokine" subunit, commonly known as IL-12α or p35, is coded for on human and mouse chromosome 3, shows homology with the IL-6-like cytokines and is not active on its own. The "soluble receptor" subunit, called IL-12β or p40, is coded for on human chromosome 5 and mouse chromosome 11, is a member of the cytokine receptor superfamily with homology to IL-6Rα/CD126 and has activity via the IL-12 receptor (IL-12R/CD212) when partnered with IL-12α. While both soluble subunits are required for biological activity, the two components are differentially regulated.[111] IL-12 is produced by APCs and has immunoregulatory effects on NK cells and T cells, two cell types that express the IL-12R.[112] IL-12 plays a critical role in cell-mediated immunity by acting as a requisite cytokine in pushing the balance between Th1 cells and Th2 cells towards Th1-type predominance. It is therefore no surprise that IL-12-deficient mice are defective in mounting an IFN-γ- or Th1-mediated immune response and/or respond with default Th2 responses when stimulated with antigen or infected with parasites or bacteria.[113-115] An interesting note on IL-12 is that a new composite cytokine has been described in mice and humans that consists of a novel α subunit, p19, that combines with IL-12β to form a unique cytokine called IL-23.[116] IL-23 has similar biological functions to IL-12 in that it can induce IFN-γ expression by T cells for example, yet it can act distinctly through an unidentified novel receptor subunit.[116]

IL-10-Like Cytokines

IL-10, IL-19 and IL-20 are members of the next related group of interleukins, those with homology to IL-10. The genes for these cytokines are closely linked on human and mouse chromosome 1.[117,118] Originally identified as human cytokine synthesis inhibitory factor (CSIF), IL-10 plays a major role in suppressing inflammatory responses. It does this by inhibiting the synthesis of IFN-γ, IL-2, IL-3, TNF-α and GM-CSF by cells such as macrophages and Th1 cells.[119,120] However, there is also evidence that IL-10 can act as a stimulator of thymocytes, mast cells and B cells.[120] Monocytes and T cells (Th2 cells) are considered to be the main sources of IL-10, although many other cell types can be made to produce IL-10 including B cells, mast cells and keratinocytes.[120] The participation of IL-10 in limiting Th1 cell responses and favoring Th2 cell development has been explored in IL-10 knockout mice. Mice that are deficient in IL-10 spontaneously develop chronic intestinal inflammation caused by uncontrolled cytokine production from dysregulated macrophages and Th1 cells.[121,122] IL-19 and IL-20 have been recently identified as IL-10 homologues. IL-19 is under patent application and not yet described while IL-20 appears to stimulate keratinocytes via its unique receptor.[118]

Interferons

The interferons are a family of cytokines that play a pivotal role in pathogen resistance. There are two types of interferons, type I and II, that signal through different receptors to produce distinct, but overlapping, cellular effects.[123] The pleiotropic cytokines IFN-α, originally referred to as leukocyte interferon, and IFN-β, originally referred to as fibroblast interferon, are type I interferons that are secreted by virus-infected cells.[124-128] Infection by most viruses causes a reaction in the host that includes innate and adaptive immune responses, such as the production of cytokines, increased expression of MHC class I and cytotoxic T cell mobilization. IFN-α and IFN-β, coded for by genes on human chromosome 9 and mouse chromosome 4, appear to be central players in innate immune responses.[128] IFN-α and IFN-β also have the unique ability to regulate adaptive T cell responses, perhaps directly by stimulating production of IFN-γ by activated T cells[129] or indirectly by inhibiting IL-4-inducible gene expression in monocytes.[130] These properties have been verified in knockout mice. Mice lacking the type I IFN receptor (CD118) exhibit impaired antiviral defenses and are deficient in promoting IFN-γ production by T cells.[129,131]

IFN-γ, also known as immune interferon or type II interferon, is secreted by activated T cells (Th1 cells) and NK cells.[123] It was originally identified as an antiviral agent and its gene was mapped to human chromosome 12 and mouse chromosome [10,123, 132, 133] IFN-γ signals through its own CDw119 receptor and has many biological functions. For example, IFN-γ can stimulate macrophages, increase antigen processing and expression of MHC molecules, promote an Ig class switch to IgG2a antibody secretion, and control the proliferation of transformed cells.[123] The immunomodulatory function of IFN-γ, however, has become a major research focus for this cytokine. IFN-γ secretion is the hallmark of proinflammatory Th1 cells but its exact role in T cell subset differentiation remains unclear. Th1 responses are associated with cell-mediated immunity and can best deal with intracellular invaders. Mice with mutations in IFN-γ[134] or IFN-γ receptor[135] expression show decreased macrophage and NK cell activity and increased susceptibility to many intracellular pathogens and viruses. Cell-mediated immune responses can still develop in IFN-γ knockout mice even though enhancements in Th2-type responses can be observed.[136, 137] As discussed above, IL-12 plays a critical role in eliciting Th1 responses. IFN-γ may act in synergy with IL-12 to accelerate development of the Th1 cell subset and also repress Th2 cells either directly or indirectly.[123]

Tumor Necrosis Factors

The TNF family is another example of a large group of interrelated cytokines that has stimulated a vast amount of scientific study.[138, 139] Most of this work has centered on the TNF family members' shared properties as cell death effectors.[140] The TNF family has been expanding a great deal recently so we have chosen five representative proteins, TNF-α, TNF-β, lymphotoxin (LT)-β, LIGHT (an acronym for homologous to lymphotoxins, exhibits inducible expression, and competes with HSV glycoprotein D for HVEM, a receptor expressed by T cells) and Fas ligand (FasL)/CD178, to describe in more detail. It appears that all the TNF family members act in trimeric form.[138, 139] Also, with the exception of TNF-β, the TNF ligands are formed as type II transmembrane proteins. Signaling by TNF family members is quite different than other cytokines we have discussed (see reference 138 and 140).

TNF-α, with its gene on human chromosome 6 and mouse chromosome 17 in close linkage to TNF-β, LT-β and MHC genes, is a pro-inflammatory cytokine that was originally identified as a tumour cell killer.[141-143] TNF-α can be found in a membrane bound or soluble form following proteolytic processing. TNF-α shares a receptor with TNF-β (CD120a, b), which is expressed on virtually all cell types except erythrocytes. TNF-α is produced mainly by activated macrophages, NK cells and T cells (mainly Th1 cells).[139] The most potent inducer of TNF-α is lipopolysaccharide (LPS), a microbial agent. TNF-α plays a role in endothelial activation and lymphocyte movement and is one of the crucial mediators in acute and chronic inflammatory conditions, such as autoimmunity, toxic shock and tuberculosis.[139, 144] It is also a direct pyrogen and can indirectly alter hormone and IL-1 secretion to induce fever. Like other members of the TNF family, TNF-α can induce apoptosis (programmed cell death) in some targets.[140]

TNF-β, also known as LT-α or LT, is derived from T and B cells and shares 30% homology at the amino acid level with TNF-α.[142, 145] TNF-β can exist as a true secreted homotrimeric protein or as a heterotrimeric membrane-associated complex with LT-β.[139] Like TNF-α, TNF-β plays a role in endothelial activation, tumour cell killing, apoptosis and mediation of inflammation. While occasional qualitative and quantitative differences have been demonstrated between the actions of TNF-α and TNF-β, the unique functions of TNF-β have not been fully elucidated.[138]

LT-β is a type II membrane protein that can anchor TNF-β in a heterotrimeric complex.[146] LT-β utilizes the LT-β receptor (LT-βR) and the herpes virus entry mediator (HVEM).[147] HVEM is a host-encoded receptor that is a member of the tumour necrosis factor receptor family and is exploited by herpes simplex virus (HSV) for entry. The same receptors can also be bound by LIGHT, a recent addition to the TNF family. LIGHT is produced by activated T cells, encoded on chromosome 16 in humans and chromosome 17 in mice and capable of both stimulating T cells and causing apoptosis depending on receptor expression.[147] LT-β, however, is produced by activated T and B cells much like TNF-β and is involved in lymph node development.[139]

As a testament to their important roles as immune mediators, TNF-α, TNF-β and LT-β knockout studies indicate that these three cytokines are required for normal lymphocyte compartmentalization in the spleen (summarized in reference 148). TNF-α- and TNFR1/CD120a-deficient mice lack follicular dendritic cells and fail to form B cell follicles. TNF-β and LT-β knockout mice exhibit similar defects in the spleen and also show impaired development of other lymphoid organs such as lymph nodes and Peyer's patches. These findings may stem from a role for the TNF-α and membrane TNF-β/LT-β heterotrimer in providing developmental cues to stromal cells to produce the chemokines necessary for lymphoid tissue organization.

FasL, newly assigned to CD178, is located on human and mouse chromosome 1.[149, 150] Like TNF-α, FasL can undergo proteolytic processing and exist as a soluble mediator. FasL is produced by T cells and is a key mediator of lymphocyte apoptosis and tolerance when associated with its receptor Fas/CD95. Most types of immune cells, as well as many nonlymphoid tissues, express Fas and/or FasL either constitutively or following activation. The Fas system is therefore very important in immune homeostasis and its powerful role must be tightly regulated or dangerous immune reactions and cancers would occur. Two very useful murine models have allowed a good dissection of the 'death' roles Fas and FasL play in immunity.[151, 152] The lpr (lymphoproliferation) mouse has a mutation in Fas that prevents Fas-induced apoptosis and causes complex defects in the B and T cell lymphoid compartments. Similarly, gld (generalized lymphoproliferative disease) mice are mutated in FasL and suffer the same immunoregulatory defects. The role of FasL in killing Fas-expressing T cells is especially evident in the testes, an area of immune privilege that can accept allografts and xenografts, where FasL expression by Sertoli cells is likely responsible for maintaining an immune barrier or immune tolerance.[153]

TGF-β

The transforming growth factor (TGF)-_ family consists of more than 30 members. TGF-β1, 2 and 3 are particularly interesting as they are remarkably multifunctional and indispensable, at least in the mouse. These homodimeric proteins are expressed by and have effects on many cell types. They are involved in development, immune regulation, immune tolerance, carcinogenesis, tissue repair and the generation and differentiation of many types of cells. As such, the TGF-β cytokine family represents an excellent example of a point of integration for multiple information networks, i.e., the immune and developmental programs. These functions cannot be completely outlined here and the reader is directed to several reviews for more details.[154-156] While the three isoforms of TGF-β are expressed under the control of unique promoters, they share a sequence identity of 70-80%, have similar cell targets and signal through the same serine-threonine kinase receptors (TGF-βR1, 2 and 3) in a manner that is unique from other cytokines.

TGF-β1 is the most abundant form of TGF-β and as such is often plainly referred to as TGF-β. It was originally identified for its ability to promote the growth of fibroblasts and assigned to chromosome 19 in humans and to chromosome 7 in mice.[157, 158] The human and mouse homologues differ by only one residue in their amino acid sequence. TGF-β1 is produced by every leukocyte lineage and has profound regulatory effects on a myriad of developmental, physiological and immune processes.[154] In general, TGF-β1 possesses both pro- and anti-inflammatory activity depending on the presence of other growth factors and the activation or differentiation state of the target cell.[154] For example, at a site of developing inflammation TGF-β1 can modulate the expression of adhesion molecules, act as a chemoattractant, and orchestrate the immune response by suppressing or activating leukocytes.[154, 159, 160] This orchestration by TGF-β1 also applies to the Th cell subset paradigm. TGF-β1 can alter the production of, and response to, cytokines of both Th subsets and can therefore skew Th1 or

Th2 immune responses as it sees fit depending on the composition of the inflammatory environment.[154] In fact, TGF-β1 secretion is a hallmark of a new candidate regulatory T cell subset called Th3 that also secrete IL-4 and IL-10.[161-163] With such widespread responsibilities, it is no surprise that TGF-β1 knockout mice exhibit immune dysregulation and succumb to a progressive wasting syndrome shortly after birth.[164-166] This mortal phenotype is characterized by changes in lymphoid organ architecture, including both the shrinking of the thymus and the swelling of lymph nodes, enhanced proliferation in vivo and defective mitogen responses in vitro. These mice also exhibit massive infiltrations of lymphocytes and macrophages in many organs resembling those found in autoimmune disorders.

TGF-β2, encoded on human and mouse chromosome 1, was originally identified as a suppressor of glioblastoma-derived T cells but is better known for its essential role in the developmental pathways of many tissues.[167] Accordingly, TGF-β2-deficient mice exhibit perinatal mortality and a wide array of tissue defects including craniofacial, skeletal, heart, eyes, ears and urogenital anomalies.[168] Likewise, TGF-β3, encoded on human chromosome 14 and mouse chromosome 12, appears to have an important role in certain developmental pathways as evidenced by TGF-β3-deficient mice that show severe defects in palate and lung morphogenesis and early death.[169-170]

Chemokines—Their Receptors and Their Genes

Chemokines are a family of low molecular weight chemotactic cytokines that regulate leukocyte migration through interactions with seven-transmembrane, rhodopsin-like G protein-coupled receptors.[172-174] Chemokines have significant structural homology and overlapping functions and can often bind to more than one receptor. In general, ligand binding results in chemokine receptor activation hallmarked by the phosphorylation of carboxyl-terminal serine/threonine residues, dissociation of heterotrimeric G proteins, generation of inositol trisphosphate, intracellular calcium release and activation of protein kinase C (PKC).[175] With additional activation of the Ras and Rho families of guanosine triphosphate (GTP)-binding proteins, chemokine receptors mediate multiple signaling pathways that regulate a wide variety of cellular responses.[175]

The chemokine field has developed at a rapid pace. This growth has caused classification headaches similar to those experienced by cytokine researchers decades ago. A classification system has been introduced to reduce confusion regarding the nomenclature of these molecules.[172, 174] Depending on the positions (or in one group the presence) of the first two cysteine residues in the primary structure of these molecules, the chemokine family can be divided into four groups as outlined in Table 2.2. Unlike the cytokines listed in Table 2.1, the human chemokines listed in Table 2.2 do not completely represent the murine chemokines because there are many differences in chemokine terminology between the two species and no matching homologues in some cases (see reference 172 and 174 for more details). The C group of chemokines (lacks cysteines one and three) has been recently described and consists of at least two ligands (XCL), namely lymphotactin/XCL1 and SCM-1β/XCL2, which both bind XCR1.[176] Lymphotactin, coded for on human chromosome 1, attracts lymphocytes but not monocytes or neutrophils. The human CC chemokine group (no intervening amino acid) includes at least 27 members (CCL), most of which are encoded on human chromosome 17, that bind at least 10 receptors (CCR). CC chemokine targets include monocytes, T cells, dendritic cells, eosinophils and NK cells. Representative CC chemokines include monocyte chemotactic protein (MCP)-1/CCL2, macrophage inflammatory protein (MIP)-1α/CCL3, MIP-1β/CCL4, regulated upon activation normally T expressed and secreted (RANTES)/CCL5 and eotaxin/CCL11. The CXC group of human chemokines (one amino acid lies between the first two cysteines) includes at least 14 ligands (CXCL). CXC chemokines are mostly encoded on human chromosome 4, bind at least five receptors (CXCR) and mediate mainly neutrophil chemotaxis. The CXC chemokine group can be divided into two main categories based on the presence of the tripeptide Glu-Leu-Arg (ELR) motif preceding the CXC motif. Representative CXC chemokines include IL-8/CXCL8 (ELR), monokine-induced by IFN-γ (MIG)/CXCL9

Table 2.2. Human chemokine and chemokine receptors[1]

Name	Synonym(s)	Amino Acids[2]	Ligand Location	Molecular Weight (Da)[2]	Chemokine Receptor(s)	Receptor Location
'C Chemokines'						
XCL1	lymphoactin α, SCM-1α, ATAC	114	1q21-q25	12517	XCR1	3p21
XCL2	lymphoactin β, SCM-1β, ATAC	114	1q23	12567	XCR1	3p21
'CC Chemokines'						
CCL1	I-309	96	17	10992	CCR8	3p22
CCL2	MCP-1, MCAF	99	17q11.2-q12	11025	CCR2	3p21
CCL3	MIP-1α, LD78α	92	17q11-q21	10085	CCR1, CCR5	3p21
CCL4	MIP-1β, LAG-1, ACT-2	92	17q11-q23	10212	CCR5	3p21
CCL5	RANTES	91	17q11.2-q12	9990	CCR1, CCR3, CCR5	3p21
CCL7	MCP-3	99	17q11.2-q12	11200	CCR1, CCR2, CCR3	3p21
CCL8	MCP-2	99	17q11.2	11246	CCR3	3p21
CCL11	eotaxin	97	17q21.1-q21.2	10732	CCR3	3p21
CCL13	MCP-4	98	17q11.2	10986	CCR2, CCR3	3p21
CCL14	HCC-1	93	17q11.2	10678	CCR1	3p21
CCL15	HCC-2, Lkn-1, MIP-1δ, MIP-5	113	17q11.2	12248	CCR1, CCR3	3p21
CCL16	HCC-4, LEC, LMC, LCC-1	120	17q11.2	13600	CCR1	3p21
CCL17	TARC	94	16q13	10507	CCR4	3p24
CCL18	DC-CK1, PARC, AMAC-α, MIP-4	89	17q11.2	9849	?	
CCL19	MIP-3β, ELC, exodus-3	98	9p13	10993	CCR7	17q12-q21.1
CCL20	MIP-3α, LARC, exodus-1	96	2q33-q37	10762	CCR6	6q27
CCL21	6Ckine, SLC, exodus-2	134	9p13	14646	CCR7	17q12-q21.2

continued on next page

Table 2.2. Cont.

CCL22	MDC, STCP-1		93	16q13	10580	CCR4	3p22
CCL23	MPIF-1, MIP-3, CKβ-8		120	17q11.2	13443	CCR1	3p21
CCL24	MPIF-2, eotaxin-2, CKβ-6		119	7q11.23	13133	CCR3	3p21
CCL25	TECK, MIP-4α		150	19p13.2	16639	CCR9	3p21
CCL26	eotaxin-3		94	7q11.2	10648	CCR3	3p21
CCL27	Eskine, CTACK, ILC		112	9p13	12618	CCR10	3p21
'CXC Chemokines'		ELR?					
CXCL1	GROα, MGSA-α	+	107	4q21	11301	CXCR1, CXCR2	2q35
CXCL2	GROβ, MGSA-β, MIP-2α	+	107	4q21	11389	CXCR2	2q35
CXCL3	GROγ, MGSA-γ, MIP-2β	+	107	4q21	11342	CXCR2	2q35
CXCL4	PF4, oncostatin A	-	101	4q12-q13	10845	?	
CXCL5	ENA-78	+	114	4q13-q21	11972	CXCR2	2q35
CXCL6	GCP-2	+	114	4q21	11897	CXCR1, CXCR2	2q35
CXCL7	NAP-2, PPBP	+	375	4q12-13	42823	CXCR2	2q35
CXCL8	IL-8, NAP-1, NAF, MDNCF	+	99	4q12-13	11098	CXCR1, CXCR2	2q35
CXCL9	Mig	-	125	4q21	14019	CXCR3	Xq13
CXCL10	IP-10	-	98	4q21	10856	CXCR3	Xq13
CXCL11	I-TAC	-	94	4q21.2	10365	CXCR3	Xq13
CXCL12	SDF-1α/β	-	93	10q11.1	10666	CXCR4	2q21
CXCL13	BLC, BCA-1	-	109	4q21	12664	CXCR5	11
CXCL14	BRAK	-	99	5q31	11722	?	
'CX3C Chemokines'							
CX3CL1	fractalkine		397	16q13	42202	CX3CR1	3p21

[1] List based on terminology from Zlotnick and Yoshie, 2000 with added annotation and data from Gene Cards (World Wide Web URL: http://genome-www.stanford.edu/genecards). Note that some of the chemokines listed are not discussed in this chapter.
[2] Data describes the unprocessed precursor.

(nonELR), IFN-γ inducible protein-10 (IP-10)/CXCL10 (nonELR) and stromal cell-derived factor-1 (SDF-1)/CXCL12 (nonELR). Lastly, the sole CX3C chemokine (three intervening amino acids), namely fractalkine/CX3CL1, is encoded on human chromosome 16, binds CX3CR1 and attracts T cells and monocytes but not neutrophils.[177]

Our understanding of the roles of chemokines in physiological and pathological processes has advanced significantly. It has become clear that in addition to wound healing, metastasis, angiogenesis/angiostasis, cell recruitment, lymphoid organ development, and lymphoid trafficking,[172,173] chemokines are fundamental in mediating innate and adaptive immune responses by their ability to activate cells of the immune system.[178-180] As with the cytokines, chemokine gene disruption studies have confirmed most of these biological functions. For example, the MIP-1α/CCL3 (a monocyte and T cell chemoattractant) knockout mouse was the first to be generated. While developmentally normal with no apparent lymphoid or myeloid defects, these mice were reduced in their ability to mount an inflammatory response to influenza infection.[181] In keeping with the role of eotaxin/CCL11 in attracting eosinophils, eotaxin/CCL11-deficient mice are reduced in their ability to mount eosinophil responses upon antigen challenge.[182] SDF-1/CXCL12-mutated mice exhibit a normal T cell compartment but have dramatic defects in B cell lymphopoiesis and myelopoiesis at the level of the bone marrow.[183] This result supports the critical role of SDF-1/CXCL12 as a modulator of progenitor cell development in the bone marrow. Knocking out the CXCR2 gene leads to impaired neutrophil migration in response to CXC chemokines, increases in circulating neutrophil numbers, and a dramatic increase in B cells.[184,185] Knocking out another CXC chemokine receptor, CXCR5, leads to perturbations in B cell colonization of secondary lymphoid tissues indicating the importance of BCA-1/CXCL13 in B cell coordination.[186,187] CCR7-deficient mice exhibit impaired lymphocyte migration, delayed antibody responses, no contact or delayed type hypersensitivity and defects in lymphoid architecture signifying an important role for CCR7 signaling in coordinating primary immune responses.[188] Lastly, mutating CCR1, CCR2 or CCR5 in mice impairs monocyte functions such as chemokine-dependent chemotaxis and alters the balance of Th1 or Th2 cytokine responses upon challenge with Th class-specific antigens or pathogens.[189-194] With this result, it is interesting to speculate that chemokines play a pivotal role in regulatory and inflammatory responses just like cytokines. For example, chemokines and their receptors have been associated with predominant Th1 or Th2 responses as eluded to earlier.[6,195-204] This association is evidenced by the linkage of MIP-1α, CXCR3 and CCR5 to Th1-type cells and MCP-1, CCR3, CCR4, and CCR8 to Th2-type cells. Along with cytokine-cytokine receptor interactions, chemokine-chemokine receptor interactions may modulate and stabilize the extent of leukocyte migration to and the nature of inflammation at a developing pathological site. Considering the power of chemokines in recruiting immune cells, these proteins may augment immune responses (helpful or dangerous) via normal immune surveillance mechanisms and may even determine the phenotype of responding cells (e.g., Th1 versus Th2 cells).

Concluding Remarks

As introduced earlier, the immune system is essentially a network supersystem utilizing specialized languages for communication between cells. This chapter focused on only the powerful language of cytokine and chemokine signaling but others exist such as hormone, neurotransmitter, complement and allergic mediator production. At their discretion, immune cells can listen to or send these signals as required to reach cells in the immediate microenvironment or throughout the organism. Just as miscommunication can crash modern electronic networks, the information super-highway of the immune system is not without its vulnerabilities. The knockout models discussed above exemplify the dramatic consequences of man-made alterations to immune network communication. The permanent absence of a particular cytokine at a whole-body level, however, may mask more subtle defects in the role of the conventional signaling component. Similarly, natural defects in communication appear to drive the pathogenesis of autoimmunity. In cases such as this, however, headway is being made in understanding how the immune system communicates and many therapies are being developed that may reset dangerous crashes and return an organism to 'online' protective status. A new understanding of the language used by cells of immune system to achieve this protection is emerging and cytokines and chemokines will certainly be at the heart of our progress in communication.

References

1. Cohen S, Bigazzi PE, Yoshida T. Similarities of T cell function in cell-mediated immunity and antibody production. Cell Immunol 1974; 12:150-159.
2. Dumonde DC, Wolstencroft RA, Panayi GS et al. Lymphokines: Nonantibody mediators of cellular immunity generated by lymphocyte activation. Nature 1969; 224:38-42.
3. Paul WE, Seder RA. Lymphocyte responses and cytokines. Cell 1994; 76:241-251.
4. Seder RA, Paul WE. Acquisition of lymphokine-producing phenotype by CD4+ T cells. Annu Rev Immunol 1994; 12:635-73:635-673.
5. Romagnani S. The Th1/Th2 paradigm. Immunol Today 1997; 18:263-266.
6. Luther SA, Cyster JG. Chemokines as regulators of T cell differentiation. Nat Immunol 2001; 2:102-107.
7. Mosmann TR, Cherwinski H, Bond MW et al. Two types of murine helper T cell clone. I. Definition according to profiles of lymphokine activities and secreted proteins. J Immunol 1986; 136:2348-2357.
8. Ivashkiv LB. Cytokines and STATs: How can signals achieve specificity? Immunity 1995; 3:1-4.
9. Murphy KM, Ouyang W, Farrar JD et al. Signaling and transcription in T helper development. Annu Rev Immunol 2000; 18:451-494.
10. Dinarello CA. Interleukin-1. Cytokine Growth Factor Rev 1997; 8:253-265.
11. Nicklin MJ, Weith A, Duff GW. A physical map of the region encompassing the human interleukin-1 alpha, interleukin-1 beta, and interleukin-1 receptor antagonist genes. Genomics 1994; 19:382-384.
12. Sims JE, Gayle MA, Slack JL et al. Interleukin 1 signaling occurs exclusively via the type I receptor. Proc Natl Acad Sci U S A 1993; 90:6155-6159.
13. Colotta F, Re F, Muzio M et al. Interleukin-1 type II receptor: A decoy target for IL-1 that is regulated by IL-4. Science 1993; 261:472-475.
14. Horai R, Asano M, Sudo K et al. Production of mice deficient in genes for interleukin (IL)-1alpha, IL-1beta, IL-1alpha/beta, and IL-1 receptor antagonist shows that IL-1beta is crucial in turpentine-induced fever development and glucocorticoid secretion. J Exp Med 1998; 187:1463-1475.
15. Kozak W, Zheng H, Conn CA et al. Thermal and behavioral effects of lipopolysaccharide and influenza in interleukin-1 beta-deficient mice. Am J Physiol 1995; 269:R969-R977.
16. Zheng H, Fletcher D, Kozak W et al. Resistance to fever induction and impaired acute-phase response in interleukin-1 beta-deficient mice. Immunity 1995; 3:9-19.
17. Alheim K, Chai Z, Fantuzzi G et al. Hyperresponsive febrile reactions to interleukin (IL) 1alpha and IL- 1beta, and altered brain cytokine mRNA and serum cytokine levels, in IL-1beta-deficient mice. Proc Natl Acad Sci USA 1997; 94:2681-2686.
18. Fantuzzi G, Dinarello CA. The inflammatory response in interleukin-1 beta-deficient mice: comparison with other cytokine-related knock-out mice. J Leukoc Biol 1996; 59:489-493.
19. Labow M, Shuster D, Zetterstrom M et al. Absence of IL-1 signaling and reduced inflammatory response in IL-1 type I receptor-deficient mice. J Immunol 1997; 159:2452-2461.
20. Arend WP, Malyak M, Guthridge CJ et al. Interleukin-1 receptor antagonist: Role in biology. Annu Rev Immunol 1998; 16:27-55.
21. Butcher C, Steinkasserer A, Tejura S et al. Comparison of two promoters controlling expression of secreted or intracellular IL-1 receptor antagonist. J Immunol 1994; 153:701-711.
22. Hirsch E, Irikura VM, Paul SM et al. Functions of interleukin 1 receptor antagonist in gene knockout and overproducing mice. Proc Natl Acad Sci USA 1996; 93:11008-11013.
23. Dinarello CA, Novick D, Puren AJ et al. Overview of interleukin-18: More than an interferon-gamma inducing factor. J Leukoc Biol 1998; 63:658-664.
24. Takeda K, Tsutsui H, Yoshimoto T et al. Defective NK cell activity and Th1 response in IL-18-deficient mice. Immunity 1998; 8:383-390.
25. Hoshino K, Tsutsui H, Kawai T et al. Cutting edge: generation of IL-18 receptor-deficient mice: Evidence for IL-1 receptor-related protein as an essential IL-18 binding receptor. J Immunol 1999; 162:5041-5044.
26. Rothe H, Hausmann A, Casteels K et al. IL-18 inhibits diabetes development in nonobese diabetic mice by counterregulation of Th1-dependent destructive insulitis. J Immunol 1999; 163:1230-1236.

27. Wild JS, Sigounas A, Sur N et al. IFN-gamma-inducing factor (IL-18) increases allergic sensitization, serum IgE, Th2 cytokines, and airway eosinophilia in a mouse model of allergic asthma. J Immunol 2000; 164:2701-2710.
28. Leite-De-Moraes MC, Hameg A, Pacilio M et al. IL-18 enhances IL-4 production by ligand-activated NKT lymphocytes: A pro-Th2 effect of IL-18 exerted through NKT cells. J Immunol 2001; 166:945-951.
29. He YW, Malek TR. The structure and function of gamma c-dependent cytokines and receptors: regulation of T lymphocyte development and homeostasis. Crit Rev Immunol 1998; 18:503-524.
30. Meissner U, Blum H, Schnare M et al. A soluble form of the murine common gamma chain is present at high concentrations in vivo and suppresses cytokine signaling. Blood 2001; 97:183-191.
31. Cao X, Shores EW, Hu-Li J et al. Defective lymphoid development in mice lacking expression of the common cytokine receptor gamma chain. Immunity 1995; 2:223-238.
32. Sugamura K, Asao H, Kondo M et al. The interleukin-2 receptor gamma chain: Its role in the multiple cytokine receptor complexes and T cell development in XSCID. Annu Rev Immunol 1996; 14:179-205.
33. Smith KA. T-cell growth factor. Immunol Rev 1980; 51:337-357.
34. Smith KA. Interleukin-2: Inception, impact, and implications. Science 1988; 240:1169-1176.
35. Schorle H, Holtschke T, Hunig T et al. Development and function of T cells in mice rendered interleukin-2 deficient by gene targeting. Nature 1991; 352:621-624.
36. Kundig TM, Schorle H, Bachmann MF et al. Immune responses in interleukin-2-deficient mice. Science 1993; 262:1059-1061.
37. Willerford DM, Chen J, Ferry JA et al. Interleukin-2 receptor alpha chain regulates the size and content of the peripheral lymphoid compartment. Immunity 1995; 3:521-530.
38. Suzuki H, Kundig TM, Furlonger C et al. Deregulated T cell activation and autoimmunity in mice lacking interleukin-2 receptor beta. Science 1995; 268:1472-1476.
39. Suzuki H, Zhou YW, Kato M et al. Normal regulatory alpha/beta T cells effectively eliminate abnormally activated T cells lacking the interleukin 2 receptor beta in vivo. J Exp Med 1999; 190:1561-1572.
40. Klein SC, Golverdingen JG, Bouwens AG et al. An alternatively spliced interleukin 4 form in lymphoid cells. Immunogenetics 1995; 41:57.
41. Paul WE. Interleukin-4: A prototypic immunoregulatory lymphokine. Blood 1991; 77:1859-1870.
42. Nelms K, Keegan AD, Zamorano J et al. The IL-4 receptor: Signaling mechanisms and biologic functions. Annu Rev Immunol 1999; 17:701-38:701-738.
43. Mosmann TR, Sad S. The expanding universe of T-cell subsets: Th1, Th2 and more. Immunol Today 1996; 17:138-146.
44. Minty A, Chalon P, Derocq JM et al. Interleukin-13 is a new human lymphokine regulating inflammatory and immune responses. Nature 1993; 362:248-250.
45. Kopf M, Le Gros G, Bachmann M et al. Disruption of the murine IL-4 gene blocks Th2 cytokine responses. Nature 1993; 362:245-248.
46. Kaplan MH, Schindler U, Smiley ST et al. Stat6 is required for mediating responses to IL-4 and for development of Th2 cells. Immunity 1996; 4:313-319.
47. Shimoda K, van Deursen J, Sangster MY et al. Lack of IL-4-induced Th2 response and IgE class switching in mice with disrupted Stat6 gene. Nature 1996; 380:630-633.
48. McKenzie GJ, Emson CL, Bell SE et al. Impaired development of Th2 cells in IL-13-deficient mice. Immunity 1998; 9:423-432.
49. Goodwin RG, Lupton S, Schmierer A et al. Human interleukin 7: Molecular cloning and growth factor activity on human and murine B-lineage cells. Proc Natl Acad Sci USA 1989; 86:302-306.
50. Lupton SD, Gimpel S, Jerzy R et al. Characterization of the human and murine IL-7 genes. J Immunol 1990; 144:3592-3601.
51. Hofmeister R, Khaled AR, Benbernou N et al. Interleukin-7: Physiological roles and mechanisms of action. Cytokine Growth Factor Rev 1999; 10:41-60.
52. Freeden-Jeffry U, Vieira P, Lucian LA et al. Lymphopenia in interleukin (IL)-7 gene-deleted mice identifies IL-7 as a nonredundant cytokine. J Exp Med 1995; 181:1519-1526.
53. Maki K, Sunaga S, Ikuta K. The V-J recombination of T cell receptor-gamma genes is blocked in interleukin-7 receptor-deficient mice. J Exp Med 1996; 184:2423-2427.
54. Maki K, Sunaga S, Komagata Y et al. Interleukin 7 receptor-deficient mice lack gammadelta T cells. Proc Natl Acad Sci USA 1996; 93:7172-7177.
55. Peschon JJ, Morrissey PJ, Grabstein KH et al. Early lymphocyte expansion is severely impaired in interleukin 7 receptor-deficient mice. J Exp Med 1994; 180:1955-1960.
56. Moore TA, Freeden-Jeffry U, Murray R et al. Inhibition of gamma delta T cell development and early thymocyte maturation in IL-7 -/- mice. J Immunol 1996; 157:2366-2373.

57. Renauld JC, Goethals A, Houssiau F et al. Human P40/IL-9. Expression in activated CD4+ T cells, genomic organization, and comparison with the mouse gene. J Immunol 1990; 144:4235-4241.
58. Townsend JM, Fallon GP, Matthews JD et al. IL-9-deficient mice establish fundamental roles for IL-9 in pulmonary mastocytosis and goblet cell hyperplasia but not T cell development. Immunity 2000; 13:573-583.
59. Grabstein KH, Eisenman J, Shanebeck K et al. Cloning of a T cell growth factor that interacts with the beta chain of the interleukin-2 receptor. Science 1994; 264:965-968.
60. Giri JG, Anderson DM, Kumaki S et al. IL-15, a novel T cell growth factor that shares activities and receptor components with IL-2. J Leukoc Biol 1995; 57:763-766.
61. Carson WE, Giri JG, Lindemann MJ et al. Interleukin (IL) 15 is a novel cytokine that activates human natural killer cells via components of the IL-2 receptor. J Exp Med 1994; 180:1395-1403.
62. Demirci G, Ferrari-Lacraz S, Groves C et al. IL-15 and IL-2: A matter of life and death for T cells in vivo. Nat Med 2001; 7:114-118.
63. Kennedy MK, Glaccum M, Brown SN et al. Reversible defects in natural killer and memory CD8 T cell lineages in interleukin 15-deficient mice. J Exp Med 2000; 191:771-780.
64. Lodolce JP, Boone DL, Chai S et al. IL-15 receptor maintains lymphoid homeostasis by supporting lymphocyte homing and proliferation. Immunity 1998; 9:669-676.
65. Miyajima A, Mui AL, Ogorochi T et al. Receptors for granulocyte-macrophage colony-stimulating factor, interleukin-3, and interleukin-5. Blood 1993; 82:1960-1974.
66. Lantz CS, Boesiger J, Song CH et al. Role for interleukin-3 in mast-cell and basophil development and in immunity to parasites. Nature 1998; 392:90-93.
67. Nishinakamura R, Nakayama N, Hirabayashi Y et al. Mice deficient for the IL-3/GM-CSF/IL-5 beta c receptor exhibit lung pathology and impaired immune response, while beta IL3 receptor-deficient mice are normal. Immunity 1995; 2:211-222.
68. Otsuka T, Miyajima A, Brown N et al. Isolation and characterization of an expressible cDNA encoding human IL-3. Induction of IL-3 mRNA in human T cell clones. J Immunol 1988; 140:2288-2295.
69. Yang YC, Ciarletta AB, Temple PA et al. Human IL-3 (multi-CSF): Identification by expression cloning of a novel hematopoietic growth factor related to murine IL-3. Cell 1986; 47:3-10.
70. Dorssers L, Burger H, Bot F et al. Characterization of a human multilineage-colony-stimulating factor cDNA clone identified by a conserved noncoding sequence in mouse interleukin-3. Gene 1987; 55:115-124.
71. Ichihara M, Hara T, Takagi M et al. Impaired interleukin-3 (IL-3) response of the A/J mouse is caused by a branch point deletion in the IL-3 receptor alpha subunit gene. EMBO J 1995; 14:939-950.
72. Mach N, Lantz CS, Galli SJ et al. Involvement of interleukin-3 in delayed-type hypersensitivity. Blood 1998; 91:778-783.
73. Kinashi T, Harada N, Severinson E et al. Cloning of complementary DNA encoding T-cell replacing factor and identity with B-cell growth factor II. Nature 1986; 324:70-73.
74. Azuma C, Tanabe T, Konishi M et al. Cloning of cDNA for human T-cell replacing factor (interleukin-5) and comparison with the murine homologue. Nucleic Acids Res 1986; 14:9149-9158.
75. Campbell HD, Tucker WQ, Hort Y et al. Molecular cloning, nucleotide sequence, and expression of the gene encoding human eosinophil differentiation factor (interleukin 5). Proc Natl Acad Sci USA 1987; 84:6629-6633.
76. Kopf M, Brombacher F, Hodgkin PD et al. IL-5-deficient mice have a developmental defect in CD5+ B-1 cells and lack eosinophilia but have normal antibody and cytotoxic T cell responses. Immunity 1996; 4:15-24.
77. Yoshida T, Ikuta K, Sugaya H et al. Defective B-1 cell development and impaired immunity against Angiostrongylus cantonensis in IL-5R alpha-deficient mice. Immunity 1996; 4:483-494.
78. Lee F, Yokota T, Otsuka T et al. Isolation of cDNA for a human granulocyte-macrophage colony-stimulating factor by functional expression in mammalian cells. Proc Natl Acad Sci USA 1985; 82:4360-4364.
79. Kaushansky K, O'Hara PJ, Berkner K et al. Genomic cloning, characterization, and multilineage growth-promoting activity of human granulocyte-macrophage colony-stimulating factor. Proc Natl Acad Sci USA 1986; 83:3101-3105.
80. Cantrell MA, Anderson D, Cerretti DP et al. Cloning, sequence, and expression of a human granulocyte/macrophage colony-stimulating factor. Proc Natl Acad Sci USA 1985; 82:6250-6254.
81. Huffman JA, Hull WM, Dranoff G et al. Pulmonary epithelial cell expression of GM-CSF corrects the alveolar proteinosis in GM-CSF-deficient mice. J Clin Invest 1996; 97:649-655.
82. Ohtani T, Ishihara K, Atsumi T et al. Dissection of signaling cascades through gp130 in vivo. Immunity 2000; 12:95-105.

83. Jones SA, Horiuchi S, Topley N et al. The soluble interleukin 6 receptor: Mechanisms of production and implications in disease. FASEB J 2001; 15:43-58.
84. Hirano T, Yasukawa K, Harada H et al. Complementary DNA for a novel human interleukin (BSF-2) that induces B lymphocytes to produce immunoglobulin. Nature 1986; 324:73-76.
85. Ferguson-Smith AC, Chen YF, Newman MS et al. Regional localization of the interferon-beta 2/ B-cell stimulatory factor 2/hepatocyte stimulating factor gene to human chromosome 7p15- p21. Genomics 1988; 2:203-208.
86. Sehgal PB, May LT, Tamm I et al. Human beta 2 interferon and B-cell differentiation factor BSF-2 are identical. Science 1987; 235:731-732.
87. Kopf M, Herren S, Wiles MV et al. Interleukin 6 influences germinal center development and antibody production via a contribution of C3 complement component. J Exp Med 1998; 188:1895-1906.
88. Fattori E, Cappelletti M, Costa P et al. Defective inflammatory response in interleukin 6-deficient mice. J Exp Med 1994; 180:1243-1250.
89. Kopf M, Baumann H, Freer G et al. Impaired immune and acute-phase responses in interleukin-6-deficient mice. Nature 1994; 368:339-342.
90. Kopf M, Ramsay A, Brombacher F et al. Pleiotropic defects of IL-6-deficient mice including early hematopoiesis, T and B cell function, and acute phase responses. Ann NY Acad Sci 1995; 762:308-318.
91. Gadient RA, Patterson PH. Leukemia inhibitory factor, interleukin 6, and other cytokines using the GP130 transducing receptor: roles in inflammation and injury. Stem cells 1999; 17:127-137.
92. McKinley D, Wu Q, Yang-Feng T et al. Genomic sequence and chromosomal location of human interleukin-11 gene (IL11). Genomics 1992; 13:814-819.
93. Paul SR, Bennett F, Calvetti JA et al. Molecular cloning of a cDNA encoding interleukin 11, a stromal cell-derived lymphopoietic and hematopoietic cytokine. Proc Natl Acad Sci USA 1990; 87:7512-7516.
94. Du XX, Williams DA. Interleukin-11: A multifunctional growth factor derived from the hematopoietic microenvironment. Blood 1994; 83:2023-2030.
95. Du X, Williams DA. Interleukin-11: Review of molecular, cell biology, and clinical use. Blood 1997; 89:3897-3908.
96. Nandurkar HH, Robb L, Tarlinton D et al. Adult mice with targeted mutation of the interleukin-11 receptor (IL11Ra) display normal hematopoiesis. Blood 1997; 90:2148-2159.
97. Gough NM, Gearing DP, King JA et al. Molecular cloning and expression of the human homologue of the murine gene encoding myeloid leukemia-inhibitory factor. Proc Natl Acad Sci USA 1988; 85:2623-2627.
98. Sutherland GR, Baker E, Hyland VJ et al. The gene for human leukemia inhibitory factor (LIF) maps to 22q12. Leukemia 1989; 3:9-13.
99. Escary JL, Perreau J, Dumenil D et al. Leukaemia inhibitory factor is necessary for maintenance of haematopoietic stem cells and thymocyte stimulation. Nature 1993; 363:361-364.
100. Chesnokova V, Auernhammer CJ, Melmed S. Murine leukemia inhibitory factor gene disruption attenuates the hypothalamo-pituitary-adrenal axis stress response. Endocrinology 1998; 139:2209-2216.
101. Zarling JM, Shoyab M, Marquardt H et al. Oncostatin M: A growth regulator produced by differentiated histiocytic lymphoma cells. Proc Natl Acad Sci USA 1986; 83:9739-9743.
102. Rose TM, Lagrou MJ, Fransson I et al. The genes for oncostatin M (OSM) and leukemia inhibitory factor (LIF) are tightly linked on human chromosome 22. Genomics 1993; 17:136-140.
103. Nagata S, Tsuchiya M, Asano S et al. Molecular cloning and expression of cDNA for human granulocyte colony-stimulating factor. Nature 1986; 319:415-418.
104. Demetri GD, Griffin JD. Granulocyte colony-stimulating factor and its receptor. Blood 1991; 78:2791-2808.
105. Yoshikawa A, Murakami H, Nagata S. Distinct signal transduction through the tyrosine-containing domains of the granulocyte colony-stimulating factor receptor. EMBO J 1995; 14:5288-5296.
106. Liu F, Wu HY, Wesselschmidt R et al. Impaired production and increased apoptosis of neutrophils in granulocyte colony-stimulating factor receptor-deficient mice. Immunity 1996; 5:491-501.
107. Lieschke GJ, Grail D, Hodgson G et al. Mice lacking granulocyte colony-stimulating factor have chronic neutropenia, granulocyte and macrophage progenitor cell deficiency, and impaired neutrophil mobilization. Blood 1994; 84:1737-1746.
108. Wolf SF, Temple PA, Kobayashi M et al. Cloning of cDNA for natural killer cell stimulatory factor, a heterodimeric cytokine with multiple biologic effects on T and natural killer cells. J Immunol 1991; 146:3074-3081.

109. Gubler U, Chua AO, Schoenhaut DS et al. Coexpression of two distinct genes is required to generate secreted bioactive cytotoxic lymphocyte maturation factor. Proc Natl Acad Sci USA 1991; 88:4143-4147.
110. Kobayashi M, Fitz L, Ryan M et al. Identification and purification of natural killer cell stimulatory factor (NKSF), a cytokine with multiple biologic effects on human lymphocytes. J Exp Med 1989; 170:827-845.
111. D'Andrea A, Rengaraju M, Valiante NM et al. Production of natural killer cell stimulatory factor (interleukin 12) by peripheral blood mononuclear cells. J Exp Med 1992; 176:1387-1398.
112. Gately MK, Renzetti LM, Magram J et al. The interleukin-12/interleukin-12-receptor system: Role in normal and pathologic immune responses. Annu Rev Immunol 1998; 16:495-521.
113. Cooper AM, Magram J, Ferrante J et al. Interleukin 12 (IL-12) is crucial to the development of protective immunity in mice intravenously infected with mycobacterium tuberculosis. J Exp Med 1997; 186:39-45.
114. Mattner F, Magram J, Ferrante J et al. Genetically resistant mice lacking interleukin-12 are susceptible to infection with Leishmania major and mount a polarized Th2 cell response. Eur J Immunol 1996; 26:1553-1559.
115. Magram J, Connaughton SE, Warrier RR et al. IL-12-deficient mice are defective in IFN gamma production and type 1 cytokine responses. Immunity 1996; 4:471-481.
116. Oppmann B, Lesley R, Blom B et al. Novel p19 protein engages IL-12p40 to form a cytokine, IL-23, with biological activities similar as well as distinct from IL-12. Immunity 2000; 13:715-725.
117. Kim JM, Brannan CI, Copeland NG et al. Structure of the mouse IL-10 gene and chromosomal localization of the mouse and human genes. J Immunol 1992; 148:3618-3623.
118. Blumberg H, Conklin D, Xu WF et al. Interleukin 20: Discovery, receptor identification, and role in epidermal function. Cell 2001; 104:9-19.
119. Vieira P, Waal-Malefyt R, Dang MN et al. Isolation and expression of human cytokine synthesis inhibitory factor cDNA clones: Homology to Epstein-Barr virus open reading frame BCRFI. Proc Natl Acad Sci USA 1991; 88:1172-1176.
120. Moore KW, O'Garra A, de Waal MR et al. Interleukin-10. Annu Rev Immunol 1993; 11:165-190.
121. Kuhn R, Lohler J, Rennick D et al. Interleukin-10-deficient mice develop chronic enterocolitis. Cell 1993; 75:263-274.
122. Berg DJ, Davidson N, Kuhn R et al. Enterocolitis and colon cancer in interleukin-10-deficient mice are associated with aberrant cytokine production and CD4(+) TH1-like responses. J Clin Invest 1996; 98:1010-1020.
123. Boehm U, Klamp T, Groot M et al. Cellular responses to interferon-gamma. Annu Rev Immunol 1997; 15:749-795.
124. Lawn RM, Adelman J, Dull TJ et al. DNA sequence of two closely linked human leukocyte interferon genes. Science 1981; 212:1159-1162.
125. Owerbach D, Rutter WJ, Shows TB et al. Leukocyte and fibroblast interferon genes are located on human chromosome 9. Proc Natl Acad Sci USA 1981; 78:3123-3127.
126. Shows TB, Sakaguchi AY, Naylor SL et al. Clustering of leukocyte and fibroblast interferon genes of human chromosome 9. Science 1982; 218:373-374.
127. Derynck R, Content J, DeClercq E et al. Isolation and structure of a human fibroblast interferon gene. Nature 1980; 285:542-547.
128. De Maeyer E, Maeyer-Guignard J. Type I interferons. Int Rev Immunol 1998; 17:53-73.
129. Cousens LP, Peterson R, Hsu S et al. Two roads diverged: Interferon alpha/beta- and interleukin 12-mediated pathways in promoting T cell interferon gamma responses during viral infection. J Exp Med 1999; 189:1315-1328.
130. Dickensheets HL, Donnelly RP. Inhibition of IL-4-inducible gene expression in human monocytes by type I and type II interferons. J Leukoc Biol 1999; 65:307-312.
131. Muller U, Steinhoff U, Reis LF et al. Functional role of type I and type II interferons in antiviral defense. Science 1994; 264:1918-1921.
132. Gray PW, Goeddel DV. Structure of the human immune interferon gene. Nature 1982; 298:859-863.
133. Naylor SL, Sakaguchi AY, Shows TB et al. Human immune interferon gene is located on chromosome 12. J Exp Med 1983; 157:1020-1027.
134. Dalton DK, Pitts-Meek S, Keshav S et al. Multiple defects of immune cell function in mice with disrupted interferon-gamma genes. Science 1993; 259:1739-1742.
135. Huang S, Hendriks W, Althage A et al. Immune response in mice that lack the interferon-gamma receptor. Science 1993; 259:1742-1745.
136. Wang ZE, Reiner SL, Zheng S et al. CD4+ effector cells default to the Th2 pathway in interferon gamma-deficient mice infected with Leishmania major. J Exp Med 1994; 179:1367-1371.

137. Graham MB, Dalton DK, Giltinan D et al. Response to influenza infection in mice with a targeted disruption in the interferon gamma gene. J Exp Med 1993; 178:1725-1732.
138. Wallach D, Varfolomeev EE, Malinin NL et al. Tumor necrosis factor receptor and Fas signaling mechanisms. Annu Rev Immunol 1999; 17:331-367.
139. Gruss HJ, Dower SK. The TNF ligand superfamily and its relevance for human diseases. Cytokines Mol Ther 1995; 1:75-105.
140. Ashkenazi A, Dixit VM. Death receptors: Signaling and modulation. Science 1998; 281:1305-1308.
141. Pennica D, Nedwin GE, Hayflick JS et al. Human tumour necrosis factor: Precursor structure, expression and homology to lymphotoxin. Nature 1984; 312:724-729.
142. Nedospasov SA, Shakhov AN, Turetskaya RL et al. Tandem arrangement of genes coding for tumor necrosis factor (TNF-alpha) and lymphotoxin (TNF-beta) in the human genome. Cold Spring Harb Symp Quant Biol 1986; 51 Pt 1:611-624.
143. Old LJ. Tumor necrosis factor (TNF). Science 1985; 230:630-632.
144. Sedgwick JD, Riminton DS, Cyster JG et al. Tumor necrosis factor: A master-regulator of leukocyte movement. Immunol Today 2000; 21:110-113.
145. Gray PW, Aggarwal BB, Benton CV et al. Cloning and expression of cDNA for human lymphotoxin, a lymphokine with tumour necrosis activity. Nature 1984; 312:721-724.
146. Browning JL, Ngam-ek A, Lawton P et al. Lymphotoxin beta, a novel member of the TNF family that forms a heteromeric complex with lymphotoxin on the cell surface. Cell 1993; 72:847-856.
147. Mauri DN, Ebner R, Montgomery RI et al. LIGHT, a new member of the TNF superfamily, and lymphotoxin alpha are ligands for herpesvirus entry mediator. Immunity 1998; 8:21-30.
148. Ngo VN, Korner H, Gunn MD et al. Lymphotoxin alpha/beta and tumor necrosis factor are required for stromal cell expression of homing chemokines in B and T cell areas of the spleen. J Exp Med 1999; 189:403-412.
149. Takahashi T, Tanaka M, Inazawa J et al. Human Fas ligand: Gene structure, chromosomal location and species specificity. Int Immunol 1994; 6:1567-1574.
150. Suda T, Takahashi T, Golstein P et al. Molecular cloning and expression of the Fas ligand, a novel member of the tumor necrosis factor family. Cell 1993; 75:1169-1178.
151. Takahashi T, Tanaka M, Brannan CI et al. Generalized lymphoproliferative disease in mice, caused by a point mutation in the Fas ligand. Cell 1994; 76:969-976.
152. Nagata S, Suda T. Fas and Fas ligand: lpr and gld mutations. Immunol Today 1995; 16:39-43.
153. Bellgrau D, Gold D, Selawry H et al. A role for CD95 ligand in preventing graft rejection. Nature 1995; 377:630-632.
154. Letterio JJ, Roberts AB. Regulation of immune responses by TGF-beta. Annu Rev Immunol 1998; 16:137-161.
155. Wahl SM, McCartney-Francis N, Mergenhagen SE. Inflammatory and immunomodulatory roles of TGF-beta. Immunol Today 1989; 10:258-261.
156. McCartney-Francis NL, Wahl SM. Transforming growth factor beta: A matter of life and death. J Leukoc Biol 1994; 55:401-409.
157. Fuji D, Brissenden JE, Derynck R et al. Transforming growth factor beta gene maps to human chromosome 19 long arm and to mouse chromosome 7. Somat Cell Mol Genet 1986; 12:281-288.
158. Roberts AB, Anzano MA, Lamb LC et al. New class of transforming growth factors potentiated by epidermal growth factor: Isolation from nonneoplastic tissues. Proc Natl Acad Sci USA 1981; 78:5339-5343.
159. Allen JB, Manthey CL, Hand AR et al. Rapid onset synovial inflammation and hyperplasia induced by transforming growth factor beta. J Exp Med 1990; 171:231-247.
160. Brandes ME, Allen JB, Ogawa Y et al. Transforming growth factor beta 1 suppresses acute and chronic arthritis in experimental animals. J Clin Invest 1991; 87:1108-1113.
161. Bridoux F, Badou A, Saoudi A et al. Transforming growth factor beta (TGF-beta)-dependent inhibition of T helper cell 2 (Th2)-induced autoimmunity by self-major histocompatibility complex (MHC) class II-specific, regulatory CD4(+) T cell lines. J Exp Med 1997; 185:1769-1775.
162. Fukaura H, Kent SC, Pietrusewicz MJ et al. Induction of circulating myelin basic protein and proteolipid protein-specific transforming growth factor-beta1-secreting Th3 T cells by oral administration of myelin in multiple sclerosis patients. J Clin Invest 1996; 98:70-77.
163. Chen Y, Kuchroo VK, Inobe J et al. Regulatory T cell clones induced by oral tolerance: Suppression of autoimmune encephalomyelitis. Science 1994; 265:1237-1240.
164. Shull MM, Ormsby I, Kier AB et al. Targeted disruption of the mouse transforming growth factor-beta 1 gene results in multifocal inflammatory disease. Nature 1992; 359:693-699.
165. Kulkarni AB, Huh CG, Becker D et al. Transforming growth factor beta 1 null mutation in mice causes excessive inflammatory response and early death. Proc Natl Acad Sci USA 1993; 90:770-774.

166. Christ M, McCartney-Francis NL, Kulkarni AB et al. Immune dysregulation in TGF-beta1-deficient mice. J Immunol 1994; 153:1936-1946.
167. de Martin R, Haendler B, Hofer-Warbinek R et al. Complementary DNA for human glioblastoma-derived T cell suppressor factor, a novel member of the transforming growth factor-beta gene family. EMBO J 1987; 6:3673-3677.
168. Sanford LP, Ormsby I, Gittenberger-de Groot AC et al. TGFbeta2 knockout mice have multiple developmental defects that are non overlapping with other TGFbeta knockout phenotypes. Development 1997; 124:2659-2670.
169. ten Dijke P, Hansen P, Iwata KK et al. Identification of another member of the transforming growth factor type beta gene family. Proc Natl Acad Sci USA 1988; 85:4715-4719.
170. Barton DE, Foellmer BE, Du J et al. Chromosomal mapping of genes for transforming growth factors beta 2 and beta 3 in man and mouse: dispersion of TGF-beta gene family. Oncogene Res 1988; 3:323-331.
171. Kaartinen V, Voncken JW, Shuler C et al. Abnormal lung development and cleft palate in mice lacking TGF-beta 3 indicates defects of epithelial-mesenchymal interaction. Nat Genet 1995; 11:415-421.
172. Zlotnik A, Yoshie O. Chemokines: A new classification system and their role in immunity. Immunity 2000; 12:121-127.
173. Baggiolini M, Dewald B, Moser B. Human chemokines: An update. Annu Rev Immunol 1997; 15:675-705.
174. Murphy PM, Baggiolini M, Charo IF et al. International union of pharmacology. XXII. Nomenclature for chemokine receptors. Pharmacol Rev 2000; 52:145-176.
175. Luster AD. Chemokines—Chemotactic cytokines that mediate inflammation. N Engl J Med 1998; 338:436-445.
176. Kelner GS, Kennedy J, Bacon KB et al. Lymphotactin: A cytokine that represents a new class of chemokine. Science 1994; 266:1395-1399.
177. Bazan JF, Bacon KB, Hardiman G et al. A new class of membrane-bound chemokine with a CX3C motif. Nature 1997; 385:640-4.
178. Strieter RM, Polverini PJ, Kunkel SL et al. The functional role of the ELR motif in CXC chemokine-mediated angiogenesis. J Biol Chem 1995; 270:27348-27357.
179. Xia Y, Pauza ME, Feng L et al. RelB regulation of chemokine expression modulates local inflammation. Am J Pathol 1997; 151:375-387.
180. Ward SG, Bacon K, Westwick J. Chemokines and T lymphocytes: more than an attraction. Immunity 1998; 9:1-11.
181. Cook DN, Beck MA, Coffman TM et al. Requirement of MIP-1 alpha for an inflammatory response to viral infection. Science 1995; 269:1583-1585.
182. Rothenberg ME, MacLean JA, Pearlman E et al. Targeted disruption of the chemokine eotaxin partially reduces antigen-induced tissue eosinophilia. J Exp Med 1997; 185:785-790.
183. Nagasawa T, Hirota S, Tachibana K et al. Defects of B-cell lymphopoiesis and bone-marrow myelopoiesis in mice lacking the CXC chemokine PBSF/SDF-1. Nature 1996; 382:635-638.
184. Cacalano G, Lee J, Kikly K et al. Neutrophil and B cell expansion in mice that lack the murine IL-8 receptor homolog. Science 1994; 265:682-684.
185. Czuprynski CJ, Brown JF, Steinberg H et al. Mice lacking the murine interleukin-8 receptor homologue demonstrate paradoxical responses to acute and chronic experimental infection with Listeria monocytogenes. Microb Pathog 1998; 24:17-23.
186. Forster R, Mattis AE, Kremmer E et al. A putative chemokine receptor, BLR1, directs B cell migration to defined lymphoid organs and specific anatomic compartments of the spleen. Cell 1996; 87:1037-1047.
187. Ansel KM, Ngo VN, Hyman PL et al. A chemokine-driven positive feedback loop organizes lymphoid follicles. Nature 2000; 406:309-314.
188. Forster R, Schubel A, Breitfeld D et al. CCR7 coordinates the primary immune response by establishing functional microenvironments in secondary lymphoid organs. Cell 1999; 99:23-33.
189. Gao JL, Wynn TA, Chang Y et al. Impaired host defense, hematopoiesis, granulomatous inflammation and type 1-type 2 cytokine balance in mice lacking CC chemokine receptor 1. J Exp Med 1997; 185:1959-1968.
190. Gerard C, Frossard JL, Bhatia M et al. Targeted disruption of the beta-chemokine receptor CCR1 protects against pancreatitis-associated lung injury. J Clin Invest 1997; 100:2022-2027.
191. Boring L, Gosling J, Chensue SW et al. Impaired monocyte migration and reduced type 1 (Th1) cytokine responses in C-C chemokine receptor 2 knockout mice. J Clin Invest 1997; 100:2552-2561.
192. Kurihara T, Warr G, Loy J et al. Defects in macrophage recruitment and host defense in mice lacking the CCR2 chemokine receptor. J Exp Med 1997; 186:1757-1762.

193. Zhou Y, Kurihara T, Ryseck RP et al. Impaired macrophage function and enhanced T cell-dependent immune response in mice lacking CCR5, the mouse homologue of the major HIV-1 coreceptor. J Immunol 1998; 160:4018-4025.
194. Sato N, Ahuja SK, Quinones M et al. CC chemokine receptor (CCR)2 is required for langerhans cell migration and localization of T helper cell type 1 (Th1)-inducing dendritic cells. Absence of CCR2 shifts the Leishmania major-resistant phenotype to a susceptible state dominated by Th2 cytokines, b cell outgrowth, and sustained neutrophilic inflammation. J Exp Med 2000; 192:205-218.
195. Karpus WJ, Lukacs NW, McRae BL et al. An important role for the chemokine macrophage inflammatory protein-1 alpha in the pathogenesis of the T cell-mediated autoimmune disease, experimental autoimmune encephalomyelitis. J Immunol 1995; 155:5003-5010.
196. Kunkel SL. Th1- and Th2-type cytokines regulate chemokine expression. Biol Signals 1996; 5:197-202.
197. Karpus WJ, Kennedy KJ. MIP-1alpha and MCP-1 differentially regulate acute and relapsing autoimmune encephalomyelitis as well as Th1/Th2 lymphocyte differentiation. J Leukoc Biol 1997; 62:681-687.
198. Sallusto F, Mackay CR, Lanzavecchia A. Selective expression of the eotaxin receptor CCR3 by human T helper 2 cells. Science 1997; 277:2005-2007.
199. Bonecchi R, Bianchi G, Bordignon PP et al. Differential expression of chemokine receptors and chemotactic responsiveness of type 1 T helper cells (Th1s) and Th2s. J Exp Med 1998; 187:129-134.
200. Loetscher P, Uguccioni M, Bordoli L et al. CCR5 is characteristic of Th1 lymphocytes. Nature 1998; 391:344-345.
201. Sallusto F, Lenig D, Mackay CR et al. Flexible programs of chemokine receptor expression on human polarized T helper 1 and 2 lymphocytes. J Exp Med 1998; 187:875-883.
202. Kunkel SL, Strieter RM, Lindley IJ et al. Chemokines: New ligands, receptors and activities. Immunol Today 1995; 16:559-561.
203. Cameron MJ, Arreaza GA, Grattan M et al. Differential expression of CC chemokines and the CCR5 receptor in the pancreas is associated with progression to type 1 diabetes. J Immunol 2000; 165:1102-1110.
204. Karpus WJ, Lukacs NW, Kennedy KJ et al. Differential CC chemokine-induced enhancement of T helper cell cytokine production. J Immunol 1997; 158:4129-4136.

CHAPTER 3

Cytokine and Cytokine Receptor Genes in the Susceptibility and Resistance to Organ-Specific Autoimmune Diseases

Hélène Coppin, Marie-Paule Roth and Roland S. Liblau

Introduction

It is beyond the scope of this Chapter to review exhaustively the research on all autoimmune diseases. We will instead focus on three highly-prevalent chronic inflammatory/autoimmune diseases, namely multiple sclerosis, rheumatoid arthritis and insulin-dependent diabetes mellitus. For each disease, an in-depth analysis of the available data regarding the potential role of pro-inflammatory cytokines, anti-inflammatory cytokine and chemokine genes in conferring genetic susceptibility or resistance will be performed.

Cytokine and Cytokine Receptor Genes in the Susceptibility to Multiple Sclerosis (MS)

Multiple sclerosis (MS) is an inflammatory disease of the central nervous system (CNS) white matter characterized by demyelination, focal T cell and macrophage infiltrates, axonal injury and loss of neurological function.[1-3] The disease typically manifests between the ages of 20 and 40 and affects women twice as often as men. It is the major cause of neurological disability in young people in the Western Hemisphere. MS is generally categorized as being either relapsing-remitting or primary-progressive in onset. The relapsing-remitting form of disease (85% of cases) is characterized by a series of attacks that result in varying degrees of disability and from which the patients recover partly or completely. The progressive form of disease (15% of cases) lacks the acute attacks and instead typically involves a gradual clinical decline. The course of disease of relapsing-remitting patients may ultimately changes to a progressive form known as secondary-progressive MS.

Experimental autoimmune encephalomyelitis (EAE) is an inflammatory condition that has been used as an animal model of MS. It can be induced in susceptible animals by immunization with myelin, myelin components or their immuno-dominant epitopes, or by transfer of CD4+ myelin antigen-specific T lymphocytes. Similarly to MS, EAE is characterized by multifocal perivascular CNS inflammatory infiltrates primarily comprised of T cells and monocytes.[4-5]

Because no cure exists for MS and because MS is known to have an underlying genetic component,[6] considerable effort has gone into mapping the predisposing loci.[7] However, the genetic, geographic, and disease heterogeneities that characterize MS make this a difficult task

Cytokines and Chemokines in Autoimmune Disease, edited by Pere Santamaria.

in human populations. Given the immunopathological similarities between EAE and MS, this animal model, which is also under genetic control, has been used to aid this undertaking.

Cytokines have been divided into proinflammatory (type 1) cytokines, such as IL-2, interferon-γ, TNF, and IL-12, and immunoregulatory (type 2) cytokines, including IL-4, IL-10, IL-13, and transforming growth factor-β. The pathogenic role of the former and the protective role of the latter have been established in animal models of MS.[8] In MS, a similar pathogenic role of proinflammatory cytokines and a down-modulating role of type 2 cytokines has been suggested.[8,9] Moreover, treatment with IFN-γ is deleterious,[10] whereas beneficial therapies have been associated with immune deviation favoring type 2 cytokine production.[11-13] A large number of polymorphisms have been reported within cytokine genes; many of them occur within known or putative regulatory regions and have been shown to influence gene expression (for review, see Bidwell et al).[14] Since differential rates of the production of cytokines may influence the susceptibility to MS, these polymorphisms are useful chromosomal markers in genetic studies of MS. Interestingly, genetic susceptibility to experimental demyelinating models of MS has been shown to be provided by several genomic regions, some of which contain cytokine or cytokine receptor genes.[15-20] Several groups have therefore tested the hypothesis that cytokine or cytokine receptor genes may influence susceptibility to MS. A review of the results of these studies is presented below.

Activated myelin-reactive T cells capable of secreting proinflammatory cytokines must migrate from the periphery into the CNS to participate in the demyelination and pathology associated with EAE and MS. Chemokines can enhance T cell and monocyte migration through direct chemoattraction and by activating leukocyte integrins to bind their adhesion receptors on endothelial cells. Certain chemokines have recently been identified which appear to selectively recruit these cells into the CNS and which are associated with EAE disease activity.[21] Similarly in MS, the expression levels of chemokines and chemokine receptors are increased in the cerebrospinal fluid of MS patients as well as in brain tissue (for review, see Ransohoff).[22] This suggests that the chemokine pathway is an important component in the etiology of EAE and MS and has recently prompted several authors to examine the influence of polymorphisms in the chemokine and chemokine receptor genes on MS susceptibility.

Pro-Inflammatory Cytokine and Cytokine Receptor Genes and Susceptibility to MS

Genes Encoding Tumor Necrosis Factors α and β and Their Receptors

TNFα is a pro-inflammatory cytokine which is toxic to myelin in vitro, causes selective injury to oligodendrocytes and induces proliferation of astrocytes.[23] It is present in active areas of demyelination in brains of patients with MS[24] and is upregulated prior to relapses.[25] Both in vivo and in vitro studies have shown that the amount of TNFα produced upon a standard stimulus varies between different individuals. The observation that the inter-individual differences in the capacity to produce TNF may be caused by genetic differences in the genes encoding the tumor necrosis factors has generated a huge amount of research on the possible involvement of polymorphisms at the TNFα and TNFβ loci, located near the HLA class III region on chromosome 6p, in susceptibility to MS.

Digestion with *Nco*I produces two restriction fragment length polymorphism (RFLP) alleles because of a single base difference in the first intron of TNFβ. No significant difference in the distribution of these alleles was found between control groups and patients with MS in Denmark[26] and France.[27]

Four dinucleotide repeat polymorphisms within the TNF complex have also been studied (TNFa, TNFb, TNFc, and TNFd), especially since the TNFa2 allele was found to correlate with higher levels of TNFα secretion by lipopolysaccharide-stimulated monocytes. Comparisons of these microsatellite allelic distributions in cases and controls from France,[27,28] Belgium,[29] Sweden,[30] Germany,[31] Japan[32] and Northern Ireland[33] revealed that none of these polymorphisms was associated with MS after controlling for the HLA-DR2 haplotype. Only two groups have found significant, but contradictory, haplotype associations, independent of HLA-DRB1*1501: the TNFa11b4 haplotype indeed was observed more commonly in Irish patients[34] but less frequently in Spanish patients than in ethnically matched controls. In the latter study, two other haplotypes, TNFa10b4 and TNFa1b5, were more common in patients than in controls.[35]

More recently, a biallelic polymorphism has been identified in the promoter region of TNFα. The less common TNF2 allele, in which guanine at the -308 position is substituted by adenosine, has been found to correlate with higher constitutive and inducible TNFα secretion. Several groups, using patients and controls from Sweden,[36] Germany,[36,37] the United States,[38] the Netherlands,[39] Japan,[32] Poland,[40] Greek Cyprus,[37] Spain,[41] Sardinia[42] and France,[28] have reported that this TNFα -308 polymorphism was not associated with MS susceptibility. Two other G→A polymorphisms at positions -376 and -238 relative to the TNFα transcription start site have been described. A significant association was found between MS susceptibility and the TNFα -376 polymorphism in Spanish patients, independent of the HLA class II association.[41] The combined inheritance of DRB1*1501 and TNFα -376A was even shown to more than additively increase susceptibility to MS. Huizinga et al[39] found a significant association with the TNFa -238 polymorphism in Dutch patients with severe disease, but this result was not confirmed in Spanish,[41] German[31] and French[28] patients. Of note, Mycko et al[40] noticed that a combined TNFα promoter/TNFβ exon 3 polymorphism contributed to the development of MS, particularly to the highest disability scores, independently of HLA-DR2. To complete this investigation, a comprehensive search for possible mutations in the four TNFα exons and the promoter region was performed, but revealed no polymorphisms associated with the course or outcome of MS.[43]

In conclusion, even though several lines of evidence suggest that TNF is involved in the pathogenesis of MS and certain TNF polymorphisms are associated with higher levels of expression, genetic variants of TNF have not been convincingly and reproducibly associated with either susceptibility to MS or disease course.

The first step in the induction of many of the biological effects elicited by TNFa is its binding to the cell surface TNF receptors I and II. Interestingly, increased serum levels of TNF-RI have been found in MS patients.[44,45] This prompted McDonnell et al.[33] to genotype patients and controls from Northern Ireland for a microsatellite marker in the vicinity of the TNF-RI gene on chromosome 12p13.2-p13.3 and another in the vicinity of the TNF-RII gene on chromosome 1p21.2-p23.1. They did not find, however, any significant association of either of these markers with MS.

Genes Encoding Interferon γ, Its Receptors and Interferon Regulatory Factors

IFNγ is known to exert an important disease-promoting and pro-inflammatory effect in MS. Systemic administration of IFNγ indeed leads to an exacerbation of the disease.[10] The IFNγ gene is a single-copy gene located on chromosome 12q14-q15. A polymorphic CA repeat element has been identified in the first intron, at the heart of a complex intronic system of transcription-regulatory activity. Goris et al[46] analyzed the potential association of this dinucleotide repeat polymorphism with MS in German, Northern Italian, Swedish, and Sardinian population-based data sets. Previously, Vandenbroeck et al[47] presented evidence for association of this polymorphism with MS in Sardinian patients not carrying the predisposing HLA-DRB1*03 or 04 alleles. In their study, Goris et al further analyzed the nature of this association by transmission disequilibrium testing and observed that the transmission bias was essentially confined to DRB1*03 and 04 negative males. They also evaluated the relevance of this association in other European populations. The only population showing a global disease association with this microsatellite polymorphism was the Swedish one. Of note, in another study on Swedish MS, He et al[48] found no association in a case-control study but reported weak indication of linkage for the same IFNγ gene polymorphism. No association or linkage with this polymorphism was found in Finnish case-control and multiplex datasets,[49] in

German MS patients and controls[39] or in French sib pairs.[50] Giedraitis et al[51] screened the IFN-γ gene promoter and part of the first intron, known to contain a c-Rel specific enhancer, for possible mutations. They found a C to T substitution in the IFNγ promoter at position -333. Screening for this mutation in 214 Swedish MS patients and 164 controls identified two patients, both heterozygous, but no controls with this mutation. The low frequency of this mutation indicates that this polymorphism is unlikely to contribute greatly to the weak indication of linkage previously observed.[48]

It can be concluded that the IFNγ gene is highly conserved and that changes in IFNγ expression may be due to the influence of regulatory factors on gene transcription, rather than gene polymorphisms. Epplen et al[31] thus investigated two microsatellites, one in intron 7 of the interferon regulatory factor 1 (IRF-1) gene on chromosome 5q31.1, and the other in the vicinity of the IRF-2 gene on chromosome 4q35. IRF-1 is a transcriptional activator of IFN and IFN-inducible genes, while IRF-2 is an antagonist of IRF-1 activity. None of these markers was statistically associated with the disease. The repeat polymorphism in the 7th intron of the IRF-1 gene was also used as a marker to test for association with MS in a case-control study including individuals from Germany, Northern Italy, and Sweden. In none of these populations was a significant allelic association with MS found. This lack of association was confirmed by testing transmission disequilibrium of individual IRF-1 alleles in a sample of Sardinian simplex MS families. No deviation from the expected 50% transmission rate was seen.[52] These studies do not provide evidence of IRF-1 being a candidate for conferring genetic susceptibility to MS.

To investigate the possible linkage between IFNγ receptors and MS, Reboul et al[50] genotyped 116 French sib pairs for three microsatellite markers surrounding the IFNγ receptor 1 on chromosome 6q23-q24 and three microsatellites surrounding the IFNγ receptor 2 on chromosome 21q22.1. They found, however, no evidence for linkage of MS with any of these markers.

Genes Encoding Interleukin-1 α and β, Their Receptors and Interleukin-1 Receptor Antagonist

IL-1 α and β are important pro-inflammatory cytokines which are produced in and at the edge of the MS lesions by macrophage and microglia[53] and could well participate in the destruction of CNS myelin. The interleukin-1 receptor antagonist (IL-1ra) is a naturally occurring competitive inhibitor of IL-1α and IL-1β and, as such, plays an important role in the regulation of the inflammatory process. In MS, circulating levels of IL-1ra have been shown to correlate with disease activity[54] and in EAE, treatment with recombinant IL-1ra is known to reduce the severity of the disease.[55] Genes encoding the three structurally related cytokines (IL-1α, IL-1β and IL-1ra) are clustered within a 430-kb region and together with the closely linked IL-1 type 1 and type 2 receptor genes form the IL-1 gene cluster on human chromosome 2 (2q12-q22). Intron 5 of the IL-1α gene contains a dinucleotide repeat, exon 5 of the IL-1β gene a biallelic polymorphism, and intron 2 of the IL-1ra gene a variable number of tandem repeats (VNTR). The number of repeats in this VNTR has been shown to influence the production of IL-1ra, with the allele A2 resulting in higher levels of IL-1ra by comparison with the A1 allele.

Disease association of the IL-1ra A2 allele with MS has been initially reported in Dutch patients.[56] Subsequently, however, attempts to replicate this finding by groups in Sweden,[57] France,[58] Spain,[59] Germany,[31] Finland,[49] the Netherlands[60] or the United States[61] have been unsuccessful, although de la Concha et al[59] found some evidence for association of the A2 allele in a subgroup of relapsing-remitting patients. More recently, Sciacca and coll.[62] in Italy studied the same polymorphism and found the opposite effect, reporting that susceptibility to the disease was associated with the A1 allele. Analysis of the polymorphism in connection with disease progression has produced equally conflicting results, with some authors claiming that carriers of IL-1ra A2 and not IL-1β A2 in the Netherlands had a more aggressive disease[60] or that carriers of IL-1ra A3 or IL-1β A2 in the United States had a favorable outcome,[61] and others finding the reverse,[62] the IL-1ra allele A1 being associated with a more rapid progression

Table 1. Cytogenetic and chromosome positions of cytokines and chemokines

Cytokine/Chemokine	Human Cytogenetic Position	Mouse Chromosome and cM Position	
TNF-α	6p21.3	17	19.06
TNF-β	6p21.3	17	19.06
TNF-RI	12p13.2-p13.3	6	60.55
TNF-RII	1p21.2-p23.1	4	75.5
IFN-α	9p22	4	42.6
IFN-β	9p22	4	42.6
INF-γ	12q14-q15	10	67
INF-γ R1	4q35	10	15
INF-γ R2	21q22.1	16	65
IL-1α	2q13	2	73
IL-1β	2q13	2	73
IL-1 Ra	2q13	2	10
IL-2	4q34-q27	3	19.2
IL-2 Rα	10p15-p14	2	6.4
IL-2 Rβ	22q13.1	15	43.3
IL-3	5q31.1	11	28.5
IL-4	5q31.1	11	29
IL-4 Rα	16p12.1	7	62
IL-5	5q31.1	11	29.2
IL-6	7p21	5	17
IL-10	1q31-q32	1	69.9
IL-10 R	11q23.3	9	26
IL-12 p35	3q25-q26.2	3	37
IL-12 p40	5q31.1-q33.1	11	19
IL-12 Rβ1	19q13.1	8	33.5
IL-12 Rβ2	1p31.3-p31.2	6	30.1
IL-13	5q31	11	29
TGF-β1	19q13.2	7	6.5
TGF-β2	1q41	1	101.5
TGF-β3	14q24	12	41
TGF-β R1	9q33	4	19.3
TGF-β R2	3p22	9	69
TGF-β R3	1p32-p33	3	n.d.
GM-CSF	5q23-q31	11	29.5
IRF1	5q31.1	11	29
IRF2	4q34.1-q35.1	8	n.d.
CCR2	3p21-p24	9	72
CCR5	3p21-p24	9	72
MCP-3	17q11.2	11	46.5

Cytogenetic and map positions are taken from http://www.ncbi.nlm.nih.gov/Homology/

in Italy. The IL-1α microsatellite was not found associated with MS in Germany[31] or Northern Ireland.[63] Against this background, Feakes et al[64] typed the IL-1ra VNTR in 536 simplex families. In order to improve the information extracted from these families they also typed a closely mapped SNP from the promoter of the IL-1β gene. Disease associations were assessed by transmission disequilibrium testing. None of the alleles from the VNTR, the SNP, or their haplotype, showed statistically significant evidence for association with MS or with disease severity. The same authors performed a crude meta-analysis by combining all the published data concerning the IL-1ra gene VNTR. This analysis suggests that any effect of this gene on susceptibility to MS or its progression is non-existent or, at best, small.

Genes Encoding Interleukin-2 and Its Receptor

IL-2 plays a key role in promoting the growth and proliferation of human T cells by its interaction with a specific surface receptor, IL-2R. This receptor is composed of two distinct subunits, the alpha and the beta chains. Both of these chains bind IL-2, alpha at low affinity and beta at intermediate affinity. When linked together non-covalently, these two chains form the physiological, high affinity binding site; signal is transduced through a third receptor chain called γc. Serum sIL-2R levels have been found to be elevated in MS.[65-67] The gene for IL-2 maps to 4q34-q27 and a polymorphic repeat has been located in its 3' flanking region. The IL-2Rα chain gene has been mapped to chromosome 10p15 and the IL-2Rβ chain gene to chromosome 22q13. A dinucleotide repeat has been identified with the 5' regulatory region of the IL-2Rβ chain gene.

To examine the influence of the IL-2 and IL-2Rβ genes on MS susceptibility and clinical course, McDonnell et al[63] genotyped for these two markers Northern Irish patients and controls. They found no significant association of any of these markers with either MS susceptibility or the clinical course of the disease. Similarly, no association or linkage of these markers with MS were observed in German cases and controls,[31] in Swedish multiplex families,[48] or in French sib pairs.[50] Of note, however, Reboul et al.[50] who analyzed two microsatellite markers in the vicinity of the IL-2Rb gene found indication of linkage in the HLA-DR15-negative subgroup. No such evidence was shown for two microsatellite markers in the vicinity of the IL-2Rα gene.

Recently, Encinas et al[17] demonstrated that an EAE-resistance gene was colocalized with the Idd3 diabetes resistance gene in a genetic interval of less than 0.15 cM on mouse chromosome 3. This interval contains the IL-2 gene. The SJL/J allele of the EAE-susceptible mouse differs from the C57BL/6 allele of the EAE-resistant mouse by a single base substitution in the sixth amino acid residue of the mature protein. The SJL/J allele also has a duplication of a 12-bp segment of DNA that results in a 4-aa insertion, and a compensatory 12-bp deletion that results in a deletion of 4 glutamines from a stretch of 12 consecutive glutamines. Compared with the IL-2 protein produced by C57BL/6 mice, NOD/SJL-produced IL-2 shows differences in glycosylation that may affect its functional half-life. This suggests a possible influence of the NOD/SJL allele of IL-2 on EAE and diabetes susceptibility.

The Interleukin-6 Gene

Several studies suggest that IL-6 plays an important role in the regulation of the inflammatory CNS response in EAE and MS. For example, IL-6 deficient mice are resistant to the induction of MOG induced EAE,[68] and the administration of neutralizing anti-IL-6 antibodies reduces the severity of clinical disease in a MBP-induced EAE model.[69] In MS patients, increased serum concentrations of soluble IL-6 receptor[70] as well as elevated levels of IL6-mRNA in peripheral blood lymphocytes[71] suggest that a dysregulation of IL-6 might contribute to MS pathogenesis. The IL-6 gene maps to chromosome 7p15-p21. Previous reports indicated that the C allele of a variable number of a tandem repeat polymorphism in the 3' flanking region of the IL-6 gene was associated with a reduced activity of IL-6 in vivo. This VNTR polymorphism was analyzed in 96 German MS patients and 106 ethnically matched healthy controls[72] and in 192 Sardinian simplex families with MS,[72] but none of the alleles was associated with susceptibility to MS. However, Vandenbroeck et al[73] found the C allele associated with a benign course of the disease and larger alleles with a malignant course, suggesting that allelic variations in the IL-6 gene might predispose to alterations in the course and initial onset of MS.

Genes Encoding the Interleukin-12 Subunits and Their Receptors

Converging studies of EAE indicate that IL-12 may be a critical factor in the pathogenesis of the disease. Segal et al[74] have demonstrated that endogenous production of IL-12 is critical for the generation of autoreactive Th1 T cells, as EAE cannot be induced in IL-12-deficient mice. Consistent with these results, several groups have demonstrated that the in vivo administration of IL-12 to mice or rats could exacerbate EAE or induce clinical relapses, while administration of anti-IL-12 monoclonal antibody ameliorated disease severity and prevented relapses.[75-77] Recently, intracellular cytokine staining confirmed that PBMCs from chronic progressive MS patients express more IL-12 upon activation than do those from normal individuals.[78] Moreover, treatment of chronic progressive MS patients with cyclophosphamide and methylprednisolone reduces the frequency of IL-12-staining monocytes to normal levels.

Reboul et al,[50] in their systematic analysis of cytokine and cytokine receptor genes in MS, genotyped 116 French affected sib pairs for two microsatellites in the vicinity of each of the genes encoding the IL-12 p40 subunit on chromosome 5q33, the IL-12 p35 subunit on chromosome 3q25-q26.2, the β1 chain of IL-12 receptor on chromosome 19q13 and the β2 chain on chromosome 1p31. None of these candidate genes, however, was significantly linked to MS.

Anti-Inflammatory Cytokine and Cytokine Receptor Genes and Susceptibility to MS

Genes Encoding Interferon α and Interferon β

The IFNα and IFNβ genes have evolved by duplication and recombination events in a gene cluster on chromosome 9q22. This cluster contains about 15 closely linked functional IFNα genes in addition to a single IFNβ gene. Treatment with recombinant IFNβ is known to reduce exacerbation rates and destruction of CNS components in MS patients.[79] IFNβ has the ability to inhibit or decrease IFNγ[80] and to augment defective suppressor cell function in MS patients.[81] Thus, the question arises whether genetically determined differences in the individual responsiveness to IFNβ production might affect susceptibility to MS. To answer this question, Miterski et al[82] screened the IFNβ gene by SSCP and sequencing and identified a single nucleotide polymorphism which was not associated with MS predisposition in German patients. Miterski et al then studied an intergenic dinucleotide polymorphism located in the IFN cluster in 505 patients and 369 controls. This association study revealed significant protection from MS for carriers of allele 2 and an increased risk for carriers of allele 7, a result previously suggested by the same authors on a smaller sample of patients.[31] Additional evidence for a candidate gene within the interferon region was obtained by analysis of the linkage disequilibrium between the IFNβ gene dimorphism and the microsatellite marker. This analysis provided evidence for a haplotype predisposing to MS. Finally, the authors extended their study to neighboring genes and analyzed several functionally relevant polymorphisms, i.e., premature stop codons in the IFN-a10 and IFN-a17 genes and an aminoacid substitution in the IFN-a17 gene. They showed that patients carrying a non-functional IFN-a17 allele had an increased risk to develop MS, suggesting that the gene predisposing to MS in that region could be in the vicinity of the IFN-a17 gene.

The intergenic microsatellite has also been analyzed by other groups, with non-significant results. Vandenbroeck et al[83] have characterized 137 unrelated simplex MS families from Sardinia for this marker. Comparison of parentally transmitted versus non-transmitted alleles revealed that none of the alleles was associated with the disease, even when patients were stratified according to HLA status or gender. Similarly, in a study on 51 Swedish MS patients belonging to 24 multiplex families, and 141 healthy controls, no genetic association or linkage of this microsatellite with MS was found.[84]

Genes Encoding Interleukin-4 and Its Receptor

Secretion of IL-4 is associated with a Th2-type immune response. IL-4 is also autostimmulatory to Th2 cells whilst being inhibitory to Th1 cells.[85,86] In EAE, administration of IL-4 has an ameliorating effect.[87,88] In MS, IL-4 secreting peripheral blood cell numbers are increased in both relapsing-remitting MS and chronic progressive MS to a similar extent.[89] It has been postulated that, as IL-4 inhibits IFNγ-secreting cells, this increase may be a marker for cells that are attempting to attenuate the IFNγ-associated tissue destructive response in MS. The human gene for IL-4 has been mapped to the long arm of chromosome 5 (5q23-q31), in close vicinity to the IL-3, IL-5, IL-13, GM-CSF and IRF1 cluster of cytokines.

In order to investigate whether IL-4 polymorphisms favor or modify clinical aspects of MS, Vandenbroeck et al[90] studied the relationship between a variable number of a 70 bp-tandem repeat located in the third intron of the IL-4 gene and clinical and physiological features of 256 sporadic MS patients from Italy and Sardinia and 146 healthy controls with similar ethnic background. No association was found between IL-4 alleles and disease susceptibility, disease progression, sex, or ethnic background of the patients. Of interest, however, the IL-4 B1 allele was shown to be associated with late onset of MS and might therefore represent a modifier that delays disease onset. So far, it has not been shown that the number of copies of the VNTR located in the third intron of the IL-4 gene affects its transcriptional activity and the resulting immune response. It can thus only be speculated that an increased responsiveness of the B1 allele to transcriptional activation might lead to overexpression of IL-4, which might in turn bias Th cell development toward the Th2 pathway and lead to down-regulation of the Th1 response needed to sustain inflammation in MS. Of note, McDonnell and coll. found no evidence of association between the IL-4 alleles and MS susceptibility[63] in a study of 277 Northern Irish patients and 216 controls. Two other groups used either the marker located in intron 3 or two microsatellite markers surrounding the IL-4 gene, as well as two other markers surrounding the IL-4Ra gene on chromosome 16p12.1, to genotype 34 Swedish multiplex families, 147 sporadic MS cases and 95 ethnically-matched healthy controls[48] or 116 French affected sib pairs.[50] Although they did not find any evidence for association or linkage between MS and these two chromosomal regions, their results do not exclude a possible association of an allele of the third intron VNTR with MS late onset, especially since no stratification of the patient population for age of onset was carried out in these studies.

Genes Encoding Interleukin-10 and Its Receptor

IL-10 is an important anti-inflammatory cytokine that inhibits the synthesis of pro-inflammatory cytokines, chemokines, and inflammatory enzymes in activated macrophages, T-cells and natural killer cells. Decreased levels of IL-10 mRNA were shown to be associated with increased disease activity[91] and increased IL-10 levels were detected in MS patients with stable disease,[92] suggesting that IL-10 plays an important role in the control of progression of multiple sclerosis. The IL-10 gene maps to the long arm of chromosome 1. Characterization of the promoter region has revealed a CA_n microsatellite repeat region 164 bp upstream of the proposed TATA box and three point mutations at positions -1082, -819, and -519 from the transcription start site. Only three haplotypic combinations, GCC, ATA, and ACC, have been observed in Caucasians. The substitution of a A for a G at position -1082, providing the GCC haplotype, was found to be related to production of significantly higher levels of IL-10 than the remaining two haplotypes.

As it was unclear whether inherited differences in the production of IL-10 could influence susceptibility to MS or disease outcome, Pickard et al[93] genotyped 185 British MS patients and 211 ethnically matched controls for each of the three dimorphisms. No association was found for any IL-10 promoter polymorphisms when MS cases were compared with controls or when patients with mild to moderate disease were compared to patients with severe disability ten years after disease onset. Moreover, no association with any allele of the microsatellite marker located 164 bp upstream of the proposed TATA box was found when the MS patient group was compared with the control panel. Three other groups used the same microsatellite marker in an investigation of cytokine candidate genes in 34 Swedish multiplex families, 147 sporadic MS cases and 95 ethnically-matched healthy controls,[48] 116 French affected sib pairs[50], or 277 Northern Irish patients and 216 controls.[63] Similar to the data obtained by Pickard et al,[93] they did not find any association or linkage of this marker to multiple sclesosis. Moreover, Mäurer et al[94] analyzed the -1082 diallelic polymorphism in 181 German MS patients and 85 healthy controls and did not find any association between this dimorphism and disease susceptibility, clinical course, or age of onset of multiple sclerosis. In conclusion from these different studies, polymorphisms in the IL-10 promoter do not appear to be significantly associated with MS or to influence disease progression. Of note, Reboul et al[50] also genotyped 116 French affected sib pairs for two microsatellite surrounding the IL-10R gene on chromosome 11q23.3 and did not find any evidence for linkage between this locus and multiple sclerosis.

Genes Encoding Transforming Growth Factors β1 and β2

TGFβ, of which three homologous isoforms exist (1, 2 and 3), is a strongly immunosuppressive cytokine, inhibiting expression of pro-inflammatory cytokines such as TNFα and IL-1, and blocking cytokine induction of adhesion molecules such as ICAM-1 and VCAM-1. Given systemically, TGFβ1 inhibits EAE,[95] whereas neutralizing antibodies against TGFβ1 enhance clinical severity of EAE.[96] High TGFβ levels exist in the blood cell cultures of MS patients during regression of a relapse,[97] and levels in cerebrospinal fluid (CSF) have been found to correlate positively with the duration and frequency of relapses in patients with a relapsing-remitting course.[98] Soluble E-selectin, whose expression on endothelial cells is inhibited by TGFβ, is found at higher levels in both the serum[99] and CSF[100] of patients with primary progressive disease compared to those following a relapsing-remitting course. From these observations it can be extrapolated that TGFβ has an important role in inducing disease remission in MS and that a lack of TGFβ may be partly responsible for the gradually progressive, unremitting course characteristic of patients with primary progressive disease.

In view of the possible important role of the TGFβ family in MS, the TGFβ gene regions have been considered a suitable area for study across the clinical spectrum of MS. Accordingly, McDonnell et al[101] undertook gene association studies using the microsatellite marker D19S223 in the region of TGFβ1 on chromosome 19q13.2 and a polymorphic CA_n repeat in the 5'-flanking region of TGFβ2 on chromosome 1q14, incorporating 151 relapsing-remitting or secondary progressive MS patients, 104 primary progressive patients and 159 normal controls from Northern Ireland. No significant differences were found in allele frequencies between either MS group and controls, indicating that TGFβ1 and TGFβ2 loci are not likely to influence either relapsing remitting/secondary progressive or primary progressive MS in this population. Of note, an affected pedigree member analysis performed with the same markers on 34 Swedish multiplex families indicated a possible linkage of MS to the TGF-b2 locus, although neither marker was significantly associated with MS in a case/control analysis of 147 Swedish sporadic MS cases and 95 healthy controls.[48] Mertens et al,[102] in their systematic analysis of oligodendrocyte growth factors in MS, genotyped 88 French affected sib pairs for two microsatellites in the vicinity of each of the genes encoding not only TGβ1 and TGFβ2, but also TGFβ3 on chromosome 14q24 and the TGFβ receptors TGFβR1 on chromosome 9q33-q34, TGFβR2 on chromosome 3p22, and TGFβR3 on chromosome 1p32-p33. None of these candidate genes, however, was significantly linked to MS, with the exception of TGFβ3 in the subgroup of sibpairs in which both affected individuals had at least one HLA-DRB1*15 allele. Because of the number of statistical tests performed in this study, this result cannot be regarded as conclusive, but it suggests a possible role for TGFβ3 in association with HLA in genetic susceptibility to MS.

Chemokine and Chemokine Receptor Genes and Susceptibility to MS

Genes Encoding the CCR5 and CC2RB Receptors

Full-genome screenings in multiplex families have identified several susceptibility regions. Among these, evidence for weak linkage was observed at 3p/3cen, suggesting the presence of a MS gene of modest effect. Encoded in this region are two chemokine receptors, CCR5, a CC-type receptor that binds RANTES, macrophage inflammatory protein (MIP)-1α and MIP-1β, and CCR2B, a receptor for the monocyte attractants MCP-1, -2, -3, and -4. Independent of their suggestive location, CCR5 and CCR2B are interesting MS candidate genes for several reasons. Aberrant expression of chemokines and chemokine receptors has been detected in both human and experimental central nervous system demyelinating lesions,[103-105] suggesting the involvement of chemokine-chemokine receptor interactions in disease pathogenesis. In addition, epidemiologic, migration, and cluster studies favor some role for an infectious agent in MS etiology.[106] Chemokine receptors have been shown to mediate the entry of microorganisms into target cells and to participate in the viral-mediated induction of type 1 cytokines, potential mediators of the encephalitogenic response.[107-108]

To clarify the genetic role of chemokine receptors in MS, Barcellos et al[109] analyzed in detail the chromosome 3p21-p24 segment in 125 families with multiple members affected with the relapsing form of MS. Genetic analyses of common variants within coding regions of both CCR5 and CCR2B loci, and two nearby microsatellite markers, were performed using linkage and association-based methodologies. Evidence for linkage to MS was not observed with any of the 3p21-associated markers in the MS families. Of interest, however, the mutant CCR5 allele which carries a 32-bp deletion (Δ32) appears to confer a moderate, yet significant, delay in age of disease onset. Bennetts and co-workers[110] also compared the frequency of CCR5Δ32 in 120 Australian unrelated relapsing-remitting MS patients with a sample of 168 control individuals and found no evidence for either a protective or predisposing effect. However, clinical variables such as age of onset were not examined. The development of inflammatory CNS lesions and detectable neurological deficits are likely the result of a multistep process that requires consecutive waves of activated lymphocytes crossing the blood-brain-barrier. Reduced CCR5 expression in heterozygous individuals, and its absence in homozygotes, could impair the efficiency of the homing process and the strength of the inflammatory response, delaying the expression of clinical signs. This hypothesis is in agreement with the increased expression of RANTES and MIP-1α in EAE prior to and during the onset of clinical signs,[111,112] and during MS acute attacks).[105] Because of the redundancy and overlapping molecules in the chemokine cascade, alternative pathways will eventually provide the necessary signaling and lymphocytic chemotaxis to initiate and perpetuate CNS inflammation. It is not surprising then that homozygosity for Δ32 fails to protect against MS. Of course, the association between CCR5Δ32 and delayed age of onset in MS may also result from linkage disequilibrium between the coding alleles and recently described polymorphisms within the CCR5 promoter region which appear to influence gene expression.[113] Analysis of these and other polymorphisms in the receptor regulatory regions and ligands in MS is therefore warranted.

Gene Encoding the Monocyte Chemotactic Protein 3 (MCP-3)

MCP-3 is α-chemokine that attracts mononuclear cells, including monocytes and lymphocytes, the inflammatory cell types that predominate in multiple sclerosis lesions. The expression of CC-chemokines like MCP-3 has recently been detected in MS lesions.[114] The MCP-3 gene, that contains a CA_n microsatellite sequence in the promoter-enhancer region, is located within a cluster of chemokine genes on chromosome 17q11.2. Fiten et al[115] studied the possible association between this microsatellite marker and the occurrence of multiple sclerosis in 192 Swedish MS patients and 129 healthy controls. The individual MCP-3 allele frequencies did not differ significantly between MS patients and control individuals. However, when MS patients and control subjects were stratified according to HLA-DR, the data suggested a protective effect of allele MCP-3*A4 in patients carrying HLA-DRB1*15 or DRB1*03, and similarly a protective effect of MCP-3*A2 allele in HLA-DRB1*15 and DRB1*03-negative individuals, although the results were not significant after correction for multiple testing. Further genetic studies of the MCP-3 gene or neighboring chemokine genes in the chromosome 17q11.2 cluster should be undertaken to confirm these results.

Genes Encoding Scya1 (TCA-3), Scya2 (Monocyte Chemoattractant Protein MCP-1) and Scya12 (MCP-5) in the Mouse

Whole genome scans in the mouse have identified a quantitative trait locus controlling susceptibility to monophasic remitting/non-relapsing EAE on mouse chromosome 11, eae7. Several candidate genes encoding chemokines are contained in the eae7 interval and cDNA sequence polymorphisms between the EAE-susceptible SJL/L and EAE-resistant B10.S/DvTe strains, resulting in significant amino acid substitutions, have been identified in the small inducible cytokines Scya1, Scya2, and Scya5.[116] Two of these chemokines have so far been implicated in EAE. In active disease, Scya1 expression is induced in the spinal cord one or two days before clinical signs appear and is expressed by activated encephalitogenic T cells.[111] Similarly, Scya1 has been associated with the encephalitogenic potential of T cell clones in passive disease.[117] Increased Scya2 expression in the CNS is also seen in EAE before onset of clinical disease and throughout the acute attack in a variety of mouse and rat models.[21] Importantly, there is a significant increase in Scya2 expression by astrocytes in the brain and spinal cord during the relapsing phase of relapsing-remitting EAE.[118] Given the role of these chemokines in EAE, the sequence polymorphisms identified in this study are promising candidates for eae7, a locus associated with severity of clinical signs and susceptibility to the shorter, less severe monophasic remitting/nonrelapsing form of disease. It is indeed conceivable that the allelic variants identified in Scya1, Scya2, and Scya3 form a quantitative genetic gradient that modulates the duration and severity of the clinical signs of EAE. However, further work to demonstrate the functional relevance of these polymorphisms is needed.

Conclusions

As reviewed here, it appears that for many cytokine or chemokine genes, disease associations with intragenic or closeby polymorphisms are initially reported, but attempts to replicate these findings by other groups are unsuccessful. Several reasons may explain these discrepancies.

First, there are cases where the discordance may be attributable to differences in ethnicity of the populations under investigation. Indeed, different genes may predispose to the same disease in groups of different ethnic origins, and thus divergent results may appear in different MS materials. For example, in Sardinia, MS has been shown to be linked to the DRB1*0405-DQA1*0501-DQB1*0301 and DRB1*0301-DQA1*0501-DQB1*201 HLA haplotypes, rather than the more typical DRB1*1501-DQA1*0102-DQB1*0602 haplotype which is associated with the disease all over continental Europe. The Sardinian population is phylogenetically and ethnically more homogeneous than the continental European populations. The advantage of studying such genetically isolated populations lies in the fact that linkage disequilibrium is likely to extend over greater distances from the susceptibility loci. This implies that the effects of even minor susceptibility loci can be uncovered more easily over a wider distance. It is therefore not surprising that certain associations with MS observed in Sardinian patients have not so far been confirmed in more heterogeneous MS groups.[46,90]

Second, the results emphasize the importance of considering clinical information in efforts to identify MS genes. At several loci, patient and control allelic distributions are statistically indistinguishable, yet a significant effect on age of onset[90,109] or disease severity[73] is observed. In addition, disease classification and the possible interaction of the candidate loci with HLA are important variables to be taken into account. Risk factors for relapsing-remitting or primary progressive MS can be different,[59] as well as risk factors for sporadic and familial MS,[48,109] or risk factors for HLA-DR2 positive or negative patients.[50,102,115] In many studies, these important variables are not considered, which may explain part of the discrepancies observed.

Finally, lack of replication can be due to methodological problems, including limited patient cohort size or reduced informativeness of single nucleotide polymorphisms to detect modest associations, or multiple testing for which a correction is not always applied when significance levels are reported. Moreover, sampling choices are often variable, as are statistical methods to analyze the data (association studies, affected sib pair studies or transmission disequilibrium tests), which makes comparisons between studies a difficult task.

In the light of these comments, the available data which have been summarized here suggest that the effect of polymorphisms in the cytokine and chemokine genes and their receptors on MS susceptibility or its progression is, at best, small.

Cytokine and Cytokine Receptor Genes in the Susceptibility to Rheumatoid Arthritis (RA)

Rheumatoid arthritis (RA) is characterized by a chronic synovial inflammation with synovium infiltration by CD4+ T cells, plasma cells and macrophages. It has been postulated that the Th1/Th2 balance could play a key role in the initiation and perpetuation of synovial inflammation and that Th1 cells secreting pro-inflammatory cytokines such as IFNγ were preferentially activated in rheumatoid synovium. In fact, other pro-inflammatory cytokines such as IL-1, TNFα and IL-6, secreted by macrophages in the synovium, are also known to be involved in synovial inflammation. The initiating event in RA has not yet been defined, nor have the factors leading to the chronicity of the disease and, except for the MHC genes, genes involved in predisposition to RA. It is possible that the set of genes predisposing to RA have subtle variations that, when present together, add up to give an overall phenotype for immune responsiveness. In addition, unknown environmental agents very likely trigger the autoimmune response, which leads to the production of mediators, particularly cytokines, that drive the pathophysiological process leading to the clinical manifestations of RA. Because of the key role of several cytokines in the autoimmune process, the genes encoding these proteins are obvious candidate gene and a large number of linkage and association studies have been performed on pro- or anti-inflammatory cytokine genes in order to better characterize disease predisposition.

Pro-Inflammatory Cytokines and Cytokines Receptor Genes and Susceptibility to RA

Genes Encoding TNFα, TFNβ and the TNF Receptor, TNFR2

In the hunt for genetic factors that contribute to RA, the TNF locus on chromosome 6 has received considerable attention. Several lines of evidence indeed suggest a pivotal role of TNFα in RA. Elevated levels are found in the synovial fluid and cartilage-pannus junction.[119] The bone erosions characteristic of RA are mediated via TNFα, acting through synoviocyte production of a cascade of proinflammatory mediators.[120] Furthermore, treatment of patients suffering from RA with anti-TNFα antibodies produces sustained clinical improvement.[121]

Five microsatellite markers have been described, that flank the entire TNFα/TNFβ locus over a 20-kb region. TNFa and TNFb are located upstream of the TNFβ gene, TNFc is in the first intron of the TNFβ gene, and TNFd and TNFe are downstream of the TNFα gene. Mulcahy et al[122] analyzed the segregation of these five markers with disease in 50 multicase RA families from Ireland, United Kingdom and Utah. Using logistic regression to assess independent effects from the TNF haplotype a6-b5-c1-d3-e3 and the HLA-DRB1 shared epitope, they provided evidence for the presence of a susceptibility gene in or nearby the TNF locus, distinct from HLA-DR. Several other studies have suggested that TNF microsatellite alleles are associated with RA susceptibility and/or severity. However, there have been differences regarding which particular alleles are associated and whether they are independent of the effects ascribed to the HLA-DRB1 shared epitope. While Hajeer et al[123] showed that the increased frequency of the TNFa6 allele in British RA patients was in fact due to linkage disequilibrium with some subtypes of DR4 carrying the shared epitope, Mattey et al[124] reported that the association of the shared epitope with disease severity was influenced by an interaction with the TNFa6 allele. Martinez et al[125] showed that the TNFa6-b5 haplotype was significantly associated with susceptibility to RA in Spanish patients, independently of the HLA-DR shared epitope. Of note, Mu et al[126] observed a striking association between another allele, TNFa11, and the shared epitope in Californian female patients, the most severe outcomes being observed among individuals who had both TNFa11 and the shared epitope, and the best outcomes among individuals who had inherited TNFa11 in the absence of the shared epitope.

Since TNF microsatellites themselves are not likely to affect TNF production, they might be markers for functionally relevant TNF gene polymorphisms. The search for genetic heterogeneity within the TNFα gene uncovered several single nucleotide polymorphisms. Disease association studies in Dutch patients revealed that the promoter -376, -308, -238, -163 and -70 polymorphisms do not contribute to disease susceptibility in RA.[127] However, disease stratification pointed to a role for the -238 promoter polymorphism in disease severity, independent of the presence of HLA-DR4.[127,128] In addition, a polymorphism at position + 489 in the first intron of the TNFa gene was shown to be associated with both susceptibility to and outcome of RA in Dutch patients.[129] Investigation of Japanese patients revealed that none of the haplotypes formed by the nucleotides at positions -1031, -863 and -857 was associated with RA, independently of DRB1*0405.[130] Conversely, systemic juvenile RA was shown to be associated with the -1031C/-863A/-857T haplotype in Japanese patients, and the -857T allele appears to enhance the effect of DRB1*0405 in predisposing to development of systemic juvenile RA.[131] Of note, the -308 variant was not found associated with either RA susceptibility or radiological progression[132] in Polish patients, a result which conforms to the previous observations by Brinkman et al.[127]

In conclusion, the available data suggest that some TNFα gene variants are markers for disease severity in RA, independent of, and additive to, the HLA shared epitope alleles. Further studies are necessary to determine whether the relevant TNF gene variants contribute directly to the pathophysiology of the disease through disturbance of the cytokine production or whether they are simply markers for additional polymorphisms in the TNF locus or neighboring genes.[133]

A genome-wide screening of affected sibpairs with RA indicated that chromosomal region 1p36 might contain a susceptibility gene.[134,135] Of interest, the gene encoding TNF receptor 2 is located in that region and can be considered as a candidate gene. Shibue and coll.[130] thus genotyped a large series of Japanese patients and controls for a TNFR2 polymorphism at position 196. They did not, however, observe an association between this polymorphism and RA.

Gene Encoding IFNγ

IFNγ is produced by T cells infiltrating the inflammed synovium and is secreted into the joint space, although its role in the progression of the articular injury remains controversial.[136] The gene consists of four exons with three intervening regions. A variable-length dinucleotide repeat polymorphism has been described in human and lower primates within the first intron of this gene, between positions 1349 and 1373.[137] Although the number of alleles reported at this marker varies according to the detection methods, evidence suggests that some of these alleles are associated with differing level of IFNγ secretion[138] and hence may be of biological importance in RA.

A case-control study was performed in 60 severe RA patients, 39 mild RA and 65 healthy individuals,[139] using this microsatellite marker. The allele frequencies in patients with severe RA differed substantially from those in controls and in patients with mild disease. Indeed, frequency of the 126 bp allele was 73% in the 60 patients with severe RA, 21% in the 39 patients with mild disease and 12% in the controls. In contrast, the 122 bp allele was detected in only 7% of patients with severe disease compared to 64% of patients with mild disease and 80% of controls. These results suggest that a polymorphism in the IFNγ gene or a gene in close linkage disequilibrium has an important role in determining the severity of RA.

Genes Encoding IL-1 α and β and IL-1 Receptor Antagonist

Interleukin 1 (IL-1) is a proinflammatory cytokine secreted by activated macrophages which is overexpressed in RA serum, synovial fluid and synovial tissue.[136,140] Together with IL-6 and TNFα, IL-1 mediates the acute phase protein response. IL-1 initiates the recruitment of immune cells, the severity of inflammation, and levels of circulating IL-1 in plasma of RA patients have been related to disease severity.[141,142] IL-1 is also involved in chondrocyte-mediated cartilage damage by inducing metalloproteinase enzymes and decreasing glycosaminoglycan synthesis in joint erosion as well as by stimulating bone resorption.[143-146] Several polymorphisms described in the IL-1 locus have been used in RA studies : a VNTR of a 46 bp-repeat within IL-1α intron 6,[147,148] two microsatellites, 222/223 and gz5/gz6, located within intron 5 of the IL-1α gene,[149] a biallelic polymorphism C/T at position -889 in the 5' regulatory region containing the IL-1α promoter[150] and a polymorphism at position +4845 within the exon 5, responsible for a shift of amino acid 112 ;[151] a transition G/A at position -511 in the promoter[152] and a transition C/T at position +3953 in exon 5 of the IL-1β gene,[153] two microsatellites, gaat. p33330 and Y31, located between the IL-1β and IL-1Ra genes[154] and a VNTR of a 86-bp element located in the intron 2 of the IL-1Ra gene.[155]

Three association studies investigated the potential role of IL-1α in RA predisposition. In the first one[147] the 46-bp repeat VNTR was analyzed in 50 patients and controls of Caucasian origin. Two groups of patients were individualized with either benign or severe RA. Although the allele coresponding to 8 repeats was over-represented in the RA population (14%) versus control (8%), the allelic distribution did not significantly differ between RA and controls or between benign and severe RA. The second study[156] was performed on 183 patients and 275 healthy controls, classified according to whether or not they possessed the known predisposing HLA-DRB1 alleles. Microsatellite polymorphisms in intron 5 of the IL12 gene did not show significant association with RA. The third study[157] was performed on 98 patients, classified in two groups: those with (57) and those without (41) destructive arthritis. The control population was composed of 94 blood donors. The IL-1α gene polymorphism was analyzed using the biallelic polymorphisms at positions -889 and +4845. The two markers were always linked. There was no difference between the allele frequencies at these sites in the patient and in the control population. However, comparison between destructive and mild disease showed an overrepresentation of allele 1 [C] in non-destructive RA patients and an overrepresentation of allele 2 [T] in destructive arthritis. The results thus suggest that these IL-1α gene polymorphisms may contribute to the pathogenesis of the disease.

An association study[157] was also performed on 108 RA patients and 128 unrelated controls using two bi-allelic polymorphisms at position -511 and +3953 in the IL-1β gene. These polymorphisms did not reveal any association with RA patients.

Two studies were performed to evaluate the association between RA and the VNTR in the second intron of the IL-1Ra gene. Forty-three RA patients and 119 unrelated controls matched for age and sex were studied in one case[159] and 108 patients and 128 unrelated controls were studied in the other case.[158] In both studies the alleles frequencies were not significantly different between RA patients and controls.

Crilly et al[160] evaluated the influence of the three IL-1 gene polymorphisms on the outcome of the disease and tested the possibility to use these IL-1 polymorphisms as predictive value for surgery. This study included 100 RA patients within a 15-year period of disease diagnosis. Among these, 50 patients had undergone major joint surgery. A group of 66 ethnically matched controls was also included. All patients were typed for HLA-DRB1. Biallelic polymorphisms at positions -889 in the IL-1α gene and –511 in the IL-1β gene were investigated as well as the VNTR in intron 2 for IL-1Ra gene. No difference in the allele or genotype frequencies was found between controls and patients with RA either with or without surgery. However IL-1β allele 2 (T) was over-represented in patients with RA who had undergone surgery compared with patients who had not (40% versus 27%).

Finally Cox et al[154] investigated several candidate genes of the IL-1 gene cluster with different markers: gaat.P33330, Y31, 222/223, gz5/gz6, IL-1a +4845, IL-1β -511 and +3953, and ILRa +2018. 195 multicase RA families (576 individuals) were sampled among which 251 subjects had an erosive RA. Their results yielded suggestive evidence for linkage of genes of the IL-1 cluster to RA in patients with severe erosive disease.

Gene Encoding Interleukin-6

IL-6 is a pleiotropic cytokine with both pro- and anti-inflammatory effects. Along with IL-1 and TNFα it is a major mediator of the acute phase response in RA. However IL-6 increases circulating levels of IL-1 receptor antagonist and soluble tumor necrosis factor receptor, both having potential anti-inflammatory effects by competing with the action of IL-1 and TNFα.[161] Several polymorphisms have been described in the IL-6 gene: a highly polymorphic AT rich repeat located in the 3' region of the gene[162-164] and two single nucleotide polymorphisms located at positions -622 and -174 (G/C) in the 5' flanking region.[165] A RFLP study was performed on DNAs of 33 European caucasians (patients and controls) using as probe a cDNA fragment containing the full length IL-6 cDNA.[166] No co-segregation between either 14.5 or 13.8 alleles and RA could be observed. Fugger et al[167] also investigated two RFLP polymorphism, located in the IL-6 gene in 24 Danish patients with RA and 72 unrelated healthy Danes. No significant association of these markers was observed with RA.

A linkage study[168] was performed on two hundred RA affected sib pairs genotyped using the microsatellite marker D7S493. Data were stratified according to the age at onset of the disease, the sex and the severity of the disease. No significant linkage of RA with the D7S493 marker was detected using either the whole population or the stratified data set.

A study was performed on 163 patients and 157 controls of the same ethnic origin. Pascual et al investigated the possible association between the IL-6 promoter polymorphisms at positions -622 and -174 and susceptibility to RA. No significant difference was observed in either the genotype or the allele distributions between patients and controls.

The most recent study investigated 95 RA patients and 55 controls.[169]Patient were split in two groups: 47 had undergone major joint replacement surgery within 15 years of disease diagnosis and 48 had a disease duration greater than 15 years without major surgery. The G7 allele of the 3' AT-rich repeat was found over represented in the patients needing surgery compared to non surgery patients and controls.[170] In contrast the G8 allele was reduced in non-surgery patients. However, despite this trend, the comparisons were not statistically significant.

Gene Encoding Interleukin-3

The IL-3 gene represents a potential candidate for RA because its product affects differentiation, proliferation, and function in three hematopoietic lineages in bone marrow and has a key role for mature myeloid cells. The human IL-3 gene is located on chromosome 5q23-31 in a cluster of cytokine genes, in particular GM-CSF, IL-4, IL-5, IL-9[171] and a polymorphism in the region of the IL-3 gene can potentially detect association with other genes nearby. A case-control study was designed using different single nucleotide polymorphisms in the vicinity of the IL-3 gene :-16 T/C in the promoter region, 131 T/C in IL-3 exon 1 and 23 C/T in exon 4 of GM-CSF.[172,173] Comparison of 254 Japanese patients and 881 matched controls indicated a strong association of these polymorphisms with the disease.[174]

Anti-Inflammatory Cytokines and Susceptibility to RA

Gene Encoding Interleukin-4

The presence of IL-4 is difficult to detect in synovial tissue of RA patients. However its strong anti-inflammatory properties probably play an important role in the course of the disease. Several polymorphisms identified in the IL-4 gene have been used in two different studies to look for association between IL-4 and RA susceptibility.

In the first work[158] two polymorphisms in the IL-4 gene were studied. The first is a VNTR located in the third intron of the gene[175,176] with three alleles consisting of 3 repeats [IL-4 (1)], 2 repeats [IL-4 (2)], or 4 repeats of a 70 bp-element. The second polymorphism results in a C to T substitution at position –590 recently described in the promoter region of IL-4.[177] The allelic distributions were investigated in 106 white patients with recent onset RA and 128 unrelated healthy white individuals who were living in the southwest of France. The patients were evaluated with a broad set of assessments, including functional measures, complete clinical evaluation, laboratory investigations, and radiographs of the hands and feet. The presence of bone erosions on radiographs or a sustained progression were considered as hallmarks of disease severity after 2 years and the patients were stratified into 2 groups: erosive RA and non-erosive RA. The patients had a significantly higher frequency of the allele IL-4 (2) as compared to controls. The carriage rate of this allele was also increased, from 17.9% in the normal population to 33.3% in the patient group. The frequency of the IL-4 -590T allele in the promoter region of IL-4 was also increased in RA patients compared with controls, but the difference was not statistically significant. However, a significantly higher frequency of the IL-4 (2) / IL-4 -590 T allelic combination was found in RA patients. The presence of both alleles was observed in 33 patients (30.8%), which compared with only 3 control subjects (2.5%), is highly significant.

The second study[178] included 335 patients with chronic polyarthritis, all from the Lyon area of France. All patients had a disease duration of at least 2 years. Arthritis patients were classified into two groups according to the type of involvement of the wrist, one of the most commonly affected sites in RA.[179,180] The first group included patients without joint destruction and the second those with significant damage.[157] The control population consisted of 104 donors of the Lyon blood bank, and had the same genetic background as the patients. The VNTR in the IL-4 gene was used in this study.[175,176] The carriage rates of the rare IL-4(2) allele and the more frequent IL-4(1) allele did not differ significantly between controls (26.0 and 99.0%) and patients (28.4 and 98.2%). However when the patients were stratified according to joint destruction, the carriage rate of the IL-4(2) allele was increased in patients with non-destructive RA (40%) and decreased in patients with destructive RA (22.3%). Allele 2 was thus overrepresented in patients with non-destructive RA. Conversely, the frequency of allele 1 was very significantly increased in destructive *vs* non-destructive RA.

Although there are important differences between the two studies, the VNTR IL-4 allele (2) seems to be associated mild RA

Gene Encoding Interleukin-10

On in vitro RA synovial tissue, endogenous IL-10 downregulates TNFα and IL-1β which are highly produced during synovial inflammation.[181-184] Several polymorphisms have been described in the IL-10 promoter or closeby. In the promoter itself, polymorphisms exist at positions -592, -819, -1082 and the presence of an A at the last position has been correlated with a low IL-10 production after stimulation of T cells in vitro.[185,186] Moreover the 5'-flanking region of the human IL-10 gene contains two very polymorphic dinucleotide repeats. These microsatellites are located 1.2 kb (IL-10.G) and 4.0 kb (IL-10.R) 5' of the transcription start site.[187,188] Four studies looked for an association between IL-10 and disease susceptibility.

The first study[189] includes 117 RA patients well characterized for HLA-class II alleles and 119 kidney donors from northwest England served as controls. Patients and controls were genotyped for the the three SNP in IL-10 promoter. No significant difference in allele or haplotype frequencies was seen between control and RA patients.

A similar study on 106 white patients with recent-onset RA and 128 unrelated healthy white individuals who were living in the southwest of France.[158] No significant association between the single nucleotide variant at position -1082 in the IL-10 promoter and RA was found.

The third study[185] was performed on 103 white patients from the Glasgow University Center for Rheumatic Disease and 148 patients from the Oxford Nuffield Orthopaedic center. They were compared to 94 and 87 white controls from the same area, respectively. Moreover,

61 African-American RA patients from Atlanta were compared to 38 African-American controls. The two microsatellites located upstream of the IL-10 gene were used as markers. No association was found between RA patients and the IL10.G microsatellite in any group. However, the overall distribution of the alleles at the IL-10.R locus differed significantly between patients and controls in all three groups. This study thus showed that the IL-10.R microsatellite was associated with RA in patients of different ethnic origins.

The last study involved 117 unrelated RA patients who were stratified based on the rheumatoid factor isotype (IgM, IgG, IgA). A subgroup of 24 patients IgA+/IgG- was identified. RA patients were compared to 119 ethnically-matched controls. Genotyping of the three polymorphic sites in the IL-10 promoter allowed the establishment of individual haplotypes. One of them (-1082A / -819C / -592C) was found significantly increased in patients who were positive for IgA rheumatoid factor but negative for IgG. Therefore, among four association studies aimed at investigating the role of IL-10, gene polymorphism in RA susceptibility, two were negative and two were positive. It can be hypothesized that either the association between IL-10 and RA is weak and then can be missed in low powered studies or that ethnic differences account for the discrepancies.

Chemokine and Chemokine Receptor Genes and Susceptibility to RA

Gene Encoding CCR5

Chemokines are an extensive family of related cytokines grouped into four subfamilies based on the positioning of conserved cysteine residues. The vast majority of chemokines fall into two of the four chemokines subfamilies, namely the CXC and CC subfamilies and are distinguished by the presence or absence of an amino acid separating the two of four conserved cysteine residues. Several human chemokines are highly homologous and contain an ELR amino acid motif located within the N terminal region of each molecule. This motif is essential for high affinity binding to the CXCR2 receptor. The ELR-containing CXC chemokines preferentially chemoattract neutrophils.[191-194] In contrast the CC chemokines with extensive homology to MCP-1 (monocyte chemotatic protein-1) do not attract neutrophils but do chemoattract monocytes and T cells. This is due to the selected expression of specific chemokine receptors on distinct cell populations such as CCR2 on monocytes, basophils and T cells or CCR5, a major receptor for HIV-1, on Th1 cells as well as monocyte lineage cells. Chemokine receptors CCR3 and CCR4 appear to be differentially expressed on Th2 cells compared to Th1 cells. RA is characterized by predominant infiltration of Th1 cells in the synovium and cells expressing CCR5 accumulate in RA synovial fluid.[195-196] Several groups described a 32-base pair deletion in the CCR5 gene, termed CCR5Δ32, which generates a nonfunctional receptor that provides nonresponsiveness to specific chemokines such as RANTES, Mip-1α or Mip-1β.[192] Since Th1 cell infiltrates are predominant in synovial joint, a nonfunctional CCR5Δ 32 could result in impaired recruitment of inflammatory cells in RA and would be protective. To test this hypothesis, several groups have evaluated RA disease prevalence and activity in individual homozygous for the CCR5Δ32 mutation.

In two studies, involving 673[197] and 580[198] Caucasian RA patients, no homozygote for the CCR5Δ32 mutation was found and the frequency of heterozygotes was lower than the control population. In a third study involving 278 western European RA patients, the frequency of the mutation was also reduced in RA patients compared with controls but homozygosity for the CCR5Δ32 mutation was present in two patients who had severe erosive disease.[199] A fourth study was performed on 160 RA patients (71 with severe and 89 with milder phenotype) and 500 healthy individuals. The frequency of the CCR5Δ32 allele was significantly higher in mild RA patients than in patients with severe RA. Finally, a last study involving 163 Danish RA patients did not find any difference in the gene frequency of the CCR5Δ32 allele between RA patients and the control group.[200]

Although in three of these studies the homozygous CCR5Δ32 mutation frequency was lower in RA patients group than in controls, the low frequency of this mutation requires larger sample sizes before a statistically significant difference can be reached. Moreover, patients homozygous for the CCR5Δ32 deletion were observed in two studies, meaning that CCR5 is dispensable for the development of the disease.

Cytokine and Cytokine Receptor Genes in the Susceptibility to Insulin-Dependent Diabetes Mellitus (IDDM)

Several studies have strongly indicated that, in men and rodents, IDDM is a polygenic disease. Genes encoding cytokine /chemokines or their receptors are reasonable candidates in the genetic susceptibility to type 1 diabetes owing to the role of their gene products in antigen-presenting cell and lymphocyte activation, migration and homeostasis. Different experimental approaches have validated this idea. Genetic manipulations leading to enforced expression (transgenesis) or experimental deletion (knockout mice) have underlined the importance of given cytokines/chemokines in the pathophysiology of spontaneous or induced mouse models of autoimmune diabetes. Another, more open, approach that has pinpointed cytokines or cytokine receptors in diabetes has been linkage analyses in human families with multiple cases of the disease or in crosses between susceptible and resistant animal strains.

The latter work has relied primarily on the non-obese diabetic (NOD) mouse that spontaneously develops a disease closely resembling human type 1 diabetes and, therefore, represents a useful model to study the genetics and pathophysiology of autoimmune diabetes. In this model diabetes ultimately results from a multistep process involving expansion and differentiation of autoreactive T cells, mononuclear cell infiltration of Langherans islets (insulitis), and destruction of insulin-producing islet β cells.

An intense and pioneer work has carried out over the last ten years to finely map the location of mouse type 1 diabetes susceptibility genes (*Idd*) in the NOD model. John Todd, Linda Wicker and their colleagues using classical linkage analyses in crosses between susceptible NOD mice and resistant C57Bl mice have revealed numerous non-MHC linked susceptibility genes. An estimated 20 genes, each probably contributing through minor functional polymorphism, are thought to be involved in genetic susceptibility in this model. The complexity of these genetic analyses has increased tremendously as linkage studies have also revealed the existence of resistance loci even in the susceptible mouse strain suggesting that disease results from a delicate interplay between susceptibility and resistance genes. Ongoing work with congenic NOD strains harbouring specific regions of B6 or B10 resistance allows the fine mapping of the *Idd* genes. One unexpected result generated by this congenic approach is that frequently, for a given region, susceptibility or resistance is contributed by several linked genes rather than by a single gene. This suggests that polymorphisms in linked genes with related functions combine to result in an experimentally detectable phenotype. For example, the B10 *Idd9* allele on chromosome 4 confers protection from type 1 diabetes, although no difference is observed in terms of kinetics and severity of the insulitis.[201] However, the NOD.B10 *Idd9* congenic mice display obvious qualitative differences in insulitis. Indeed, whereas in control NOD mice IFNγ and TNFβ are preferentially expressed in the pancreas, IL-4, IL-13 and TGFβ predominate in the tissue of NOD.B10 *Idd9* congenics. Analysis of substrains of congenic mice have in fact revealed that the *Idd9* effect results from the combined action of at least 3 loci. This region of mouse chromosome 4 harbours a cluster of genes encoding for members of the TNF receptor superfamily including CD30, TNF-R2 and CD137 (4-1BB) and variations in the coding sequence of these three genes exist between NOD and B10 mice.[201]

Pro-Inflammatory Cytokine and Cytokine Receptor Genes and Susceptibility to IDDM

Genes Encoding Tumor Necrosis Factors α and β and Their Receptors

TNFα and TNFβ are clearly expressed in pancreatic islets from prediabetic and recently diabetic NOD female mice.[202] Moreover, TNFα enhances the cytotoxicity of β islet cells mediated by IL-1 and IFNγ.[203] However, treatment of NOD mice with TNFα prevents or exacerbates insulitis and diabetes depending on the timing of the cytokine treatment.[204] NOD mice overexpressing TNFα specifically in β islet cells display accelerated diabetes as a consequence of increased β cell death, local recruitment of antigen-presenting cells and enhanced presentation of β cell autoantigens to self-reactive lymphocytes. Therapeutic modulation of NOD diabetes with anti-TNFα strategies also indicate that TNFα plays a deleterious role in the disease process. To date, however, this knowledge has not led to the identification of mutation in the TNF/TNF-R signaling pathway in patients with autoimmune diabetes.

Several groups studying patients from Denmark,[205] the United Kingdom[206] and the United States[207] have revealed strong association between the TNF2 allele of the –308 TNFα promoter polymorphism and IDDM. However, this association was secondary to linkage disequilibrium between the TNF allele and the closely linked HLA-DRB1*0301 allele and no independent contribution of the TNF promoter allele was found. Similarly, five biallelic polymorphisms in the 5'-flanking region of the TNFα gene (-1031T/C; -863C/A; -857C/T; -308G/A and -238G/A) did not significantly increase the risk in Japanese when two-locus analyses were performed to take into account the risk conferred by the HLA-B and/or HLA-DRB1 alleles.[208] However, allele 9 of the TNFα VTNR (which is associated with higher LPS-induced TNFα production) is associated with disease in young-onset, but not adult-onset, Japanese patients with IDDM.[209] This association appeared to be independent of HLA-DR and HLA-B genes. If confirmed, these results would suggest that the age of onset is, at least in part, determined by a genetic factor influencing the magnitude of the inflammatory response.

To date, the data concerning young onset patients along with the above-mentioned results on *Idd9* in NOD mice are the most suggestive evidence that the TNF/TNF-R genes could play a role in the genetic susceptibility to IDDM.

Gene Encoding Interferon-γ

An important pathogenic role has been proposed for IFNγ in mouse models of IDDM based on experiments involving transgenic expression of this cytokine in β islet cells, treatment of NOD mice with neutralizing monoclonal antibodies against IFNγ, and use of IFNγ receptor-deficient mice.[210-213] Beyond its activating effect on macrophages and T cells, IFNγ might directly contribute to β cell death. However, recent data indicate that resistance to spontaneous diabetes in NOD IFNγ–R knockout mice is not directly due to the defective IFNγ–R gene but is rather dependent on, as yet unidentified, closely linked gene(s) on chromosome 10 derived from the 129 genome.[214] Moreover IFNγ/IFNγ-R interactions do not play an obligatory pathogenic role in NOD mice as mice knockout for either IFNγ or the IFNγ-Rβ chain (responsible for signal transduction) do not exhibit significant protection from diabetes but only a slight delay in diabetes onset.[215, 216] Nevertheless, IFNγ/IFNγ–R interactions in NOD mice play a central role in the acceleration of diabetes induced by cyclophosphamide since the NOD IFNγ-R-/- subline which is susceptible to spontaneous diabetes remains refractory to the cyclophosphamide effect.

In humans, the polymorphic CA repeat in the first intron of the IFNγ gene has been tested in case-control studies in different populations. In Japanese, evidence for an association was found between this polymorphism and IDDM. This association was even more obvious in the subgroup of patients with young-onset diabetes with the "3/6" genotype conferring a relative risk of 5.7.[217] In patients from England an overrepresentation of allele 3 was also present but no difference was found when the patients were stratified according to age of onset.[218] The fact that allele 3 of the IFNγ gene is associated with high levels of IFNγ production by PHA-stimulated blood mononuclear cells provides a plausible functional link between the genotype and the phenotype. However, an extensive analysis involving patients from Denmark and Finland with both case-control and TDT approaches could not confirm an association of this polymorphism with IDDM.[219]

Genes Encoding Interleukin-1α and β, Their Receptors and IL1RA

A large body of evidence indicates that, in vitro, IL-1β causes pancreatic β cell death either directly or by rendering them susceptible to IFNγ-induced apoptosis.[220-221]

In the NOD strain a type 1 diabetes locus, *Idd5* has been mapped to a proximal region of chromosome 1 including the IL-1-R genes.[222-223] This has led to thorough work both in mice and men to determine whether IL-1-R is a plausible candidate gene in IDDM. In congenic mice in which a 69 cM portion of diabetes-resistant B10 strain was introduced in the NOD genetic background, diabetes frequency was largely reduced. However, refining further the interval in additional congenic mice ruled out IL-1-R1 and IL-1-R2 (as well as IL-10 and CXCR4 genes) as being responsible for the *Idd5*-mediated effect.[224]

In humans, linkage analyses have mapped a type 1 diabetes susceptibility region to chromosome 2q31-q35 a region synthenic to the region of mouse chromosome 1 harbouring *Idd5,* and containing the human IL-1 gene cluster. Case-control studies involving patients with type 1 diabetes have suggested association between polymorphisms within the IL-1 gene cluster and disease,[225] or disease phenotype.[226] However, family-based genetic analyses, which prevent artifacts due to population stratification effects, have not revealed increased transmission of a given IL-1 gene cluster haplotype to diabetic siblings. Indeed, when transmission analyses of markers near or within the IL-1 gene cluster were performed in 352 diabetic families from the UK, biased transmission to diabetic siblings was not found.[227] Transmission disequilibrium was also tested in 245 Danish multiplex IDDM families using the four well-characterized intragenic IL-1 gene cluster polymorphisms (RFLP or VNTR). No linkage or intrafamilial association with IDDM was revealed in this study even when data were stratified according to HLA type (DR3/4 heterozygous versus non-DR3/4 heterozygous patients).[228] A similar result was generated in a TDT analysis of 91 Indian IDDM families.[229] Recently SNPs have been identified in the IL-1-R1 promoter region, one of which (G/A at position 1622) might have functional relevance since it is associated with differences in IL-1-R1 plasma levels.[230] A trend towards preferential transmission of the allele associated with higher IL-1-R1 plasma levels in affected siblings was found. This observation obviously needs replication in other data sets.

Although no strong linkage or association of IDDM has been found with the IL-1 gene cluster, it remains plausible that such an association exists with specific complications of diabetes in which an inflammatory component is involved.[226]

Genes Encoding Interleukin-2 and Its Receptor

Congenic mapping between NOD mice and diabetes-resistant B6 mice has assigned the location of *Idd3* to a 780-kb fragment of mouse chromosome 3.[231] In this narrow fragment resides the IL-2 gene, a very likely candidate as it displays allelic variations between B6 and NOD mice leading to a proline to serine substitution and a different number of polyglutamine repeats in the N-terminus of the protein. As a result, the glycosylation of the IL-2 protein differs substantially between the two strains.[232] Although this does not lead to detectable differences in the half-life of the molecule, in its affinity for the IL2R, and in its proliferative activity, it might affect the diffusion of the molecule from the tissue. As a consequence major immunological processes such as the generation of regulatory T cell populations (CD4+ CD25+) or induction of activation-induced T cell death could be perturbed in the NOD mice, both having a key role in induction of T cell tolerance and development of autoimmunity.

Gene Encoding Interleukin-6

Several studies indicate that IL-6 has an essential role in the pathogenesis of IDDM in the NOD mice[233] and that IL-6 is present in the islets of recently diagnosed patients with IDDM.[234-235]

A SNP at position -174 in the 5' flanking region of the IL-6 gene is associated with the level of the cytokine in the plasma of normal controls. The G allele (associated with higher IL-6 plasma levels) is significantly more frequent in patients as compared to the control population. However no excess transmission of the G allele to the affected sibling was found in a limited number (n=53) of parent-proband trios.[236] The IL-6 promoter polymorphism has also been shown to influence insulin sensitivity [237] and raises the interesting possibility that this polymorphism controls, in part, the slope of disease progression.

Genes Encoding Interleukin-12 and Interleukin-18

Evidence that IL-12 plays an important role in the induction of insulitis and diabetes comes from experiments in NOD mice in which administration of IL-12 accelerates onset of diabetes whereas pharmacological inhibition of IL-12 before establishment of insulitis prevents development of disease.[238] In addition, IL-12 mRNA expression in pancreatic islets of NOD mice increases with β cell destruction.[239] Moreover NOD macrophages produce far more IL-12 in response to various stimuli in vitro than macrophages from normal mouse strains.[240] This intrinsic property of NOD macrophages to produce high levels of IL-12 may bias autoreactive T cell differentiation towards the Th1 pathway and, thereby, favor destructive autoimmunity.

Recently, Morahan et al[241] have uncovered a significant linkage between polymorphisms in the vicinity of the IL-12p40 gene on chromosome 5q33-34 and human type 1 diabetes. After stratification according to HLA haplotype sharing, a significant linkage was found only for the HLA-identical diabetic sibpairs. Identification of several polymorphisms within the IL-12p40 gene led the authors to evaluate whether particular IL-12p40 gene alleles were preferentially transmitted to diabetic siblings. One of these polymorphisms, located in the 3' untranslated region of the IL-12p40 gene, was in strong linkage disequilibrium with an IDDM susceptibility locus as indicated by biased transmission in two independent cohorts of IDDM families. Some experimental data even suggest that the IL-12 3' untranslated region polymorphism itself might represent the susceptibility variant. First, linkage disequilibrium was limited to a 30 kb region in which IL-12p40 is the only known gene. Second, the preferentially transmitted 3' untranslated region allele was associated with higher IL-12p40 mRNA expression levels than the non-transmitted allele consistent with the biased Th1 response in infiltrated islets of diabetic patients.[241] These data therefore suggest that the IL-12p40 gene itself, rather than an unknown linked gene, is involved in the genetic susceptibility to type 1 diabetes in a subgroup of patients.

IL-18, previously called IFNγ-inducing factor, is a recently discovered cytokine mostly produced by antigen presenting cells with biological effects closely related to, and synergistic with, those of IL-12 in promoting IFNγ production, Th1 differentiation and NK cell cytotoxicity. IL-18 mRNA is present in NOD mouse pancreas in the early stages of disease[242] and IL-18 mRNA can be expressed by rodent β islet cells.[243] IL-18 may indirectly contribute to β cell death through upregulation of IFNγ, IL-1β, TNFα and FasL production by inflammatory cells.

Although genetic analyses of IL-18/IL-18-R in diabetes are scarce, it has been shown that the mouse IL-18 gene lies within the 20 cM-large *Idd2* region on chromosome 9 and represents an attractive candidate as a NOD susceptibility locus.[242] However, the human IL-18 gene on chromosome 11q22.2-22.3 appears distinct from any mapped IDDM locus.[244]

Anti-Inflammatory Cytokine and Cytokine Receptor Genes and Susceptibility to IDDM

Genes Encoding Interleukin-4 and Its Receptor

It has been suggested based on observations in transgenic NOD mice overexpressing IL-4 in β islet cells[245, 246] and on IL-4 transcription in NOD pancreas[247] that IL-4 prevents destruction of β islet cells by influencing homing and activation of autoreactive T cells to become autodestructive. Treatment with IL-4 prevents occurrence of IDDM in the NOD mice.

The IL-4 promoter SNP at position -590 (C/T) has been investigated in a case-control study from England and no differences in the frequency of the two alleles were found.[218] Similarly, the IL-4 and IL-4-Rα genes were tested as candidates for susceptibility genes in diabetic families from the United States using the affected sibpair and TDT statistics and no evidence of linkage or association was found.[248]

Genes Encoding Interleukin-10 and Transforming Growth Factor β

IL-10 has a complex role in IDDM. Administration of IL-10 at the appropriate times in NOD mice blocks disease development.[249] However, early transgenic expression of IL-10 in β islet cells accelerates diabetes in NOD mice[250] whereas transgenic expression of Epstein-Barr viral IL-10 has the opposite effect.[251] TGFβ has potent anti-inflammatory properties and has been shown to be a key regulator of pathogenic autoimmune responses in several animal models. Transgenic expression of TGFβ1 either in pancreatic α cells or in β cells has a major suppressive effect on NOD diabetes.[252, 253]

Little work has been devoted to IL-10 and TGFβ genes in human IDDM. In a case-control study from Denmark, two polymorphisms in the TGFβ1 gene on chromosome 19q13.2 (one resulting in a Thr to Ile substitution at position 263, and the second being a single base deletion in an intron) were evaluated. No association of the TGFβ1 polymorphisms was found with diabetes although a possible association with development of diabetic nephropathy was suggested.[254] Since no correlations between TGFβ1 mRNA levels and genotypes were observed, the functional importance of these TGFβ1 gene polymorphisms remains to be established.

Chemokine and Chemokine Receptor Genes and Susceptibility to IDDM

Evaluation of chemokine protein content in pancreata from NOD mice indicates that early insulitis correlates with local expression of MIP-1α. Similarly higher intrapancreatic levels of MIP-1α were found in diabetic female NOD mice as compared to their non-diabetic littermates, whereas elevated MIP-1β expression was associated with protection from diabetes. CCR5 (receptor for MIP-1α) mRNA levels are also higher in diabetic mice relative to protected animals.[255]

NOD mice T cells have been shown to respond weakly to TCR engagement. This hyporesponsiveness is under genetic control that mapping to a central region of chromosome 11 containing the CC chemokine gene cluster and the *Idd4* susceptibility locus.[256]

In humans the frequency of alleles carrying a 32 bp deletion in the CCR5 gene was comparable in 115 children with IDDM and 280 non-diabetic controls indicating a lack of association of the CCR5Δ32 variant with IDDM.[257] Confirming these data, a French study found a similar frequency of the CCR5Δ32 mutation in Caucasian IDDM patients and controls. Furthermore, the CCR5 allelic frequencies did not differ according to age of diabetes onset.[258] Interestingly, in the pediatric population, a G to A susbtitution at position 190 of the CCR2 gene (leading to an amino acid change at residue 64) was significantly associated with diabetes.[257]

A G to A substitution at position 801 of the SDF-1 3' untranslated region, albeit as frequent in IDDM patients as in controls, was significantly associated with a younger age at onset of diabetes.[258] Although the functional consequences of this variant are at the present time unknown, this variant has been shown to influence the course of HIV infection. These data have yet to be replicated but it is worth noting that the SDF-1 gene is located in a region of chromosome 10q11 that has been shown in linkage analyses to harbour the *IDDM10* susceptibility gene.

In conclusion, the importance of cytokine/chemokine or cytokine/chemokine receptor genes in determining susceptibility to autoimmune diseases or in modifying the course of disease remains to be firmly established in most cases. Owing to the clustering of non-MHC loci in a number of autoimmune diseases both in experimental models and in humans, the confirmed linkage/association in one disease should promote testing of the implicated markers in the other autoimmune diseases.[17,259-261] New tools such as the large scale identification of closely spaced SNP for both mouse and human genomes and accurate automated methods for their analysis should simplify and accelerate genotyping for complex traits such as autoimmune diseases.

References

1. Adams CW, Poston RN, Buk SJ. Pathology, histochemistry and immunocytochemistry of lesions in acute multiple sclerosis. J Neurol Sci 1989; 92:291-306.
2. Prineas JW, Raine CS. Electron microscopy and immunoperoxidase studies of early multiple sclerosis lesions. Neurology 1976; 26:29-32.
3. Prineas JW, Wright RG. Macrophages, lymphocytes, and plasma cells in the perivascular compartment in chronic multiple sclerosis. Lab Invest 1978; 38:409-421.
4. Mokhtarian F, McFarlin DE, Raine CS. Adoptive transfer of myelin basic protein-sensitized T cells produces chronic relapsing demyelinating disease in mice. Nature 1984; 309:356-358.
5. Zamvil S, Nelson P, Trotter J et al. T-cell clones specific for myelin basic protein induce chronic relapsing paralysis and demyelination. Nature 1985; 317:355-358.
6. Ebers GC, Sadovnick AD, Risch NJ. A genetic basis for familial aggregation in multiple sclerosis. Canadian Collaborative Study Group. Nature 1995; 377:150-151.
7. Sawcer S, Goodfellow PN, Compston A. The genetic analysis of multiple sclerosis. Trends Genet 1997; 13:234-239.
8. Olsson T. Critical influences of the cytokine orchestration on the outcome of myelin antigen-specific T-cell autoimmunity in experimental autoimmune encephalomyelitis and multiple sclerosis. Immunol Rev 1995; 144:245-268.
9. Martin R, Ruddle NH, Reingold S et al. T helper cell differentiation in multiple sclerosis and autoimmunity. Immunol Today 1998; 19:495-498.
10. Panitch HS, Hirsch RL, Schindler J et al. Treatment of multiple sclerosis with IFNγ: Exacerbations associated with activation of the immune system. Neurology 1987; 37: 1097-1102.
11. Rudick RA, Ransohoff RM, Lee JC et al. In vivo effects of interferon beta-1a on immunosuppressive cytokines in multiple sclerosis. Neurology 1998; 50:1294-1300.
12. Rep MH, Schrijver HM, van Lopik T et al. Interferon (IFN)-beta treatment enhances CD95 and interleukin 10 expression but reduces interferon-gamma producing T cells in MS patients. J Neuroimmunol 1999; 96:92-100.
13. Smith DR, Balashov KE, Hafler DA et al. Immune deviation following pulse cyclophosphamide/methylprednisolone treatment of multiple sclerosis: Increased interleukin-4 production and associated eosinophilia. Ann Neurol 1997; 42:313-318.
14. Bidwell J, Keen L, Gallagher G et al. Cytokine gene polymorphism in human disease: On-line databases. Genes and Immunity 1999; 1:3-19
15. Sundvall M, Jirholt J, Yang HT et al. Identification of murine loci associated with susceptibility to chronic experimental autoimmune encephalomyelitis. Nat Genet 1995; 10:313-317.
16. Encinas JA, Lees MB, Sobel RA et al. Genetic analysis of susceptibility to experimental autoimmune encephalomyelitis in a cross between SJL/J and B10.S mice. J Immunol 1996; 157:2186-2192.
17. Encinas JA, Wicker LS, Peterson LB et al. QTL influencing autoimmune diabetes and encephalomyelitis map to a 0.15-cM region containing Il2. Nat Genet 1999; 21:158-160.
18. Teuscher C, Rhein DM, Livingstone KD et al. Evidence that Tmevd2 and eae3 may represent either a common locus or members of a gene complex controlling susceptibility to immunologically mediated demyelination in mice. J Immunol 1997; 159:4930-4934.
19. Butterfield RJ, Sudweeks JD, Blankenhorn EP et al. New genetic loci that control susceptibility and symptoms of experimental allergic encephalomyelitis in inbred mice. J Immunol 1998; 161:1860-1867.

20. Roth MP, Viratelle C, Dolbois L et al. A genome-wide search identifies two susceptibility loci for experimental autoimmune encephalomyelitis on rat chromosomes 4 and 10. J Immunol 1999; 162:1917-1922.
21. Karpus WJ, Ransohoff RM. Chemokine regulation of experimental autoimmune encephalomyelitis: Temporal and spatial expression patterns govern disease pathogenesis. J Immunol 1998; 161:2667-2671.
22. Ransohoff RM. Mechanisms of inflammation in MS tissue: Adhesion molecules and chemokines. J Neuroimmunol 1999; 98:57-68.
23. Selmaj KW, Raine CS. Tumor necrosis factor mediates myelin and oligodendrocyte damage in vitro. Ann Neurol 1988; 23:339-346.
24. Selmaj K, Raine CS, Cannella B et al. Identification of lymphotoxin and tumor necrosis factor in multiple sclerosis lesions. J Clin Invest 1991; 87:949-954.
25. Beck J, Rondot P, Catinot L et al. Increased production of interferon gamma and tumor necrosis factor precedes clinical manifestation in multiple sclerosis: do cytokines trigger off exacerbations? Acta Neurol Scand 1988; 78:318-323.
26. Fugger L, Morling N, Sandberg-Wollheim M et al. Tumor necrosis factor alpha gene polymorphism in multiple sclerosis and optic neuritis. J Neuroimmunol 1990; 27:85-88.
27. Roth MP, Nogueira L, Coppin H et al. Tumor necrosis factor polymorphism in multiple sclerosis: No additional association independent of HLA. J Neuroimmunol 1994; 51:93-99.
28. Lucotte G, Bathelier C, Mercier G. TNF alpha polymorphisms in multiple sclerosis: No association with -238 and -308 promoter alleles, but the microsatellite allele a11 is associated with the disease in French patients. Mult Scler 2000; 6:78-80.
29. Vandevyver C, Raus P, Stinissen P et al. Polymorphism of the tumour necrosis factor beta gene in multiple sclerosis and rheumatoid arthritis. Eur J Immunogenet 1994; 21:377-382.
30. Sandberg-Wollheim M, Ciusani E, Salmaggi A et al. An evaluation of tumor necrosis factor microsatellite alleles in genetic susceptibility to multiple sclerosis. Mult Scler 1995; 1:181-185.
31. Epplen C, Jackel S, Santos EJ et al. Genetic predisposition to multiple sclerosis as revealed by immunoprinting. Ann Neurol 1997; 41:341-352.
32. Ma JJ, Nishimura M, Mine H et al. HLA-DRB1 and tumor necrosis factor gene polymorphisms in Japanese patients with multiple sclerosis. J Neuroimmunol 1998; 92:109-112.
33. McDonnell GV, Kirk CW, Middleton D et al. Genetic association studies of tumour necrosis factor alpha and beta and tumour necrosis factor receptor 1 and 2 polymorphisms across the clinical spectrum of multiple sclerosis. J Neurol 1999; 246:1051-1058.
34. Kirk CW, Droogan AG, Hawkins SA et al. Tumour necrosis factor microsatellites show association with multiple sclerosis. J Neurol Sci 1997; 147:21-25.
35. Allcock RJ, de la Concha EG, Fernandez-Arquero M et al. Susceptibility to multiple sclerosis mediated by HLA-DRB1 is influenced by a second gene telomeric of the TNF cluster. Hum Immunol 1999; 60:1266-1273.
36. He B, Navikas V, Lundahl J et al. Tumor necrosis factor alpha-308 alleles in multiple sclerosis and optic neuritis. J Neuroimmunol 1995; 63:143-147.
37. Maurer M, Kruse N, Giess R et al. Gene polymorphism at position -308 of the tumor necrosis factor alpha promotor is not associated with disease progression in multiple sclerosis patients. J Neurol 1999; 246:949-954.
38. Wingerchuk D, Liu Q, Sobell J et al. A population-based case-control study of the tumor necrosis factor alpha-308 polymorphism in multiple sclerosis. Neurology 1997; 49:626-628.
39. Huizinga TW, Westendorp RG, Bollen EL et al. TNF-alpha promoter polymorphisms, production and susceptibility to multiple sclerosis in different groups of patients. J Neuroimmunol 1997; 72:149-153.
40. Mycko M, Kowalski W, Kwinkowski M et al. Multiple sclerosis: The frequency of allelic forms of tumor necrosis factor and lymphotoxin-alpha. J Neuroimmunol 1998; 84:198-206.
41. Fernandez-Arquero M, Arroyo R et al. Primary association of a TNF gene polymorphism with susceptibility to multiple sclerosis. Neurology 1999; 53:1361-1363.
42. Sotgiu S, Pugliatti M, Serra C et al. Tumor necrosis factor 2 allele does not contribute to increased tumor necrosis factor-alpha production in Sardinian multiple sclerosis. Ann Neurol 1999; 46:799-800.
43. Weinshenker BG, Wingerchuk DM, Liu Q et al. Genetic variation in the tumor necrosis factor alpha gene and the outcome of multiple sclerosis. Neurology 1997; 49:378-385.
44. Rieckmann P, Martin S, Weichselbraun I et al. Serial analysis of circulating adhesion molecules and TNF receptor in serum from patients with multiple sclerosis: cICAM-1 is an indicator for relapse. Neurology 1994; 44:2367-2372.

45. Matsuda M, Tsukada N, Miyagi K et al. Increased levels of soluble tumor necrosis factor receptor in patients with multiple sclerosis and HTLV-1-associated myelopathy. J Neuroimmunol 1994; 52:33-40.
46. Goris A, Epplen C, Fiten P et al. Analysis of an IFN-gamma gene (IFNG) polymorphism in multiple sclerosis in Europe: Effect of population structure on association with disease. J Interferon Cytokine Res 1999; 19:1037-1046.
47. Vandenbroeck K, Opdenakker G, Goris A et al. Interferon-gamma gene polymorphism-associated risk for multiple sclerosis in Sardinia. Ann Neurol 1998; 44:841-842.
48. He B, Xu C, Yang B et al. Linkage and association analysis of genes encoding cytokines and myelin proteins in multiple sclerosis. J Neuroimmunol 1998; 86:13-19.
49. Wansen K, Pastinen T, Kuokkanen S et al. Immune system genes in multiple sclerosis: Genetic association and linkage analyses on TCR beta, IGH, IFN-gamma and IL-1ra/IL-1 beta loci. J Neuroimmunol 1997; 79:29-36.
50. Reboul J, Mertens C, Levillayer F et al. Cytokines in genetic susceptibility to multiple sclerosis: A candidate gene approach. French Multiple sclerosis Genetics Group. J Neuroimmunol 2000; 102:107-112.
51. Giedraitis V, He B, Hillert J. Mutation screening of the interferon-gamma gene as a candidate gene for multiple sclerosis. Eur J Immunogenet 1999; 26:257-259.
52. Vandenbroeck K, Hardt C, Louage J et al. Lack of association between the interferon regulatory factor-1 (IRF1) locus at 5q31.1 and multiple sclerosis in Germany, Northern Italy, Sardinia and Sweden. Genes and Immunity 2000; 1:290-292.
53. Cannella B, Raine CS. The adhesion molecule and cytokine profile of multiple sclerosis lesions. Ann Neurol 1995; 37:424-435.
54. Nicoletti F, Patti F, DiMarco R et al. Circulating serum levels of IL-1ra in patients with relapsing remitting multiple sclerosis are normal during remission phases but significantly increased either during exacerbations or in response to IFN-beta treatment. Cytokine 1996; 8:395-400.
55. Martin D, Near SL. Protective effect of the interleukin-1 receptor antagonist (IL-1ra) on experimental allergic encephalomyelitis in rats. J Neuroimmunol 1995; 61:241-245.
56. Crusius JB, Pena AS, Van Oosten BW et al. Interleukin-1 receptor antagonist gene polymorphism and multiple sclerosis. Lancet 1995; 346:979.
57. Huang WX, He B, Hillert J. An interleukin 1-receptor-antagonist gene polymorphism is not associated with multiple sclerosis. J Neuroimmunol 1996; 67:143-144.
58. Semana G, Yaouanq J, Alizadeh M et al. Interleukin-1 receptor antagonist gene in multiple sclerosis. Lancet 1997; 349:476.
59. de la Concha EG, Arroyo R, Crusius JB et al. Combined effect of HLA-DRB1*1501 and interleukin-1 receptor antagonist gene allele 2 in susceptibility to relapsing/remitting multiple sclerosis. J Neuroimmunol 1997; 80:172-178.
60. Schrijver HM, Crusius JB, Uitdehaag BM et al. Association of interleukin-1beta and interleukin-1 receptor antagonist genes with disease severity in MS. Neurology 1999; 52:595-599.
61. Kantarci OH, Atkinson EJ, Hebrink DD et al. Association of two variants in IL-1beta and IL-1 receptor antagonist genes with multiple sclerosis. J Neuroimmunol 2000; 106:220-227.
62. Sciacca FL, Ferri C, Vandenbroeck K et al. Relevance of interleukin 1 receptor antagonist intron 2 polymorphism in Italian MS patients. Neurology 1999; 52:1896-1898.
63. McDonnell GV, Kirk CW, Hawkins SA et al. An evaluation of interleukin genes fails to identify clear susceptibility loci for multiple sclerosis. J Neurol Sci 2000; 176:4-12.
64. Feakes R, Sawcer S, Broadley S et al. Interleukin 1 receptor antagonist (IL-1ra) in multiple sclerosis. J Neuroimmunol 2000; 105:96-101.
65. Greenberg SJ, Marcon L, Hurwitz BJ et al. Elevated levels of soluble interleukin-2 receptors in multiple sclerosis. N Engl J Med 1988; 319:1019-1020.
66. Adachi K, Kumamoto T, Araki S. Elevated soluble interleukin-2 receptor levels in patients with active multiple sclerosis. Ann Neurol 1990; 28:687-691.
67. Capra R, Mattioli F, Marciano N et al. Significantly higher levels of soluble interleukin 2 in patients with relapsing-remitting multiple sclerosis compared with healthy subjects. Arch Neurol 1990; 47:254.
68. Eugster HP, Frei K, Kopf M et al. IL-6-deficient mice resist myelin oligodendrocyte glycoprotein-induced autoimmune encephalomyelitis. Eur J Immunol 1998; 28:2178-2187.
69. Gijbels K, Brocke S, Abrams JS et al. Administration of neutralizing antibodies to interleukin-6 (IL-6) reduces experimental autoimmune encephalomyelitis and is associated with elevated levels of IL-6 bioactivity in central nervous system and circulation. Mol Med 1995; 1:795-805.

70. Padberg F, Feneberg W, Schmidt S et al. CSF and serum levels of soluble interleukin-6 receptors (sIL-6R and sgp130), but not of interleukin-6 are altered in multiple sclerosis. J Neuroimmunol 1999; 99:218-223.
71. Navikas V, Matusevicius D, Soderstrom M et al. Increased interleukin-6 mRNA expression in blood and cerebrospinal fluid mononuclear cells in multiple sclerosis. J Neuroimmunol 1996; 64:63-69.
72. Schmidt S, Papassotiropoulos A, Bagli M et al. No association of serum levels of interleukin-6 and its soluble receptor components with a genetic variation in the 3'flanking region of the interleukin-6 gene in patients with multiple sclerosis. Neurosci Lett 2000; 294:139-142.
73. Vandenbroeck K, Fiten P, Ronsse I et al. High-resolution analysis of IL-6 minisatellite polymorphism in Sardinian multiple sclerosis: Effect on course and onset of disease. Genes and Immunity 2000; 1:460-463
74. Segal BM, Dwyer BK, Shevach EM. An interleukin (IL)-10/IL-12 immunoregulatory circuit controls susceptibility to autoimmune disease. J Exp Med 1998; 187:537-546.
75. Bright JJ, Musuro BF, Du C et al. Expression of IL-12 in CNS and lymphoid organs of mice with experimental allergic encephalitis. J Neuroimmunol 1998; 82:22-30.
76. Constantinescu CS, Wysocka M, Hilliard B et al. Antibodies against IL-12 prevent superantigen-induced and spontaneous relapses of experimental autoimmune encephalomyelitis. J Immunol 1998; 161:5097-5104.
77. Leonard JP, Waldburger KE, Schaub RG et al. Regulation of the inflammatory response in animal models of multiple sclerosis by interleukin-12. Crit Rev Immunol 1997; 17:545-553.
78. Comabella M, Balashov K, Issazadeh S et al. Elevated interleukin-12 in progressive multiple sclerosis correlates with disease activity and is normalized by pulse cyclophosphamide therapy. J Clin Invest 1998; 102: 671-678.
79. Paty DW, Li DK. Interferon beta-1β is effective in relapsing-remitting multiple sclerosis. II. MRI analysis results of a multicenter, randomized, double-blind, placebo-controlled trial. UBC MS/MRI Study Group and the IFNB Multiple sclerosis Study Group. Neurology 1993; 43:662-667.
80. Noronha A, Toscas A, Jensen MA. Interferon beta decreases T cell activation and interferon gamma production in multiple sclerosis. J Neuroimmunol 1993; 46:145-153.
81. Noronha A, Toscas A, Jensen MA. Interferon beta augments suppressor cell function in multiple sclerosis. Ann Neurol 1990; 27:207-210.
82. Miterski B, Jaeckel S, Epplen JT et al. The interferon gene cluster: A candidate region for MS predisposition ? Genes and Immunity 1999; 1:37-44
83. Vandenbroeck K, Goris A, Murru R et al. A dinucleotide repeat polymorphism located in the IFN-alpha/beta gene cluster at chromosome 9p22 is not associated with multiple sclerosis in Sardinia. Exp Clin Immunogenet 1999; 16:26-29.
84. Bergkvist M, Martinsson T, Aman P et al. No genetic linkage between multiple sclerosis and the interferon alpha/beta locus. J Neuroimmunol 1996; 65:163-165.
85. Street NE, Mosmann TR. Functional diversity of T lymphocytes due to secretion of different cytokine patterns. Faseb J 1991; 5:171-177.
86. Peleman R, Wu J, Fargeas C et al. Recombinant interleukin 4 suppresses the production of interferon gamma by human mononuclear cells. J Exp Med 1989; 170:1751-1756.
87. Kuchroo VK, Das MP, Brown JA et al. B7-1 and B7-2 costimulatory molecules activate differentially the Thl/Th2 developmental pathways: Application to autoimmune disease therapy. Cell 1995; 80:707-718.
88. Racke MK, Bonomo A, Scott DE et al. Cytokine-induced immune deviation as a therapy for inflammatory autoimmune disease. J Exp Med 1994; 180:1961-1966.
89. Lu CZ, Jensen MA, Arnason BG. Interferon gamma- and interleukin-4-secreting cells in multiple sclerosis. J Neuroimmunol 1993; 46:123-128.
90. Vandenbroeck K, Martino G, Marrosu M et al. Occurrence and clinical relevance of an interleukin-4 gene polymorphism in patients with multiple sclerosis. J Neuroimmunol 1997; 76:189-192.
91. van Boxel-Dezaire AH, Hoff SC, van Oosten BW et al. Decreased interleukin-10 and increased interleukin-12p40 mRNA are associated with disease activity and characterize different disease stages in multiple sclerosis. Ann Neurol 1999; 45:695-703.
92. Rieckmann P, Albrecht M, Kitze B et al. Cytokine mRNA levels in mononuclear blood cells from patients with multiple sclerosis. Neurology 1994; 44:1523-1526.
93. Pickard C, Mann C, Sinnott P et al. Interleukin-10 (IL10) promoter polymorphisms and multiple sclerosis. J Neuroimmunol 1999; 101:207-210.
94. Maurer M, Kruse N, Giess R et al. Genetic variation at position -1082 of the interleukin 10 (IL10) promotor and the outcome of multiple sclerosis. J Neuroimmunol 2000; 104:98-100.

95. Santambrogio L, Hochwald GM, Saxena B et al. Studies on the mechanisms by which transforming growth factor-beta (TGF-beta) protects against allergic encephalomyelitis. Antagonism between TGF-beta and tumor necrosis factor. J Immunol 1993; 151:1116-1127.
96. Johns LD, Sriram S. Experimental allergic encephalomyelitis: Neutralizing antibody to TGF beta 1 enhances the clinical severity of the disease. J Neuroimmunol 1993; 47:1-7.
97. Beck J, Rondot P, Jullien P et al. TGF-beta-like activity produced during regression of exacerbations in multiple sclerosis. Acta Neurol Scand 1991; 84:452-455.
98. Rollnik JD, Sindern E, Schweppe C et al. Biologically active TGF-beta 1 is increased in cerebrospinal fluid while it is reduced in serum in multiple sclerosis patients. Acta Neurol Scand 1997; 96:101-105.
99. Giovannoni G, Thorpe JW, Kidd D et al. Soluble E-selectin in multiple sclerosis: Raised concentrations in patients with primary progressive disease. J Neurol Neurosurg Psychiatry 1996; 60:20-26.
100. McDonnell GV, McMillan SA, Douglas JP et al. Raised CSF levels of soluble adhesion molecules across the clinical spectrum of multiple sclerosis. J Neuroimmunol 1998; 85:186-192.
101. McDonnell GV, Kirk CW, Hawkins SA et al. Lack of association of transforming growth factor (TGF)-beta 1 and beta 2 gene polymorphisms with multiple sclerosis (MS) in Northern Ireland. Mult Scler 1999; 5:105-109.
102. Mertens C, Brassat D, Reboul J, -Darpoux F, Baron-Van Evercooren A, Lyon-Caen O, Liblau R, Fontaine B et al. A systematic study of oligodendrocyte growth factors as candidates for genetic susceptibility to MS. French Multiple sclerosis Genetics Group. Neurology 1998; 51:748-753.
103. Hvas J, McLean C, Justesen J et al. Perivascular T cells express the pro-inflammatory chemokine RANTES mRNA in multiple sclerosis lesions. Scand J Immunol 1997; 46:195-203.
104. Jiang Y, Salafranca MN, Adhikari S et al. Chemokine receptor expression in cultured glia and rat experimental allergic encephalomyelitis. J Neuroimmunol 1998; 86:1-12.
105. Sorensen TL, Tani M, Jensen J et al. Expression of specific chemokines and chemokine receptors in the central nervous system of multiple sclerosis patients. J Clin Invest 1999; 103:807-815.
106. Johnson RT. The virology of demyelinating diseases. Ann Neurol 1994; 36 Suppl:S54-60.
107. Alkhatib G, Combadiere C, Broder CC et al. CC CKR5: A RANTES, MIP-1alpha, MIP-1beta receptor as a fusion cofactor for macrophage-tropic HIV-1. Science 1996; 272:1955-1958.
108. Choe H, Farzan M, Sun Y et al. The beta-chemokine receptors CCR3 and CCR5 facilitate infection by primary HIV-1 isolates. Cell 1996; 85:1135-1148.
109. Barcellos LF, Schito AM, Rimmler JB et al. CC-chemokine receptor 5 polymorphism and age of onset in familial multiple sclerosis. Multiple sclerosis Genetics Group. Immunogenetics 2000; 51:281-288.
110. Bennetts BH, Teutsch SM, Buhler MM et al. The CCR5 deletion mutation fails to protect against multiple sclerosis. Hum Immunol 1997; 58:52-59.
111. Godiska R, Chantry D, Dietsch GN et al. Chemokine expression in murine experimental allergic encephalomyelitis. J Neuroimmunol 1995; 58:167-176.
112. Karpus WJ, Lukacs NW, McRae BL et al. An important role for the chemokine macrophage inflammatory protein-1 alpha in the pathogenesis of the T cell-mediated autoimmune disease, experimental autoimmune encephalomyelitis. J Immunol 1995; 155:5003-5010.
113. Martin MP, Dean M, Smith MW et al. Genetic acceleration of AIDS progression by a promoter variant of CCR5. Science 1998; 282:1907-1911.
114. McManus C, Berman JW, Brett FM et al. MCP-1, MCP-2 and MCP-3 expression in multiple sclerosis lesions: an immunohistochemical and in situ hybridization study. J Neuroimmunol 1998; 86:20-29.
115. Fiten P, Vandenbroeck K, Dubois B et al. Microsatellite polymorphisms in the gene promoter of monocyte chemotactic protein-3 and analysis of the association between monocyte chemotactic protein-3 alleles and multiple sclerosis development. J Neuroimmunol 1999; 95:195-201.
116. Teuscher C, Butterfield RJ, Ma RZ et al. Sequence polymorphisms in the chemokines Scya1 (TCA-3), Scya2 (monocyte chemoattractant protein (MCP)-1), and Scya12 (MCP-5) are candidates for eae7, a locus controlling susceptibility to monophasic remitting/nonrelapsing experimental allergic encephalomyelitis. J Immunol 1999; 163:2262-2266.
117. Kuchroo VK, Martin CA, Greer JM et al. Cytokines and adhesion molecules contribute to the ability of myelin proteolipid protein-specific T cell clones to mediate experimental allergic encephalomyelitis. J Immunol 1993; 151:4371-4382.
118. Glabinski AR, Tani M, Strieter RM etal. Synchronous synthesis of alpha- and beta-chemokines by cells of diverse lineage in the central nervous system of mice with relapses of chronic experimental autoimmune encephalomyelitis. Am J Pathol 1997; 150:617-630.

119. Brennan FM, Maini RN, Feldmann M. TNF alpha—A pivotal role in rheumatoid arthritis? Br J Rheumatol 1992; 31:293-298.
120. Griffiths RJ, Pettipher ER, Koch K et al. Leukotriene B4 plays a critical role in the progression of collagen-induced arthritis. Proc Natl Acad Sci USA 1995; 92:517-521.
121. Elliott MJ, Maini RN, Feldmann M et al. Treatment of rheumatoid arthritis with chimeric monoclonal antibodies to tumor necrosis factor alpha. Arthritis Rheum 1993; 36:1681-90.
122. Mulcahy B, Waldron-Lynch F, McDermott MF et al. Genetic variability in the tumor necrosis factor-lymphotoxin region influences susceptibility to rheumatoid arthritis. Am J Hum Genet 1996; 59:676-683.
123. Hajeer AH, Worthington J, Silman AJ et al. Association of tumor necrosis factor microsatellite polymorphisms with HLA-DRB1*04-bearing haplotypes in rheumatoid arthritis patients. Arthritis Rheum 1996; 39:1109-1114.
124. Mattey DL, Hassell AB, Dawes PT et al. Interaction between tumor necrosis factor microsatellite polymorphisms and the HLA-DRB1 shared epitope in rheumatoid arthritis: influence on disease outcome. Arthritis Rheum 1999; 42:2698-2704.
125. Martinez A, Fernandez-Arquero M, Pascual-Salcedo D et al. Primary association of tumor necrosis factor-region genetic markers with susceptibility to rheumatoid arthritis. Rheum Arthritis 2000; 43:1366-1370.
126. Mu H, Chen JJ, Jiang Y et al. Tumor necrosis factor a microsatellite polymorphism is associated with rheumatoid arthritis severity through an interaction with the HLA-DRB1 shared epitope. Arthritis Rheum 1999; 42:438-442.
127. Brinkman BM, Huizinga TW, Kurban SS et al. Tumour necrosis factor alpha gene polymorphisms in rheumatoid arthritis: Association with susceptibility to, or severity of, disease? Br J Rheumatol 1997; 36:516-521.
128. Kaijzel EL, van Krugten MV, Brinkman BM et al. Functional analysis of a human tumor necrosis factor alpha (TNF-alpha) promoter polymorphism related to joint damage in rheumatoid arthritis. Mol Med 1998; 4:724-733.
129. van Krugten MV, Huizinga TW, Kaijzel EL et al. Association of the TNF +489 polymorphism with susceptibility and radiographic damage in rheumatoid arthritis. Genes Immun 1999; 1:91-96.
130. Shibue T, Tsuchiya N, Komata T et al. Tumor necrosis factor alpha 5'-flanking region, tumor necrosis factor receptor II, and HLA-DRB1 polymorphisms in Japanese patients with rheumatoid arthritis. Arthritis Rheum 2000; 43:753-757.
131. Date Y, Seki N, Kamizono S et al. Identification of a genetic risk factor for systemic juvenile rheumatoid arthritis in the 5'-flanking region of the TNFalpha gene and HLA genes. Arthritis Rheum 1999; 42:2577-2582.
132. Lacki JK, Moser R, Korczowska I et al. TNF-alpha gene polymorphism does not affect the clinical and radiological outcome of rheumatoid arthritis. Rheumatol Int 2000; 19:137-140.
133. Verweij CL. Tumour necrosis factor gene polymorphisms as severity markers in rheumatoid arthritis. Ann Rheum Dis 1999; 58:I20-26.
134. Cornelis F, Faure S, Martinez M et al. New susceptibility locus for rheumatoid arthritis suggested by a genome-wide linkage study. Proc Natl Acad Sci USA 1998; 95:10746-10750.
135. Shiozawa S, Hayashi S, Tsukamoto Y, et al. Identification of the gene loci that predispose to rheumatoid arthritis. Int Immunol 1998; 10:1891-1895.
136. Feldmann M, Brennan FM, Maini RN. Role of cytokines in rheumatoid arthritis. Annu Rev Immunol 1996; 14:397-440.
137. Ruiz-Linares A. Dinucleotide repeat polymorphism in the IFN-γ gene. Hum Mol Genet 1993; 2:1508.
138. Pravica V, Asderakis A, Perrey C et al. In vitro production of IFN-γ correlates with CA repeat polymorphism in the human IFN-γ gene. Eur J Immunogenet 1999; 26:1-3.
139. Khani-Hanjani A, Lacaille D, Hoar D et al. Association between dinucleotide repeat in non-coding region of IFN-γ gene and susceptibility to, and severity of, rheumatoid arthritis. Lancet 2000; 356:820-825.
140. Dinarello CA. Biologic basis for interleukin-1 in disease. Blood 1996; 87:2095-2147.
141. Eastgate JA, Symons JA, Wood NC et al. Plasma levels of interleukin-1-alpha in rheumatoid arthritis. Br J Rheumatol 1991; 30:295-297.
142. Eastgate JA, Symons JA, Wood NC et al. Correlation of plasma interleukin 1 levels with disease activity in rheumatoid arthritis. Lancet 1988; 2:706-709.
143. Gowen M, Wood DD, Ihrie EJ et al. An interleukin 1 like factor stimulates bone resorption in vitro. Nature 1983; 306:378-380.

144. Mino T, Sugiyama E, Taki H et al. Interleukin-1alpha and tumor necrosis factor alpha synergistically stimulate prostaglandin E2-dependent production of interleukin-11 in rheumatoid synovial fibroblasts. Arthritis Rheum 1998; 41:2004-2013.
145. van de Loo AA, van den Berg WB. Effects of murine recombinant interleukin 1 on synovial joints in mice: Measurement of patellar cartilage metabolism and joint inflammation. Ann Rheum Dis 1990; 49:238-245.
146. Zhang Y, McCluskey K, Fujii K et al. Differential regulation of monocyte matrix metalloproteinase and TIMP-1 production by TNF-alpha, granulocyte-macrophage CSF, and IL-1 beta through prostaglandin-dependent and -independent mechanisms. J Immunol 1998; 161:3071-3076.
147. Bailly S, di Giovine FS, Blakemore AI et al. Genetic polymorphism of human interleukin-1 alpha. Eur J Immunol 1993; 23:1240-1245.
148. Bailly S, di Giovine FS, Duff GW. Polymorphic tandem repeat region in interleukin-1 alpha intron 6. Hum Genet 1993; 91:85-86.
149. Cox A, Camp NJ, Nicklin MJ et al. An analysis of linkage disequilibrium in the interleukin-1 gene cluster, using a novel grouping method for multiallelic markers. Am J Hum Genet 1998; 62:1180-1188.
150. McDowell TL, Symons JA, Ploski R et al. A genetic association between juvenile rheumatoid arthritis and a novel interleukin-1 alpha polymorphism. Arthritis Rheum 1995; 38:221-228.
151. van den Velden PA, Reitsma PH. Amino acid dimorphism in IL1A is detectable by PCR amplification. Hum Mol Genet 1993; 2:1753.
152. di Giovine FS, Takhsh E, Blakemore AI et al. Single base polymorphism at -511 in the human interleukin-1 beta gene (IL1 beta). Hum Mol Genet 1992; 1:450.
153. Pociot F, Molvig J, Wogensen L et al. A TaqI polymorphism in the human interleukin-1 beta (IL-1 beta) gene correlates with IL-1 beta secretion in vitro. Eur J Clin Invest 1992; 22:396-402.
154. Cox A, Camp NJ, Cannings C et al. Combined sib-TDT and TDT provide evidence for linkage of the interleukin-1 gene cluster to erosive rheumatoid arthritis. Hum Mol Genet 1999; 8:1707-1713.
155. Tarlow JK, Blakemore AI, Lennard A et al. Polymorphism in human IL-1 receptor antagonist gene intron 2 is caused by variable numbers of an 86-bp tandem repeat. Hum Genet 1993; 91:403-404.
156. Gomolka M, Menninger H, Saal JE et al. Immunoprinting: Various genes are associated with increased risk to develop rheumatoid arthritis in different groups of adult patients. J Mol Med 1995; 73:19-29.
157. Jouvenne P, Chaudhary A, Buchs N et al. Possible genetic association between interleukin-1alpha gene polymorphism and the severity of chronic polyarthritis. Eur Cytokine Netw 1999; 10:33-36.
158. Cantagrel A, Navaux F, Loubet-Lescoulie P et al. Interleukin-1beta, interleukin-1 receptor antagonist, interleukin-4, and interleukin-10 gene polymorphisms: Relationship to occurrence and severity of rheumatoid arthritis. Arthritis Rheum 1999; 42:1093-1100.
159. Perrier S, Coussediere C, Dubost JJ et al. IL-1 receptor antagonist (IL-1RA) gene polymorphism in Sjogren's syndrome and rheumatoid arthritis. Clin Immunol Immunopathol 1998; 87:309-313.
160. Crilly A, Maiden N, Capell HA et al. Predictive value of interleukin 1 gene polymorphisms for surgery. Ann Rheum Dis 2000; 59:695-699.
161. Tilg H, Trehu E, Atkins MB et al. Interleukin-6 (IL-6) as an anti-inflammatory cytokine: induction of circulating IL-1 receptor antagonist and soluble tumor necrosis factor receptor p55. Blood 1994; 83:113-118.
162. Bowcock AM, Ray A, Erlich H et al. Rapid detection and sequencing of alleles in the 3' flanking region of the interleukin-6 gene. Nucleic Acids Res 1989; 17:6855-6864.
163. Murray RE, McGuigan F, Grant SF et al. Polymorphisms of the interleukin-6 gene are associated with bone mineral density. Bone 1997; 21:89-92.
164. Gyapay G, Morissette J, Vignal A et al. The 1993-94 Genethon human genetic linkage map. Nat Genet 1994; 7:246-339.
165. Fishman D, Faulds G, Jeffery R et al. The effect of novel polymorphisms in the interleukin-6 (IL-6) gene on IL-6 transcription and plasma IL-6 levels, and an association with systemic-onset juvenile chronic arthritis. J Clin Invest 1998; 102:1369-76.
166. Blankenstein T, Volk HD, Techert-Jendrusch C et al. Lack of correlation between BglII RFLP in the human interleukin 6 gene and rheumatoid arthritis. Nucleic Acids Res 1989; 17:8902.
167. Fugger L, Morling N, Bendtzen K et al. IL-6 gene polymorphism in rheumatoid arthritis, pauciarticular juvenile rheumatoid arthritis, systemic lupus erythematosus, and in healthy Danes. J Immunogenet 1989; 16:461-465.
168. John S, Myerscough A, Marlow A et al. Linkage of cytokine genes to rheumatoid arthritis. Evidence of genetic heterogeneity. Ann Rheum Dis 1998; 57:361-365.

169. Crilly A, Bartlett JM, White A et al. Investigation of novel polymorphisms within the 3' region of the Il-6 gene in patients with rheumatoid arthritis using genescan analysis. Cytokine 2001; 13:109-112.
170. Bartlett JM, Crilly A, White A et al. Modification of the GeneScan 2500 fluorescent dye standard for accurate product sizing. Mol Biotechnol 1999; 13:185-189.
171. Yang YC, Kovacic S, Kriz R et al. The human genes for GM-CSF and IL 3 are closely linked in tandem on chromosome 5. Blood 1988; 71:958-961.
172. Jeong MC, Navani A, Oksenberg JR. Limited allelic polymorphism in the human interleukin-3 gene. Mol Cell Probes 1998; 12:49-53.
173. Yamada R, Tanaka T, Ohnishi Y, Suematsu K, Minami M, Seki T et al. Identification of 142 single nucleotide polymorphisms in 41 candidate genes for rheumatoid arthritis in the Japanese population. Hum Genet 2000; 106:293-297.
174. Yamada R, Tanaka T, Unoki M et al. Association between a single-nucleotide polymorphism in the promoter of the human interleukin-3 gene and rheumatoid arthritis in Japanese patients, and maximum-likelihood estimation of combinatorial effect that two genetic loci have on susceptibility to the disease. Am J Hum Genet 2001; 68:674-685.
175. Arai N, Nomura D, Villaret D et al. Complete nucleotide sequence of the chromosomal gene for human IL-4 and its expression. J Immunol 1989; 142:274-282.
176. Mout R, Willemze R, Landegent JE. Repeat polymorphisms in the interleukin-4 gene (IL4). Nucleic Acids Res 1991; 19:3763.
177. Rosenwasser LJ, Klemm DJ, Dresback JK et al. Promoter polymorphisms in the chromosome 5 gene cluster in asthma and atopy. Clin Exp Allergy 1995; 25:74-8; discussion 95-96.
178. Buchs N, Silvestri T, Di Giovine FS, Chabaud M, Vannier E, Duff GW, Miossec P. IL-4 VNTR gene polymorphism in chronic polyarthritis. The rare allele is associated with protection against destruction. Rheumatology 2000; 39:1126-1131.
179. Scott DL, Coulton BL, Popert AJ. Long term progression of joint damage in rheumatoid arthritis. Ann Rheum Dis 1986; 45:373-8.
180. van der Heijde DM, van Leeuwen MA, van Riel PL et al. Radiographic progression on radiographs of hands and feet during the first 3 years of rheumatoid arthritis measured according to Sharp's method (van der Heijde modification). J Rheumatol 1995; 22:1792-1796.
181. de Waal Malefyt R, Abrams J, Bennett B et al. Interleukin 10(IL-10) inhibits cytokine synthesis by human monocytes: An autoregulatory role of IL-10 produced by monocytes. J Exp Med 1991; 174:1209-1220.
182. Katsikis PD, Chu CQ, Brennan FM et al. Immunoregulatory role of interleukin 10 in rheumatoid arthritis. J Exp Med 1994; 179:1517-1527.
183. Lacraz S, Nicod LP, Chicheportiche R et al. IL-10 inhibits metalloproteinase and stimulates TIMP-1 production in human mononuclear phagocytes. J Clin Invest 1995; 96:2304-2310.
184. Walmsley M, Katsikis PD, Abney E, Parry S, Williams RO, Maini RN et al. Interleukin-10 inhibition of the progression of established collagen-induced arthritis. Arthritis Rheum 1996; 39:495-503.
185. Eskdale J, McNicholl J, Wordsworth P et al. Interleukin-10 microsatellite polymorphisms and IL-10 locus alleles in rheumatoid arthritis susceptibility. Lancet 1998; 352:1282-1283.
186. Turner DM, Williams DM, Sankaran D et al. An investigation of polymorphism in the interleukin-10 gene promoter. Eur J Immunogenet 1997; 24:1-8.
187. Eskdale J, Gallagher G. A polymorphic dinucleotide repeat in the human IL-10 promoter. Immunogenetics 1995; 42:444-5.
188. Eskdale J, Kube D, Gallagher G. A second polymorphic dinucleotide repeat in the 5' flanking region of the human IL10 gene. Immunogenetics 1996; 45:82-83.
189. Coakley G, Mok CC, Hajeer AH et al. Interleukin-10 promoter polymorphisms in rheumatoid arthritis and Felty's syndrome. Br J Rheumatol 1998; 37:988-991.
190. Eskdale J, Gallagher G, Verweij CL et al. Interleukin 10 secretion in relation to human IL-10 locus haplotypes. Proc Natl Acad Sci USA 1998; 95:9465-9470.
191. DeVries ME, Ran L, Kelvin DJ. On the edge: The physiological and pathophysiological role of chemokines during inflammatory and immunological responses. Semin Immunol 1999; 11:95-104.
192. Patel DD, Zachariah JP, Whichard LP. CXCR3 and CCR5 ligands in rheumatoid arthritis synovium. Clin Immunol 2001; 98:39-45.
193. Bonecchi R, Bianchi G, Bordignon PP et al. Differential expression of chemokine receptors and chemotactic responsiveness of type 1 T helper cells (Th1s) and Th2s. J Exp Med 1998; 187:129-134.
194. Allusto F, Lenig D, Mackay CR et al. Flexible programs of chemokine receptor expression on human polarized T helper 1 and 2 lymphocytes. J Exp Med 1998; 187:875-883.

195. Loetscher P, Uguccioni M, Bordoli L et al. CCR5 is characteristic of Th1 lymphocytes. Nature 1998; 391:344-345.
196. Suzuki N, Nakajima A, Yoshino S et al. Selective accumulation of CCR5+ T lymphocytes into inflamed joints of rheumatoid arthritis. Int Immunol 1999; 11:553-559.
197. Gomez-Reino JJ, Pablos JL, Carreira PE et al. Association of rheumatoid arthritis with a functional chemokine receptor, CCR5. Arthritis Rheum 1999; 42:989-992.
198. Pablos JL, Carreira PE, Serrano L et al. The homozygous D32 deletion of CC chemokine receptor CCR5 protects against rheumatoid arthritis. Arthritis Rheum 1997; 40:S157.
199. Cooke SP, Forrest G, Venables PJ et al. The delta32 deletion of CCR5 receptor in rheumatoid arthritis. Arthritis Rheum 1998; 41:1135-1136.
200. Garred P, Madsen HO, Petersen J et al. CC chemokine receptor 5 polymorphism in rheumatoid arthritis. J Rheumatol 1998; 25:1462-1465.
201. Lyons PA, Hancock WW, Denny P et al. The NOD Idd9 genetic interval influences the pathogenicity of insulitis and contains molecular variants of Cd30, Tnfr2, and Cd137. Immunity 2000; 13:107-115.
202. Hirai H, Kaino Y, Ito T, Kida K. Analysis of cytokine mRNA expression in pancreatic islets of nonobese diabetic mice. J Pediatr Endocrinol Metab 2000; 13:91-98.
203. Rabinovitch A, Suarez-Pinzon WL. Cytokines and their roles in pancreatic islet beta-cell destruction and insulin-dependent diabetes mellitus. Biochem Pharmacol 1998; 55:1139-1149.
204. Yang XD, Tisch R, Singer SM et al. Effect of tumor necrosis factor α on insulin-dependent diabetes mellitus in NOD mice. I- The early development of autoimmunity and the diabetogenic process. J Exp Med 1994; 180:995-1004.
205. Pociot F, Wilson AG, Nerup J, Duff GW. No independent association between a tumor necrosis factor-alpha promotor region polymorphism and insulin-dependent diabetes mellitus. Eur J Immunol. 1993; 3:3050-3053.
206. Cox A, Gonzalez AM, Wilson AG, Wilson RM, Ward JD, Artlett CM, Welsh K, Duff GW. Comparative analysis of the genetic associations of HLA-DR3 and tumour necrosis factor alpha with human IDDM. Diabetologia 1994; 37:500-503.
207. Deng GY, Maclaren NK, Huang HS, Zhang LP, She JX. No primary association between the 308 polymorphism in the tumor necrosis factor alpha promoter region and insulin-dependent diabetes mellitus. Human Immunol 1996; 45:137-142.
208. Hamaguchi K, Kimura A, Seki N et al. Analysis of tumor necrosis factor-α promoter polymorphism in type 1 diabetes: HLA-B and -DRB1 alleles are primarily associated with the disease in Japanese. Tissue Antigens 2000; 55:10-16.
209. Obayashi H, Nakamura N, Fukui M et al. Influence of the TNF microsatellite polymorphisms (TNFa) on age-at-onset of insulitis-dependent diabetes mellitus. Human Immunol 1999; 60:974-978.
210. Sarvetnick N, Shizuru J, Liggitt D et al. Loss of pancreatic islet tolerance induced by beta-cell expression of interferon-gamma. Nature 1990; 346:844-847.
211. Campbell IL, Kay TW, Oxbrow I et al. Essential role for interferon-gamma and interleukin-6 in autoimmune insulin-dependent diabetes in NOD/wehi mice. J Clin Invest 1991; 87:739-742.
212. Debray-Sachs M, Carnaud C, Boitard C et al. Prevention of diabetes in NOD mice treated with antibody to murine IFN-γ. J Autoimmunity 1991; 4:237-248.
213. Wang B, Andre I, Gonzalez A et al. IFN-γ impacts at multiple points during the progression of autoimmune diabetes. Proc Natl Acad Sci USA 1997; 94:13844-13849.
214. Kanagawa O, Xu G, Tevaarwerk A et al. Protection of nonobese diabetic mice from diabetes by gene(s) closely linked to IFN-γ receptor loci. J Immunol 2000; 164:3919-3923.
215. Hultgren B, Huang X, Dybdal N et al. Genetic Absence of gamma-interferon delays but does not prevent diabetes in NOD mice. Diabetes 1996; 45:812-817.
216. Serreze DV, Post CM, Chapman HD et al. Interferon-γ receptor signaling is dispensable in the development of autoimmune type 1 diabetes in NOD mice. Diabetes 2000; 49:2007-2011.
217. Awata T, Matsumoto C, Urakami T et al. Association of polymorphism in the interferon gamma gene with IDDM. Diabetologia 1994; 37:1159-1162.
218. Jahromi M, Millward A, Demaine A. A CA repeat polymorphism of the IFN-γ gene is associated with susceptibility to type 1 diabetes. J Interferon Cytokine Research 2000; 20:187-190.
219. Pociot F, Veijola R, Johannesen J et al. Analysis of an IFN-γ gene polymorphism in Danish and Finnish insulin-dependent diabetes mellitus (IDDM) patients and control subjects. Danish Study Group of Diabetes in Childhood. J Interferon Cytokine Res 1997; 17:87-93.
220. Mandrupt-Poulsen T. The role of interleukin-1 in the pathogenesis of IDDM. Diabetologia 1996; 39:1005-1029.
221. Hoorens A, Strangé G, Pavlovic D et al. Distinction between interleukin-1-induced necrosis and apoptosis of islet cells. Diabetes 2001; 50:551-557.

222. Cornall RJ, Prins JB, Todd JA, Pressey A, DeLarato NH, Wicker LS et al. Type 1 diabetes in mice is linked to the interleukin-1 receptor and Lsh/Ity/Bcg genes on chromosome 1. Nature 1991; 353:262-265.
223. Garchon HJ, Bedossa P, Eloy L et al. Identification and mapping to chromosome 1 of a susceptibility locus for periinsulitis in nonobese diabetic mice. Nature 1991; 353:260-262.
224. Hill NJ, Lyons PA, Armitage N et al. NOD Idd5 locus controls insulitis and diabetes and overlaps the orthologous CLTA4/IDDM12 and NRAMP1 loci in humans. Diabetes 2000; 49:1744-1747.
225. Metcalfe KA, Hitman GA, Pociot F, Bergholdt R, Tuomilehto-Wolf E, Tuomilehto J, Viswanathan M, Ramachandran A, Nerup J. An association between type 1 diabetes and the interleukin-1 receptor type 1 gene. Hum Immunol. 1996; 51:41-48.
226. Blakemore AI, Cox A, Gonzalez AM et al. Interleukin-1 receptor antagonist allele (IL1RN*2) associated with nephropathy in diabetes mellitus. Hum Genet 1996; 97:369-374.
227. Esposito L, Hill NJ, Pritchard LE et al. Genetic analysis of chromosome 2 in type 1 diabetes: Analysis of putative loci IDDM7, IDDM12, and IDDM13 and candidate genes NRAMP1 and IA-2 and interleukin-1 gene cluster. Diabetes 1998; 47:1797-1799.
228. Kristiansen OP, Pociot F, Johannesen J et al. Linkage disequilibrium testing of four interleukin-1 gene-cluster polymorphisms in Danish multiplex families with insulin-dependent diabetes mellitus. Cytokine 2000; 12:171-175.
229. Ogunkolade WB, Ramachandran A, McDermott MF et al. Family association studies of markers on chromosome 2q and type 1 diabetes in subjects from South India. Diabetes Metab Res Rev 2000; 16:276-280.
230. Bergholdt R, Larsen ZM, Andersen NA et al. Characterization of new polymorphisms in the 5' UTR of the human interleukin-1 receptor type 1 (IL1RI) gene: Linkage to type 1 diabetes and correlation to IL-1RI plasma level. Genes Immunity 2000; 8:495-500.
231. Lyons PA, Armitage N, Argentina F et al. Congenic mapping of the type 1 diabetes locus Idd3, to a 780-kb region of mouse chromosome 3: Identification of a candidate segment of ancestral DNA by haplotype mapping. Genome Res 2000; 10:446-453.
232. Podolin PL, Wilusz MB, Cubbon RM et al. Differential glycosylation of interleukin 2, the molecular basis for the NOD Idd3 type 1 diabetes gene? Cytokine 2000; 12:477-482.
233. Campbell IL, Kay TW, Oxbrow L et al. Essential role for interferon-gamma and interleukin-6 in autoimmune insulin-dependent diabetes in NOD/Wehi mice. J Clin Invest 1991; 87:739-742.
234. Foulis AK, Farquhaarson MA, Meaher H. Immunoreactive alpha-interferon in insulitis-secreting beta cells in type 1 diabetes mellitus. Lancet 1987, 2:1423-1427.
235. Somoza N, Vargas F, Roura-Mir C, Vives-Pi M, Fernandez-Figueras MT, Ariza A et al. Pancreas in recent onset insulin-dependent diabetes mellitus changes in HLA, adhesion molecules and autoantigens, restricted T cell receptor V beta usage, and cytokine profile. J Immunol 1994; 153:1360-1377.
236. Jahromi MM, Millward BA, Demaine AG. A polymorphism in the promoter region of the gene for interleukin-6 associated with susceptibility to type 1 diabetes mellitus. J Interfer Cytok Res 2000; 20:885-888.
237. Fernandez-Real JM, Broch M, Vendrell J et al. Interleukin-6 gene polymorphism and insulin sensitivity. Diabetes 2000; 49:517-520.
238. Trembleau S, Penna G, Gregori S, Gately M, Adorini L. Deviation of pancreas-infiltrating cells to Th2 by interleukin-12 antagonist administration inhibits autoimmune diabetes. Eur J Immunol; 27:2330-2239.
239. Rabinovitch A, Suarez-Pinzon WL, Sorensen O. Interleukin 12 mRNA expression in islets correlates with β-cell destruction in NOD mice. J Autoimmunity 1996; 9:645-651.
240. Alleva DG, Pavlovitch RP, Grant C et al. Aberrant macrophage cytokine production is a conserved feature among autoimmune-prone mouse strains. Diabetes 2000; 49:1106-111.
241. Morahan G, Huang D, Ymer SI et al. Linkage disequilibrium of a type 1 diabetes susceptibility locus with a regulatory IL12B allele. Nature Genetics 2001; 27:218-221.
242. Rothe H, Jenkins NA, Copeland NG et al. Active stage of autoimmune diabetes is associated with the expression of a novel cytokine, IGIF, which is located near Idd2. J Clin Invest 1997; 99:469-474.
243. Hong TP, Andersen NA, Nielsen K et al. Interleukin-18 mRNA, but not interleukin-18 receptor mRNA, is constitutively expressed in islet beta-cells and up-regulated by interferon-γ. Eur Cytokine Netw 2000; 11:193-205.
244. Nolan KF, Greaves DR, Waldmann H. The human interleukin 18 gene IL18 maps to 11q22.2-22.3, closely linked to the DRD2 gene locus and distinct from mapped IDDM loci. Genomics 1998; 51:161-163.
245. Mueller R, Krahl T, Sarvetnick N. Pancreatic expression of interleukin-4 abrogates insulitis and autoimmune diabetes in nonobese diabetic (NOD) mice. J Exp Med 1996; 184:1093-1099.

246. Mueller R, Bradley LM, Krahl T et al. Mechanism underlying counterregulation of autoimmune diabetes by IL4. Immunity 1997; 7:411-418.
247. Fox CJ, Danska JS. IL-4 expression at the onset of islet inflammation predicts nondestructive insulitis in nonobese diabetic mice. J Immunol 1997, 158:2414-2424.
248. Reimsnider SK, Eckenrode SE, Marron MP et al. IL4 and IL4alpha genes are not linked or associated with type 1 diabetes. Pediatr Res 2000; 47:246-249.
249. Pernnline KJ, Roquegaffney E, Monahan M. Human IL-10 prevents the onset of diabetes in the NOD mouse. Clin Immunol Immunopathol 1994; 71:169-175.
250. Balasa B, Van Gunst K, Jung N et al. Islet-specific expression of IL-10 promotes diabetes in nonobese diabetic mice independent of Fas, perforin, TNF receptor-1, and TNF receptor-2 molecules. J Immunol 2000; 165:2841-2849.
251. Kawamoto S, Nitta Y, Tashiro F et al. Suppression of T(h)1 cell activation and prevention of autoimmune diabetes in NOD mice by local expression of viral 1L-10. Int Immunol 2001; 13:685-694.
252. King C, Davies J, Mueller R et al. TGF-β1 alters APC preference, polarizing islet antigen responses toward a Th2 phenotype. Immunity 1998; 8:601-613.
253. Moritani M, Yoshimoto K, Wong SF et al. Abrogation of autoimmune diabetes in nonobese diabetic mice and protection against effector lymphocytes by transgenic paracrine TGF-beta1. J Clin Invest 1998; 102:499-506.
254. Pociot F, Hansen PM, Karlsen AE et al. TGF-β1 gene mutations in insulin-dependent diabetes mellitus and diabetic nephropathy. J Am Soc Nephrol 1998; 9:2302-2307.
255. Cameron MJ, Arreaza GA, Grattan M et al. Differential expression of cc chemokines and the CCR5 receptor in the pancreas is associate with progression to type I diabetes. J Immunol 2000; 165:1102-1110.
256. Gill BM, Jaramillo A, Ma L, Laupland KB, et al. Genetic linlage of thymic T-cell proliferative unresponsiveness to mouse chromosome 11 in NOD mice. Diabetes 1995; 44:614-619.
257. Szalai C, Csaszar A, Czinner A et al. Chemokine receptor CCR2 and CCR5 polymorphisms in children with insulin-dependent diabetes mellitus. Pediatr Res 1999; 46:82-84.
258. Dubois-Laforgue D, Hendel H, Caillat-Zucman S et al. A common stromal cell-derived factor-1 chemokine gene variant is associated with the early onset of type 1 diabetes. Diabetes 2001; 50:1211-1213.
259. Vyse TJ, Todd JA. Genetic analysis of autoimmune disease. Cell 1996; 85:311-318.
260. Becker KG, Simon RM, Bailey-Wilson JE et al. Clustering of non-major histocompatibility complex susceptibility candidate loci in human autoimmune diabetes. Proc Natl Acad Sci USA 1998, 95:9979-9984.
261. Merriman TR, Cordell HJ, Eaves IA et al. Suggestive evidence for association of human chromosome 18q12-q21 and its orthologue on rat and mouse chromosome 18 with several autoimmune diseases. Diabetes 2001; 50:184-194.

CHAPTER 4

Cytokines, Lymphocyte Homeostasis and Self Tolerance

Yiguang Chen and Youhai Chen

Introduction

Cytokines play pivotal roles in maintaining lymphocyte homeostasis and self tolerance. Cytokines are required for activating and inactivating as well as deleting cells of the immune system during immune responses. Mutations in cytokine genes in humans and animals can lead to the breakdown of self tolerance and the development of autoimmune diseases. In this chapter, we will discuss the important roles of cytokines in lymphocyte homeostasis and self tolerance. Since many cytokines are involved, albeit to different degrees, in maintaining lymphocyte homeostasis, a complete review of all cytokines involved is beyond the scope of this chapter. Instead, we will focus on those cytokines whose gene mutations cause autoimmune problems. These include transforming growth factor-β, interleukin-10, interleukin-2, interferon-γ and the tumor necrosis factor family of proteins.

Our immune system is capable of generating a diverse repertoire of lymphocyte antigen receptors with literally unlimited number of specificities. This is achieved through a random gene rearrangement process during lymphocyte development using a limited number of antigen receptor genes. Although such a strategy confers the immune system with the capacity to specifically recognize and respond to a vast range of foreign antigens, it also inevitably generates receptors capable of recognizing self antigens. Therefore, in order to prevent immune attack against self tissues, the immune system must eliminate or suppress lymphocytes that express self-reactive antigen receptors. On the other hand, although immune attacks against foreign antigens are required for eliminating infectious pathogens, they can also cause 'bystander' injury to self tissues. Once the foreign antigens are removed, the immune response must be down-regulated and the activated effector cells be eliminated to ensure the homeostasis of the immune system and the wellbeing of the host. In the past decade, we have learned that the immune homeostasis and self tolerance are orchestrated by a highly complex and intricate network of membrane and secretory proteins. Prominent among these are antigen receptors, costimulatory molecules and cytokines. In this chapter, we will start with a general review of self tolerance and lymphocyte homeostasis. We will then focus on the roles of several key cytokines in maintaining lymphocyte homeostasis and self tolerance.

Self Tolerance and Lymphocyte Homeostasis

Our immune system does not normally attack self tissues. This state of immune unresponsiveness to self antigens is called self tolerance. Self tolerance is a learned process, which can be divided into two types based on the sites of the tolerance induction, i.e., central tolerance and peripheral tolerance. Central tolerance is established during the early stage of lymphocyte development when immature lymphocytes expressing high affinity self-reactive receptors undergo clonal deletion inside the primary (central) lymphoid organs. This 'negative' selection

Cytokines and Chemokines in Autoimmune Disease, edited by Pere Santamaria.

process may eliminate the majority of self-reactive lymphocytes. However, not all self antigens are present in primary lymphoid organs. Some antigens are exclusively expressed in the peripheral tissues while others are only expressed at certain developmental stages outside the lymphoid organs. Thus, self-reactive lymphocytes can escape the negative selection process and migrate into peripheral lymphoid organs. Indeed, there is mounting evidence that all healthy individuals harbor potentially pathologic self-reactive lymphocytes in their peripheral lymphoid tissues. For example, self-reactive T cell clones can be easily isolated from the peripheral blood of normal healthy individuals by repeated in vitro stimulation with self antigens. Similarly, immunization of normal animals with self antigens in adjuvant can induce activation and expansion of self-reactive lymphocytes and elicit inflammatory autoimmune responses. Studies of self-reactive TCR transgenic mice revealed that most, if not all, lymphocytes recognizing self antigens expressed in nonlymphoid tissues develop normally and are present in the periphery as naïve 'ignorant' cells. However, the fact that most individuals remain healthy despite of the presence of the self-reactive lymphocytes suggests that these lymphocytes must be under a state of immune tolerance.

In the past decade, much has been learned about the mechanisms whereby peripheral tolerance (i.e., immune tolerance induced in the periphery) is generated and maintained. Ironically, most information came from studies of experimentally induced immune tolerance to foreign antigens. For many years, immunologists were amazed by the fact that peripheral tolerance can be induced by certain modes of antigen administration. For example, soluble peptides or monomeric proteins delivered by the intraperitoneal (i.p.), subcutaneous (s.c.), oral, nasal or intravenous (i.v.) route induces antigen-specific T cell hyporesponsiveness. It must be emphasized that this type of immune tolerance is not a total unresponsiveness. Rather, it is characterized by a reduced immune response upon specific antigen challenge. Different arms of the immune responses may be affected to different degrees. For example, in the case of tolerance induced by oral feeding of antigens, single administration of high dose antigens (>0.5 mg antigen per gram of body weight) induces suppression of virtually all arms of the immune responses. On the other hand, multiple low doses (<0.1 mg per gram of body weight per feeding) are more likely to cripple Th1 type response while stimulating other (e.g., regulatory Th2 or Th3 cell) responses. Tolerance induced in the periphery can be mediated by both deletion and nondeletion mechanisms.[1-6] Although apoptotic cells can not be directly demonstrated in conventional animals (due to the extreme low frequency of antigen reactive cells), a role for deletion has been strongly suggested by the decrease in T cell frequency and the lack of a complete reversal of the tolerance, especially when high dose antigens were used.[7] Using TcR transgenic mice and conventional mice injected with transgenic T cells, we have demonstrated directly that mucosal exposure of antigen induced apoptosis and deletion of antigen specific T cells.[6, 8] Because deletion eliminates specific cells from the system, this pathway of peripheral tolerance is usually not reversible and therefore the most effective. Nondeletion mechanisms of peripheral tolerance may involve clonal anergy of TH1 cells and immune deviation of TH2 cells. Clonal anergy is characterized by deficiency in IL-2 production and the reversibility of hyporesponsiveness by IL-2.[9] By contrast, deviation is characterized by preferential activation of T cells producing TGF-β and TH2 cytokines. The deviated T cells may be not only nonpathogenic but also capable of down-regulating TH1 responses through releasing their cytokines.[3, 10] This may explain the suppressor phenomenon described in some early tolerance models.[11, 12] It should be emphasized that the deviation pathway leads to tolerance of TH1 but not TH2 cells and that the end result of immune deviation is partial or 'split' tolerance rather than full tolerance. In the oral tolerance model, the deviation pathway is favored by low dose antigen administration. When high doses of antigens are used, both TH1 and TH2 cells are tolerized, and this may be mediated by deletion and/or anergy pathways.

While development of central tolerance may not require cytokines, development and maintenance of peripheral tolerance are dependent on the roles of cytokines. This is because peripheral tolerance is achieved through an active process in which activation, inactivation, and deletion of antigen-specific lymphocytes are involved. Indeed, cytokine requirement for lymphocyte activation during the inductive phase of peripheral tolerance may resemble that of an active immune response, and lymphocytes undergoing tolerance induction exhibit many of the early phenotypic characteristics of activated lymphocytes that are capable of supporting a productive immune response. As will be discussed in detail later in this chapter, inactivation and deletion of lymphocytes require participation of such cytokines as transforming growth factor (TGF)-β, interleukin (IL)-10, Fas ligand (FasL) and tumor necrosis factor (TNF).

On the other hand, the immune system must specifically recognize foreign pathogens and generate a productive immune response against them. This is usually accomplished by specific activation of T and B lymphocytes in the presence of inflammatory signals. During a typical immune response, antigen-specific lymphocytes divide extensively, generating a large number of effector cells. Depending on the priming condition, a complex network of cytokines is generated. As a result, cellular and/or humoral immune responses emerge which eventually lead to the elimination of foreign antigens or cells expressing them. Once the antigens are cleared, the vast majority of the effector cells die by apoptosis. Only a small fraction of them survive as long lived memory cells. This ensures that the total number of lymphocytes in the body remains constant, thus maintaining lymphocyte homeostasis of the immune system. Cytokines are essential for these processes because they dictate not only the degree of lymphocyte expansion during the inductive phase of the immune response, but also the degree of lymphocyte contraction after the antigen removal.

Thus, cytokines can be involved in maintaining both self tolerance and immune homeostasis. Cytokines can serve as immune modulators by regulating lymphocyte activation, proliferation, differentiation, survival and apoptosis. Dysfunction of cytokines can lead to the breakdown of self tolerance and the development of autoimmnue diseases. Dysfunction of cytokines can also lead to the breakdown of immune homeostasis which in turn causes self tissue injury either by heightened immune attacks against foreign antigens or by activating self-reactive naïve lymphocytes that are otherwise nonpathogenic. Below, we will discuss several of the cytokines that are crucial for self tolerance and immune homeostasis.

TGF-β and IL-10

Many cytokines exhibit anti-inflammatory activities. The two most intensively studied anti-inflammatory cytokines are TGF-β1 and IL-10. Both TGF-β1 and IL-10 can deliver negative signals to lymphocytes, preventing their activation and proliferation. The generation of TGF-β1 and IL-10 gene knockout mice led to the recognition that these two cytokines are essential for immune homeostasis and self tolerance.

Belonging to a superfamily of >20 distinct dimeric proteins that share a similar structure, TGF-β1 is one of the three isoforms of the TGF-β expressed in mammalian species.[13] It is produced by a variety of cell types, mostly of the lymphoid origin, and is found in large amounts in platelets and bones, and circulates in the plasma. TGF-β1 is synthesized as an inactive precursor and requires activation before exerting its function. The active molecule is a 25-kd homodimer linked by disulfide bonds. There are three types of TGF-β receptors, which are designated as type I, II, and III receptors. The type I and type II receptors are transmembrane serine-threonine kinases that interact with each other to facilitate intracellular signaling. The type III receptor is a membrane proteoglycan that has no signaling role but acts to present TGF-β to the other TGF-β receptors.

Although initially identified as a growth factor for fibroblasts, TGF-β has been found to play important roles in a number of biological processes including embryonic development, tissue repair, wound healing, inflammation as well as immune regulation.[13] Although all three isoforms of the TGF-β bind to the same set of receptors, each of the TGF-β isoforms may perform distinct functions. Thus, germ line disruption of TGF-β1 gene has little effect on the development and function of nonimmune systems, but its effect on the immune system is dramatic and fatal.[14] Days after the birth, TGF-β1 deficient mice develop systemic inflammatory diseases, presumably of autoimmune origin, and die within 2-6 weeks of age.[15] By contrast, disruption of TGF-β2 gene leads to perinatal mortality and a wide range of developmental defects. Similarly, germ line disruption of TGF-β3 gene does not result in any overlapping phenotype with TGF-β1- or TGF-β2-deficient mice, but leads to the development of cleft palate, presumably as a result of impaired adhesion of apposing medialmedial edge epithelia of the palatal shelves and subsequent elimination of the mid-line epithelial seam. The lack of phenotypic overlap in mice deficient in different isoforms of TGF-β suggests that they be endowed with distinct nonoverlapping functions.

TGF-β1 is a potent inhibitor of immune responses. It down-regulates the functions of virtually all immune cells including B cells, CD4+ Th1 and Th2 cells, CD8+ cytotoxic T lymphocytes (CTLs), natural killer (NK) cells, and macrophages. It suppresses the production of many cytokines including interferon (IFN)-γ, TNF-α and IL-2. It inhibits IL-2 receptor and IL-12 receptor expression, and can induce apoptosis in T cells. In macrophages, TGF-β1 antagonizes the activities of TNF-α and IFN-γ, inhibits inducible nitric oxide synthase (iNOS) activity, suppresses the production of both nitric oxide (NO) and superoxide ion, and alters the expression of costimulatory molecules. TGF-β1 also down-regulates the MHC class I and II expression in a variety of cell types including B cells and macrophages. It alters the expression of adhesion molecules such as E-selectin, and thus interferes with the adhesion of neutrophils and lymphocytes to the vascular endothelial cells. Additionally, TGF-β1 inhibits the secretion of IgG and IgM by B lymphocytes, but promotes the production of IgA by activating the Cα gene promoter.[16] Some of the inhibitory activities of TGF-β1 may result from its suppression of tyrosine phosphorylation and activation of Jak-1, STAT-5 and Tyk-2.

Numerous studies revealed that increased TGF-β1 production correlates with the resolution of inflammatory responses, particularly in organ-specific antoimmune diseases. Regulatory T cells induced by oral feeding of antigens secrete TGF-β1 and have been designated as Th3 cells.[10] The in vivo relevance of TGF-β1 in oral tolerance was confirmed by the demonstration that injection of anti-TGF-β1 mAb into animals reversed oral tolerance induced by low dose antigen. The TGF-β1-mediated by-stander suppression plays important roles in oral tolerance in a number of models including diabetes, experimental granulomatous colitis, adjuvant arthritis and experimental tracheal eosinophilia as well as autoimmune encephalomyelitis. A recent study showed that TGF-β1 could be up-regulated in T lymphocytes by cross-linking CTLA-4, suggesting that CTLA-4 may work through TGF-β1 to down-regulate T cell function.[17]

IL-10 is another important anti-inflammatory cytokine that is crucial for immune homeostasis.[18, 19] It is produced primarily by CD4+ Th2 cells, monocytes, and B cells, and circulates as a homodimer consisting of two tightly packed 160-amino-acid polypeptides. IL-10 is a potent inhibitor of Th1 cells, suppressing both IL-2 and IFN-γ production. This was the reason why IL-10 was initially designated as cytokine synthesis inhibition factor. In addition to its effect on TH1 cells, IL-10 is also a potent deactivator of pro-inflammatory cytokines produced by monocytes/macrophages. Upon engaging its high-affinity 110-kD receptor on monocytes/ macrophages, IL-10 inhibits the secretion of TNF-α, IL-1, IL-6, IL-8, IL-12, granulocyte colony-stimulating factor, MIP-1α, and MIP-2α. IL-10 also inhibits cell surface expression of MHC class II molecules, B7, and the LPS recognition and signaling molecule CD14. Furthermore, IL-10 inhibits cytokine production by neutrophils and natural killer cells, and attenuates surface expression of TNF receptors. Some of the inhibitory activities of IL-10 may result from its suppression of the nuclear factor κB translocation and its enhancement of the cytokine messenger RNA degradation.

Not surprisingly, IL-10-/- mice spontaneously develop chronic inflammatory enteritis.[20] The animals show dysregulated production of pro-inflammatory cytokines in the inflamed tissues and uncontrolled expansion of IFN-γ producing T cells. Conversely, systemic or local administration of IL-10 inhibits organ-specific autoimmune diseases. And regulatory Tr1 cells that inhibit autoimmune inflammation produce abundant amount of IL-10.[21, 22]

IL-2 and IFN-γ

Although IL-2 and IFN-γ are normally considered to be pro-inflammatory cytokines, recent studies in IL-2-/- and IFN-γ-/- mice revealed that they both play important roles in limiting and terminating immune responses.[23, 24] For example, IL-2-/- mice or mice lacking high affinity IL-2 receptors (IL-2Rα-/- and IL-2Rβ-/-) develop lymphoid hyperplasia and autoimmune diseases. While the postoperative blockade of CD28 and/or CD40L induces long-term survival of allogeneic cardiac grafts in wild-type mice, it failed to induce long term cardiac allograft acceptance in IL-2-/- mice or mice injected with anti-IL-2 neutralizing antibodies. Alloantigen-induced T cell apoptosis is impaired in IL-2-/- mice, which leads to increased accumulation of alloreactive T lymphocytes. In addition, Fas-mediated activation-induced T cell apoptosis is also severely defective in IL-2-/- mice. These findings suggest that the principle immunoregulatory function of IL-2 might be to program activated T lymphocytes for apoptosis. In fact, IL-2 has been shown to up-regulate Fas ligand expression while down-regulate FLIP, a protein that inhibits Fas-mediated cell death.[25]

Similarly, T cell costimulation blockade failed to induce long-term cardiac allograft acceptance in IFN-γ-/- mice or in wild type recipients treated with anti-IFN-γ neutralizing antibodies.[26] CTLA-4-Ig administration blocks alloantigen-induced T cell proliferation in wild type mice but fails to do so in IFN-γ-/- mice. Compared to wild type T cells, IFN-γ-/- T lymphocytes display heightened proliferation and CTL activity upon allostimulation in vitro. Heightened T lymphocyte proliferation is also observed in IFN-γ-/- mice after injection of bacterial superantigen or allogeneic splenocytes.[27] The cutaneous delayed-type hypersensitivity (DTH) response to allogeneic splenocytes is also increased in the absence of IFN-γ.

The TNF Superfamily

The TNF/TNF receptor superfamily of proteins consists of approximately 20 ligands and 30 receptors. Most of these proteins are involved in regulating lymphocyte activation and/or apoptosis, and therefore, are important for immune homeostasis and self tolerance. Below we will discuss the important roles of several TNF-related proteins in immune homeostasis and autoimmunity.

One of the most extensively studied members of the TNF family that regulate immune homeostasis and self tolerance is FasL (Apo-1L, CD95L). FasL is normally expressed by a small number of cell types including activated lymphocytes and cells of the immune privileged organs (such as eye, testis, brain and spinal cord). Its receptor Fas (CD95) is a type I membrane protein of the TNF-receptor family. Unlike FasL, Fas is expressed constitutively in most tissues and is dramatically up-regulated at sites of inflammation. Fas/FasL interaction activates FADD, which in turn triggers the activation of the IL-1 converting enzyme (ICE) family of caspases, leading to DNA fragmentation and cell death. However, Fas/FasL interaction does not always lead to apoptosis. Under certain conditions, Fas/FasL interaction can also activate target cells, presumably through the nuclear factor (NF)-κB pathway. In this case, Fas may transmit similar activating signals as TNF-receptors, leading to secretion of pro-inflammatory cytokines such as IL-1 and IL-8.

Fas/FasL have been reported to both inhibit and promote autoimmune inflammation. Mutations of genes encoding Fas or FasL lead to lymphocytic proliferation and autoimmune inflammatory diseases in both humans and mice.[28, 29] Under these conditions, T cells of presumably autoimmune origin accumulate in extremely large numbers and exhibit a peculiar phenotype, i.e., CD4-CD8-B220+ or CD4+CD8-B220+. In the late stages of the disease, these aberrant cells become functionally inactive, or anergic. While these observations have led to the recognition that Fas and FasL are essential for maintaining self tolerance, presumably by deleting autoreactive cells through activation-induced cell death (AICD), recent studies suggest that Fas/FasL interaction can also contribute to autoimmune inflammation. Thus, unlike FasL expressed in the eye, testis, joints and certain tumors or transplants that confers immune privilege, FasL expressed in the thyroid gland, pancreatic islets and some other tumors or transplants enhances inflammation. It is not known whether these opposing effects of Fas/FasL are due to the intrinsic differences of the Fas signals generated (i.e., apoptotic versus activating or chemotactic signals), or differences in the way that the target tissues respond to apoptosis. Nor is it clear what factor(s) determines whether Fas/FasL interaction will promote or inhibit inflammation in a given tissue.

Similarly, other members of the TNF family are also crucial in regulating immune homeostasis and autoimmunity. For example, TNF is not only required for AICD of T lymphocytes, but also capable of promoting autoimmune diseases. In fact, anti-TNF therapy is effective in preventing autoimmune arthritic inflammation both in humans and animals. Other members of the TNF family that are capable of inhibiting autoimmune inflammation are TRAIL (TNF-related apoptosis-inducing ligand) and CD30 ligand (CD30L). Chronic blockade of TRAIL in mice exacerbated autoimmune arthritis, and intra-articular TRAIL gene transfer ameliorated the disease.[30] In vivo, TRAIL-blockade led to profound hyper-proliferation of synovial cells and arthritogenic lymphocytes, and heightened the production of cytokines and autoantibodies. In vitro, TRAIL inhibited DNA synthesis and prevented cell cycle progression of lymphocytes.[30] Similarly, CD30L may also play an anti-inflammatory role in autoimmune diseases. Autoreactive CD8+ T cells deficient in CD30L elicited more severe autoimmune insulitis in mice.[31]

Thus, both pro-inflammatory and anti-inflammatory cytokines can regulate lymphocyte homeostasis and autoimmunity. This regulation may determine whether an autoreactive cell is to become activated, inactivated, deleted or remain as a harmless naïve cell. A challenge for immunologists working in the post-genome era is to determine how many players are involved in regulating immune homeostasis and how many molecular pathways are used to modulate self tolerance and autoimmunity.

Acknowledgments

This work was supported by grants from the National Institutes of Health (NS40188, NS36581, NS40447, AR44914 and AI41060).

References

1. Singer GG, Abbas AK. The Fas antigen is involved in peripheral but not thymic deletion of T lymphocytes in T cell receptor transgenic mice. Immunity 1994; 1:365-371.
2. Burstein HJ, Shea CM, Abbas AK. Aqueous antigens induce in vivo tolerance selectivity in IL-2 and IFN-g-producing (Th_1) cells. J Immunol 1992; 148(12):3687-3691.
3. Weiner HL, Friedman A, Miller F et al. Oral tolerance: Immunologic mechanisms and treatment of murine and human organ specific autoimmune diseases by oral administration of autoantigens. Annu Rev Immunol 1994; 12:809.
4. Whitacre CC, Gienapp IE, Orosz CG et al. Oral tolerance in experimental autoimmune encephalomyelits. III. Evidence for clonal anergy. J Immunol 1991; 147:2155-2163.
5. Rizzo LV, Miller-Rivero NE, Chan CC et al. Interleukin-2 treatment potentiates induction of oral tolerance in a murine model of autoimmunity. J Clin Invest 1994; 94(4):1668-72.
6. Chen Y, Inobe J, Marks R et al. Peripheral deletion of antigen-reactive T cells in oral tolerance. Nature 1995; 376:177-180.
7. Whitacre CC, Gienapp IE, Meyer A et al. Oral tolerance in experimental autoimmune encephalomyelitis. Ann NY Acad Sci 1996; 778:217-227.
8. Chen Y, Inobe J-i, Weiner HL. Inductive events in oral tolerance in the TCR transgenic adoptive transfer model. Cell Immunol 1997; 178:62-68.
9. Friedman A, Weiner HL. Induction of anergy or active suppression following oral tolerance is determined by antigen dosage. Proc Nat Acad Sci USA 1994; 91:6688-6692.
10. Chen Y, Kuchroo VK, Inobe J-I et al. Regulatory T cell clones induced by oral tolerance: suppression of autoimmune encephalomyelitis. Science 1994; 265:1237-1240.
11. Tomasi T, Jr. Oral tolerance. [Review]. Transplantation 1980; 29(5):353-6.
12. MacDonald TT. Immunosuppression caused by antigen feeding. I. Evidence for the activation of a feedback suppressor pathway in the spleens of antigen-fed mice. Eur J Immunol 1982; 12(9):767-773.
13. Roberts AB, Sporn MB. Physiological actions and clinical applications of transforming growth factor beta (TGFb). Growth Factors 1993; 8(1):1-9.

14. Letterio JJ, Geiser AG, Kulkarni AB et al. Autoimmunity associated with TGF-beta1-deficiency in mice is dependent on MHC class II antigen expression. J Clin Invest 1996; 98(9):2109-19.
15. Christ M, McCartney-Francis NL, Kulkarni AB et al. Immune dysregulation in TGF-beta 1-deficient mice. J Immunol 1994; 153(5):1936-46.
16. Coffman RL, Lebman DA, Shrader B. Transforming growth factor beta (TGFb) specifically enhances IgA production by lipopolysaccharide-stimulated murine B lymphocytes. J Exp Med 1989; 144:3411-16.
17. Chen W, Jin W, Wahl SM. Engagement of cytotoxic T lymphocyte-associated antigen 4 (CTLA-4) induces transforming growth factor beta (TGF-beta) production by murine CD4(+) T cells. J Exp Med 1998; 188(10):1849-57.
18. Mosmann TR, Coffman RT. Heterogeneity of cytokine secretion patterns and functions of helper T cells. Adv Immunol 1989; 46:111-47.
19. de Vries JE. Immunosuppressive and anti-inflammatory properties of interleukin 10. Ann Med 1995; 27(5):537-41.
20. Kuhn R, Lohler J, Rennick D et al. Interleukin-10-deficient mice develop chronic enterocolitis. Cell 1993; 75(2):263-74.
21. Asseman C, Powrie F. Interleukin 10 is a growth factor for a population of regulatory T cells. Gut 1998; 42(2):157-8.
22. Kitani A, Chua K, Nakamura K et al. Activated self-MHC-reactive T cells have the cytokine phenotype of Th3/T regulatory cell 1 T cells. J Immunol 2000; 165(2):691-702.
23. MacDonald TT. Gastrointestinal inflammation. Inflammatory bowel disease in knockout mice. Curr Biol 1994; 4(3):261-3.
24. Ma A, Datta M, Margosian E et al. T cells, but not B cells, are required for bowel inflammation in interleukin 2-deficient mice. J Exp Med 1995; 182(5):1567-72.
25. Refaeli Y, Van Parijs L, London CA et al. Biochemical mechanisms of IL-2-regulated Fas-mediated T cell apoptosis. Immunity 1998; 8(5):615-23.
26. Konieczny BT, Dai Z, Elwood ET et al. IFN-gamma is critical for long-term allograft survival induced by blocking the CD28 and CD40 ligand T cell costimulation pathways. J Immunol 1998; 160(5):2059-64.
27. Dalton DK, Pitts-Meek S, Keshav S et al. Multiple defects of immune cell function in mice with disrupted interferon-gamma genes. Science 1993; 259(5102):1739-42.
28. Cohen PL, Eisenberg RA. The lpr and gld genes in systemic autoimmunity: Life and death in the Fas lane. Immunology Today 1992; 13(11):427-8.
29. Nagata S, Golstein P. The Fas death factor. Science 1995; 267:1449-56.
30. Song K, Chen Y, Goke R et al. Tumor necrosis factor-related apoptosis-inducing ligand (TRAIL) is an inhibitor of autoimmune inflammation and cell cycle progression. J Exp Med 2000; 191(7):1095-104.
31. Kurts C, Carbone FR, Krummel MF et al. Signalling through CD30 protects against autoimmune diabetes mediated by CD8 T cells. Nature 1999; 398(6725):341-4.

CHAPTER 5

The Role of Cytokines as Effectors of Tissue Destruction in Autoimmunity

Thomas W.H. Kay, Rima Darwiche, Windy Irawaty, Mark M.W. Chong, Helen L. Pennington and Helen E. Thomas

Introduction

Target cell damage in autoimmune disease is likely to be mediated by multiple effector pathways only some of which are cytokines. Recent progress in cell death research has dramatically changed ideas of how target cells might be destroyed and new effector pathways have been discovered. Multiple extra-cellular effector mechanisms may converge on a limited number of increasingly well-characterized intracellular cell death pathways. This increases the possibility that blockade of the damaging effects of inflammatory cytokines, cell death receptors that trigger caspase activation and noncytokine mechanisms such as the contents of the cytotoxic T cell granule may be a realistic and logical point of intervention in autoimmune disease.

Death of target cells is the culmination of the immunological events that cause many autoimmune diseases. When T cells specific for target cell autoantigens are activated they express a range of effector mechanisms analogous to those used by the immune system to clear infectious micro-organisms. The issue is to understand which are critical in target cell death in autoimmunity and how cells can be protected from them. This is a realistic and logical potential point of intervention.

Effector mechanisms of target organ damage have been most extensively studied in the pancreatic beta cell because of the realistic prospect of beta cell replacement therapy and the availability of many animal models of type 1 diabetes (T1DM). Beta cell destruction will therefore be the main focus of this review. Prospects for beta cell replacement as a treatment for T1DM have improved recently with progress in both islet cell transplantation[1] and differentiation of pancreatic precursor cells[2] or embryonal stem cells[3] to insulin-producing cells[4] Whatever forms of beta cell replacement are eventually most useful clinically they will require concurrent immunosuppression or, preferably, genetic modification for protection against the highly active anti-beta cell immune response found in people with established T1DM. A detailed knowledge of how beta cells are destroyed will indicate the most effective forms of immunoprotection and could eventually allow beta cell replacement without systemic immunosuppression. In other autoimmune diseases treatment with pharmacological replacement therapy is already successful (pernicious anemia, hypothyroidism) and knowledge of how target cells are killed is unlikely to be directly applied but is of scientific interest. In multiple sclerosis, replacement of damaged tissue seems unlikely in the short term but knowledge of mechanisms of disease may guide and refine systemic therapy.

A problematic issue in protecting cells from immune attack is the very high likelihood that multiple immune effector mechanisms are involved in target cell destruction.[5] If multiple pathways of cell destruction are to be overcome then either cells will have to carry multiple

Cytokines and Chemokines in Autoimmune Disease, edited by Pere Santamaria.

exogenous protective genes or points of convergence in the pathways must be found so that one added gene will protect against several mechanisms.

The most compelling evidence for a pathogenic role of a single molecule in target cell destruction is for perforin, a noncytokine that is a key constituent of the granules of cytotoxic T cells. These data come from the nonobese diabetic (NOD) mouse that is recognised to be the best available mouse model of T1DM (reviewed in ref. 6). Perforin-deficient NOD mice have a much reduced frequency of diabetes (16% compared with 77% in wild-type NOD females) that comes on later in age.[7] This result has been confirmed in independently derived perforin-deficient NOD mice.[8] Perforin, in contrast to most other immune effector mechanisms including most cytokines, has a limited range of functions. Changes in disease progression in perforin-deficient mice are likely to indicate a role for perforin-dependent cytotoxicity, presumably of beta cells, although some argue that this is not yet proven.[9] Because some perforin-deficient NOD mice develop diabetes, these data also indicate that nonperforin-dependent mechanisms can also destroy beta cells.

Cytokines, especially those that bind to receptors with a death domain are important candidate nonperforin mechanisms of cell death. The death domain is a conserved region in the cytoplasmic domain of members of the TNF receptor family that allows binding and aggregation of adaptor proteins.[10] These link the receptors to the caspase pathway, a cascade of proteases that leads to cell death by apoptosis.[11] This is a morphologically defined mode of cell death that appeared in evolution perhaps for cells to respond to virus infection but is known also to occur in response to inflammatory stimuli such as cytokines. This is the most direct way that cytokines might act as effector molecules and cause death of the target cells of autoimmunity.[12] Clearly if this pathway is important in autoimmune disease it should be possible to observe the histological hallmarks of apoptosis in target organs. Apoptosis of thyrocytes has been observed in Hashimoto's disease[13-15] and, with some difficulty, in rodent models of diabetes[16-18] and in human diabetes.[19] The very small amount of pathological material available from humans with type 1 diabetes, the rapid clearance of apoptotic cells and the problems of colocalizing staining make these studies difficult to interpret. One major study has failed to observe apoptosis in human diabetes using the TUNEL method.[20] On balance, apoptosis is probably the mode of target cell death in autoimmune disease but further confirmation would be useful. Triggering of receptors with death domains is not the only way to cause apoptosis. Perforin causes cell death by apoptosis through a less well defined pathway, as do stresses such as irradiation. Intracellular free radical generation for example by formation of nitric oxide (NO) also causes DNA fragmentation.[21] Beta cell death induced by NO is not blocked by caspase inhibitors.[22]

All of the cytokines proposed as effectors of target cell death in autoimmune diseases have many different effects on different cell types. Few have yet been subjected to a rigorous test of their effects on the beta cell alone. While NOD mice with interruptions of cytokine pathways (such as TNF receptor knockouts[23] and Fas-deficient NOD-*lpr* mice[24,25]) are protected from diabetes, controversy still persists about whether this is due to cytotoxic effects on beta cells or to effects of these cytokines on lymphoid or other nonbeta cells. Some of these issues can be addressed by making chimeric animals with either transplanted islets, transplanted bone marrow or transferred mature T cells. Ideally mice should be produced with targeted interference with these pathways in beta cells only.[26] A further important problem that arises with genetically manipulated mice should also be acknowledged. When a genetic locus of interest is backcrossed onto an in-bred strain other than the one in which it was created (usually the 129 strain), there is significant potential for artefacts to ensue unless recombinants close to the locus of interest are specifically chosen.[27] Within the genetic interval surrounding the locus of interest there may be genes that differ between 129 and NOD, for example, that can affect disease progression. This is an important issue that will only be overcome by the routine use of embryonic stem cells from the strain of interest, e.g., NOD.[28]

Interleukin-1

IL-1 was originally identified as the active constituent of supernatant from concanavalin-A-activated spleen cells able to impair the function of rat islets in vitro and eventually cause them to die.[29,30] The receptor for IL-1 is expressed at high levels on beta cells.[31,32] IL-1 is expressed in intra-islet macrophages and at times by beta cells themselves[33] and IL-1 expression has been observed in islets of NOD mice[34] and BB rats.[35] It is produced in response to TNF, bacterial products such as lipopolysaccharide and viral infection amongst other stimuli. IL-1 produced by intra-islet macrophages and other nonbeta cells is responsible for some of the effects of TNF on beta cells since some of these can be blocked by IL-1 antagonists (Fig. 5.1).[36] IL-1 damages beta cells mainly by stimulating expression of inducible nitric oxide synthase (iNOS).[37-39] In mouse and human islets, IFNγ, as well as IL-1, is required for iNOS induction. Exposure to cytokines need not be simultaneous as IFNγ-induction of the signaling molecule STAT1 is persistent in beta cells and can prime beta cells for IL-1 induced iNOS expression up to 7 days ahead of time.[40] INOS production induced by IL-1 after IFNγ priming is much less in alpha cells than beta cells[40] perhaps contributing to the beta-cell specificity of type 1 diabetes. IL-1 stimulates iNOS expression by activating pathways of signal transduction including the NF-κB pathway and the MAP kinase pathway both in beta cells[41] and other cell types (Fig. 5.2).[42] Inactivation of JNK MAP kinase signaling using cell-permeable peptide inhibitors of JNK renders cells less sensitive to IL-1 mediated cell death although cells treated in this way appear to make similar levels of NO.[43-45] Less differentiated islet cell lines make less iNOS in response to cytokines and appear to have less IL-1-induced activation of the JNK signal transduction pathway.[44,46] The details of the relative importance of different members of the MAP kinase pathway in mediating effects including iNOS induction, beta cell death and beta cell differentiation are complex but are being actively explored and may be a possible opportunity to protect beta cells from IL-1.

Production of NO in response to IL-1 results in beta cell dysfunction and death that does not have the classical appearance of apoptosis . DNA laddering is not a prominent feature[47] and it is not inhibited by caspase-inhibitors; this is consistent with the IL-1 receptor not directly activating the caspase pathway. Nevertheless a component of typical apoptosis may also occur in response to IL-1 in combination with other cytokines[48-50] and this is not reduced in iNOS deficient islets.[51] These are protected from IL-1-induced apoptosis after 6 days but not at 9 days.[51] Beta cells appear to be particularly sensitive to IL-1 because they have low levels of anti-oxidant enzymes that protect cells from free radical stress. Transfection of beta cell lines with anti-oxidant enzymes can reduce their susceptibility to cytokine-induced cell death by blocking oxidation of cellular proteins and nucleic acids.[47, 52] IL-1 can also have beta cell protective effects in that it induces several molecules such as anti-oxidant enzymes and heat shock protein 70 (hsp70) that are cyto-protective against the effects of alloxan, streptozotocin and nitric oxide but not against cytokine combinations.[53] Therefore changes in beta cells after exposure to IL-1 may be protective as well as deleterious and the effects in vivo of this balance remain uncertain. Overexpression of hsp70 can protect against cytokine mediated beta cell destruction[54] and the effects seem similar in cells treated with troglitazone or J series prostaglandins, ligands of the peroxisome proliferator-activated receptors.[55] Additionally reduction of hsp70 in a human beta cell line increased its sensitivity to NO-induced necrosis and apoptosis.[56] A further way of protecting beta cell lines was developed by selecting cells that survived treatment with IL-1 and IFNγ.[52] When steps in cytokine signaling were analyzed in these cells it was found that the cells overexpressed the transcription factor STAT1a that normally mediates IFN signaling.[57] The mechanism by which STAT1a overexpression causes unresponsiveness to IL-1 and IFNγ is currently unclear. These studies show the potential for inhibiton of cytokine signaling pathways in producing beta cells protected from the immune system. Effective protection of beta cells from cytokines may be achievable before their role in beta cell destruction is finally clarified.

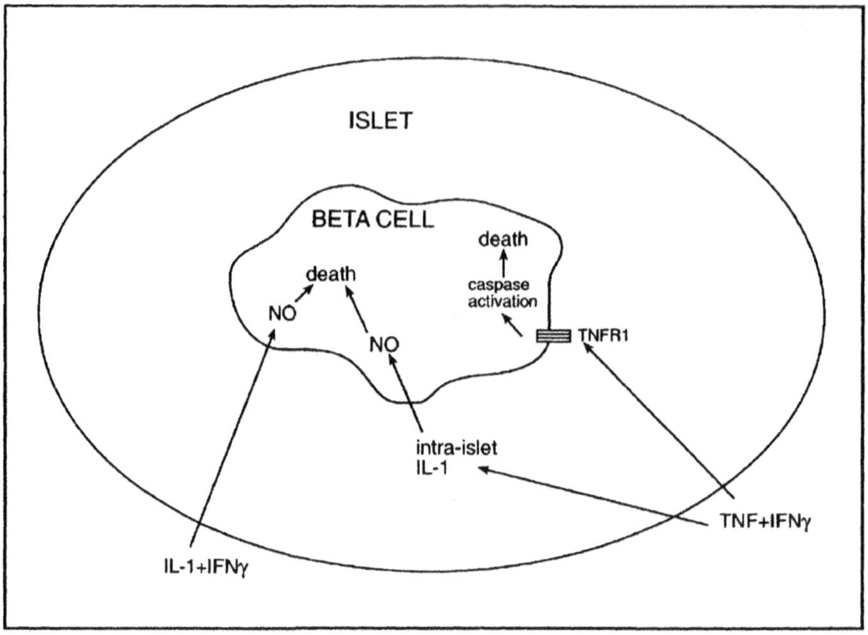

Fig. 5.1. How IL-1, TNF and NO kill beta cells in vitro. Treatment of mouse islets with IL-1 and IFNγ in vitro results in NO production and beta cell death. NO is produced by many cells within the islet (endothelial cells, macrophages, ductal and beta cells), however NO production by the beta cell itself is responsible for IL-1 and IFNγ-mediated beta cell damage. TNF and IFNγ on the other hand also induce intra-islet IL-1 production that then stimulates NO production and beta cell death. TNF may also directly damage beta cells via TNFR1-mediated activation of caspases.

IL-1 therefore remains an important candidate effector mechanism for target cell destruction by NO-dependent and independent cell death and by "marking" cells for Fas ligand mediated killing.[58] The role of IL-1 is less well established in vivo although neutralisation of IL-1 has been reported to prevent accelerated diabetes in NOD mice treated with cyclophosphamide[59] or transplanted with syngeneic islets.[60] NOD mice deficient in iNOS are known to develop diabetes normally. Genetic manipulation of IL-1 or its receptor in NOD mice has not been described to date and translation of the vast literature on IL-1 effects on beta cells in vitro to clinical testing at least in rodent models needs to be more thorough. IL-1 has similar effects on other autoimmune targets such as the thyrocyte.[61] It has been implicated in the pathogenesis of inflammatory arthritis,[62] perhaps in the same pathway as TNF. Both IL-1 and TNF are capable of increasing the expression of molecules that damage joints such as matrix metalloproteases, including collagenase.[63] Whether the effect of these cytokines in arthritis is by altering inflammatory cell trafficking, by directly damaging synovial cells or by activating immune responses is unclear.[64] Neutralization of IL-1 with IL-1 receptor antagonist in patients with rheumatoid arthritis appears most effective at decreasing cartilage and bone destruction[65] and so IL-1 may be a particularly important effector of joint damage. These molecules may also play a role in inflammatory colitis and demyelination.

Interferon-gamma

IFNγ cooperates with TNF and IL-1 to stimulate the expression of many immune inflammatory genes including iNOS, adhesion molecules, caspases and major histocompatability

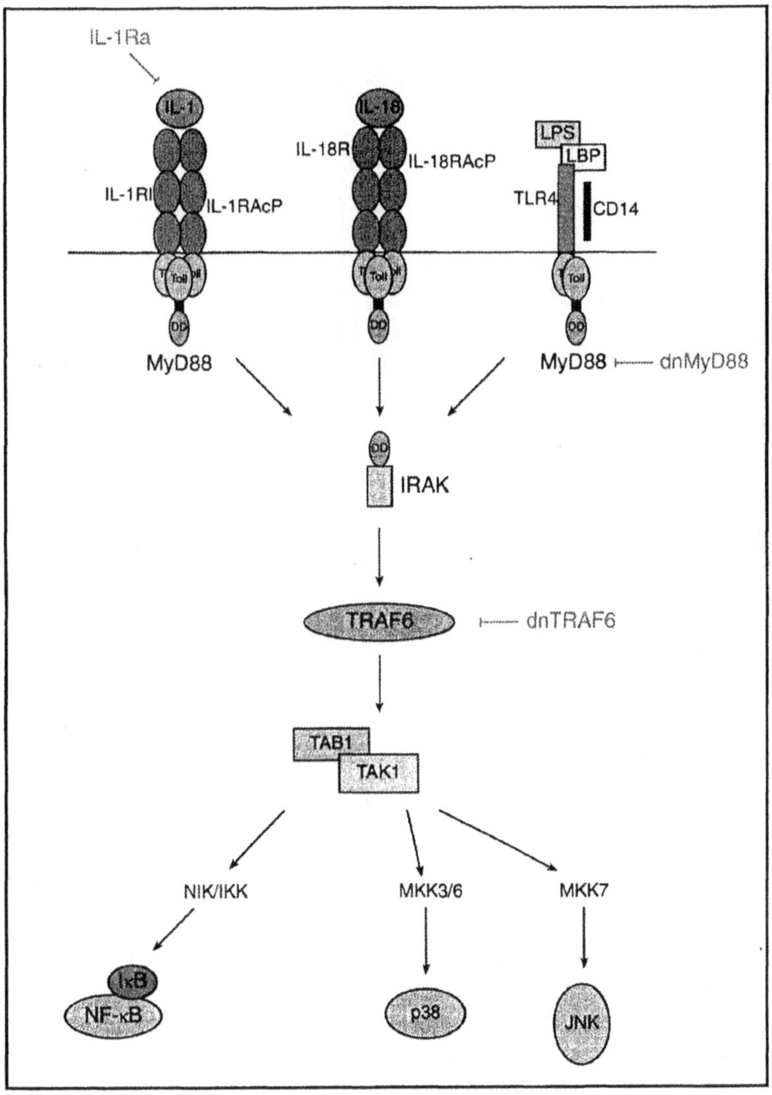

Fig. 5.2. IL-1R/Toll-like receptor signaling. After ligand binding, the adaptor protein MyD88 associates with the receptor via a Toll domain, and with IL-1R-associated kinase (IRAK) via a death domain (DD). Interaction of autophosphorylated IRAK with TNFR-associated factor 6 (TRAF6) leads to activation of kinase cascades and the transcription factors NF-κB and AP-1. This pathway can be inhibited at many stages with dominant-negative forms of most of the signaling molecules (including MyD88, IRAK, TRAF6 and I-κB); and also by synthetic inhibitors of MAPK proteins (for example the p38 inhibitor SB-203580).

genes (Fig. 5.3).[66] It therefore potentially enhances both immune recognition of cells and cytotoxicity. To explore the direct effects of IFNγ on beta cells, we took advantage of an elegant series of experiments carried out by Schreiber's group using a dominant negative mutant of the interferon-gamma receptor.[67] Dighe et al showed that, when overexpressed in transgenic mice,

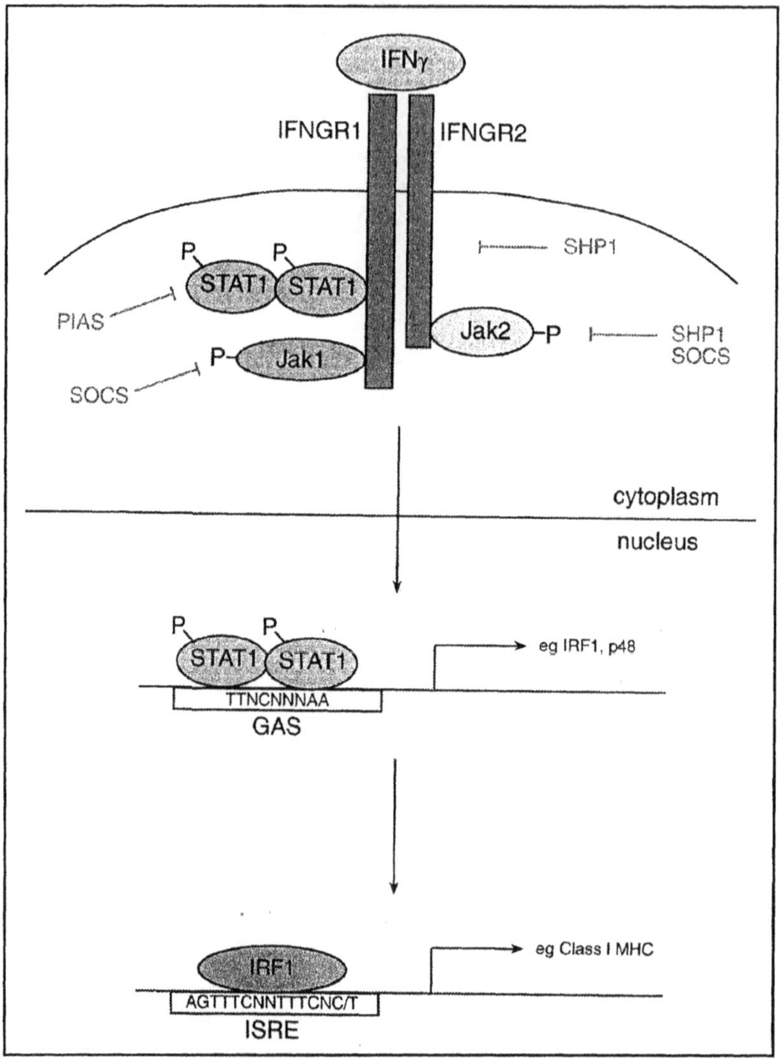

Fig. 5.3. IFNγ signal transduction. IFNγ binds to the IFNγR, leading to phosphorylation of Jak1 and Jak2, IFNGR1 and STAT1. STAT1 homodimers then translocate to the nucleus where they bind to an IFNγ-activated sequence (GAS) to activate gene expression. IFNγ activates the transcription factors IRF1 and p48, which in turn bind to an IFN-specific responsive element (ISRE) to further activate gene expression. Molecules which inhibit IFNγ signaling include the tyrosine phosphatase SHP1, the protein inhibitor of activated STAT (PIAS) family and the suppressors of cytokine signaling (SOCS) family of molecules.

this mutant could inhibit signaling from endogenous wild-type receptors in fibroblasts, lymphocytes and macrophages.[68] When we overexpressed the mutant IFNγ receptor (ΔγR) in beta cells under the control of the rat insulin promoter, the beta cells were unresponsive to at least 100U/ml of IFNγ whereas other cells responded to <10U/ml.[69] Beta cells from these transgenic mice were also unresponsive in vivo because the rise with age of beta cell class I MHC protein

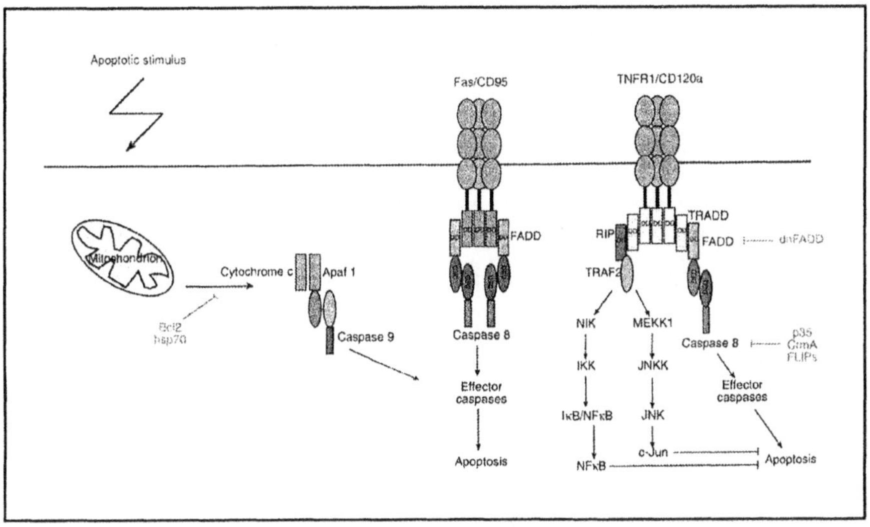

Fig. 5.4. Caspase-dependent apoptosis. Fas and TNFR1 form homodimers upon ligand binding. With both Fas and TNFR1, this can lead to association of Fas-associated death domain (FADD) and TNFR-associated death domain (TRADD) via a death domain (DD), and recruitment of caspase 8 by a death effector domain (DED), resulting in activation of downstream effector caspases and apoptosis. In the case of TNFR1, ligand binding can also lead to activation of the transcription factors AP-1 and NF-κB through association of receptor interacting protein (RIP) and TNFR associated factor 2 (TRAF2) with the receptor, followed by activation of MAPK and NIK/IKK kinase cascades. Activation of these pathways is thought to protect cells from TNF-mediated apoptosis by inducing expression of anti-apoptotic genes.
Stimuli such as γ-irradiation or chemotherapeutic drugs induce apoptosis through activation of cytochrome c in the mitochondria. Cytochrome c then interacts with Apaf 1 and caspase 9 is recruited, thus activating downstream caspases and apoptosis of the cell.
Caspase-dependent apoptosis can be blocked by inhibition of mitochondrial release of cytochrome c (with anti-apoptotic members of the Bcl2 family or hsp70), inhibition of caspases (with the viral molecules p35 and CrmA or with FLICE-inhibitory proteins (FLIPs)), and dominant negative forms of adaptor molecules in the pathway (such as FADD).

expression that is normally seen in NOD mice did not occur. In the NOD model there was no evidence of protection from diabetes and this is consistent with other recent studies in IFNγ or IFNγR knockout NOD mice[27,70,71] casting doubt on a major role for IFNγ in diabetes in NOD mice. We have also tested beta cell expression of ΔγR in other models of diabetes with results strikingly different to those in NOD mice. In the transgenic model of diabetes in which the glycoprotein (gp) from lymphocytic choriomeningitis virus (LCMV) is expressed in beta cells (RIP-LCMV mice) and diabetes induced by LCMV infection, class I MHC rose on beta cells but diabetes was significantly reduced in mice with the ΔγR transgene.[72] Our interpretation of these findings is that following LCMV infection there is a high level of IFNγ in the insulitis lesion compared with NOD insulitis that causes beta cell damage. When the ΔγR transgene is present this is blunted but some IFNγ action remains that is sufficient to increase class I expression in the ΔγR transgenic beta cells. The data imply that there is a difference in the concentration of IFNγ required to increase class I MHC gene expression and that needed for beta cell damage. They also highlight the dramatic differences in pathogenesis of different mouse models of diabetes and the problems inherent in extrapolating findings in one model to another or to human diabetes.

If as we have proposed, IFNγ causes or contributes to beta cell destruction in the RIP-LCMV model, how would this occur? By increasing class I MHC expression and enhancing intracellular pathways of antigen presentation in beta cells IFNγ may make target cells more easily recognised by cytotoxic T cells. IFNγ is known to enhance the effects of other cytotoxic molecules such as IL-1, FasL or TNF. The addition of IL-1 and IFNγ to beta cells causes damage mediated predominantly by induction of iNOS and production of NO from both beta cells and other cells within the islet that are capable of responding to cytokines. NO produced by these nonbeta cells may be an important contributor to beta cell damage. Islets from the ΔγR mice are chimeras of beta cells responsive to IL-1 but not IFNγ, and non beta cells responsive to both. Addition of IL-1 and IFNγ to islets from ΔγR mice in vitro resulted in significantly reduced iNOS expression than in normal islets and expression was shown by confocal microscopy to be principally in nonbeta cells. The ΔγR beta cells were protected from IL-1 and IFNγ—indicating that the NO produced by neighbouring cells in response to the cytokines is not sufficient to cause beta cell death. IFNγ is also known to sensitize cells to FasL-mediated cell death. The ΔγR beta cells were not killed by FasL after Fas upregulation by IL-1 and IFNγ in vitro—in part because Fas upregulation is reduced without IFNγ and also because IFNγ may enhance the expression of factors within the caspase pathway downstream of Fas aggregation. These data show that blockade of IFNγ is able to affect IL-1 and Fas-dependent beta cell damage indicating the possibility of blocking several pathways with one intervention. This is sufficient to block diabetes in RIP-LCMV but not NOD mice. Whether Fas or IL-1 are involved in the RIP-LCMV model is unknown.

TNF and TNF Family Members

TNF is expressed in islets of NOD mice and BB rats[35, 73] and is potentially cytotoxic because one of its receptors, TNFR1, activates the caspase pathway, a major signal transduction pathway leading to cell death (Fig. 5.4).[74] TNF-induced cell death is tightly regulated in most cells. Multiple inhibitors of the caspase pathway are found in cells and the expression of these is upregulated by TNF itself.[75] Apart from activation of caspases, TNF stimulates similar signal transduction pathways to IL-1 including the NF-κB and MAP kinase pathways. TNF acts via two receptors[76]—TNFR1 is constitutively expressed on most cells and is found at low level on beta cells.[77] TNFR2 is less widely expressed and is not found constitutively on beta cells, but may be upregulated on beta cells during inflammation.[78] TNF is both secreted and produced as a transmembrane molecule that is more efficient than the soluble form.[79] When expressed transgenically in beta cells (probably mainly soluble), TNF is not potently cytotoxic for beta cells.[80, 81] Transgenic TNF does however accelerate diabetes in the neonatal NOD mouse and other models of diabetes, perhaps by enhancing antigen presentation.[82] NOD mice deficient in TNFR1[23] or expressing high levels of soluble TNFR1 as a TNF antagonist[83] do not develop diabetes although insulitis develops normally. Similarly TNFR1-deficient islets are not destroyed in a transfer system by islet-specific $CD4^+$ T cells from BDC2.5 T cell receptor transgenic mice.[84] When TNFR1 knockout islets and wild-type islets were simultaneously present and exposed in vivo to BDC2.5 cells, both wild-type and knockout islets were destroyed, indicating that once these T cells are fully activated they can kill TNFR1 deficient islets.[85] This experiment suggests that TNF is required for T cell activation not beta cell cytotoxicity. It would be good to see this elegant experiment reproduced perhaps using polyclonal T cells and islets from NOD mice before fully accepting that TNF does not cause beta cell cytotoxicity. The balance of evidence suggests that TNF is more likely to act in chemotaxis and activation of antigen presentation than as an effector mechanism that destroys target cells. Similar conclusions have been reached in EAE in which the kinetics of lymphocytic infiltration and course of disease but neither paralysis nor demyelination are prevented by TNF deficiency[86,87] even though TNF can cause damage to oligodendrocytes in vitro.[88] Clearly TNF neutralization has a major role in treatment of rheumatoid arthritis and colitis, but again it may be that steps other than effector damage to target cells are important.

Fas Ligand

Fas ligand (FasL) is the member of the TNF superfamily that has attracted most attention as a potential effector molecule particularly in Hashimoto's disease[89] and in type 1 diabetes. Interestingly Fas is expressed on most beta cells from islets with lymphocytic infiltration in humans with type 1 diabetes[19,20] but on hardly any beta cells in the NOD mouse.[22] This indicates either significant differences in how Fas is regulated in mouse and humans or differences in the cytokines present in the local islet microenvironment. Fas is not detectable on normal beta cells in mice or in humans.[22, 51] Fas expression on mouse beta cells can be induced by IL-1 and the addition of IFNγ further enhances expression.[22, 90, 91] Human beta cells have been reported to express Fas in response to NO production [19] but this is not the case in mice in which induction of Fas is unaffected by iNOS deficiency or inhibition.[22,51,92] It has been reported that Fas is significantly up-regulated by IFNγ alone in human beta cells.[51] This is consistent with the observed pattern of widespread Fas expression on islet cells in human diabetes because other molecules such as class I MHC that are increased by IFNγ are expressed similarly. We have been unable to detect FasL on beta cells from NOD mice by flow cytometry. This was confirmed functionally by showing that beta cells that express Fas due to treatment in vitro with IL-1/IFNγ were not susceptible to fratricidal killing[22] FasL is most likely expressed by islet infiltrating inflammatory cells and not by beta cells.

Fas has been observed on oligodendrocytes in the brains of patients with multiple sclerosis[93] although the significance of this finding remains unclear. The role of Fas in clearance of activated T cells in the CNS may be more important than its role in death of oligodendrocytes.[94, 95] Fas has been observed on thyroid epithelial cells from normal as well as inflamed thyroids.[96-98] Normal thyroid cells are resistant to killing by FasL due to the presence of inhibitors, but the influence of these can be overcome by inflammatory cytokines including IL-1, IFNγ and TNFα.[99] Whether or not FasL is expressed by nonmalignant thyroid cells and if so whether it plays a role in "fratricidal" killing of thyroid cells or killing of infiltrating lymphocytes remains highly controversial.[97,100] This is also the case for FasL expression on beta cells and diabetes. Clarification of whether FasL expression on epithelial cells plays a significant role in autoimmune disease remains an important priority.

The possibility that FasL is a major effector mechanism of beta cell destruction in diabetes was raised by observations that NOD*lpr* mice that lack Fas expression were not susceptible to diabetes[24] and were free of insulitis.[25] It was subsequently found that NOD*lpr* islets were destroyed after transplantation into diabetic NOD mice with relatively minor differences from wild-type transplanted islets.[101,102] This, however, is a stern test because diabetic NOD mice have a highly activated anti-beta cell immune response even capable in one report of destroying class I MHC deficient islets.[103] Based on the experimental data and current ideas of lymphocyte-mediated cytotoxicity being due to perforin, FasL and perhaps other members of the TNF super-family including TNF and TRAIL, it is likely that FasL is involved in beta cell destruction even if it is not the dominant mechanism. FasL appears to be the mechanism of cytolysis used by at least one beta-cell specific $CD8^+$ T cell clone, 8.3, that is able to kill beta cells when rendered perforin deficient but is not able to kill Fas-deficient beta cells in vitro or in vivo.[8] Perforin deficient 8.3 TCR transgenic mice develop diabetes normally. Similarly 4.1 TCR transgenic CD4 cells appear to primarily use FasL to kill beta cells and cytolysis is dependent on beta cell expression of Fas.[58] Both the 4.1 and 8.3 clones are derived from early insulitis and these data support the concept that early insulitis is dependent on beta cell apoptosis mediated through Fas. Enhancement of Fas-mediated killing is one of the ways that cytokines like IFNγ may enhance diabetes in some models. Given it's likely involvement and it's easily detectable expression in islets from humans with diabetes, it is unclear why Fas expression on beta cells has been so difficult to detect in vivo in NOD mice. Further understanding of the role of Fas in beta cell destruction will be aided by development of transgenic NOD mice that express inhibitors of Fas function in beta cells.

TRAIL

It is clear that currently tested mechanisms cannot between them explain all of beta cell destruction—for example the mechanism used by the highly cytotoxic T cells from BDC2.5 transgenic mice remains obscure.[84,104] Other TNF family members may also be involved and there is particular interest in the role of a member of this family termed TNF-related apoptosis-inducing ligand (TRAIL).[105,106] TRAIL has a complex receptor system that includes two apparently positive signaling receptors, DR4 (TRAIL-R1) and DR5 (TRAIL-R2) as well as two inhibitory receptors DcR1 (TRAIL-R3) and DcR2 (TRAIL-R4).[74] The effects of TRAIL on beta cells have not yet been described but TRAIL effects on thyroid cells have been studied. Primary thyroid cells express the DR4 and DR5 receptors for TRAIL[107]and are susceptible to TRAIL mediated killing in the presence of cycloheximide, implying the presence of inhibitors of cytotoxicity. Intra-thyroidal lymphocytes from thyroiditis specimens also showed TRAIL expression and interestingly cytokine-treated thyroid cells also expressed TRAIL,[107] again raising the possibility of "fratricidal" killing.

A Blueprint for Protection

The obvious question is how feasible is it to inhibit the multiple immune effector pathways, both cytokine and noncytokine. Caspase-mediated apoptosis by death receptors such as TNFR1, Fas, DR4 should be inhibited by overexpression of inhibitory molecules from viruses (e.g., crmA), mammalian cells (e.g., cFLIP) or dominant negative mutations of caspase pathway members eg dominant negative FADD. Also promising are viral inhibitory molecules such as those within the E3 region of adenovirus that decrease Fas and TRAIL expression.[108] The E3-gp19K gene also inhibits antigen presentation by blocking class I MHC transport to the cell surface.[109] Similarly, inhibition of IL-1 and IFNγ effects should be feasible by overexpression of inhibitors that preferably affect several cytokines simultaneously. These include members of the suppressors of cytokine signaling family,[110] Stat1a[57] and perhaps hsp70. The most challenging problem remains inhibition of perforin-mediated apoptosis. This is the least defined cell death pathway and the one for which no effective inhibitors exist. At the time of beta cell replacement in a patient with established diabetes, protection from perforin would be vital. While removal of class I MHC protein from the beta cell surface would be one way to exclude the effects of $CD8^+$ T cells, practical steps to achieve this for human islets have not yet been devised. In the meantime blockade of cytokine receptors and death receptors may be refined by testing them in conjunction with encapsulation of the transplanted cells. In this way effector mechanisms requiring direct cell contact such as perforin may be neutralized physically and the efficacy of other measures tested.

Acknowledgments

This work was supported by the National Health and Medical Research Council (Regkey 973002) and the Juvenile Diabetes Research Foundation via a Career Development Award to TWHK.

References

1. Shapiro AM, Lakey JR, Ryan EA et al. Islet transplantation in seven patients with type 1 diabetes mellitus using a glucocorticoid-free immunosuppressive regimen. N Engl J Med 2000; 343(4):230-238.
2. Ramiya VK, Maraist M, Arfors KE et al. Reversal of insulin-dependent diabetes using islets generated in vitro from pancreatic stem cells [see comments]. Nat Med 2000; 6(3):278-282.
3. Soria B, Roche E, Berna G et al. Insulin-secreting cells derived from embryonic stem cells normalize glycemia in streptozotocin-induced diabetic mice. Diabetes 2000; 49(2):157-162.
4. Serup P, Madsen OD, Mandrup-Poulsen T. Islet and stem cell transplantation for treating diabetes. BMJ 2001; 322(7277):29-32.
5. Kay TWH, Thomas HE, Harrison LC et al. The beta cell in autoimmune diabetes: Many mechanisms and pathways of loss. Trends Endocrinol Metab 2000; 11(1):11-15.

6. Atkinson MA, Leiter EH. The NOD mouse model of type 1 diabetes: As good as it gets? Nat Med 1999; 5(6):601-604.
7. Kagi D, Odermatt B, Seiler P et al. Reduced incidence and delayed onset of diabetes in perforin-deficient nonobese diabetic mice. J Exp Med 1997; 186(7):989-997.
8. Amrani A, Verdaguer J, Anderson B et al. Perforin-independent beta-cell destruction by diabetogenic CD8(+) T lymphocytes in transgenic nonobese diabetic mice. J Clin Invest 1999; 103(8):1201-1209.
9. Dilts SM, Solvason N, Lafferty KJ. The role of CD4 and CD8 T cells in the development of autoimmune diabetes. J Autoimmun 1999; 13(3):285-290.
10. Smith CA, Farrah T, Goodwin RG. The TNF receptor superfamily of cellular and viral proteins: Activation, costimulation, and death. Cell 1994; 76(6):959-962.
11. Nicholson DW, Thornberry NA. Caspases: Killer proteases. Trends in Biochem Sci 1997; 22(8):299-306.
12. Vaux DL, Flavell RA. Apoptosis genes and autoimmunity. Curr Opin Immunol 2000; 12(6):719-724.
13. Kotani T, Aratake Y, Hirai K et al. Apoptosis in thyroid tissue from patients with Hashimoto's thyroiditis. Autoimmunity 1995; 20(4):231-236.
14. Tanimoto C, Hirakawa S, Kawasaki H et al. Apoptosis in thyroid diseases: A histochemical study. Endocr J 1995; 42(2):193-201.
15. Hammond LJ, Lowdell MW, Cerrano PG et al. Analysis of apoptosis in relation to tissue destruction associated with Hashimoto's autoimmune thyroiditis. J Pathol 1997; 182(2):138-44.
16. Kurrer MO, Pakala SV, Hanson HL et al. Beta cell apoptosis in T cell-mediated autoimmune diabetes. Proc Natl Acad Sci USA 1997; 94(1):213-218.
17. O'Brien BA, Harmon BV, Cameron DP et al. Apoptosis is the mode of beta-cell death responsible for the development of IDDM in the nonobese diabetic (NOD) mouse. Diabetes 1997; 46(5):750-757.
18. Augstein P, Stephens LA, Allison J et al. Beta-cell apoptosis in an accelerated model of autoimmune diabetes. Mol Med 1998; 4(8):495-501.
19. Stassi G, Maria RD, Trucco G et al. Nitric oxide primes pancreatic beta cells for Fas-mediated destruction in insulin-dependent diabetes mellitus. J Exp Med 1997; 186(8):1193-1200.
20. Moriwaki M, Itoh N, Miyagawa J et al. Fas and Fas ligand expression in inflamed islets in pancreas sections of patients with recent-onset Type I diabetes mellitus. Diabetologia 1999; 42:1332-1340.
21. Schulze-Osthoff K, Krammer PH, Droge W. Divergent signalling via APO-1/Fas and the TNF receptor, two homologous molecules involved in physiological cell death. EMBO J 1994; 13(19):4587-4596.
22. Thomas HE, Darwiche R, Corbett JA et al. Evidence that beta cell death in the nonobese diabetic mouse is Fas independent. J Immunol 1999; 163(3):1562-1569.
23. Kagi D, Ho A, Odermatt B et al. TNF receptor 1-dependent beta cell toxicity as an effector pathway in autoimmune diabetes. J Immunol 1999; 162(8):4598-4605.
24. Chervonsky AV, Wang Y, Wong FS et al. The role of Fas in autoimmune diabetes. Cell 1997; 89(1):17-24.
25. Itoh N, Imagawa A, Hanafusa T et al. Requirement of Fas for the development of autoimmune diabetes in nonobese diabetic mice. J Exp Med 1997; 186(4):613-618.
26. Benoist C, Mathis D. Cell death mediators in autoimmune diabetes—No shortage of suspects. Cell 1997; 89(1):1-3.
27. Kanagawa O, Xu G, Tevaarwerk A et al. Protection of nonobese diabetic mice from diabetes by gene(s) closely linked to IFN-gamma receptor loci. J Immunol 2000; 164:3919-3923.
28. Nagafuchi S, Katsuta H, Kogawa K et al. Establishment of an embryonic stem (ES) cell line derived from a nonobese diabetic (NOD) mouse: in vivo differentiation into lymphocytes and potential for germ line transmission. FEBS Lett 1999;4 55(1-2):101-104.
29. Mandrup-Poulsen T, Bendtzen K, Nielsen JH et al. Cytokines cause functional and structural damage to isolated islets of Langerhans. Allergy 1985; 40(6):424-429.
30. Mandrup-Poulsen T, Bendtzen K, Nerup J et al. Affinity-purified human interleukin I is cytotoxic to isolated islets of Langerhans. Diabetologia 1986; 29(1):63-67.
31. Deyerle KL, Sims JE, Dower SK et al. Pattern of IL-1 receptor gene expression suggests role in noninflammatory processes. J Immunol 1992; 149(5):1657-1665.
32. Jafarian-Tehrani M, Amrani A, Homo-Delarche F et al. Localization and characterization of interleukin-1 receptors in the islets of Langerhans from control and nonobese diabetic mice. Endocrinology 1995; 136(2):609-613.
33. Heitmeier MR, Arnush M, Scarim AL et al. Pancreatic beta-cell damage mediated by beta-cell production of IL-1: A novel mechanism for virus-induced diabetes. J Biol Chem 2000.

34. Rabinovitch A, Suarez-Pinzon WL, Sorensen O et al. Inducible nitric oxide synthase (iNOS) in pancreatic islets of nonobese diabetic mice: Identification of iNOS- expressing cells and relationships to cytokines expressed in the islets. Endocrinology 1996; 137(5):2093-2099.
35. Jiang Z, Woda BA. Cytokine gene expression in the islets of the diabetic Biobreeding/Worcester rat. J Immunol 1991; 146(9):2990-2994.
36. Corbett JA, McDaniel ML. Intraislet release of interleukin 1 inhibits beta cell function by inducing beta cell expression of inducible nitric oxide synthase. J Exp Med 1995; 181(2):559-568.
37. Corbett JA, Lancaster JR, Jr., Sweetland MA et al. Interleukin-1 beta-induced formation of EPR-detectable iron-nitrosyl complexes in islets of Langerhans. Role of nitric oxide in interleukin-1 beta-induced inhibition of insulin secretion. J Biol Chem 1991; 266(32):21351-4.
38. Southern C, Schulster D, Green IC. Inhibition of insulin secretion by interleukin-1 beta and tumour necrosis factor-alpha via an L-arginine-dependent nitric oxide generating mechanism. FEBS Lett 1990; 276(1-2):42-44.
39. Welsh N, Eizirik DL, Bendtzen K et al. Interleukin-1 beta-induced nitric oxide production in isolated rat pancreatic islets requires gene transcription and may lead to inhibition of the Krebs cycle enzyme aconitase. Endocrinology 1991; 129(6):3167-3173.
40. Heitmeier MR, Scarim AL, Corbett JA. Prolonged STAT1 activation is associated with interferon-gamma priming for interleukin-1-induced inducible nitric-oxide synthase expression by islets of Langerhans. J Biol Chem 1999; 274(41):29266-29273.
41. Larsen CM, Wadt KA, Juhl LF et al. Interleukin-1beta-induced rat pancreatic islet nitric oxide synthesis requires both the p38 and extracellular signal-regulated kinase 1/2 mitogen-activated protein kinases. J Biol Chem 1998; 273(24):15294-15300.
42. Kristof AS, Marks-Konczalik J, Moss J. Mitogen-activated protein kinases mediate AP-1-dependent human inducible nitric oxide synthase promoter activation. J Biol Chem 2000; 8:8.
43. Bonny C, Oberson A, Steinmann M et al. IB1 reduces cytokine-induced apoptosis of insulin-secreting cells. J Biol Chem 2000; 275(22):16466-16472.
44. Ammendrup A, Maillard A, Nielsen K et al. The c-Jun amino-terminal kinase pathway is preferentially activated by interleukin-1 and controls apoptosis in differentiating pancreatic beta-cells. Diabetes 2000; 49(9):1468-1476.
45. Bonny C, Oberson A, Negri S et al. Cell-permeable peptide inhibitors of JNK: novel blockers of beta-cell death. Diabetes 2001; 50(1):77-82.
46. Nielsen K, Karlsen AE, Deckert M et al. Beta-cell maturation leads to in vitro sensitivity to cytotoxins. Diabetes 1999; 48(12):2324-32.
47. Lortz S, Tiedge M, Nachtwey T et al. Protection of insulin-producing RINm5F cells against cytokine-mediated toxicity through overexpression of antioxidant enzymes. Diabetes 2000; 49(7):1123-1130.
48. Rabinovitch A, Suarez-Pinzon WL, Strynadka K et al. Human pancreatic islet beta-cell destruction by cytokines is independent of nitric oxide production. J Clin Endocrinol Metab 1994; 79(4):1058-1062.
49. Eizirik DL, Sandler S, Welsh N et al. Cytokines suppress human islet function irrespective of their effects on nitric oxide generation. J Clin Invest 1994; 93(5):1968-1974.
50. Delaney CA, Pavlovic D, Hoorens A et al. Cytokines induce deoxyribonucleic acid strand breaks and apoptosis in human pancreatic islet cells. Endocrinology 1997; 138(6):2610-2614.
51. Liu D, Pavlovic D, Chen MC et al. Cytokines induce apoptosis in beta-cells isolated from mice lacking the inducible isoform of nitric oxide synthase (iNOS-/-). Diabetes 2000; 49(7):1116-1122.
52. Hohmeier HE, Thigpen A, Tran VV et al. Stable expression of manganese superoxide dismutase (MnSOD) in insulinoma cells prevents IL-1beta-induced cytotoxicity and reduces nitric oxide production. J Clin Invest 1998; 101(9):1811-1820.
53. Ling Z, Van de Casteele M, Eizirik DL et al. Interleukin-1beta-induced alteration in a beta-cell phenotype can reduce cellular sensitivity to conditions that cause necrosis but not to cytokine-induced apoptosis. Diabetes 2000; 49(3):340-345.
54. Bellmann K, Jaattela M, Wissing D et al. Heat shock protein hsp70 overexpression confers resistance against nitric oxide. FEBS Lett 1996; 391(1-2):185-188.
55. Maggi LB, Jr., Sadeghi H, Weigand C et al. Anti-inflammatory actions of 15-deoxy-delta 12,14-prostaglandin J2 and troglitazone: evidence for heat shock-dependent and -independent inhibition of cytokine-induced inducible nitric oxide synthase expression. Diabetes 2000; 49(3):346-355.
56. Burkart V, Liu H, Bellmann K et al. Natural resistance of human beta cells toward nitric oxide is mediated by heat shock protein 70. J Biol Chem 2000; 275(26):19521-19528.
57. Chen G, Hohmeier HE, Newgard CB. Expression of the transcription factor STAT-1alpha in insulinoma cells protects against cytotoxic effects of multiple cytokines. J Biol Chem 2001; 276(1):766-772.

58. Amrani A, Verdaguer J, Thiessen S et al. IL-1-alpha, IL-1-beta, and IFN-gamma mark beta cells for Fas-dependent destruction by diabetogenic CD4(+) T lymphocytes. J Clin Invest 2000; 105:459-468.
59. Nicoletti F, Di Marco R, Barcellini W et al. Protection from experimental autoimmune diabetes in the nonobese diabetic mouse with soluble interleukin-1 receptor. Eur J Immunol 1994; 24(8):1843-1847.
60. Sandberg JO, Eizirik DL, Sandler S. IL-1 receptor antagonist inhibits recurrence of disease after syngeneic pancreatic islet transplantation to spontaneously diabetic nonobese diabetic (NOD) mice. Clin Exp Immunol 1997; 108(2):314-317.
61. Reimers JI, Rasmussen AK, Karlsen AE et al. Interleukin-1 beta inhibits rat thyroid cell function in vivo and in vitro by an NO-independent mechanism and induces hypothyroidism and accelerated thyroiditis in diabetes-prone BB rats. J Endocrinol 1996; 151(1):147-157.
62. van den Berg WB, Bresnihan B. Pathogenesis of joint damage in rheumatoid arthritis: Evidence of a dominant role for interleukin-1. Baillieres Best Pract Res Clin Rheumatol 1999; 13(4):577-597.
63. Dinarello CA. Biologic basis for interleukin-1 in disease. Blood 1996; 87(6):2095-2147.
64. Dinarello CA. The role of the interleukin-1-receptor antagonist in blocking inflammation mediated by interleukin-1. N Engl J Med 2000; 343(10):732-734.
65. Bresnihan B, Alvaro-Gracia JM, Cobby M et al. Treatment of rheumatoid arthritis with recombinant human interleukin-1 receptor antagonist. Arthritis Rheum 1998; 41(12):2196-2204.
66. Boehm U, Klamp T, Groot M et al. Cellular responses to interferon-gamma. Annu Rev Immunol 1997; 15:749-795.
67. Dighe AS, Farrar MA, Schreiber RD. Inhibition of cellular responsiveness to interferon-gamma (IFN gamma) induced by overexpression of inactive forms of the IFN gamma receptor. J Biol Chem 1993; 268(14):10645-10653.
68. Dighe AS, Campbell D, Hsieh CS et al. Tissue-specific targeting of cytokine unresponsiveness in transgenic mice. Immunity 1995; 3(5):657-666.
69. Thomas HE, Parker JL, Schreiber RD et al. IFN-gamma action on pancreatic beta cells causes class I MHC upregulation but not diabetes. J Clin Invest 1998; 102(6):1249-1257.
70. Hultgren B, Huang X, Dybdal N et al. Genetic absence of gamma-interferon delays but does not prevent diabetes in NOD mice. Diabetes 1996; 45(6):812-817.
71. Serreze DV, Post CM, Chapman HD et al. Interferon-gamma receptor signaling is dispensable in the development of autoimmune type 1 diabetes in NOD mice. Diabetes 2000; 49(12):2007-2011.
72. Seewaldt S, Thomas HE, Ejrnaes M et al. Virus-induced autoimmune diabetes: most beta cells die through inflammatory cytokines and not perforin from autoreactive (anti-viral) CTL. Diabetes 2000; 49:1801-1809.
73. Held W, MacDonald HR, Weissman IL et al. Genes encoding tumor necrosis factor alpha and granzyme A are expressed during development of autoimmune diabetes. Proc Natl Acad Sci USA 1990; 87(6):2239-2243.
74. Ashkenazi A, Dixit VM. Death receptors: Signaling and modulation. Science 1998; 281(5381):1305-1308.
75. Baichwal VR, Baeuerle PA. Activate NF-kappa B or die? Curr Biol 1997; 7(2):R94-R96.
76. Tartaglia LA, Goeddel DV. Two TNF receptors. Immunol Today 1992; 13(5):151-153.
77. Stephens LA, Thomas HE, Ming L et al. Tumor necrosis factor-alpha-activated cell death pathways in NIT-1 insulinoma cells and primary pancreatic beta cells. Endocrinology 1999; 140(7):3219-3227.
78. Walter U, Frantzke A, Sarukhan A et al. Monitoring gene expression of TNFR family members by beta-cells during development of autoimmune diabetes. Eur J Immunol 2000; 30(4):1224-1232.
79. Grell M, Douni E, Wajant H et al. The transmembrane form of tumor necrosis factor is the prime activating ligand of the 80 kDa tumor necrosis factor receptor. Cell 1995; 83(5):793-802.
80. Higuchi Y, Herrera P, Muniesa P et al. Expression of a tumor necrosis factor alpha transgene in murine pancreatic beta cells results in severe and permanent insulitis without evolution towards diabetes. J Exp Med 1992; 176(6):1719-1731.
81. Picarella DE, Kratz A, Li CB et al. Insulitis in transgenic mice expressing tumor necrosis factor beta (lymphotoxin) in the pancreas. Proc Natl Acad Sci USA 1992; 89(21):10036-10040.
82. Green EA, Flavell RA. Tumor necrosis factor alpha and the progression of diabetes in nonobese diabetic mice. Immunol Rev 1999; 169:11-22.
83. Hunger RE, Carnaud C, Garcia I et al. Prevention of autoimmune diabetes mellitus in NOD mice by transgenic expression of soluble tumor necrosis factor receptor p55. Eur J Immunol 1997; 27(1):255-261.

84. Pakala SV, Chivetta M, Kelly CB et al. In autoimmune diabetes the transition from benign to pernicious insulitis requires an islet cell response to tumor necrosis factor alpha. J Exp Med 1999; 189(7):1053-1062.
85. Pakala S, Sarvetnick N. Beta cells play a TNF-alpha dependent role in their own demise. In: Juvenile Diabetes Foundation Fellows and their Research; 2000; Leesburg, VA 2000.
86. Korner H, Riminton DS, Strickland DH et al. Critical points of tumor necrosis factor action in central nervous system autoimmune inflammation defined by gene targeting. J Exp Med 1997; 186(9):1585-1590.
87. Sean Riminton D, Korner H, Strickland DH et al. Challenging cytokine redundancy: Inflammatory cell movement and clinical course of experimental autoimmune encephalomyelitis are normal in lymphotoxin-deficient, but not tumor necrosis factor-deficient, mice. J Exp Med 1998; 187(9):1517-1528.
88. Raine CS. The Norton Lecture: A review of the oligodendrocyte in the multiple sclerosis lesion. J Neuroimmunol 1997; 77(2):135-152.
89. Baker JR. Dying (apoptosing?) for a consensus on the Fas death pathway in the thyroid. J Clin Endocrinol Metab 1999; 84(8):2593-2595.
90. Yamada K, Takane-Gyotoku N, Yuan X et al. Mouse islet cell lysis mediated by interleukin-1-induced Fas. Diabetologia 1996; 39(11):1306-1312.
91. Harrison M, Dunger AM, Berg S et al. Growth factor protection against cytokine-induced apoptosis in neonatal rat islets of Langerhans: Role of Fas. FEBS Lett 1998; 435(2-3):207-210.
92. Zumsteg U, Frigerio S, Hollander GA. Nitric oxide production and Fas surface expression mediate two independent pathways of cytokine-induced murine beta cell damage. Diabetes 2000; 49(1):39-47.
93. D'Souza SD, Bonetti B, Balasingam V et al. Multiple sclerosis: Fas signaling in oligodendrocyte cell death. J Exp Med 1996; 184(6):2361-2370.
94. Bonetti B, Pohl J, Gao YL et al. Cell death during autoimmune demyelination: Effector but not target cells are eliminated by apoptosis. J Immunol 1997;159(11):5733-5741.
95. Pender MP. Activation-induced apoptosis of autoreactive and alloreactive T lymphocytes in the target organ as a major mechanism of tolerance. Immunol Cell Biol 1999; 77(3):216-223.
96. Kawakami A, Eguchi K, Matsuoka N et al. Thyroid-stimulating hormone inhibits Fas antigen-mediated apoptosis of human thyrocytes in vitro. Endocrinology 1996; 137(8):3163-3169.
97. Giordano C, Stassi G, De Maria R et al. Potential involvement of Fas and its ligand in the pathogenesis of Hashimoto's thyroiditis. Science 1997; 275(5302):960-963.
98. Arscott PL, Knapp J, Rymaszewski M et al. Fas (APO-1, CD95)-mediated apoptosis in thyroid cells is regulated by a labile protein inhibitor. Endocrinology 1997; 138(11):5019-5027.
99. Bretz JD, Arscott PL, Myc A et al. Inflammatory cytokine regulation of Fas-mediated apoptosis in thyroid follicular cells. J Biol Chem 1999; 274(36):25433-25438.
100. Baker J, Bretz JD. Specificity questions concerning the clone 33 anti-fas ligand antibody. Cell Death Differ 2000; 7(1):8-9.
101. Allison J, Strasser A. Mechanisms of beta cell death in diabetes: a minor role for CD95. Proc Natl Acad Sci USA 1998; 95(23):13818-13822.
102. Kim YH, Kim S, Kim KA et al. Apoptosis of pancreatic beta-cells detected in accelerated diabetes of NOD mice: No role of Fas-Fas ligand interaction in autoimmune diabetes. Eur J Immunol 1999; 29(2):455-465.
103. Osorio RW, Ascher NL, Melzer JS et al. beta-2 Microglobulin gene disruption prolongs murine islet allograft survival in NOD mice. Transplant Proc 1994; 26(2):752.
104. Dobbs CM, Haskins K. Comparison of a T Cell Clone and of T Cells from a TCR Transgenic Mouse: TCR Transgenic T Cells Specific for Self-Antigen Are Atypical. J Immunol 2001; 166(4):2495-2504.
105. Pitti RM, Marsters SA, Ruppert S et al. Induction of apoptosis by Apo-2 ligand, a new member of the tumor necrosis factor cytokine family. J Biol Chem 1996; 271(22):12687-12690.
106. Wiley SR, Schooley K, Smolak PJ et al. Identification and characterization of a new member of the TNF family that induces apoptosis. Immunity 1995; 3(6):673-682.
107. Bretz JD, Rymaszewski M, Arscott PL et al. TRAIL death pathway expression and induction in thyroid follicular cells. J Biol Chem 1999; 274(33):23627-23632.
108. Benedict CA, Norris PS, Prigozy TI et al. Three Adenovirus E3 Proteins Cooperate to Evade Apoptosis by Tumor necrosis factor-related Apoptosis-inducing Ligand Receptor-1 and -2. J Biol Chem 2001; 276(5):3270-3278.
109. Horwitz MS. Adenovirus immunoregulatory genes and their cellular targets. Virology 2001; 279(1):1-8.
110. Starr R, Willson TA, Viney EM et al. A family of cytokine-inducible inhibitors of signalling. Nature 1997; 387(6636):917-921.

CHAPTER 6

Cytokines in the Treatment and Prevention of Autoimmune Responses—A Role of IL-15

Xin Xiao Zheng, Wlodzmierz Maslinski, Sylvie Ferrari-Lacraz and Terry B. Strom

Introduction

Cytokines are important protein mediators of immunity, inflammation, cell proliferation, differentiation, and fibrosis.[1] These are the major biological processes underlying autoimmunity. Hence, it is not surprising that there is now convincing evidence that the dysregulation of cytokines plays an important role in the pathogenesis of autoimmunity.[2,3]

Despite the existence of more than 100 cytokines, it has been shown that only very few of them, namely TNF-α, IL-1β, and recently added IL-12, IL-15 and IL-18, are consistently linked to the pathogenesis of rheumatoid arthritis (RA) and other autoimmune diseases.[4] Acting alone or in concert TNF-α, IL-1β, IL-12, IL-15, IL-18 and other cytokines and chemokines induce inflammation.[1,4] Based on these observations, it has been hypothesized that blocking the production or biological activities of these cytokines may help to control chronic autoimmune diseases like rheumatoid arthritis (RA).[4-6] This hypothesis was tested in in vitro experiments, where neutralizing anti-TNF-α antibodies diminish the induction of other proinflammatory cytokines (e.g., IL-1β and IL-6).[1-4] Further experiments showed that anti-TNF-α Abs diminish the incidence and reduce the severity of collagen-induced arthritis (CIA), an animal model of RA.[7] Encouraged by these results, an open-label trial on 20 RA patients treated with chimeric (human/mouse) anti-TNF-α cA2 Ab (presently known as infliximab) was conducted in 1992.[8] The results show remarkable efficacy of the treatment, further proof that blocking TNF-α may represent a new strategy to control RA.[8] Based on these successful results several other biological agents aimed to neutralize TNF-α were developed. FDA recently approved etanercept and infliximab for the treatment of RA. Although both drugs could be used as a monotherapy, higher efficacy is achieved in combination with methotrexate.[9-11] Etanercept and infliximab treatment blocks (i) the early inflammatory autoimmunity of RA and (ii) joint degradation as measured by radiographic analysis during one year treatment period.[10,11] Several excellent reviews describing both basic and clinical aspects of anti-TNF-α therapy have recently been published.[12,13] Despite the success of anti-TNF-α therapy, there are some limitations. First, a group of 20-40% of RA patients does not respond to the treatment.[9-11] Second, the therapy does not cure the disease; clinical symptoms of RA recurred after the cessation of the treatment.[9-13]

Another candidate cytokine for participation in the pathogenesis of RA is IL-1. IL-1 exerts many proinflammatory properties of TNF-α.[34] While the inflammatory process of RA depends more on TNF-α,[14,15] the IL-1 plays a more important role in the pathogenesis of articular cartilage degradation.[14,15] Based on the results of in vitro and in vivo experiments on animal models,[14,15] clinical trials using natural IL-1 receptor antagonist (IL-1Ra) were conducted. Despite the encouraging decrease of cartilage degradation of RA reported from the natural

Cytokines and Chemokines in Autoimmune Disease, edited by Pere Santamaria.

interleukin-1 receptor antagonist (IL-1Ra) clinical trial, the overall clinical improvement of RA patients was less satisfactory compared with that treated with anti-TNF-α therapy.[16,17] At present, clinical trials using combined anti-TNF-α and IL-1Ra therapy are in progress.

Progress made during recent years highlights IL-15 as a key cytokine contributing to autoimmunity. In this chapter, we will focus on the biology and clinical applications of this interesting cytokine.

IL-15 and IL-15Rα

Since the cloning of interleukin (IL)-15 six years ago, there have been numerous studies examining the molecular and cellular biology of this cytokine. Owing to shared receptor (R) signaling components (IL-2Rβ and γ), the in vitro biologic activities of IL-15 are similar to those of IL-2.[18-20] However, specificity for IL-15 versus IL-2 is provided by unique private α-chain receptors that complete the IL-15R and IL-2R heterimeric high-affinity receptor complexes and allow differential responsiveness depending on the ligand and high-affinity receptor expressed.[21] In contrast to the IL-2Rα, which has low affinity for IL-2 (Kd approximately 10^{-8} M) in the absence of IL-2Rβγ,[22] the IL-15Rα alone was sufficient for high-affinity (Kd greater than or equal to 10^{-11} M) binding of IL-15. Similar to IL-2Rα, IL-15Rα seemed play no role in signal transduction.[21] Intriguingly, both IL-15 and IL-15R transcripts have a much broader tissue distribution than IL-2/IL-2R. IL-15 mRNA is produced by multiple tissues (placenta, skeletal muscle, kidney, lung, heart, monocytes/macrophages).[23] IL-15Rα is expressed in activated T cells, activated NK cells, activated B cells, activated macrophages, activated vascular endothelial cells, as well as thymic and bone marrow stromal cells. IL-15Rα mRNA also has wide range of tissue expression including liver, heart, spleen, lung, and skeletal muscle.[21,24] Further, multi-level complex regulatory mechanisms tightly control IL-15 expression.[25] Although transcriptional control of IL-15 is important, the principal level of IL-15 regulation appears to be posttranscriptional.[25] In addition, there are two IL-15 isoforms that differ only by the signal sequence.[26-28] The isoform having a 48 amino acid (aa) long signal sequence is directed to the plasma membrane or may be secreted.[27,28] Surface expressed IL-15 exerts biological activities: stimulates T-cell proliferation[25] and induces granzyme B and perforin.[30] The protease, cleaving mature protein from the surface, has not been identified yet. The other IL-15 isoform containing a 21 aa short signal sequence is found in the cell cytoplasm and nucleus. The biological role of this isoform has to be elucidated. Thus, based upon complex regulation, as well as differential patterns of IL-15 and IL-15R expression, it is consistent with evidence that in vivo functions of this receptor/ligand pair differ from those of IL-2 and IL-2R. Studies to date examining the biology of IL-15 have identified several key nonredundant roles, such as the importance roles in the development and homeostasis of natural killer (NK) cells, NK-T cells, $CD8^+$ T cells, and intestinal intraepithelial lymphocytes (Fig. 6.1).

IL-15/IL-15R System Is Critical for NK Cell Development and Function

Recently generated mice with targeted disruption of the IL-15R and IL-15 (IL-15RKO and IL-15KO) provide direct evidence that the IL-15/IL-15R system is critical for murine NK cell development.[31] The IL-15RKO mice contain multiple defects in innate immune effectors, including an absence of splenic NK cells and NK cytotoxic activity. It is significant that IL-15KO mice also lack any phenotypic or functional NK cells in the spleen and liver, a defect that is reversible upon administration of exogenous IL-15 for 1 week.[32] Moreover, exogenous IL-15 treatment of normal mice enhances NK cell activity and increases both the percentage and absolute number of splenic NK cells.[32-34] Transgenic mice that overexpress murine IL-15 manifest a striking early expansion in NK cells.[35] Thus, in vivo evidence demonstrates that IL-15 is requisite for murine NK cell development, and exogenous IL-15 supports the differentiation of human NK cells in BM culture systems. These basic observations provide invaluable insight into the critical, nonredundant role of IL-15 during NK cell development and suggest the potential utility of IL-15 therapy to expand NK cells in patients.

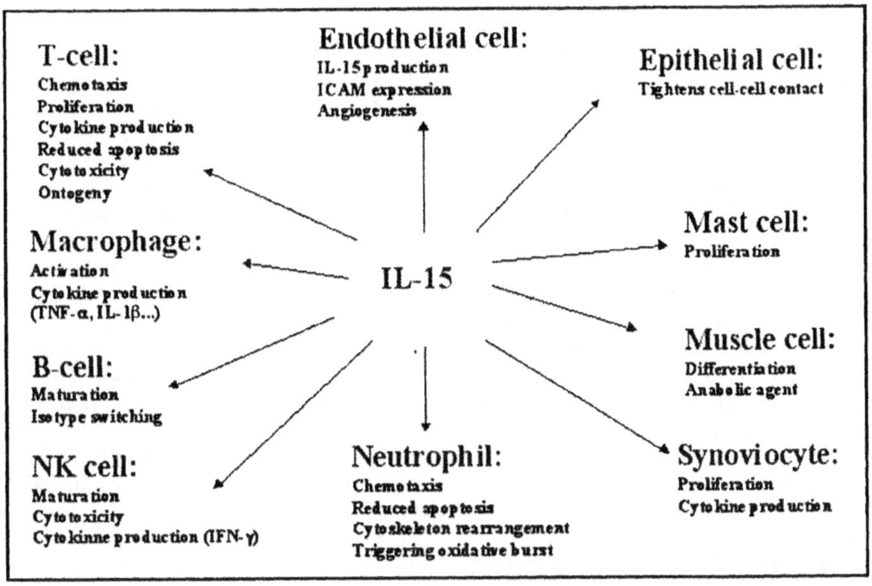

Fig. 6.1. Biological effects of IL-15 on immune and non-immune cells.

IL-15 is a potent stimulus for GM-CSF production by resting human and murine NK cells.[19, 36] In concert with IL-12, IL-15 costimulates NK cells to produce the macrophage-activating factors IFN-γ and TNF-α.[19, 37] Therefore, in both mice and humans, activated macrophages and NK cells interact through a paracrine feedback loop, with macrophages producing monokines (e.g., IL-15 and IL-12) that bind to surface receptors constitutively present on NK cells, resulting in the production of macrophage-activating factors (e.g., IFN-γ). NK cell-derived macrophage-activating factors in turn feed back upon the macrophages to further augment their activation. Thus, macrophage-derived IL-15 contributes with other monokines (especially IL-12) to the proinflammatory cascade leading to IFN-γ production.

In addition, IL-15 stimulates NK cells to produce the C-C chemokines, macrophage inflammatory protein (MIP)-1, which is augmented with the addition of IL-12.[36, 38] Because C-C chemokines also serve as chemoattractants for NK cells,[39] MIP-1 production may direct trafficking of additional NK cells to the site of inflammation. Moreover, chemokine production may have implications in the interactions between macrophages and NK cells, as MIP-1 has been shown to potentiate IFN-γ inducible secretion of inflammatory cytokines by macrophages such as IL-1.[40] Therefore, IL-15 seems to play a critical role in the proinflammatory cascade. Thus blockade of the IL-15/IL-15R signal pathway may represent a novel strategy to interrupt the proinflammatory cascade of autoimmune responses.

Function of IL-15 on TCR T Cells

The studies from IL-15RKO and IL-15KO mice reveal a pivotal role of IL-15 for the homeostasis and proliferation of memory $CD8^+$ T-cells. IL-15RKO mice have a selective deficit in both thymic and peripheral $CD8^+$ T-cells.[31] IL-15KO mice have reduced numbers of memory-phenotype $CD8^+$ T cells in the spleen and lymph nodes that were reversible upon provision of exogenous IL-15.[32] Because IL-15KO mice had normal numbers of single-positive CD8 thymocytes, IL-15 may not be requisite for the development of $CD8^+$ T cells but may be critical for their expansion or survival.[32] The subtle differences in the thymic CD8 single-positive cells between IL-15R and IL-15 knockout mice warrant additional investigation.

IL-15KO mice maintained good health when housed under specific pathogen-free conditions; however, they demonstrated a dramatic lethal sensitivity to vaccinia virus infection compared with control mice.[32] Because both NK cells and $CD8^+$ T cells are important for protection against vaccinia, failure to mount a protective host response is likely due to the deficiencies in these lymphocyte populations. IL-15 stimulates the proliferation of human memory ($CD45RO^+$) CD4 and CD8 and naive (CD45RO) $CD8^+$ human T cells in vitro, while having no effect on naive CD4 T lymphocytes.[41] Further, in mice transgenic for the long signal peptide of IL-15 (LSP-IL-15) cDNA or with IL-2 signal peptide replacing LSP of IL-15 under the control of an MHC class I promoter, CD44hiLy-$6C^+CD8^+$ memory-phenotype T cells were increased in peripheral lymphoid tissues. Moreover, unlike IL-2, IL-15 closely mimics the effects of type I IFN in causing strong and selective stimulation of memory-phenotype CD44hi $CD8^+$ (but not $CD4^+$) cells in vivo; similar specificity applies to purified T cells in vitro.[42]

The therapeutic implication of the pivotal role of IL-15 for the homeostasis and proliferation of memory $CD8^+$ T-cells is obvious. As T-cell costimulation blockade has proved to be a potent therapeutic strategy to blunt certain T-cell dependent autoimmunity and induce allograft tolerance,[43-46] clinical trials using CTLA4/Ig fusion protein which blocks B7/CD28 costimulation pathway and anti-CD40L mAb which blocks CD40/CD40L costimulation pathway have been conducted. Despite the potent immunosuppressive effects of costimulation blockade, costimulation resistant $CD8^+$ T cells are responsible for the failure of these therapies in certain auto-and alloimmune responses.[47-49] Since IL-15 plays a pivotal role in the homeostasis and proliferation of memory $CD8^+$ T cells, targeting IL-15/IL-15R signal pathway in concert with costimulation blockade may represent a novel strategy to prevent and treat T-cell mediated autoimmune diseases and induce allograft tolerance.

Although IL-15 and IL-2 share properties as T-cell growth factors, their effects on T-cell apoptosis are poles apart. It is notable that IL-2, IL-2Rα, and IL-2Rβ deficient mice share a similar phenotype, impaired activation induced apoptosis, lymphoproliferative disorders, and autoimmunity,[50-52] indicating that IL-2 plays a unique and irreplaceable role in activation induced T-cell death (AICD). In striking contrast to IL-2, IL-15 has been shown to prevent rather than promote T-cell apoptosis.[53,54] As AICD of lymphocytes is an important homeostatic mechanism in the immune system and is involved in the induction and maintenance of peripheral tolerance to auto- and alloantigens,[55,56] targeting IL-15/IL-15R signal pathway may foster AICD and facilitate tolerance induction.

Moreover, Li X et al recently reported the distinct roles of IL-15 and IL-2 for primary T-cell expansion in vivo.[57] IL-15 seems to be a critical growth factor in initiating T-cell division in vivo, whereas the unique role of IL-2 in vivo is to control the magnitude of clonal expansion by regulating γ-c expression on cycling T cells. Thus blockade of IL-15/IL-15R signaling pathway may prevent early T-cell activation and the T-cell mediated cytopathic autoimmune process.

Role of IL-15 in Autoimmune and Inflammatory Disease

Rheumatoid arthritis (RA) is a chronic degenerative condition of synovial membranes mediated in part by aberrant cytokine regulation that ultimately results in abnormally high levels of proinflammatory cytokines, such as TNF-α, within the joints. Lymphocytes and TNF-α are strongly implicated in the pathogenesis of CIA and clinical RA.[4] Furthermore, the success of therapeutic strategies that neutralize TNF-α in the murine CIA model[58,59] and in clinical RA[60,61] underscore the crucial role played by TNF-α in disease pathogenesis. McInnes et al reported that TNF-α production is increased by a direct T cell to macrophage contact through an IL-15 dependent process that also results in activation of T cells.[62] This effect was cell-contact dependent, and antibodies to CD69, lymphocyte function-associated antigen (LFA)-1, and intercellular adhesion molecule (ICAM)-1 inhibited the T-cell-induced production of TNF-α by macrophages. Moreover, IL-15, but not LPS or TNF-α, triggers IL-17 production by T-cells.[63] It is noteworthy that high level of IL-15 protein was demonstrated in the synovial fluids and synovial membranes of patients with active RA. Coincidently, high levels of IL-17 are present in the synovial fluid of RA patients.[63] Moreover, elevated levels of synovial fluid

IL-15 and IL-17 are correlated, suggesting that IL-15 may activate inert synovial fluid T cells and trigger IL-17 production.[63] In addition, the presence of IL-15 in peripheral blood of patients with RA has also been reported. Interestingly there are no significant differences between levels of IL-15 in serum and synovial fluid from the same patients.[63] This may suggest that IL-15 is produced both in the joints and in the circulating blood. On the other hand, in the joints, fibroblast-like synoviocytes of RA patients may represent a major source of IL-15.[64] These cells increase IL-15 production and release in response to TNF-α and IL-1β.[64] Therefore, the loop between IL-15 triggered TNF-α production and TNF-α induced IL-15 production by synoviocytes may well contribute to the proinflammatory cytokine overproduction seen in RA. In addition, synovial fluids from patients with active RA were found to promote not only activation of T cells but also chemoattraction. These effects were partially abrogated by the addition of anti-IL-15 antibodies. In addition, injection of a single dose of IL-15 resulted in a lymphocytic inflammatory infiltrate in vivo.[62] Current hypotheses suggest that abnormal T-cell trafficking to the joints may be a key early step in this process.[4]

IL-15/IL-15R Targeting Strategies

On the basis of work describing IL-15 as a potent T-cell attractant,[65] McInnes et al suggested that IL-15 might play a primary role in the development of RA.[66] Thus, the development of agents blocking IL-15 or targeting the receptor and signaling elements of IL-15 may provide a new perspective for treatment of diseases associated with expression of IL-15/IL-15R.

Soluble IL-15R alpha

The soluble form of IL-15Rα neutralizes the biological functions of IL-15 in vitro.[67] Moreover, administration of soluble IL-15Rα prevents collagen-induced arthritis in a murine model, suggesting that development of effective IL-15-blocking agents such as soluble receptors may be useful in the treatment of RA.[67] However, treated mice developed acute CIA soon after the discontinuation of sIL-15Rα administration.[67] These experiments further prove the important role of IL-15 in the pathogenesis of CIA and warrant the development of more effective IL-15/IL-15R targeting agents.

Mutant Cytolytic IL-15/Fc Fusion Protein

The development of agents targeting the receptor and signaling elements of IL-15 may provide a new perspective for treatment of diseases associated with expression of IL-15/IL-15R. As many other cytokines, IL-15 possesses a very high affinity for its receptor. However, its short circulating half-life and agonist activity, triggering activation of receptor-bearing target cells, limit or preclude its utility as a receptor site antagonist or a vehicle for targeting cytocidal agents to cytokine receptor bearing cells without transiently stimulating the target cells. The impetus for creating mutant IL-15/Fc fusion protein to be used as IL-15R site antagonist stems from the homologues of glutamine residues within the C terminus of the four helix structure shared by IL-2 and IL-15 (Fig. 6.2), as the 141 glutamine residue of IL-2 has been reported to be crucial for IL-2 binding to IL-2Rγ.[68] To overcome the problem associated with the short $t_{1/2}$ and agonist activity of IL-15, we designed, genetically constructed, and expressed a receptor site-specific IL-15 antagonist by mutating glutamine residues within the C terminus of IL-15 to aspartic acid and genetically linked this mutant IL-15 to murine Fcγ2a (Fig. 6.3).[69] This immunoligand retains the properties of murine Fcγ2a fragment, i.e., prolonged circulating $t_{1/2}$ and the ability to direct ADCC and CDC activities to target the cells recognized by IL-15 moiety.[69, 70] These mutant IL-15 proteins specifically bind to the IL-15R, competitively inhibit IL-15-triggered cell proliferation, and do not activate the STAT signaling pathway. Because the receptor site-specific antagonist IL-15 mutant/Fcγ2a fusion proteins had a prolonged $t_{1/2}$ in vivo and the potential for destruction of IL-15R$^+$ leukocytes, we examined the immunosuppressive activity of this agent in mice with methylated BSA-induced delayed type hypersensitivity (DTH) and collagen-induced arthritis (CIA), a murine model for RA. In a DTH model treatment with this unique IL-15 mutant/Fcγ2a fusion protein markedly attenuated DTH responses and decreased leukocyte infiltration within DTH sites.[69] In a CIA model treatment of mutant/Fcγ2a

					101							108							
Human - 97	K	E	F	L	Q	S	F	V	H	I	V	Q	M	F	I	N	T	S	114
Simian- 97	K	E	F	L	Q	S	F	V	H	I	V	Q	M	F	I	N	T	S	114
Mouse -97	T	E	F	L	Q	S	F	I	R	I	V	Q	M	F	I	N	T	S	114

Fig. 6.2. Homology between C-terminal alpha-helix of the human, simian and mouse IL-15. (mutation residues in the human are boxed)

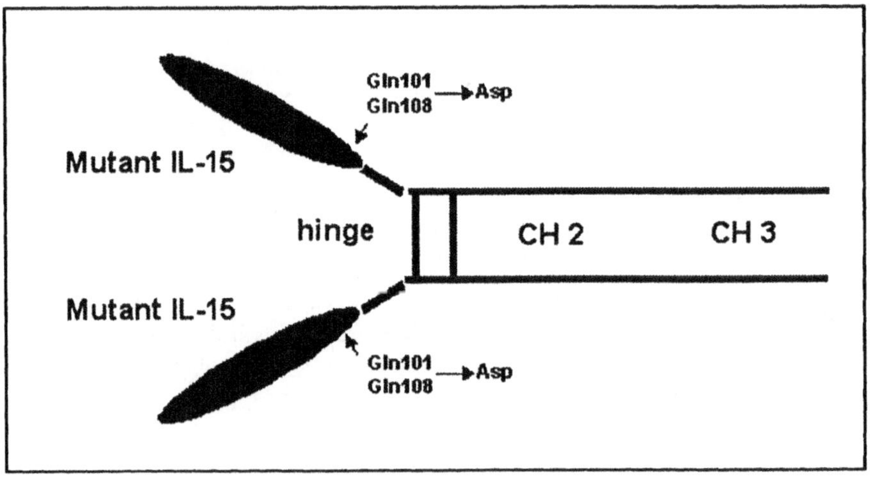

Fig. 6.3. Mutant IL-15/Fc fusion protein gene construction. Scheme of the genetic fusion of mutant human IL-15 and murine Fcγ2a cDNA to create a mutant IL-15/Fc immunoligand. Mutations were made in the fourth helix of IL-15 by using site-directed mutagenesis to replace Gln101 and Gln108 with Asp.

fusion protein markedly decreased the incidence and severity of arthritis in DBA/1 mice. Treatment was associated with a dramatic decrease in intra-articular gene and protein expression of the proinflammatory cytokines, IL-1β, TNF-α and IL-17 in mice with CIA (Ferrari-Lacraz S, in press). As high affinity IL-15Rα are present on activated, but not resting, mononuclear leukocytes, it is notable that marked reductions in T-cell receptor transcripts and frequency of proliferating $CD4^+$ T cells occur in mutant IL-15/Fc treated hosts. Histologic analysis confirms that this treatment is remarkably effective in protecting the joint from $CD4^+$, $CD8^+$ and $CD11^+$ cellular infiltration. Moreover, after cessation of treatment, remission of the arthritis persists throughout the observation period. Mutant IL-15/Fc treatment was also effective in blocking the progression of ongoing arthritis, as a significant therapeutic effect was observed with the treatment initiated after the onset of arthritis in this murine CIA model.

In summary, the IL-15/IL-15R system plays an important role in the development and maintenance of immunity. The dysregulation of IL-15/IL-15R system results in autoimmunity, zealous lymphocyte proliferation and IL-15 dependent inflammation observed in the RA. Targeting IL-15/IL-15R may represent a novel and effective strategy to control these pathogenesis processes.

References

1. Oppenheim JJ, Feldmann M. The Cytokine Reference—A Comprehensive Guide to the Role of Cytokine in Health and Disease. London: Academic Press, 2000.
2. Brennan FM, Feldmann M. Cytokines in autoimmunity. Curr Opin Immunol 1996; 8:872-877.
3. Feldmann M. The cytokine network in rheumatoid arthritis: Definition of TNF alpha as a therapeutic target. J R Coll Physicians Lond 1996; 30:560-570.
4. Feldmann M, Brennan FM, Maini RN. Role of cytokines in rheumatoid arthritis. Annu Rev Immunol 1996; 14:397-440.
5. Brennan FM, Chantry D, Jackson A et al. Inhibitory effect of TNF alpha antibodies on synovial cell interleukin-1 production in rheumatoid arthritis. Lancet 1989; 2:244-247.
6. Maini RN, Elliott MJ, Brennan FM et al. Monoclonal anti-TNF alpha antibody as a probe of pathogenesis and therapy of rheumatoid disease. Immunol Rev 1995; 144:195-223.
7. Williams RO, Feldmann M, Maini RN. Anti-tumor necrosis factor ameliorates joint disease in murine collagen-induced arthritis. Proc Natl Acad Sci USA 1992; 89:9784-9788.
8. Elliott MJ, Maini RN, Feldmann M et al. Treatment of rheumatoid arthritis with chimeric monoclonal antibodies to tumor necrosis factor alpha. Arthritis Rheum 1993; 36:1681-1690.
9. Weinblatt ME, Kremer JM, Bankhurst AD et al. A trial of etanercept, a recombinant tumor necrosis factor receptor: Fc fusion protein, in patients with rheumatoid arthritis receiving methotrexate [see comments]. N Engl J Med 1999; 340:253-259.
10. Maini R, St Clair EW, Breedveld F et al. Infliximab (chimeric anti-tumour necrosis factor alpha monoclonal antibody) versus placebo in rheumatoid arthritis patients receiving concomitant methotrexate: A randomised phase III trial. ATTRACT Study Group. Lancet 1999; 354:1932-1939.
11. Lipsky PE, van der Heijde DM, St Clair EW et al. Infliximab and Methotrexate in the treatment of rheumatoid arthritis. N Engl J Med 2000; 343:1594-1602.
12. Markham A, Lamb HM. Infliximab: A review of its use in the management of rheumatoid arthritis. Drugs 2000; 59:1341-1359.
13. Editors G, Breedveld FC, Kalden JR et al. Advances in targeted therapies II [In Process Citation]. Ann Rheum Dis 2000; 59 Suppl 1:I.
14. van den Berg WB, Joosten LA, Kollias G et al. Role of tumour necrosis factor alpha in experimental arthritis: separate activity of interleukin 1beta in chronicity and cartilage destruction. Ann Rheum Dis 1999; 58 Suppl 1:I40.
15. Joosten LA, Helsen MM, van de Loo FA et al. Anticytokine treatment of established type II collagen-induced arthritis in DBA/1 mice. A comparative study using anti-TNF alpha, anti-IL-1 alpha/beta, and IL-1Ra. Arthritis Rheum 1996; 39:797-809.
16. Bresnihan B, Alvaro-Gracia JM, Cobby M et al. Treatment of rheumatoid arthritis with recombinant human interleukin-1 receptor antagonist [see comments]. Arthritis Rheum 1998; 41:2196-2204.
17. Jiang Y, Genant HK, Watt I et al. A multicenter, double-blind, dose-ranging, randomized, placebo-controlled study of recombinant human interleukin-1 receptor antagonist in patients with rheumatoid arthritis: Radiologic progression and correlation of Genant and Larsen scores. Arthritis Rheum 2000; 43:1001-1009.
18. Bamford RN, Grant AJ, Burton JD et al. The interleukin 2 receptor beta chain is shared by IL-2 and a cytokine, provisionally designated IL-T, that stimulates T-cell proliferation and the induction of lymphokine-activated killer cells. Proc Natl Acad Sci USA 1994; 91:4940-4944.
19. Carson WE. Interleukin (IL) 15 is a novel cytokine that activates human natural killer cells via components of IL-2 receptor. J Exp Med 1994; 180:1395-1403.
20. Giri JG. Utilization of the β and γ chains of the IL-2 receptor by the novel cytokine IL-15. EMBOJ 1994; 13:2822-2830.
21. Giri JG, Kumaki S, Ahdieh M et al. Identification and cloning of a novel IL-15 binding protein that is structurally related to the alpha chain of the IL-2 receptor. Embo J 1995; 14:3654-3663.
22. Smith KA. Interleukin-2: Inception, impact, and implications. Science 1988; 240:1169-1176.
23. Grabstein KH, Eisenman J, Shanebeck K et al. Cloning of a T cell growth factor that interacts with the β chain of the interleukin-2 receptor. Nature 1994; 264:965-968.
24. Anderson DM, Kumaki S, Ahdieh M et al. Functional characterization of the human interleukin-15 receptor alpha chain and close linkage of IL15RA and IL2RA genes. J Biol Chem 1995; 270:29862-29869.
25. Waldmann TA, Tagaya Y. The multifaceted regulation of interleukin-15 expression and the role of this cytokine in NK cell differentiation and host response to intracellular pathogens. Annu Rev Immunol 1999; 17:19-49.

26. Meazza R, Gaggero A, Neglia F et al. Expression of two interleukin-15 mRNA isoforms in human tumors does not correlate with secretion: role of different signal peptides. Eur J Immunol 1997; 27:1049-1054.
27. Onu A, Pohl T, Krause H et al. Regulation of IL-15 secretion via the leader peptide of two IL-15 isoforms. J Immunol 1997; 158:255-262.
28. Tagaya Y, Kurys G, Thies TA et al. Generation of secretable and nonsecretable interleukin 15 isoforms through alternate usage of signal peptides. Proc Natl Acad Sci USA 1997; 94:14444-14449.
29. Musso T, Calosso L, Zucca M et al. Human monocytes constitutively express membrane-bound, biologically active, and interferon-gamma-upregulated interleukin-15. Blood 1999; 93:3531-3539.
30. Chae DW, Nosaka Y, Strom TB at al. Distribution of IL-15 receptor alpha-chains on human peripheral blood mononuclear cells and effect of immunosuppressive drugs on receptor expression. J Immunol 1996; 157:2813-2819.
31. Lodolce JP, Boone DL, Chai S et al. IL-15 receptor maintains lymphoid homeostasis by supporting lymphocyte homing and proliferation. Immunity 1998; 9:669-676.
32. Kennedy MK, Glaccum M, Brown SN et al. Reversible defects in natural killer and memory CD8 T cell lineages in interleukin 15-deficient mice [see comments]. J Exp Med 2000; 191:771-780.
33. Munger W, DeJoy SQ, Jeyaseelan R et al. Studies evaluating the antitumor activity and toxicity of interleukin-15, a new T cell growth factor: Comparison with interleukin-2. Cell Immunol 1995; 165:289-293.
34. Evans R, Fuller JA, Christianson G et al. IL-15 mediates anti-tumor effects after cyclophosphamide injection of tumor-bearing mice and enhances adoptive immunotherapy: The potential role of NK cell subpopulations. Cell Immunol 1997; 179:66-73.
35. Fehniger TA, Suzuk K, Ponnappan A. Fetal leukemia in interleukin-15 transgenic mice follows early expansion in NK and memory-phenotype CD8+ T cells. J Exp Med 2001; 193:219-232.
36. Fehniger TA, Shah MH, Turner MJ et al. Differential cytokine and chemokine gene expression by human NK cells following activation with IL-18 or IL-15 in combination with IL-12: Implications for the innate immune response. J Immunol 1999; 162:4511-4520.
37. Ross ME, Caligiuri MA. Cytokine-induced apoptosis of human natural killer cells identifies a novel mechanism to regulate the innate immune response. Blood 1997; 89:910-918.
38. Bluman EM, Bartynski KJ, Avalos BR et al. Human natural killer cells produce abundant macrophage inflammatory protein-1 alpha in response to monocyte-derived cytokines. J Clin Invest 1996; 97:2722-2727.
39. Maghazachi AA, al-Aoukaty A, Schall TJ. C-C chemokines induce the chemotaxis of NK and IL-2-activated NK cells. Role for G proteins. J Immunol 1994; 153:4969-4977.
40. Fahey TJd, Tracey KJ, Tekamp-Olson P et al. Macrophage inflammatory protein 1 modulates macrophage function. J Immunol 1992; 148:2764-2769.
41. Kanegane H, Tosato G. Activation of naive and memory T cells by interleukin-15. Blood 1996; 88:230-235.
42. Zhang X, Sun S, Hwang I et al. Potent and selective stimulation of memory-phenotype CD8+ T cells in vivo by IL-15. Immunity 1998; 8:591-599.
43. Finck BK, Linsley PS, Wofsy D. Treatment of murine lupus with CTLA4Ig. Science 1994; 265:1225-1227.
44. Lenschow DJ, Zeng Y, Thistlethwaite JR et al. Long-term survival of xenogeneic pancreatic islet grafts induced by CTLA4Ig. Science 1992; 257:789-792.
45. Buhlmann JE, Noelle RJ. Therapeutic potential for blockade of the CD40 ligand, gp39. J Clin Immunol 1996 ;16:83-89.
46. Mohan C, Shi Y, Laman JD et al. Interaction between CD40 and its ligand gp39 in the development of murine lupus nephritis. J Immunol 1995; 154:1470-1480.
47. Jones ND, Van Maurik A, Hara M et al. CD40-CD40 ligand-independent activation of CD8+ T cells can trigger allograft rejection. J Immunol 2000; 165:1111-1118.
48. Newell KA, He G, Guo Z et al. CTLA4Ig fails to prevent intestinal allograft rejection due to an inability to inhibit CD8 T cell responses. Transplantation 1999; 67:s45.
49. Trambley J, Bingaman AW, Lin A et al. Asialo GM1(+) CD8(+) T cells play a critical role in costimulation blockade-resistant allograft rejection. J Clin Invest 1999; 104:1715-1722.
50. Schorle H, Holtschke T, Hünig T et al. Development and function of T cells in mice rendered interleukin-2 deficient by gene targeting. Nature 1991; 352:621-623.
51. Willerford DM, Chen J, Ferry JA et al. Interleukin-2 receptor alpha chain regulates the size an content of the perpheral lymphoid compartment immunity. Immunity 1995; 3:521-530.
52. Suzuki H, Kunding TM, Furlonger C et al. Deregulated T cell activation and autoimmunity in mice lacking interleukin receptor beta. Science 1995; 268:1472-1476.

53. Dooms H, Desmedt M, Vancaeneghem S et al. Quiescence-inducing and antiapoptotic activities of IL-15 enhance secondary CD4+ T cell responsiveness to antigen. J Immunol 1998; 161:2141-2150.
54. Bulfone-Paus S, Ungureanu D, Pohl T et al. Interleukin-15 protects from lethal apoptosis in vivo. Nat Med 1997; 3:1124-1128.
55. Wells AD, Li XC, Li Y et al. Requirement for T-cell apoptosis in the induction of peripheral transplantation tolerance. Nat Med 1999; 5:1303-1307.
56. Li Y, Li XC, Zheng XX et al. Blocking both signal 1 and signal 2 of T-cell activation prevents apoptosis of alloreactive T cells and induction of peripheral allograft tolerance. Nat Med 1999; 5:1298-1302.
57. Li XC, Demirci G, Ferrari-Lacraz S et al. IL-15 and IL-2: a matter of life and death for T cells in vivo. Nat Med 2001; 7:114-118.
58. Wooley PH, Dutcher J, Widmer MB et al. Influence of a recombinant human soluble tumor necrosis factor receptor FC fusion protein on type II collagen-induced arthritis in mice. J Immunol 1993; 151:6602-6607.
59. Williams RO, Ghrayeb J, Feldmann M et al. Successful therapy of collagen-induced arthritis with TNF receptor-IgG fusion protein and combination with anti-CD4. Immunology 1995; 84:433-439.
60. Elliott MJ, Maini RN, Feldmann M et al. Randomised double-blind comparison of chimeric monoclonal antibody to tumour necrosis factor alpha (cA2) versus placebo in rheumatoid arthritis. Lancet 1994; 344:1105-1110.
61. Moreland LW, Baumgartner SW, Schiff MH et al. Treatment of rheumatoid arthritis with a recombinant human tumor necrosis factor receptor (p75)-Fc fusion protein [see comments]. N Engl J Med 1997; 337:141-147.
62. McInnes IB, Leung BP, Sturrock RD et al. Interleukin-15 mediates T cell-dependent regulation of tumor necrosis factor-α production in rheumatoid arthritis. Nat Med 1997; 3:189-195.
63. Ziolkowska M, Koc A, Luszczykiewicz G et al. High levels of IL-17 in rheumatoid arthritis patients: IL-15 triggers in vitro IL-17 production via cyclosporin A-sensitive mechanism. J Immunol 2000; 164:2832-2838.
64. Harada S, Yamamura M, Okamoto H et al. Production of interleukin-7 and interleukin-15 by fibroblast-like synoviocytes from patients with rheumatoid arthritis. Arthritis Rheum 1999; 42:1508-1516.
65. Wilkinson PC, and Liew FY. Chemoattraction of human blood T lymphocytes by interleukin-15. J Exp Med 1995; 181:1255-1259.
66. McInnes IB, al-Mughales J, Field M et al. The role of interleukin-15 in T-cell migration and activation in rheumatoid arthritis [see comments]. Nat Med 1996; 2:175-182.
67. Ruchatz H, Leung BP, Wei XQ et al. Soluble IL-15 receptor alpha-chain administration prevents murine collagen-induced arthritis: A role for IL-15 in development of antigen-induced immunopathology. J Immunol 1998; 160:5654-5660.
68. Zurawski SM, and Zurawski G. Receptor antagonist and selective agonist derivatives of mouse interleukin-2. EMBO J 1992; 11:3905-3910.
69. Kim YS, Maslinski W, Zheng XX et al. Targeting the IL-15 receptor with an antagonist IL-15 mutant/Fc gamma2a protein blocks delayed-type hypersensitivity. J Immunol 1998; 160:5742-5748.
70. Zheng XX, Steele AW, Hancock WW et al. IL-2 receptor-targeted cytolytic IL-2/Fc fusion protein treatment blocks diabetogenic autoimmunity in nonobese diabetic mice. J Immunol 1999; 163:4041-4048.

CHAPTER 7

Cytokines in the Pathogenesis and Therapy of Autoimmune Encephalomyelitis and Multiple Sclerosis

David O. Willenborg and Maria A. Staykova

Introduction

In the inflammatory diseases autoimmune encephalomyelitis (EAE) and multiple sclerosis (MS) the occurrence, severity, course and resolution of disease are dependent on a complex interaction of cells, cytokines, chemokines and myriad other mediators. This chapter describes the role of what, in the authors' opinions, are the most influential cytokines in these two diseases. To understand the effect of the various cytokines it is important to bear in mind that most are highly pleiotrophic and like most biological molecules, their effects are dose, time and site dependent. Targeting cytokines as therapy for these diseases is a feasible approach but careful consideration must be given not only to the cytokine targeted but the stage of the disease process being targeted. A 'ready reckoner' of cytokine function in relation to CNS inflammation is presented at the end of the chapter.

Multiple Sclerosis

Multiple sclerosis (MS) is an inflammatory demyelinating disease of the central nervous system (CNS) which may present clinically in one of five varieties: primary progressive (PP-MS), relapsing remitting (RR-MS), secondary progressive (SP-MS), benign and rapidly fatal.[1] Whether these various forms of disease represent different pathological features, a different pathogenesis and in fact a different etiology is also unknown.

Extensive epidemiological and genetic studies have established a complex interaction between the environment and genes in the expression of MS.[2, 3] Contemporary twin studies have described a concordance rate of 3-5% in dizygotic and 25-32% in monozygotic pairs, the excess of monozygotic concordance providing powerful support for a genetic contribution to disease etiology. At the same time the fact the rate is less than 100% suggests incomplete penetrance or most likely an obligatory interplay with the environment. [4, 5]

Linkage with an MHC allelle, the presence of inflammatory cells of the immune system within CNS lesions and extensive evidence of immune reactivity of lymphoid cells from MS patients against a number of CNS antigens has led to the widely accepted conclusion that MS has, at least in part, an autoimmune pathology.[6] Further evidence for the possible autoimmune nature of MS derives from the pathological and immunological similarities with an extensively studied animal model of the disease termed experimental autoimmune encephalomyelitis (EAE).

Cytokines and Chemokines in Autoimmune Disease, edited by Pere Santamaria.

Experimental Autoimmune Encephalomyelitis

EAE is an organ specific, T cell driven autoimmune disease of the CNS. Its pathology is characterised by lymphocytic and monocytic perivascular infiltration, an increase of blood brain barrier (BBB) permeability, astrocytic hypertrophy and demyelination.[7, 8] Three major encephalitogenic CNS proteins and their relevant peptides have been described: myelin basic protein (MBP),[9, 10] proteolipid protein (PLP)[11, 12] and myelin oligodendrocyte glycoprotein (MOG)[13, 14] all of which induce EAE in a number of species and strains of animals.

It is important to note that EAE presents clinically and pathologically in several different ways dependent on the species and strain of animal and encephalitogenic inoculum used. Therefore EAE, as MS, should be thought of as being possibly more than a single disease entity and as such, note should always be made of which model is being described in any given experimental study.

Irrespective of the different EAE models there is a generally accepted unifying factor, the activated CNS antigen-reactive T lymphocyte which drives disease. The exact sequence of events orchestrated by the activated CNS antigen-reactive T lymphocytes as well as the molecular mediators of the inflammatory process are far from completely understood. However, multiple steps in disease induction have been identified:

1. Migration of cells to the CNS and their transit across or through the tight-junctioned endothelium and the underlying basement membrane;
2. Recognition of endogenous myelin epitopes presented to the T cells by CNS-resident antigen presenting cells (APC);
3. Production of chemokines and cytokines by both the infiltrating T cells and resident cells which combine to direct the further influx of cells into CNS parenchyma;
4. Activation of both, recruited and resident cells;
5. Edema, conduction block and demyelination and
6. In some cases repair and recovery.

Which chemokines, cytokines and/or products of activated cells are required for disease development and regulation is the focus of this chapter and the chapter by Trevor Owens. The prevailing dogma depicts the inducer of EAE (and by inference MS) as polarised Th1 cells and causally links their cytokines (IL-2, IFNγ and TNF) with pathology, while recovery from disease results from a shift to Th2 cells and their cytokines.

This chapter sets out to present our most recent understanding of the role of cytokines in pathology and immunoregulation of MS and EAE. The focus will be on those cytokines for which some controversy exists such as TNFα, IFNγ, IL-4, IL-10 and IL-12. Because of constraints of space, interferon beta (IFNβ), which was in some respects the first big advance in MS therapy, will not be discussed. The reader is referred to recent reviews.[15-19]

Interleukin-1

IL-1 and EAE

IL-1, both as mRNA[20] and protein[21] has been demonstrated in the CNS of animals with EAE and a correlation with disease severity was reported.[21]

Administering IL-1 to Lewis rats one week prior to induction of MBP-EAE resulted in markedly suppressed neurological symptoms of EAE without affecting the onset or duration of the disease.[22] Since IL-1 is known to sensitise the hypothalmic-pituatary-adrenal (HPA) axis to various stressors[23] the result suggested that increased corticosterone levels may be responsible for amelioration of disease.

Another member of the IL-1 gene family is IL-1 receptor antagonist (IL-1Rα) which inhibits IL-1 activity by binding to IL-1 α and β receptors. IL-1Rα has been exploited in three studies of EAE in the Lewis rat to inhibit the function of IL-1. All studies reported a modest but significant beneficial effect on disease,[24-26] and the suppressive effect was thought to be on both, the afferent and efferent arm of the immune response.

IL-1R knockout mice were examined for susceptibility to MOG-EAE and found to be totally refractory both, with respect to clinical disease and CNS inflammation.[27] Unfortunately, it was not tested whether the IL-1R was essential for induction of the immune response or for expression of disease in the target tissue.

IL-1 and MS

Recent studies have found no significant differences in genotypes, allele frequencies, or carrier frequencies between MS patients and controls with respect to IL-1 genes. However, a specific IL-1Rα/ IL-1β combination was associated with disease severity.[28-30] Evidence to date therefore seems to indicate that IL-1Rα and IL-1β may be disease severity genes rather than disease susceptibility genes.

Therapeutic studies would seem to support the idea that IL-1 may contribute to disease severity. Corticosteroid pulse therapy is still the most widely used and effective treatment for acute MS exacerbations. In 18 patients given methylprednisolone for an acute episode of MS, IL-1 levels were significantly decreased from the elevated pretreatment levels.[31] IFNγ reduced relapse rate in RR-MS and studies have suggested that it may function to either decrease IL-1 or increase IL-1Rα (as well as sTNF-αR1 and sTNF-αR2).[32,33]

IL-1α and β are both synthesised as precursor proteins and their processing to 'mature' forms with optimal biological activity requires the cysteine protease, caspase-1.[34] Caspase-1 mRNA levels were measured in PBMNC of seven patients with RR-MS every 15 days over a one year period. Brain MRI was also performed each month. Caspase-1 mRNA levels were significantly increased in MS patients when compared to controls. Furthermore, the increases were found in the week preceeding an acute attack. There was also a correlation between the level of caspase-1 and the number of new MRI lesions.[35] If this intriguing finding can be confirmed caspase-1 could conceivably be used as a surrogate marker for MS disease activity.

The evidence would indicate that IL-1 does not determine disease susceptibility but may contribute to disease severity. Because of the pleotrophic effects of IL-1 and the fact that IL-1 is produced in CNS not only by inflammatory cells but by resident cells under the influence of many stimuli, e.g., stress, it seems reasonable that IL-1 may, at the very least, contribute to the maintenance of a chronic disease such as MS.

A point that should be made with respect to not only IL-1 but all cytokines is the fact that the effect a cytokine has is not only time and dose dependent but also site dependent, and site dependent not only whether in the periphery or the target tissue but also where in the target tissue. By way of example, the injection of IL-1β into the rat brain parenchyma failed to induce BBB breakdown and gave rise to only minimal neutrophil recruitment, whereas injection into the spinal cord induced significant BBB breakdown and recruitment of neutrophils and lymphocytes.[36] A similar result was obtained with the injection of TNFα. This observation combined with the fact that the distribution of lesions in EAE, and perhaps MS, varies considerably depending on the encephalitogen used for immunisation,[37] means that a given cytokine may have varying effects in the same disease process dependent on where the lesions form.

Tumor Necrosis Factor alpha and Lymphotoxin alpha

TNFα/LTα and EAE

There is much compelling evidence that TNFα and LTα are critical in the development of EAE. Thus mRNA as well as protein of TNFα and LTα can be found in the CNS of animals with EAE.[38] In some cases the expression of TNFα appears to be disease specific in that that there was no significant difference in the severity of inflammation in the spinal cord lesions between CR-EAE and acute EAE in the DA rat, but there was significantly more TNFα mRNA in the lesions at the first attack of CR-EAE than at the peak stage of the acute disease.[39] The fact that there is considerable demyelination in the CR-EAE suggests the high levels of cytokine may contribute to both the demyelination and the chronicity of disease. There is also evidence for a continuum of production of TNFα in EAE with first CD4⁺ T cells infiltrating the CNS producing it followed by microglia and then macrophages.[40]

The clinical manifestation of EAE can be blocked by treating animals with antibodies to TNFα[41-43] or by using soluble forms of the p55 TNF receptor.[44-46] EAE can also be reversed by giving altered peptide ligands of MBP which apparently reduces the production of TNFα by the T cell lines;[47,48] by giving phosphodiesterase inhibitors, which also decreases production of TNFα[49] or by treating with Dexanabinol (HU-211), a synthetic nonpsychotropic cannabinoid which suppresses TNFα production in the brain and peripheral blood.[50] Multiple doses of pentoxyfilline that inhibit IL-1 and TNFα failed to have any significant effect on CR-EAE in Lewis rats.[51]

Studies using gene targeted mice have, however, questioned the pathogenic role of TNFα and LTα in EAE. Kroner et al[52] found that MOG-EAE developed with the same incidence and severity in the TNFα$^{-/-}$ C57BL mouse as it did in the wild type littermates though with a significant delay in clinical signs suggestive of altered leukocyte movement into CNS. Another study also found TNFα was not essential for disease and described TNFα as perhaps a protective molecule in MOG-EAE in that in mice lacking the gene, disease was more severe and chronic than in the wild type and giving exogenous TNFa inhibited the development of disease.[53]

Depending on the background and the encephalitogen used, TNFα and LTα double knockout mice developed either a mild disease similar to their wild type littermates or an earlier onset and more severe disease indicating that TNFα and LTα are not essential for the development of EAE.[54] In contrast, Suen et al[55] reported that C57BL/6 and wild-type littermates of LTα$^{-/-}$ mice are susceptible to MOG-EAE and develop a chronic, sustained paralysis with CNS inflammation and demyelination. The LTα$^{-/-}$ mice, however, were somewhat more resistant to both clinical and histopathological signs of disease even though they generated lymphocytes which proliferated to MOG35-55 and produced antibodies to the peptide. The great difficulty with these experiments as well as those with the double knockouts is that in the absence of LTα the mice are severely immunocompromised, with greatly disrupted splenic architecture and an absence of lymph nodes and Peyer's patches.[56]

An elegant study by Riminton et al[57] may have partly overcome the problem of immunocompetence in LTα$^{-/-}$ mice. These investigators took advantage of the fact that LTα is solely a product of lymphoid cells[58] and that bone marrow cells from LTα$^{-/-}$ mice have the capacity to repopulate lymph nodes in irradiated recipients.[59] Having disrupted the LTα gene directly in the C57BL/6 strain of mouse, this allowed LTα$^{-/-}$ bone marrow to be used to establish chimeras with irradiated wild-type B57BL/6. These animals should then have the inability to produce LTα, but on the background of a normal functional immune system. Such chimeras proved to be perfectly susceptible to MOG35-55-EAE in the absence of LTα.

From the five studies with transgenic animals just described, there is no consensus on the respective roles of TNFα and LTα in the pathogenesis of EAE. We have conducted studies with TNFα receptor transgenic mice in an attempt to further elucidate the role of TNF in EAE.[60]

The TNF receptors, p55 (TNF-R1) and p75 (TNF-R2) are both high affinity receptors that bind TNFα and LTα with almost identical affinities.[58] Consistent with the different structure of their intracellular domains, the two receptors appear to fulfill different functions in vivo. Using MOG35-55-EAE in C57BL/6 x 129 mice we found that wild type animals readily developed severe EAE whereas p55$^{-/-}$ mice were resistant. p55$^{-/-}$ mice also failed to develop tEAE with effector cells from wild type mice but they did generate effector cells following active immunisation which could transfer disease to wild-type mice. These data point to the essential nature of an intact p55 TNF receptor for the development of EAE and that it is essential for the expression of pathology but not for the generation of effector cells. Similar studies with the p75 receptor indicated it was not essential for disease and in point of fact the p75$^{-/-}$ mice were equally if not more susceptible to EAE than the wild type.[60] Several recent papers have confirmed and extended the finding that the p55 receptor is essential for expression of MOG-EAE[61-63] and that the p75 receptor appears to downregulate the immune response.[64]

If one accepts that TNFα is not essential for EAE pathology (as the majority of studies showed) and accepts the essential nature of the p55 (TNF-R1) for the expression (but not the induction) of disease then one logical conclusion is that LTα, which binds with equal affinity as TNFα to both p55 and p75 receptors, is the effector molecule in disease pathology. Furthermore, the fact that the p75 receptor was not essential for EAE indicates that LTα is exerting its effect through its interaction with the p55 receptor.

The LTα data are more difficult to reconcile. The two papers using the $LT\alpha^{-/-}$ mouse model disturbingly disagree with each other. This could be due to the fact that different antigens (MOG vs SCH or PLP), different adjuncts (*Bordetella pertussis* organisms vs pertussis toxin) as well as different background strains of mice were used. On the other hand, this discrepency could simply highlight the pitfalls of using such profoundly immunocompromised animals as an experimental model. If, for the purpose of argument, one discounts the results from these studies there remains the recent study in immunocompetent mice of Riminton et al[57] indicating that LTα is not essential for EAE development. Accepting this, then we are left with the conclusion that neither TNFα nor LTα is essential for EAE development but that the p55 (TNF-R1) is. Since no other currently described ligands are known to bind to the p55 receptor, these data taken together would be consistent with an interpretation that another related ligand exists that binds to TNF-R1 and mediates disease.

TNFα/LTα and MS

In MS, TNFα levels have been shown to correlate with disease progression and LTα has been localized in MS lesions[42, 65, 66] TNF receptors have also been examined in MS brain and interestingly oligodendrocytes around active MS lesions frequently expressed TNF-R molecules belonging to the apoptotic cascade. However, these cells did not undergo apoptosis as judged by TUNEL.[67] On the other hand, lymphocytes (and a few microglial cells) in the same tissue displayed apoptosis. This might suggest TNFα as a down-regulating molecule in MS.

Serial analysis of in vitro TNFα production by leukocytes from MS patients suggested that TNFα may be a useful predictor of relapses.[68] Another longitudinal study of 40 patients followed every 2-3 months also suggested TNFα as a marker of relapses in that TNFα was individually increased during the patients' relapses.[69] The serum levels of TNFα p55 and p75 soluble receptors in 65 RR-MS patients at different stages of disease, however, did not correlate with the clinical relapses or MRI lesions.[69]

Soluble TNF-R as well as TNF were measured in CSF and plasma in another study of active MS and determinations made both before and after a six day treatment with methylprednisolone.[70] The most significant finding was that CSF sTNF-Rp55 levels were higher in acute MS patients than in controls and that post-treatment these levels were still higher than in the active phase of the disease. That the MS patients, who clinically improved, tended to have the highest CSF sTNF-Rp55 levels again suggests that TNF may be down-regulating disease (supra vidae).

TNFα directed immune therapies in humans have been carried out and have been shown to be quite beneficial in the case of rheumatoid arthritis (see chapter by Lubberts). The situation with MS is less encouraging. Treatment of a small number of MS patients with phosphodiesterase inhibitors (pentoxyfylline) was reported to either reduce the mean relapse rate[71] or to worsen disease as judged by clinical, MRI or visual evoked potential criteria, despite the reduction of in vitro produced TNFα by PBMNC.[72] Since pentoxyfylline inhibits IL-1 and IL-6 as well as TNFα these results are difficult to interpret. We also hasten to add that both the above studies were uncontrolled and extreme care must be exercised in interpretation of the results of such studies. (See section on IFNγ for further discussion of uncontrolled clinical trials).

Lenercept, a TNF-R immunoglobulin G fusion protein that had been shown to inhibit EAE[45] has been tested in a double-blind, placebo-controlled phase II study of 168 patients with RR-MS.[73] The primary outcome was reduction of new lesions on MRI but clinical exacerbations were also recorded. There were no significant differences between groups on any MRI measurements and the number of lenercept-treated patients experiencing exacerbations was significntly increased compared with patients receiving placebo.

On the basis of both, the animal studies and the studies in man it would seem that wisdom dictates a reanalysis of the strategy of targeting TNFα as a therapy in MS. The data would in fact suggest that delivering TNFα rather than inhibiting it may be of some benefit. The dose, site and timing of delivery are of course questions that must first be answered.

Interleukin-6

IL-6 and EAE and MS

IL-6 is a multifunctional cytokine with important roles in host defense, hematopoeisis and immune responses. There is no doubt that IL-6 is involved in EAE development since four studies have described resistance of IL-6 $^{-/-}$ mice to EAE.[61,74-76] Different experimental designs led to differing conclusion but on balance it would appear that IL-6 is essential for efficient and appropriate generation of effector cells and has little role in the actual pathology of EAE.

IL-6 has been described as up-regulated in the brains of MS patients using both immunohistochemistry[77] and transcriptional analysis.[78] It has also been measured in plasma and CSF of MS patients and found to be increased.[69,79,80] In none of the studies was there any good correlation between disease activity and IL-6 levels.

An interesting work by Schonrock et al[81] sought to correlate the numbers of IL-6 expressing cells in 36 MS patients to the stage of demyelinating activity and the pattern of oligodendrocyte pathology. Highest numbers of IL-6 positive cells, identified as macrophages and astrocytes by morphological criteria, were found in inactive demyelinated lesions with oligodendrocyte preservation, whereas absence of IL-6 expression correlated with oligodendrocyte loss. This suggests a possible involvement of IL-6 in oligodendrocyte protection and survival in MS lesions.

It seems likely that IL-6 plays an insignificant role in the pathology of MS and its increase during disease activity probably reflects the global activity of the immune system. The influence of IL-6 early in the development of MS may be considerable, as for IL-4, but determining this will be very difficult.

Interferon γ

IFNγ and EAE

Being produced by encephalitogenic CD4+ Th1 cells and in view of its myriad pro-inflammatory effects, IFNγ has been considered a pivotal and usually pathogenic cytokine in the context of CNS inflammation. Nonetheless, many studies have provided data suggesting that IFNγ acts to down-regulate EAE. Treatment of animals with antibodies to IFNγ resulted in enhanced disease in both rats and mice[82-85] while giving exogenous IFNγ ameliorated disease.[86]

Experiments with transgenic mice have clearly defined IFNγ as a down-regulating molecule in EAE. Work from our lab has shown that mice lacking the ligand binding chain of the IFNγ receptor (IFNγR$^{-/-}$) develop severe and usually fatal EAE when immunised with MOG_{35-55} whereas wild type mice are resistant, indicating that IFNγ is not essential for disease induction.[87] Furthermore, passive transfer of disease with MOG_{35-55} specific lymphoid cells from IFNγR$^{-/-}$ mice produces in knockout (IFNγR$^{-/-}$) mice severe EAE from which the recipients fail to recover. The same cells produce equally severe disease in IFNγR$^{+/+}$ control mice but importantly all the recipients recover fully. These results provide definitive evidence that IFNγ is not necessary for the generation or function of anti-$MOG_{35\text{-}55}$ effector cells but that it is essential in down-regulating disease. Other studies using IFNγ cytokine knock out mice have produced comparable results and conclusions.[88,89]

Further to our own studies on IFNγ we have suggested that down-regulation of disease by IFNγ must act indirectly through a secondary mediator. IFNγ$R^{-/-}$ effector cells produce extremely high levels of IFNγ[87] but cannot respond to it because of the lack of the receptor. When these cells are transferred into wild type recipients the recipient cells can and do respond to IFNγ with the production of some mediator(s) which ultimately feeds back and down-regulates the effector cells. We have recently demonstrated that IFNγ induced nitric oxide (NO) production by macrophages (and by inference microglia) is most likely the key down-regulating molecule, acting probably through induction of apoptotic cell death of effector cells.[90] Furthermore the use of bone marrow chimeric mice indicated that down-regulation occurs not only systemically but also at the level of the target tissue, the CNS. Three subsequent studies support the hypothesis that NO regulates EAE through its effect on encephalitogenic effector cells[91-93] The importance of NO as a down-stream molecule in immunoregulation should not be underestimated and has been highlighted in other studies.[94, 95]

The role of IFNγ in CR-EAE is somewhat less clear than in the acute model. Spontaneous relapses in Biozzi ABH mice as well as induced relapses in SJL/J mice were facilitated by administration of neutralizing mAb against IFNγ in the disease-free interval. Administration of IFNγ in Biozzi mice provided partial protection not only against the first attack, but also against subsequent relapses. Administration of IFNγ during the remission phase provided some protection against subsequent relapses.[84] Thus in both types of relapses, IFNγ is produced and does provide a certain degree of protection against disease progression. In CR-EAE in the rat induced by immunisation with SCH and treatment with CsA, IFNγ mRNA was not expressed at the first attack of disease but peaked at the second (and last) attack.[96] These findings suggested that suppression of IFNγ by CsA in the first attack leads to relapses, however, intraventricular injection of IFNγ before the first attack led to more relapses or more severe disease. As pointed out by the authors themselves the result could reflect the dose and/or site of delivery of IFNγ with respect to the type of effect the cytokine has.

The body of evidence on IFNγ in EAE supports the idea that rather than a pro-inflammatory cytokine it is in fact a critical immunmodulator in this disease.

IFNγ and MS

The involvement of the IFNγ gene loci (chromosome 12q14-q15) in MS is problematic. Genetic association and linkage analyses on this region in a Finnish population found no evidence for a contribution by the IFNγ loci to the genetic susceptibility to MS.[97] Using a candidate gene strategy in a study of polymorphic markers within or close to the IFNγ loci, He et al[98] found no evidence for linkage in two-point linkage analysis. They did report, however, what they interpreted to be a slightly positive LOD score for IFNγ (0.88). As an isolated finding the result falls far short of establishing a linkage.

The relation between clinical parameters of MS such as disability, exacerbation frequency, disease duration, course of disease, and IFNγ-producing blood lymphocytes was determined in 41 consecutive, clinically stable MS patients with a primary relapsing course of disease and without immunomodulatory or immunosuppressive treatment in the last three months. Petereit et al[99] found a significant positive correlation between IFNγ-producing PBMNC and disability. A 12 month study of eight RR-MS patients revealed that serum levels of IFNγ and in vitro induced cellular IFNγ production increased prior to the onset of a relapse whereas IFNγ production showed a temporal delayed increase which was related to clinical remission.[100] Even more indirect evidence for IFNγ involvement in the pathology of MS is a study by Balashov et al[101] who measured seasonal serum IFNγ levels in PP-MS patients and found significantly increased production in the autumn and winter months compared with the spring and summer months. Autumn and winter are the times of increased viral infections and there has been a study establishing a positive correlation between concomitant viral infections and MS relapses.[102] The above studies were mostly interpreted as involvement of IFNγ in MS pathology.

A positive correlation between MS attacks/severity/relapses and IFNγ is of course far from an universal finding. Serum samples from nine SP-MS patients who were treated with monthly intravenous infusions of the interferon inducer polyinosinic acid polycytidylic acid polylysine in carboxymethylcellulose were assayed for IFNγ at various times.[103] It was demonstrated that the greatest IFNγ induction did not correlate with clinical worsening and in fact the IFNγ levels were not higher in the two patients who worsened on treatment while the highest levels were found in the patient who remained stable. Van Oosten et al[104] performed a study to determine if IFNγ production (and TNFα) by stimulated PBMNC precedes or accompanies clinical and MRI signs of disease activity in MS patients. One month preceding exacerbations there was a shift toward increased TNFα but there was no significant increase in IFNγ production. Another study found the IFNγ-producing cells to be significantly less in active RR-MS than in stable patients.[105]

Though there seems to be an overriding opinion that IFNγ is a "bad" molecule in the context of MS it could be reasonably argued that the above results in fact support the idea that IFNγ has a protective role in MS. The demonstration of IFNγ at peak disease, for example, could indicate an increase in response to exacerbation with the 'intent' of limiting the attack. The prevaling consensus of IFNγ as a bad molecule seems to be the result of a single study in which IFNγ was used to treat MS patients. We comment here on this in some detail.

The study in question[106] involved 18 patients who received IFNγ. There were no controls who received a placebo and were similarly manipulated. The manipulations were certainly not insignificant. The patients were lumbar punctured before the treatment and at its end. They were given infusions of IFNγ over a two hour period twice a week for four weeks, which required hospitalisation, and were bled every time for haematological work. The treatment was with three different doses of IFNγ ranging over three orders of magnitude (1000 fold). The results reported 7 of 18 patients showing worsening. There was worsening in all three dose groups and the degree of worsening was not related to the dose. This is a tremendously flat dose response curve and must suggest that something other than the substance being given brought about the effect. Could it be the two lumbar punctures, the eight two hour infusions in hospital and eight bleedings? There are no controls with which to compare. Account must also be taken of the clinical assessment which relied, in part, on comparison with retrospectively documented pretreatment exacerbations as determined from referring physician's records —not totally reliable. We hasten to state that this is not meant as a criticism of the study per se. It was carried out as a pilot without controls and was clearly stated to be so. However, over time references to the work have simply been to the conclusion drawn that IFNγ makes MS worse. IFNγ certainly may make MS worse but it should be obvious that the data presented in this study simply do not support that statement. If it is possible to extrapolate from the mouse data to human it may be that in dismissing IFNγ as an immunomodulator we are 'throwing the baby out with the bath water'. Time will tell.

Interleukin-18

IL-18 and EAE

IL-18 mRNA expression as well as that for caspase-1 has been found to increase in CNS during the acute phase of EAE.[107] However, protein analysis is necessary to determine if caspase-1 is actually converting IL-18 to an active protein in situ. The authors point out that because of the role of IL-18 in Fas/Fas ligand mediated apoptosis, IL-18 may in fact be acting to terminate rather than enhance the immune response in CNS.

Antibodies to IL-18 have been used in Lewis rat EAE and were found to inhibit disease when given on days 8, 10 and 12 after immunisation.[108] Treating tEAE was also said to inhibit development of disease suggesting a role for IL-18 in pathology. The data from this tEAE experiment, however, were based on the supposed ability to delineate 10 different gradations of the score of 1, which represents a limp tail. Fairly unreliable data. The investigators did show that T cells from anti-IL-18 Ab treated rats produced increased levels of IL-4 when stimulated with antigen in vitro and suggested a perturbation of the Th1/Th2 balance as the mechanism of anti-IL-18 Ab activity. This is most likely due to the synergy between IL-18 and IL-12. IFNγ drives the differentiation of Th1 cells and IFNγ stimulation by IL-12 is in fact a co-induction.[109] IL-12 by itself induces only small amounts of IFNγ. Costimulants such as ligation of CD3, mitogens or specific antigens apparently induce the production of IL-18 which synergises with IL-12 for IFNγ production. Therefore anti-IL-18 Ab removes the costimulant needed for IL-12 activity and results in a block in the afferent immune response.

IL-18$^{-/-}$ mice immunised with MOG_{35-55} do not develop EAE.[110] They do develop tEAE with encephalitogenic effector cells from wild type mice but do not generate effectors capable of transferring disease to either wild type or knock out mice. NK cell function was found to be compromised in the IL-18$^{-/-}$ mice and transfer of NK cells from RAG$^{-/-}$ mice rescued the defective Th1 responses in IL-18$^{-/-}$ mice and resulted in EAE induction.

It would appear therefore that IL-18 is capable of playing a significant role in the active induction of EAE through its role of synergising, or not, with IL-12 in the production of IFNγ and the subsequent differentiation of Th1 EAE effector cells. There is no good evidence for role for IL-18 in EAE pathology.

IL-18 and MS

The human IL-18 promoter has been cloned and screened and three single nucleotide polymorphisms were detected. Two of them were analyzed in 208 MS patients and 139 healthy controls, however, no significant differences were found.[111] IL-18 mRNA has been identified in MS demyelinating lesions[112] and IL-18 protein found in the CSF of meningitis patients but only a few MS patients along with normal controls.[113]

If it were to be established that exacerbations of MS are in fact associated with repeated rounds of activation of Th1 cells, then targeting of IL-18 might be considered a possible therapeutic approach.

Transforming Growth Factor β

TGFβ and EAE

With perhaps two exceptions evidence overwhelmingly supports TGFβ as a down-regulator of EAE as well as other autoimmune diseases. In acute EAE TGFγ mRNA appears first at peak disease and increases further during recovery.[114, 115] In chronic EAE in the DA rat it was noted that TGFβ (and IL-10) were absent during the entire course of disease, such absence perhaps accounting for or contributing to chronicity.[20]

Tolerance induction in EAE whether induced by encephalitogen in incomplete Freund's adjuvant,[116] oral or nasal instillation of antigen[117-120] or treatment with altered peptide ligand (APL)[121, 122] have all been linked to TGFβ production. Mice immunised with APL had higher levels of TGFβ mRNA, and lower levels of TNFα and IFNγ mRNA in the CNS tissue than mice immunised with normal peptide. Such mice were protected from clinical disease induced with normal peptide and protection with the APL was partially abrogated by treatment with anti-TGFβ Ab.[121]

Therapeutic approaches have supported a down-regulating role for TGFβ in EAE with anti-TGFβ Ab treatment exacerbating EAE.[123,124] Administration of TGFβ either directly[93, 125-127] or via genes[46,128] or cells[129,130] all inhibited clinical disease. Treatment of animals with compounds such as omega-6 fatty acids[131] or 1,25-dihydroxyvitamin D3[132] both of which inhibited or slowed progression of EAE was shown to be linked to the increased production of TGFβ.

Transgenic mice have been generated that over-express bioactive TGFβ1 in astrocytes.[133] These TGFβ1 transgenics showed an earlier onset of clinical symptoms, more severe disease and increased mononuclear cell infiltration in the spinal cords compared with wild type littermate controls with EAE. Hence, local expression of TGFβ within the CNS parenchyma appears to enhance immune cell infiltration and intensify the CNS impairment resulting from the autoimmune responses. Whether such enhancement results from the known chemotactic effect of TGFβ[134] or is in some way related to the strong up-regulation of extracellular matrix proteins such as laminin and fibronectin[133] remains to be determined.

TGFβ and MS

Gene association studies using separate polymorphic microsatellite markers for TGFβ1 and TGFβ2 were performed on 151 RR- or SP-MS patients, 104 PP-MS patients and 159 controls.[135] The data indicate that TGFβ1 and β2 genes are not loci influencing MS susceptibility. Using a candidate gene strategy in a study of polymorphic markers within or close to the TGFβ1 or β2 loci, He et al[98] found no evidence for linkage in two-point linkage analysis in 34 Swedish multiplex MS families, 147 sporadic MS patients and 95 healthy controls.

Numerous descriptive studies have been done showing TGFβ to be present in MS brain tissue[136,137] and to be increased in CSF, serum and PBMNC.[136,138,205,139] Comparisons have been made between MBP specific T cell clones from MS patients and healthy controls with respect to TGFβ and other cytokine production in an attempt to dissect differences between the two types of cells, which in point of fact occur at similar frequencies in the two populations. The general conclusions were that the presence of Th1 secreting autoreactive T cells in healthy individuals may be counterbalanced by the presence of cells secreting Th2 cytokines and by the augmented production of the immunosuppressive cytokine TGFβ, whereas in MS there is a decrease in these anti-inflammatory agents.

Studies of several potential therapeutics in MS have been followed with respect to their influence on TGFβ production. While IFNγ reduces the number and severity of MS relapses examination of the role that TGFβ might play in betterment of disease has led to conflicting results. The serum levels of TGFβ were significantly increased in patients with RR- and SP-MS compared with sex and age matched healthy controls.[140] Moreover, in RR-MS patients, the blood levels of the cytokine were further augmented either during relapses or, in a rapid but reversible fashion, by s.c. injection with IFNγ1b. On the other hand, the numbers of PBMNC spontaneously expressing TGFβ and IFNγ) mRNA were reported to be unaffected by IFNβ1b for either 3-6 weeks or 3-6 months treatment of RR-MS patients.[141]

Oral tolerance has been successfully used in models of EAE and linked to TGFβ production. Clinical trials of myelin feeding (containing both MBP and PLP) have also been done in MS. In such a trial the investigators found a marked increase in the relative frequencies of both, MBP- and PLP-specific TGFβ-secreting T cell lines, as compared to nontreated MS patients.[142] The tacit interpretation was that this increase may contribute to the beneficial effects of oral tolerance on disease. Unfortunately, as is now well known, the large clinical trial of oral myelin showed no beneficial effect on disease. What the increased numbers of TGFβ secreting cells or TGFβ protein in MS patients means in the context of pathology or protection is still unknown.

A single study of TGFβ as a therapy for MS has been done in a phase 1 trial of 11 patients with secondary progressive disease.[143] There was no change in expanded disability status scale or MRI lesions during treatment. Unfortunately, five patients had a reversible decline in the glomerular filtration rate sounding a caution about further clinical trials of TGFβ.

Interleukin-4

IL-4 and EAE

Low levels of IL-4 in the CNS and peripheral lymphoid tissue of rats, mice and marmosets have been reported by many investigators[20,40,144-147] but there has been no good correlation with disease severity, progression or remission.

Administering IL-4 by injection[46,148,149] or by over-expression in T cells[150] had little effect on actively induced EAE. IL-4 gene delivery into CNS, however, both delayed the onset of MOG35-55-EAE when injected prior to immunisation and ameliorated disease when given at clinical onset.[35, 46, 151] IL-4 delivery via MBP-reactive encephalitogenic T cells, transduced with a retroviral gene to express IL- 4 were also found to ameliorate rEAE[152] and to delay the onset and reduce the severity of MBP-EAE.[153]

EAE inhibition by IL-4 produced by a 'bystander' effect was nicely demonstrated by immunising mice with keyhole limpet hemocyanin (KLH) in incomplete Freund's adjuvant which generates Th2 memory cells that make IL-4 upon recognition of antigen. Immunising such mice some weeks later for EAE (guinea pig myelin and CFA) with or without inclusion of KLH in the inoculum led to significantly decreased disease expression in those mice receiving KLH.[154]

The importance of IL-4 in determining susceptibility to disease was demonstrated by Constantinescu et al[155] who showed that treating EAE resistant BALB/c mice with anti-IL-4 Ab at the time of immunisation resulted in disease induction. Similarly treatment of susceptible SJL/J mice with anti-IL-12 Ab protected against disease pointing to the cytokine milleu during immunisation as crucial in directing disease susceptibility or resistance. Finally, IL-4$^{-/-}$ C57BL/6 mice have been shown to be more susceptible to MOG_{35-55}-EAE than wild type litter mates.[150, 154, 156] The difference was not striking and IL-4$^{-/-}$ mice recovered in a normal fashion.[157] Interestingly knocking out the IL-4 gene on PL/J background had no effect on incidence, severity or duration of disease.[158]

IL-4 therefore would seem to be involved in both, the afferent and efferent arms of the immune response in EAE. In the afferent limb IL-4 can direct the immune response away from a cell-mediated type response to an antibody Th2 type response. In the effector phase IL-4 could act on resident microglia and recruited macrophages to suppress the production of IL-1, IL-6, IL-8 and TNFα while enhancing the production of IL-1Ra, all acting to limit disease. In those studies above where failure to alter disease by giving IL-4 was reported it is likely that such failure is due to the inability of delivering the cytokine at the right concentration to the right site at the right time. The importance of this is perhaps highlighted in the bystander study of Falcone and Bloom [154] where delivery of KLH along with the encephalitogen led to disease suppression. It would be interesting to determine if injection of KLH at a distant site from the encephalitogen has a similar effect. We think not.

IL-4 and MS

The human IL-4 gene is on chromosome 5, the murine gene maps to chromosome 11. A recent genetic study shows that the IL-4R variant R551 may influence the predisposition for PP-MS but does not represent a general genetic factor for MS susceptibility.[159]

PBMNC and myelin antigen-specific T cell clones and lines from MS patients demonstrate heterogeneity in cytokine secretion, typical for Th0 phenotype, rather than distinct Th1 or Th2 subsets.[160-162] This is perhaps why results examining IL-4 production by PBMNC from MS patients following mitogen or antigen stimulation vary from undetectable to very high.[163-165] In one case significantly lower percentages of IL-4-producing T cells were found in stable MS patients than in controls, and in active than in stable patients.[105]

A significant correlation between the number of cells present in the CSF and the number of contrast enhancing lesions was reported in a recent study on 40 MS patients.[166] These CSF cells consistently expressed TNFα, IFNγ and IL-10 but not IL-4. The lack of expression of IL-4 in all but two most stable patients raises the possibility that IL-4 could be a regulatory cytokine in MS. Also in the studies by another group the CSF of MS patients contained four- to eight fold more myelin antigen-reactive IL-4 (and IFNγ) expressing cells.[138]

IL-4 could certainly contribute to regulation of disease in MS but, if for example, it is important in the initiating event of the disease whereby it directs the type of immune response, then studies of existing MS patients will by nature fail to reveal its previous influence.

Interleukin-10

IL-10 and EAE

With a few exceptions the majority of descriptive studies concerning the presence or absence of IL-10 during the course of EAE in both, rats and mice, point to a contribution by IL-10 in the rapid and permanent recovery from clinical disease.[39,75,107,114,167,168]

An interesting study of Marmosets with CR-EAE described the presence of immunocytochemically detected IL-10 in cells with astrocyte like morphology.[147] Importantly, treatment with nerve growth factor that resulted in a decreased demyelination was accompanied by a significantly increased percentage of IL-10 positive cells per infiltrate, and there was a distinct expression of IL-10 at high levels by glial cells, not only in the inflammatory regions but also in normal-appearing CNS white matter.[169]

Administration of recombinant IL-10, systemically or mucosally (intranasally), during the induction phase of EAE was effective in suppressing acute actively induced EAE in rats and mice[170-172] and in preventing development of relapse in CR-EAE in DA rats.[171] Other routes of delivery (subcutaneous, intracranial) and times (during the effector phase) were ineffective[46,171] as was an attempt to inhibit tEAE in mice.[173] While these results support a role for IL-10 in regulation of the afferent immune response, others suggest a more global role.

PLP-specific T cell clones expressing IL-10 under control of the IL-2 promoter and therefore producing IL-10 upon antigen recognition in the CNS were able to reverse PLP-EAE either just prior to or after onset of clinical signs.[174,175] Transgenic mice expressing IL-10 under the control of CD2 promoter in their T cells were totally resistant to PLP-EAE[150] as were transgenic mice expressing IL-10 under the control of class II MHC promoter.[176] The former study also demonstrtated that such mice, though resistant to active EAE were capable of generating cells which could transfer disease to wild type mice suggesting the block was not at the afferent end of the immune response but in the CNS.[150]

In vivo expression of IL-10 by injection of a replication-defective adenovirus vector prevented completely the development of EAE in mice but only if the vector was inoculated into the CNS.[177] Peripheral inoculation led to high circulating levels of IL-10 but no protection from disease. IL-10 gene transfer during remission in a CR-EAE model also prevented subsequent relapses. All strong evidence for a regulatory role for IL-10 in disease expression.

Taking advantage of the observation that young male SJL/J mice develop predominately a Th2 response whereas females generate a Th1, Stohlman et al[178] demonstrated that male mice immunised with KLH generate cells which upon stimulation with cognate antigen make copious amounts of IL-10. Transfer of these cells along with female PLP-reactive T cells resulted in disease. However, if the recipients were challenged subcutaneously with KLH then no EAE developed demonstrating the nonspecific bystander effect of in vivo IL-10 production. The induction of high levels of IL-10 by immunisation with PLP conjugated to Ig also led to reduced disease severity, quicker recovery and no relapses in PLP- and SCH-immunised animals.[179]

Experiments inhibiting IL-10 with antibody treatment also support an important role for IL-10 in disease regulation. Monoclonal anti-IL-10 increased the incidence and the severity of relapses in a model of Staphylococcal enterotoxin B and TNF-induced EAE in AKR and BALB/c mice,[124] caused worsening of passively induced CR-EAE when given immediately before onset of signs in SJL/J mice[173] and reversed the Th2-mediated protection in female SJL/J mice.[178] Experiments with IL-$10^{-/-}$ mice also strongly support a down-regulating role for IL-10 (see section on IL-12).

IL-10 is produced mainly by lymphoid cells but also by cells in the pituitary, hypothalamic and neural tissues and therefore, like IL-1, can act as a regulator of the HPA[180] inducing increased corticosterone synthesis.

IL-10 and MS

PBMNC and myelin antigen-specific T cell clones upon mitogen (ConA and PHA) or antigen (PLP and MBP) activation, were found to produce more IL-10 when isolated from MS patients than control subjects and when isolated from MS patients in stable phase than in acute phase.[181-183] In a detailed study by Rieckmann et al[184] RR-MS patients were followed every four weeks for 12 months and in 85% of the 27 relapses, decreased IL-10 mRNA expression in PBMNC preceded clinical symptoms.

Therapeutic studies in MS have, as might be expected, revealed disparate results concerning the contribution made by IL-10 to clinical improvement. Chronic administration of IFNγ did not result in an up-regulation of IL-10[185] and in fact IL-10 levels were lower during IFNγ treatment [186] suggesting little role for IL-10. Copaxone treatment on the other hand, which was found to significantly reduce the mean annual relapse rate, was accompanied by an elevation of serum IL-10 levels.[187] Similarly, seven out of nine patients suffering an acute relapse of MS and treated with methylprednisolone for four days displayed increased PBMNC IL-10 mRNA expression as well as higher serum IL-10 concentration.[188] In the latter study it is perhaps not surprising that IL-10 was increased since corticosteroids directly increase IL-10 transcription.[189] This does not necessarily indicate that the anti-inflammatory effect of methylprednisolone is mediated by IL-10.

Interleukin-12

IL-12 and EAE

In actively induced acute EAE in Lewis rats and CR-EAE in DA rats, the IL-12 mRNA expression in CNS appears early and peaks at the height of clinical disease, consistent with a disease-promoting role for this cytokine.[20, 114, 168] The IL-12p40 expression in SJL/J mice with tEAE also paralleled the clinical course of EAE: an increase of about 30 times in the spinal cord before the onset of clinical signs, a sharp drop at the disease peak and a slight further reduction at recovery.[190]

Administration of IL-12 during the induction phase of MBP-EAE in Lewis rats not only exacerbated EAE, but administration up to one week after recovery from disease resulted in an immediate relapse.[191] These investigators suggested that IL-12 reactivates effector cells in the periphery rather than in the target tissue since giving IL-12 to rats recovered from tEAE, which early after recovery still have inflammatory infiltrates in their CNS, did not result in a second episode of disease. Administering IL-12 to recipients of EAE effector cells has been reported to enhance disease severity.[192] In this work IL-12 was given immediately following transfer, and in the context of the work by Smith et al we would interpret as IL-12 enhancing the proliferation and differentiation of the transferred donor cells in the periphery of the recipient rather than having any effect on the CNS target tissue.

IL-12 has been shown to be indispensible for the induction of EAE in that IL-12$^{-/-}$ mice are totally resistant to MBP-EAE.[193] In IFNγ$^{-/-}$ mice that are also highly susceptible to disease, EAE was ameliorated by anti-IL-12 Ab indicating that IL-12 promotes EAE by an IFNγ independent mechanism. This study also demonstrated a very important interaction between IL-12 and IL-10 in development of disease. IL-10$^{-/-}$ mice were found to be highly susceptible to EAE as compared to controls (also shown by Bettelli et al[150]). Treatment of wild type mice with anti-IL-12 Ab resulted in generation of antigen nonspecific T cells producing IL-10 which could counter-regulate the development of autoimmune effector cells normally driven by IL-12.

Further evidence of an important role for IL-12 in potentiating disease are studies in which treatment of immunised animals with anti-IL-12 Ab inhibited both acute and relapsing EAE.[192-197]

IL-12 and MS

A number of studies have described an increase in serum levels of IL-12 as well as IL-12-producing PBMNC in MS patients when compared to controls[198,199] although in some cases IL-12 was also elevated in other neurological disease.[200] Balashov et al[201] have shown a 7 and 16 fold increase in T cell receptor mediated IL-12 secretion in RR-MS patients in acute attack and SP-MS patients, respectively. Van Boxel et al[202] reported that both RR- and SP-MS patients had increased levels of IL-12p40 mRNA compared with controls during the development of active lesions and that in RR-MS the increase was before a relapse. The same group found the baseline levels of IL-12p35 mRNA to be lower in RR-MS patients responding to therapy with IFNγ than in nonresponders[203] and reported that the IL-12p35 level correctly predicted the clinical outcome ie, benefit or no benefit, in 81% of the 26 patients under investigation.

Measurement of IL-12 in CSF of RR-MS patients demonstrated an increased (up to 1,000-fold) compartmentalized release of the p40 subunit but not of the heterodimer p70. Release of IL-12p40 correlated with classic markers of CNS inflammation and was significantly increased in patients with gadolinium-enhancing plaques on MRI.[204]

Not all studies, however, show IL-12 correlated with disease activity. The antigen stimulated IL-12 production was reduced in MS patients compared with that in healthy controls.[181] Furthermore production of the metabolically active p70 heterodimer and the p40 chains (total IL-12) was increased in stable MS compared with that in acute MS and healthy controls. Finally PBMNC from untreated MS patients produced normal amounts of IL-12 p70 but significantly less free IL-12 p40 heavy chain than PBMNC from both healthy and disease controls.[205]

In assessing the above results the reader should be aware of a number of facts concerning the IL-12 molecule. IL-12p70 is a heterodimer composed of a p40 and a p35 subunit. Both subunits must be produced by the same cell to generate functional IL-12. The cells secrete a large excess of free p40 over p70. IL12p40 exists as both, homodimer and monomer, and the homodimer binds to the IL-12Rb1 chain with equal affinity to p70 and therefore can act as a natural antagonist of the IL-12p70 heterodimer.[206] The p40 homodimer can also act as a potent chemotactic molecule for macrophages.[207] Antagonist activity has been demonstrated for both mouse and human p40 although human p40 binds with a lower affinity to the IL-12 receptor than p70 and therefore requires higher concentrations to act as antagonist. Macrophage chemotaxis by human p40 has not been shown; however, human IL-12 has been shown to be chemotactic for NK cells.[208] p40 homodimer has also been shown to induce iNOS and hence NO production in microglial cells and peritoneal macrophages,[209] a potentially important observation in the context of MS.

In the context of EAE and MS a number of scenarios can be described depending on the concentrations of the various IL-12 components. One: IL-12 accumulates in excess resulting in binding to its receptor with the production of IFNγ which along with IL-12 then promotes the differentiation and expansion of Th1 cells. This could be considered acute inflammation. Two: p40 homodimer is produced in excess resulting in competitive inhibition of IL-12 for receptor binding, decreased IFNγ production and little Th1 expansion. This will also, however, lead to accumulation of macrophages, the result of which may be chronic inflammation. The work by Tanuma et al[39] did in fact describe increased IL-12p40 and demyelination in the CNS of rats with CR-EAE which was accompanied by an increased number of macrophages. Three: IL-12 and p40 are both produced in amounts such that the p40 homodimer drives accumulation of macrophages and IL-12 binds to its receptor and promotes the production of IFNγ. This in turn activates the macrophages with increased NO production leading to apoptotic cell death of effector cells and down-regulation of the immune response. This latter scenario would result in termination of both the acute and chronic inflammation. These three outcomes, all a function of the ratio between the IL-12 components, could occur in the periphery during the afferent arm of the response or in the target tissue during the effector phase.

Concluding Observations

Where to from here? Can we target cytokines therapeutically in the context of CNS inflammation and how do we do it? We suggest the following: first one needs to determine which stage of the inflammatory process is to be addressed. Are we attempting to stop the initiation of disease (currently not feasible in MS), prevent relapses, or treat ongoing chronic disease? These concerns in turn will dictate which cytokine we target, where we deliver it and perhaps how much we deliver.

The next step should be to define a cytokine thought to be central and then ask which other molecules are its agonists and antagonists. By way of example, IL-12 is certainly a central cytokine in driving a Th1 cell response, the type of response essential for EAE and perhaps MS. IL-18 is a costimulant with IL-12 for optimal IFNγ production and Th1 differentiation. If one is speaking of naive T cells then IL-4 becomes important because the Th1-inducing effect of IL-12 and IL-18 can be overcome by the concomitant presence of IL-4.[210] IL-12 p40 homodimer can antagonise IL-12 or may be agonistic via its ability to recruit macrophages. IL-10 is a dominant inhibitor of IL-12. If we were fortunate enough one day to be able to identify the autoantigen(s) and most importantly the people at risk of developing MS, then inhibiting IL-12/IL-18 and/or administering IL-10/IL-4 or p40 systemically might be desirable. Until then, targeting RR-MS with this same strategy may be effective if we assume relapses are due to recruitment of new effectors from a naive precursor pool (epitope spreading). If epitope spreading occurs within the CNS (still problematical), then we should direct our IL-10/IL-4 or p40 to the target tissue. Progressive disease, if inflammatory, might best be treated by delivering IL-10 (to antagonise IL-12) or IL-4 (to induce IL-1Ra) or p40 to the CNS with subsequent down-regulation of inflammation.

Thus inflammation may be attacked by delivering cytokines or antagonists at various times to various places. As a further illustration, altering the dose of a given cytokine may dramatically alter its effect. Tarrant et al[211] treated mice immunised for experimental autoimmune uveitis (a Th1 cell mediated disease) with high doses of IL-12. Instead of augmenting disease as anticipated it was ameliorated, apparently due to the hyper-induction of IFNγ which led to the production of NO which down-regulated the response by inducing apoptosis in effector cells.

Chasing cytokine function 'down-stream' as this latter and other studies[90] illustrate could result in identification of molecules, such as NO, which might then be targeted with molecules other than cytokines, e.g., NO donors, thus leaving the complex society of cytokines undisturbed. This is certainly one way forward.

A Ready Reckoner to Cytokine Function in Relation to CNS Inflammation

- IL-1 is pro-inflammatory yet produced locally can stimulate the HPA axis resulting in corticosterone production
- IL-6 is pro-inflammatory and essential for EAE induction. It can also stimulate the initial production of IL-4 by $CD4^+$ T cells (and can therefore deviate the response)
- IL-4 is the single most important cytokine in the differentiation of Th2 cells
- IL-12 is indispensible in differentiation of Th1 cells and induces IFNγ
- IL-12 is essential for EAE induction
- IL-18 is necessary as a costimulant for optimal IL-12 induction of IFNγ
- IFNγ is not essential for EAE induction but is essential for disease regulation
- IFNγ activates macrophages (dendritic cells) to make IL-12
- IFNγ is a major inducer of NO which can act as a down-regulator of immune responses
- IL-10 is a dominant inhibitor of IL-12 and can also influence corticosterone synthesis
- TNFα can suppress CNS inflammation but one of its receptors (p55) is essential in promoting disease
- TGFβ can alone or with other factors initiate inflammation but its major role is in resolution of inflammatory damage.

Acknowledgments

This work was supported by grants from the National Health and Medical Research Council (NH&MRC) of Australia (to D.O.W.), Multiple Sclerosis Australia (MSA) (to D.O.W. and M.A.S.) and the Canberra Hospital Private Practice Fund (to D.O.W. and M.A.S.). D.O.W. is a Senior Research Fellow of the NH&MRC

References

1. Mcdonald IW, Thompson AJ. How many kinds of multiple scelrosis are there? In: Thompson AJ, Polman C, Hohlfeld R, eds. Multiple sclerosis. Clinical Challenges and Controversies. London: Martin Dunitz Ltd., 1997:35-42.
2. Dean G. Annual incidence, prevalence, and mortality of multiple sclerosis in white South African-born and in white immigrants to South Africa. Br Med J 1967; 2:724-730.
3. Hammond SR, McCloud JG, Milligan KS et al. The epidemiology of multiple sclerosis in three Australian cities: Perth, Newcastle and Hobart. Brain 1987; 111:1-25.
4. Sadnovick AD, Baird PA, Ward RH. Multiple sclerosis: Updated risks for relatives. Am J Med Gent 1988; 29:533-541.
5. Mumford CJ, Wood NW, Keller-Wood HF et al. The British Isles survey of multiple sclerosis in twins. Neurology 1994; 44:11-15.
6. Ewing CAAB. Insights into the aetiology and pathogenesis of multiple sclerosis. Immunol Cell Biol 1998; 76:47-54.
7. Raine CS, Barnett LB, Brown A et al. Neuropathology of experimental allergic encephalomyelitis in inbred strains of mice. Lab Invest 1980; 43:150-157.
8. Ludowyk PA, Hughes W, Hugh A et al. Astrocytic hypertrophy: An important pathological feature of chronic experimental autoimmune encephalomyelitis in aged rats. J Neuroimmunol 1993; 48:121-134.
9. Kies MW, Murphy JB, Alvord EC, Jr. Studies of the encephalitogenic factor in guinea pig central nervous system. In: Floch J, ed. Chemical Pathology of the Nervous System. New York: Pergamon Press, 1961:197-204.
10. Kono DH, Urban JL, Horvath DG, Ando DG, Saavedra AG et al. Two minor determinants of myeloin basic protein induce experimental autoimmune encephalomyelitis in SJL/J mice. J Exp Med 1988; 168:213-227.
11. Tuohy VK, Lu Z, Sobel RA, Laursen RA, Lees MB. Identification of an encephalitogenic determinant of myeli proteolipid protein for SJL/J mice. J Immunol 1989; 142:1523-1527.
12. Greer JM, Kuchroo VK, Sobel RA, Lees MB. Identification and characterisation of a second encephalitogenic determinant of myelin proteolipid protein (residues 178-191) for SJL mice. J Immunol 1992; 149:783-788.
13. Mendel I, Kelero de Rosbo N, Ben-Nun A. A myelin oligodendrocyte glycoprotein peptide induces typical chronic experimental autoimmune encephalomyelitis in H2b mice: Fine specificity and T cell receptor V beta expression of encephalitogenic T cells. Eur J Immunol 1995; 25:1951-1959.
14. Bernard CCA, Johns TG, Salvin A, Ichikawa M, Ewing C et al. Myelin oligodendrocyte glycoprotein: A novel candidate autoantigen in multiple sclerosis. J Mol Med 1997; 75:77-88.
15. Hohlfeld R. Biotechnological agents for the immunotherapy of multiple sclerosis. Principles, problems and perspectives. Brain 1997; 120:865-916.
16. Chofflon M. Recombinant human interferon beta in relapsing-remitting multiple sclerosis: A review of the major clinical trials. Eur J Neurol 2000; 7:369-380.
17. Karp CL, Biron CA, Irani DN. Interferon beta in multiple sclerosis: Is IL-12 suppression the key? Immunol Today 2000; 21:24-28.
18. Comi G. Why treat early multiple sclerosis patients? Curr Opin Neurol 2000; 13:235-240.
19. Weinstock-Guttman B, Jacobs LD. What is new in the treatment of multiple sclerosis? Drugs 2000; 59:401-410.
20. Issazadeh S, Lorentzen JC, Mustafa MI et al. Cytokines in relapsing experimental autoimmune encephalomyelitis in DA rats: persistent mRNA expression of proinflammatory cytokines and absent expression of interleukin-10 and transforming growth factor-beta. J Neuroimmunol 1996; 69:103-115.
21. Bauer J, Berkenbosch F, Van Dam AM et al. Demonstration of interleukin-1 beta in Lewis rat brain during experimental allergic encephalomyelitis by immunocytochemistry at the light and ultrastructural level. J Neuroimmunol 1993; 48:13-21.

22. Huitinga I, Schmidt ED, Van Der Cammen MJ et al. Priming with interleukin-1beta suppresses experimental allergic encephalomyelitis in the Lewis rat. J Neuroendocrinol 2000; 12:1186-1193.
23. Schmidt ED, Janszen AW, Wouterlood FG et al. Interleukin-1-induced long-lasting changes in hypothalamic corticotropin-releasing hormone (CRH)—Neurons and hyperresponsiveness of the hypothalamus-pituitary-adrenal axis. J Neurosci 1995; 15:7417-7426.
24. Jacobs CA, Baker PE, Roux ER et al. Experimental autoimmune encephalomyelitis is exacerbated by IL-1 alpha and suppressed by soluble IL-1 receptor. J Immunol 1991; 146:2983-2989.
25. Martin D, Near SL. Protective effect of the interleukin-1 receptor antagonist (IL-1ra) on experimental allergic encephalomyelitis in rats. J Neuroimmunol 1995; 61:241-245.
26. Wiemann B, Van GY, Danilenko DM et al. Combined treatment of acute EAE in Lewis rats with TNF-binding protein and interleukin-1 receptor antagonist. Exp Neurol 1998; 149:455-463.
27. Schiffenbauer J, Streit WJ, Butfiloski E et al. The induction of EAE is only partially dependent on TNF receptor signaling but requires the IL-1 type I receptor. Clin Immunol 2000; 95:117-123.
28. Schrijver HM, Crusius JB, Uitdehaag BM et al. Association of interleukin-1beta and interleukin-1 receptor antagonist genes with disease severity in MS. Neurology 1999; 52:595-599.
29. McDonnell GV, Kirk CW, Hawkins SA et al. An evaluation of interleukin genes fails to identify clear susceptibility loci for multiple sclerosis [see comments]. J Neurol Sci 2000; 176:4-12.
30. Kantarci OH, Atkinson EJ, Hebrink DD et al. Association of two variants in IL-1beta and IL-1 receptor antagonist genes with multiple sclerosis. J Neuroimmunol 2000; 106:220-227.
31. Wandinger KP, Wessel K, Trillenberg P et al. Effect of high-dose methylprednisolone administration on immune functions in multiple sclerosis patients. Acta Neurol Scand 1998; 97:359-365.
32. Sciacca FL, Canal N, Edoardo Grimaldi LM. Induction of IL-1 receptor antagonist by interferon beta: Implication for the treatment of multiple sclerosis. J Neurovirol 2000; 6 Suppl 2:S33-37.
33. Perini P, Tiberio M, Sivieri S et al. Interleukin-1 receptor antagonist, soluble tumor necrosis factor-alpha receptor type I and II, and soluble E-selectin serum levels in multiple sclerosis patients receiving weekly intramuscular injections of interferon-beta1a. Eur Cytokine Netw 2000; 11:81-86.
34. Dinarello CA. Biologic basis for interleukin-1 in disease. Blood 1996; 87:2095-2147.
35. Furlan R, Filippi M, Bergami A et al. Peripheral levels of caspase-1 mRNA correlate with disease activity in patients with multiple sclerosis: A preliminary study. J Neurol Neurosurg Psychiatry 1999; 67:785-788.
36. Schnell L, Fearn S, Schwab ME et al. Cytokine-induced acute inflammation in the brain and spinal cord. J Neuropathol Exp Neurol 1999; 58:245-254.
37. Berger T, Weerth S, Kojima K et al. Experimental autoimmune encephalomyelitis: The antigen specificity of T lymphocytes determines the topography of lesions in the central and peripheral nervous system. Lab Invest 1997; 76:355-364.
38. Renno T, Krakowski M, Piccirillo C et al. TNFα expression by resident microglia and infiltrating leukocytes in the central nervous system of mice with experimental allergic encephalomyelitis. J Immunol 1995; 154:944-953.
39. Tanuma N, Shin T, Matsumoto Y. Characterization of acute versus chronic relapsing autoimmune encephalomyelitis in DA rats. J Neuroimmunol 2000; 108:171-180.
40. Juedes AE, Hjelmstrom P, Bergman CM et al. Kinetics and cellular origin of cytokines in the central nervous system: Insight into mechanisms of myelin oligodendrocyte glycoprotein-induced experimental autoimmune encephalomyelitis. J Immunol 2000; 164:419-426.
41. Ruddle NH, Bergman CM, McGrath KM et al. An antibody to lymphotoxin and tumour necrosis factor prevents transfer of experimental allergic encephalomyelitis. J Exp Med 1990; 172:1193-1200.
42. Selmaj K, Raine CS, Cross AH. Anti-tumour necrosis factor therapy abrogates autoimmune demyelination. Ann. Neurol 1991; 30:694-697.
43. Willenborg DO, Fordham SA, Cowden WB et al. Cytokines and murine autoimmune encephalomyelitis: Inhibition or enhancement of disease with antibodies to select cytokines, or by delivery of exogeneous cytokines using a recombinant vaccinia virus system. Scand J Immunol 1995; 41:31-41.
44. Selmaj K, Paplerz W, Glabinski A et al. Prevention of chronic relapsing experimental autoimmune encephalomyelitis by soluble tumour necrosis factor receptor 1. J Neuroimmunol 1995; 56:135-141.
45. Klinkert WEF, Kojima K, Lesslaur W et al.TNFα receptor fusion protein prevents experimental autoimmune encephalomyelitis and demyelination in Lewis rats: An overview. J Neuroimmunol 1997; 72:163-168.
46. Croxford JL, Triantaphyllopoulos K, Podhajcer OL et al. Cytokine gene therapy in experimental allergic encephalomyelitis by injection of plasmid DNA-cationic liposome complex into the central nervous system. J Immunol 1998; 160:5181-5187.
47. Karin N, Mitchell DJ, Brocke S et al. Reversal of experimental autoimmune encephalomyelitis by a soluble peptide variant of a myelin basic protein epitope: T cell receptor antagonism and reduction of interferon gamma and tumor necrosis factor alpha production. J Exp Med 1994; 180:2227-2237.

48. Brocke SK, Gijbels K, Allegretta M et al. Treatment of experimental encephalomyelitis with a peptide analogue of myelin basic protein. Nature 1996; 379:343-345.
49. Sommer N, Loschmann PA, Northoff GH et al. The anti-depressant Rolipram suppresses cytokine production and prevents autoimmune encephalomyelitis. Nat Med 1995; 1:244-248.
50. Achiron A, Miron S, Lavie V et al. Dexanabinol (HU-211) effect on experimental autoimmune encephalomyelitis: Implications for the treatment of acute relapses of multiple sclerosis. J Neuroimmunol 2000; 102:26-31.
51. Grassin M, Brochet B, Coussemacq M et al. Controlled therapeutic trials of pentoxifylline in relapsing-experimental auto-immune encephalomyelitis. Acta Neurol Scand 1998; 97:404-408.
52. Korner H, Riminton DS, Strickland DH et al. Critical points of tumour necrosis factor action in central nervous system autoimmune inflammation defined by gene targeting. J Exp Med 1997; 186:1585-1590.
53. Liu J, Marino MW, Wong G et al. TNF is a potent anti-inflammatory cytokine in autoimmune-mediated demyelination. Nat Med 1998; 4:78-82.
54. Frei K, Eugster H-P, Bopst M et al. Tumour necrosis factor α and lymphotoxin α are not required for induction of autoimmune encephalomyelitis. J Exp Med 1997; 185:2177-2182.
55. Suen WE, Bergman CM, Hjelstrom P et al. A critical role for lymphotoxin in experimental allergic encephalomyelitis. J Exp Med 1997; 186:1233-1240.
56. DeTogni P, Goellner J, Ruddle NH et al. Abnormal develoment of peripheral lymphoid organs in mice deficient in lymphotoxin. Science 1994; 264:703-707.
57. Riminton DS, Korner H, Strickland DH et al. Challenging cytokine redundancy: Inflammatory cell movement and clinical course of experimental autoimmune encephalomyelitis are normal in lymphotoxin-deficient mice, but not tumour necrosis factor-deficient, mice. J Exp Med 1998; 187:1517-1528.
58. Bazzoni F, Beutler B. The tumor necrosis factor ligand and receptor families. N Engl J Med 1996; 334:1717-1725.
59. Mariathasan S, Matsumoto M, Baranyay F et al. Absence of lymph nodes in lymphotoxin-alpha (LTα)-deficient mice is due to abnormal organ development not defective lymphocyte migration. J Inflamm 1995; 45:72-78.
60. Willenborg DO, Fordham SA, O'Brien NC et al. Tumour necrosis factor-alpha and lymphotoxin-alpha in the pathollgy of experimental autoimmune encephalomyelitis: Is either one responsible or is there another ligand mediating disease? Res Immunol 1998; 149:805-810.
61. Eugster HP, Frei K, Bachmann R et al. Severity of symptoms and demyelination in MOG-induced EAE depends on TNFR-1. Eur J Immunol 1999; 29:626-632.
62. Probert L, Eugster HP, Akassoglou K et al. TNFR1 signalling is critical for the development of demyelination and the limitation of T-cell responses during immune-mediated CNS disease. Brain 2000; 123:2005-2019.
63. Kassiotis G, Kollias G. Uncoupling the proinflammatory from the immunosuppressive properties of tumor necrosis factor (TNF) at the p55 receptor level: Implications for pathogenesis and therapy of autoimmune demyelination. J Exp Med 2001; 193:427-434.
64. Suvannavejh GC, Lee HO, Padilla J et al. Divergent roles for p55 and p75 tumor necrosis factor receptors in the pathogenesis of MOG(35-55)-induced experimental autoimmune encephalomyelitis Cell Immunol 2000; 205:24-33.
65. Sharief MK, Hentgre R. Assocation between tumor necrosis factor and disease progression in patients with multiple sclerosis. N Eng J Med 1991; 325:467-472.
66. Maimone D, Gregory S, Arnason BG. Cytokine levels in the cerebrospinal fluid and serum of patients with multiple sclerosis. J Neuroimmunol 1991; 32:67-74.
67. Raine CS, Bonetti B, Cannella B. Multiple sclerosis: Expression of molecules of the tumor necrosis factor ligand and receptor families in relationship to the demyelinated plaque. Rev Neurol (Paris) 1998; 154:577-585.
68. Debruyne J, Philippe J, Dereuck J et al. Relapse markers in multiple sclerosis: are in vitro cytokine production changes reflected by circulatory T-cell phenotype alterations? Mult Scler 1998; 4:193-197.
69. Hautecoeur P, Forzy G, Gallois P et al. Variations of IL2, IL6, TNF alpha plasmatic levels in relapsing remitting multiple sclerosis. Acta Neurol Belg 1997; 97:240-243.
70. Franciotta D, Bergamaschi R, Martino G et al. Tumor necrosis factor-alpha and its soluble receptors in plasma and cerebrospinal fluid of multiple sclerosis patients treated with methylprednisolone. Eur Cytokine Netw 1999; 10:431-436.
71. Suzumura A, Nakamuro T, Tamaru T et al. Drop in relapse rate of MS by combination therapy of three different phosphodiesterase inhibitors. Mult Scler 2000; 6:56-58.
72. Myers LW, Ellison GW, Merrill JE et al. Pentoxifylline is not a promising treatment for multiple sclerosis in progression phase. Neurology 1998; 51:1483-1486.

73. group LM. TNF neutralization in MS: Results of a randomized, placebo-controlled multicenter study. The Lenercept Multiple sclerosis Study Group and The University of British Columbia MS/ MRI Analysis Group [see comments]. Neurology 1999; 53:457-465.
74. Mendel I, Katz A, Kozak N et al. Interleukin-6 functions in autoimmune encephalomyelitis: A study in gene-targeted mice. Eur J Immunol 1998; 28:1727-1737.
75. Okuda Y, Sakoda S, Bernard CC et al. IL-6-deficient mice are resistant to the induction of experimental autoimmune encephalomyelitis provoked by myelin oligodendrocyte glycoprotein. Int Immunol 1998; 10:703-708.
76. Samoilova EB, Horton JL, Hilliard B et al. IL-6-deficient mice are resistant to experimental autoimmune encephalomyelitis: Roles of IL-6 in the activation and differentiation of autoreactive T cells. J Immunol 1998; 161:6480-6486.
77. Maimone D, Guazzi GC, Annunziata P. IL-6 detection in multiple sclerosis brain. J Neurol Sci 1997; 146:59-65.
78. Baranzini SE, Elfstrom C, Chang SY et al. Transcriptional analysis of multiple sclerosis brain lesions reveals a complex pattern of cytokine expression J Immunol 2000; 165:6576-6582.
79. Hauser SL, Doolittle TH, Lincoln R et al. Cytokine accumulations in CSF of multiple sclerosis patients: frequent detection of interleukin-1 and tumor necrosis factor but not interleukin-6. Neurology 1990; 40:1735-1739.
80. Navikas V, Matusevicius D, Soderstrom M et al. Increased interleukin-6 mRNA expression in blood and cerebrospinal fluid mononuclear cells in multiple sclerosis. J Neuroimmunol 1996; 64:63-69.
81. Schonrock LM, Gawlowski G, Bruck W. Interleukin-6 expression in human multiple sclerosis lesions Neurosci Lett 2000; 294:45-48.
82. Billiau A, Heremans H, Vandekerckhove F et al. Enhancement of experimental allergic encephalomyelitis in mice by antibodies against IFN-gamma. J Immunol 1988; 140:1506-1510.
83. Duong TT, St. Louis J, Gilbert JJ et al. Effect of anti-interferon-gamma and anti-interleukin-2 monoclonal antibody treatment on the development of actively and passively induced experimental allergic encephalomyelitis in the SJL/J mouse. J Neuroimmunol 1992; 36:105-115.
84. Heremans H, Dillen C, Groenen M et al. Chronic relapsing experimental autoimmune encephalomyelitis (CREAE) in mice: Enhancement by monoclonal antibodies against interferon-gamma. Eur J Immunol 1996; 26:2393-2398.
85. Lublin FD, Knobler RL, Kalman B et al. Monoclonal anti-gamma interferon antibodies enhance experimental allergic encephalomyelitis. Autoimmunity 1993; 16:267-274.
86. Voorthuis JAC, Uitdehaag BMJ, deGroot CJA et al. Suppression of experimental allergic encephalomyelitis by intraventricular administration of interferon-gamma. Clin Exp Immunol 1990; 81:183-188
87. Willenborg DO, Fordham S, Bernard CC et al. IFN-gamma plays a critical down-regulatory role in the induction and effector phase of myelin oligodendrocyte glycoprotein-induced autoimmune encephalomyelitis. J Immunol 1996; 157:3223-3227.
88. Ferber IA, Brocke S, Taylor-Edwards C et al. Mice with a disrupted IFN-gamma gene are susceptible to the induction of experimental autoimmune encephalomyelitis (EAE). J Immunol 1996; 156:5-7.
89. Krakowski M, Owens T Interferon-gamma confers resistance to experimental allergic encephalomyelitis. Eur J Immunol 1996; 26:1641-1646.
90. Willenborg DO, Fordham SA, Staykova MA et al. IFN-gamma is critical to the control of murine autoimmune encephalomyelitis and regulates both in the periphery and in the target tissue: A possible role for nitric oxide. J Immunol 1999; 163:5278-5286.
91. Xiao BG, Huang YM, Xu LY et al. Mechanisms of recovery from experimental allergic encephalomyelitis induced with myelin basic protein peptide 68-86 in Lewis rats: A role for dendritic cells in inducing apoptosis of CD4+ T cells. J Neuroimmunol 1999; 97:25-36.
92. Chu CQ, Wittmer S, Dalton DK. Failure to suppress the expansion of the activated CD4 T cell population in interferon gamma-deficient mice leads to exacerbation of experimental autoimmune encephalomyelitis. J Exp Med 2000; 192:123-128.
93. Jin YX, Xu LY, Guo H et al. TGF-beta1 inhibits protracted-relapsing experimental autoimmune encephalomyelitis by activating dendritic cells. J Autoimmun 2000; 14:213-220.
94. O'Brien NC, Charlton B, Cowden WB et al. Nitric oxide plays a critical role in the recovery of Lewis rats from experimental autoimmune encephalomyelitis and the maintenance of resistance to reinduction. J Immunol 1999; 163:6841-6847.
95. Willenborg DO, Staykova MA, Cowden WB. Our shifting understanding of the role of nitric oxide in autoimmune encephalomyelitis: A review. J Neuroimmunol 1999; 100:21-35.

96. Tanuma N, Shin T, Kogure K et al. Differential role of TNF-alpha and IFN-gamma in the brain of rats with chronic relapsing autoimmune encephalomyelitis. J Neuroimmunol 1999; 96:73-79.
97. Wansen K, Pastinen T, Kuokkanen S et al. Immune system genes in multiple sclerosis: Genetic association and linkage analyses on TCR beta, IGH, IFN-gamma and IL-1ra/IL-1 beta loci. J Neuroimmunol 1997; 79:29-36.
98. He B, Xu C, Yang B et al. Linkage and association analysis of genes encoding cytokines and myelin proteins in multiple sclerosis. J Neuroimmunol 1998; 86:13-19.
99. Petereit HF, Richter N, Pukrop R et al. Interferon gamma production in blood lymphocytes correlates with disability score in multiple sclerosis patients. Mult Scler 2000; 6:19-23.
100. Dettke M, Scheidt P, Prange H et al. Correlation between interferon production and clinical disease activity in patients with multiple sclerosis. J Clin Immunol 1997; 17:293-300.
101. Balashov KE, Olek MJ, Smith DR et al. Seasonal variation of interferon-gamma production in progressive multiple sclerosis. Ann Neurol 1998; 44:824-828.
102. Sibley WA, Bamford CR, Clark K. Clinical viral infections and multiple sclerosis. Lancet 1985; 1:1313-1315.
103. Bever CT, Jr., Panitch HS, Levy HB et al. Gamma-interferon induction in patients with chronic progressive MS. Neurology 1991; 41:1124-1127.
104. van Oosten BW, Barkhof F, Scholten PE et al. Increased production of tumor necrosis factor alpha, and not of interferon gamma, preceding disease activity in patients with multiple sclerosis. Arch Neurol 1998; 55:793-798.
105. Franciotta D, Zardini E, Bergamaschi R et al. Interferon-γ and interleukin-4-producing T cells in peripheral blood of multiple sclerosis patients. Eur Cytokine Netw 2000; 11:677-681.
106. Panitch HS, Haley AS, Hirsch RI, Johnson KP. Exacerbations of multiple sclerosis in patients treated with gamma interferon. The Lancet 1987; ii:893-895.
107. Jander S, Stoll G. Differential induction of interleukin-12, interleukin-18, and interleukin-1beta converting enzyme mRNA in experimental autoimmune encephalomyelitis of the Lewis rat. J Neuroimmunol 1998; 91:93-99.
108. Wildbaum G, Youssef S, Grabie N et al. Neutralizing antibodies to IFN-gamma-inducing factor prevent experimental autoimmune encephalomyelitis. J Immunol 1998; 161:6368-6374.
109. Dinarello CA. IL-18: A Th1-inducing, proinflammatory cytokine and new member of the IL-1 family. J Allergy Clin Immunol 1999; 103:11-22.
110. Shi FD, Takeda K, Akira S et al. IL-18 directs autoreactive T cells and promotes autodestruction in the central nervous system via induction of IFN-gamma by NK cells. J Immunol 2000; 165:3099-3104.
111. Giedraitis V, He B, Huang W et al. Cloning and mutation analysis of the human IL-18 promoter: A possible role of polymorphisms in expression regulation [In Process Citation]. J Neuroimmunol 2001; 112:146-152.
112. Balashov KE, Rottman JB, Weiner HL et al. CCR5(+) and CXCR3(+) T cells are increased in multiple sclerosis and their ligands MIP-1alpha and IP-10 are expressed in demyelinating brain lesions. Proc Natl Acad Sci USA 1999; 96:6873-6878.
113. Fassbender K, Mielke O, Bertsch T et al. Interferon-gamma-inducing factor (IL-18) and interferon-gamma in inflammatory CNS diseases. Neurology 1999; 53:1104-1106.
114. Issazadeh S, Mustafa M, Ljungdahl A et al. Interferon gamma, interleukin 4 and transforming growth factor beta in experimental autoimmune encephalomyelitis in Lewis rats: Dynamics of cellular mRNA expression in the central nervous system and lymphoid cells. J Neurosci Res 1995; 40:579-590.
115. Issazadeh S, Navikas V, Schaub M et al. Kinetics of expression of costimulatory molecules and their ligands in murine relapsing experimental autoimmune encephalomyelitis in vivo. J Immunol 1998; 161:1104-1112.
116. Karpus WJ, Swanborg RH. CD4+ suppressor cells inhibit the function of effector cells of experimental autoimmune encephalomyelitis through a mechanism involving transforming growth factor-beta. J Immunol 1991; 146:1163-1168.
117. Chen Y, Inobe J, Kuchroo VK et al. Oral tolerance in myelin basic protein T-cell receptor transgenic mice: suppression of autoimmune encephalomyelitis and dose-dependent induction of regulatory cells. Proc Natl Acad Sci USA 1996; 93:388-391.
118. Fukaura H, Kent SC, Pietrusewicz MJ et al. Induction of circulating myelin basic protein and proteolipid protein-specific transforming growth factor-beta1-secreting Th3 T cells by oral administration of myelin in multiple sclerosis patients. J Clin Invest 1996; 98:70-77.
119. Bebo BF, Jr., Adlard K, Schuster JC et al. Gender differences in protection from EAE induced by oral tolerance with a peptide analogue of MBP-Ac1-11. J Neurosci Res 1999; 55:432-440.

120. Bai XF, Shi FD, Xiao BG et al. Nasal administration of myelin basic protein prevents relapsing experimental autoimmune encephalomyelitis in DA rats by activating regulatory cells expressing IL-4 and TGF-beta mRNA. J Neuroimmunol 1997; 80:65-75.
121. Santambrogio L, Lees MB, Sobel RA. Altered peptide ligand modulation of experimental allergic encephalomyelitis: immune responses within the CNS. J Neuroimmunol 1998; 81:1-13.
122. Young DA, Lowe LD, Booth SS et al. IL-4, IL-10, IL-13, and TGF-beta from an altered peptide ligand-specific Th2 cell clone down-regulate adoptive transfer of experimental autoimmune encephalomyelitis. J Immunol 2000; 164:3563-3572.
123. Johns LD, Sriram S. Experimental allergic encephalomyelitis: neutralizing antibody to TGF beta 1 enhances the clinical severity of the disease. J Neuroimmunol 1993; 47:1-7.
124. Crisi GM, Santambrogio L, Hochwald GM et al. Staphylococcal enterotoxin B and tumor-necrosis factor-alpha-induced relapses of experimental allergic encephalomyelitis: Protection by transforming growth factor-beta and interleukin-10. Eur J Immunol 1995; 25:3035-3040.
125. Johns LD, Flanders KC, Ranges GE et al. Successful treatment of experimental allergic encephalomyelitis with transforming growth factor-beta 1. J Immunol 1991; 147:1792-1796.
126. Racke MK, Dhib-Jalbut S, Cannella B et al. Prevention and treatment of chronic relapsing experimental allergic encephalomyelitis by transforming growth factor-beta 1. J Immunol 1991; 146:3012-3017.
127. Racke MK, Sriram S, Carlino J et al. Long-term treatment of chronic relapsing experimental allergic encephalomyelitis by transforming growth factor-beta 2. J Neuroimmunol 1993; 46:175-183.
128. Piccirillo CA, Prud'homme GJ. Prevention of experimental allergic encephalomyelitis by intramuscular gene transfer with cytokine-encoding plasmid vectors. Hum Gene Ther 1999; 10:1915-1922.
129. Inobe JI, Chen Y, Weiner HL. In vivo administration of IL-4 induces TGF-beta-producing cells and protects animals from experimental autoimmune encephalomyelitis. Ann NY Acad Sci 1996; 778:390-392.
130. Thorbecke GJ, Umetsu DT, deKruyff RH et al. When engineered to produce latent TGF-beta1, antigen specific T cells down regulate Th1 cell-mediated autoimmune and Th2 cell-mediated allergic inflammatory processes. Cytokine Growth Factor Rev 2000; 11:89-96.
131. Harbige LS, Layward L, Morris-Downes MM et al. The protective effects of omega-6 fatty acids in experimental autoimmune encephalomyelitis (EAE) in relation to transforming growth factor-beta 1 (TGF-beta1) up-regulation and increased prostaglandin E2 (PGE2) production Clin Exp Immunol 2000; 122:445-452.
132. Cantorna MT, Woodward WD, Hayes CE et al. 1,25-dihydroxyvitamin D3 is a positive regulator for the two anti-encephalitogenic cytokines TGF-beta 1 and IL-4. J Immunol 1998; 160:5314-5319.
133. Wyss-Coray T, Borrow P, Brooker MJ et al. Astroglial overproduction of TGF-beta 1 enhances inflammatory central nervous system disease in transgenic mice. J Neuroimmunol 1997; 77:45-50.
134. Wahl SM, Allen JB, McCartney-Francis N et al. Macrophage- and astrocyte-derived transforming growth factor beta as a mediator of central nervous system dysfunction in acquired immune deficiency syndrome. J Exp Med 1991; 173:981-991.
135. McDonnell GV, Kirk CW, Hawkins SA et al. Lack of association of transforming growth factor (TGF)-beta 1 and beta 2 gene polymorphisms with multiple sclerosis (MS) in Northern Ireland. Mult Scler 1999; 5:105-109.
136. Woodroofe MN, Cuzner ML. Cytokine mRNA expression in inflammatory multiple sclerosis lesions: Detection by nonradioactive in situ hybridization. Cytokine 1993; 5:583-588.
137. De Groot CJ, Montagne L, Barten AD et al. Expression of transforming growth factor (TGF)-beta1, beta2, and beta3 isoforms and TGF-beta type I and type II receptors in multiple sclerosis lesions and human adult astrocyte cultures. J Neuropathol Exp Neurol 1999; 58:174-187.
138. Link J, Soderstrom M, Olsson T et al. Increased transforming growth factor-beta, interleukin-4, and interferon-gamma in multiple sclerosis. Ann Neurol 1994; 36:379-386.
139. Bennett AL, Chao CC, Hu S et al. Elevation of bioactive transforming growth factor-beta in serum from patients with chronic fatigue syndrome. J Clin Immunol 1997; 17:160-166.
140. Nicoletti F, Di Marco R, Patti F et al. Blood levels of transforming growth factor-beta 1 (TGF-beta1) are elevated in both relapsing remitting and chronic progressive multiple sclerosis (MS) patients and are further augmented by treatment with interferon-beta 1b (IFN-beta1b). Clin Exp Immunol 1998; 113:96-99.
141. Matusevicius D, Kivisakk P, Navikas V et al. Influence of IFN-beta1b (Betaferon) on cytokine mRNA profiles in blood mononuclear cells and plasma levels of soluble VCAM-1 in multiple sclerosis. Eur J Neurol 1998; 5:265-275.

142. Hafler DA, Kent SC, Pietrusewicz MJ et al. Oral administration of myelin induces antigen-specific TGF-beta 1 secreting T cells in patients with multiple sclerosis. Ann NY Acad Sci 1997; 835:120-131.
143. Calabresi PA, Fields NS, Maloni HW et al. Phase 1 trial of transforming growth factor beta 2 in chronic progressive MS. Neurology 1998; 51:289-292.
144. Diab A, Michael L, Wahren B et al. Linomide suppresses acute experimental autoimmune encephalomyelitis in Lewis rats by counter-acting the imbalance of pro-inflammatory versus anti-inflammatory cytokines. J Neuroimmunol 1998; 85:146-154.
145. Begolka WS, Miller SD. Cytokines as intrinsic and exogenous regulators of pathogenesis in experimental autoimmune encephalomyelitis. Res Immunol 1998; 149:771-781; discussion 843-774, 855-760.
146. Di Rosa F, Francesconi A, Di Virgilio A et al. Lack of Th2 cytokine increase during spontaneous remission of experimental allergic encephalomyelitis. Eur J Immunol 1998; 28:3893-3903.
147. Laman JD, van Meurs M, Schellekens MM et al. Expression of accessory molecules and cytokines in acute EAE in marmoset monkeys (*Callithrix jacchus*). J Neuroimmunol 1998; 86:30-45.
148. Nagelkerken L, Blauw B, Tielemans M. IL-4 abrogates the inhibitory effect of IL-10 on the development of experimental allergic encephalomyelitis in SJL mice. Int Immunol 1997; 9:1243-1251.
149. Inobe J, Slavin AJ, Komagata Y et al. IL-4 is a differentiation factor for transforming growth factor-beta secreting Th3 cells and oral administration of IL-4 enhances oral tolerance in experimental allergic encephalomyelitis. Eur J Immunol 1998; 28:2780-2790.
150. Bettelli E, Das MP, Howard ED et al. IL-10 is critical in the regulation of autoimmune encephalomyelitis as demonstrated by studies of IL-10- and IL-4-deficient and transgenic mice. J Immunol 1998; 161:3299-3306.
151. Martino G, Poliani PL, Furlan R et al. Cytokine therapy in immune-mediated demyelinating diseases of the central nervous system: a novel gene therapy approach. J Neuroimmunol 2000; 107:184-190.
152. Dal Canto RA, Costa G, Shaw MD et al. Local delivery of cytokines by retrovirally transduced antigen-specific TCR+ hybridoma cells in experimental autoimmune encephalomyelitis. Eur Cytokine Netw 1998; 9:83-91.
153. Shaw MK, Lorens JB, Dhawan A et al. Local delivery of interleukin 4 by retrovirus-transduced T lymphocytes ameliorates experimental autoimmune encephalomyelitis. J Exp Med 1997; 185:1711-1714.
154. Falcone M, Bloom BR. A T helper cell 2 (Th2) immune response against nonself antigens modifies the cytokine profile of autoimmune T cells and protects against experimental allergic encephalomyelitis. J Exp Med 1997; 185:901-907.
155. Constantinescu CS, Hilliard B, Ventura E et al. Modulation of susceptibility and resistance to an autoimmune model of multiple sclerosis in prototypically susceptible and resistant strains by neutralization of interleukin-12 and interleukin-4, respectively. Clin Immunol 2001; 98:23-30.
156. Pal E, Yamamura T, Tabira T. Autonomic regulation of experimental autoimmune encephalomyelitis in IL-4 knockout mice. J Neuroimmunol 1999; 100:149-155.
157. Samoilova EB, Horton JL, Chen Y. Acceleration of experimental autoimmune encephalomyelitis in interleukin-10-deficient mice: Roles of interleukin-10 in disease progression and recovery. Cell Immunol 1998; 188:118-124.
158. Liblau R, Steinman L, Brocke S. Experimental autoimmune encephalomyelitis in IL-4-deficient mice. Int Immunol 1997; 9:799-803.
159. Hackstein H, Bitsch A, Bohnert A et al. Analysis of interleukin-4 receptor alpha chain variants in multiple sclerosis. J Neuroimmunol 2001; 113:240-248.
160. Hemmer B, Vergelli M, Calabresi P et al. Cytokine phenotype of human autoreactive T cell clones specific for the immunodominant myelin basic protein peptide (83-99). J Neurosci Res 1996; 45:852-862.
161. Hermans G, Stinissen P, Hauben L et al. Cytokine profile of myelin basic protein-reactive T cells in multiple sclerosis and healthy individuals. Ann Neurol 1997; 42:18-27.
162. Kil K, Zang YC, Yang D et al. T cell responses to myelin basic protein in patients with spinal cord injury and multiple sclerosis. J Neuroimmunol 1999; 98:201-207.
163. Bongioanni P, Fioretti C, Vanacore R et al. Lymphocyte subsets in multiple sclerosis. A study with two-colour fluorescence analysis. J Neurol Sci 1996; 139:71-77.
164. Crucian B, Dunne P, Friedman H et al. Alterations in peripheral blood mononuclear cell cytokine production in response to phytohemagglutinin in multiple sclerosis patients. Clin Diagn Lab Immunol 1995; 2:766-769.

165. Khademi M, Wallstrom E, Andersson M et al. Reduction of both pro- and anti-inflammatory cytokines after 6 months of interferon beta-1α treatment of multiple sclerosis. J Neuroimmunol 2000; 103:202-210.
166. Calabresi PA, Tranquill LR, McFarland HF et al. Cytokine gene expression in cells derived from CSF of multiple sclerosis patients. J Neuroimmunol 1998; 89:198-205.
167. Olsson T. Critical influences of the cytokine orchestration on the outcome of myelin antigen-specific T cell autoimmunity in experimental autoimmune encephalomyelitis and multiple sclerosis. Immunol Rev 1995; 144:245-268.
168. Diab A, Zhu J, Xiao BG et al. High IL-6 and low IL-10 in the central nervous system are associated with protracted relapsing EAE in DA rats. J Neuropathol Exp Neurol 1997; 56:641-650.
169. Villoslada P, Hauser SL, Bartke I et al. Human nerve growth factor protects common marmosets against autoimmune encephalomyelitis by switching the balance of T helper cell type 1 and 2 cytokines within the central nervous system. J Exp Med 2000; 191:1799-1806.
170. Rott O, Fleischer B, Cash E. Interleukin-10 prevents experimental allergic encephalomyelitis in rats. Eur J Immunol 1994; 24:1434-1440.
171. Xiao BG, Bai XF, Zhang GX et al. Suppression of acute and protracted-relapsing experimental allergic encephalomyelitis by nasal administration of low-dose IL-10 in rats. J Neuroimmunol 1998; 84:230-237.
172. Xu LY, Yang JS, Huang YM et al. Combined nasal administration of encephalitogenic myelin basic protein peptide 68-86 and IL-10 suppressed incipient experimental allergic encephalomyelitis in Lewis rats. Clin Immunol 2000; 96:205-211.
173. Cannella B, Gao YL, Brosnan C et al. IL-10 fails to abrogate experimental autoimmune encephalomyelitis. J Neurosci Res 1996; 45:735-746.
174. Mathisen PM, Yu M, Johnson JM et al. Treatment of experimental autoimmune encephalomyelitis with genetically modified memory T cells. J Exp Med 1997; 186:159-164.
175. Tuohy VK, Yu M, Yin L et al. Modulation of the IL-10/IL-12 cytokine circuit by interferon-beta inhibits the development of epitope spreading and disease progression in murine autoimmune encephalomyelitis. J Neuroimmunol 2000; 111:55-63.
176. Cua DJ, Groux H, Hinton DR et al. Transgenic interleukin 10 prevents induction of experimental autoimmune encephalomyelitis. J Exp Med 1999; 189:1005-1010.
177. Cua DJ, Hutchins B, LaFace DM et al. Central nervous system expression of IL-10 inhibits autoimmune encephalomyelitis. J Immunol 2001; 166:602-608.
178. Stohlman SA, Pei L, Cua DJ et al. Activation of regulatory cells suppresses experimental allergic encephalomyelitis via secretion of IL-10. J Immunol 1999; 163:6338-6344.
179. Legge KL, Min B, Bell JJ et al. Coupling of peripheral tolerance to endogenous interleukin 10 promotes effective modulation of myelin-activated T cells and ameliorates experimental allergic encephalomyelitis. J Exp Med 2000; 191:2039-2052.
180. Smith EM, Cadet P, Stefano GB et al. IL-10 as a mediator in the HPA axis and brain. J Neuroimmunol 1999; 100:140-148.
181. Ferrante P, Fusi ML, Saresella M et al. Cytokine production and surface marker expression in acute and stable multiple sclerosis: Altered IL-12 production and augmented signaling lymphocytic activation molecule (SLAM)-expressing lymphocytes in acute multiple sclerosis. J Immunol 1998; 160:1514-1521.
182. Rohowsky-Kochan C, Molinaro D, Cook SD. Cytokine secretion profile of myelin basic protein-specific T cells in multiple sclerosis. Mult Scler 2000; 6:69-77.
183. Inoges S, Merino J, Bandres E et al. Cytokine flow cytometry differentiates the clinical status of multiple sclerosis (MS) patients. Clin Exp Immunol 1999; 115:521-525.
184. Rieckmann P, Albrecht M, Kitze B et al. Tumor necrosis factor-alpha messenger RNA expression in patients with relapsing-remitting multiple sclerosis is associated with disease activity. Ann Neurol 1995; 37:82-88.
185. Gayo A, Mozo L, Suarez A et al. Long-term effect of IFNbeta1β treatment on the spontaneous and induced expression of IL-10 and TGFbeta1 in MS patients. J Neurol Sci 2000; 179:43-49.
186. Waubant E, Gee L, Bacchetti P et al. Relationship between serum levels of IL-10, MRI activity and interferon beta-1α therapy in patients with relapsing remitting MS. J Neuroimmunol 2001; 112:139-145.
187. Miller A, Shapiro S, Gershtein R et al. Treatment of multiple sclerosis with copolymer-1 (Copaxone): Implicating mechanisms of Th1 to Th2/Th3 immune-deviation. J Neuroimmunol 1998; 92:113-121.
188. Gayo A, Mozo L, Suarez A et al. Glucocorticoids increase IL-10 expression in multiple sclerosis patients with acute relapse. J Neuroimmunol 1998; 85:122-130.
189. Barnes PJ. Anti-inflammatory actions of glucocorticoids: Molecular mechanisms. Clin Sci 1998; 94:557-572.

190. Bright JJ, Musuro BF, Du C et al. Expression of IL-12 in CNS and lymphoid organs of mice with experimental allergic encephalitis. J Neuroimmunol 1998; 82:22-30.
191. Smith T, Hewson AK, Kingsley CI et al. Interleukin-12 induces relapse in experimental allergic encephalomyelitis in the Lewis rat. Am J Pathol 1997; 150:1909-1917.
192. Leonard JP, Waldburger KE, Goldman SJ. Prevention of experimental autoimmune encephalomyelitis by antibodies against interleukin 12. J Exp Med 1995; 181:381-386.
193. Segal BM, Dwyer BK, Shevach EM. An interleukin (IL)-10/IL-12 immunoregulatory circuit controls susceptibility to autoimmune disease. J Exp Med 1998; 187:537-546.
194. Bright JJ, Du C, Coon M et al. Prevention of experimental allergic encephalomyelitis via inhibition of IL-12 signaling and IL-12-mediated Th1 differentiation: An effect of the novel anti-inflammatory drug lisofylline. J Immunol 1998; 161:7015-7022.
195. Heremans H, Dillen C, Groenen M et al. Role of endogenous interleukin-12 (IL-12) in induced and spontaneous relapses of experimental autoimmune encephalomyelitis in mice. Eur Cytokine Netw 1999; 10:171-180.
196. Constantinescu CS, Wysocka M, Hilliard B et al. Antibodies against IL-12 prevent superantigen-induced and spontaneous relapses of experimental autoimmune encephalomyelitis. J Immunol 1998; 161:5097-5104.
197. Ichikawa M, Koh CS, Inoue A et al. Anti-IL-12 antibody prevents the development and progression of multiple sclerosis-like relapsing—Remitting demyelinating disease in NOD mice induced with myelin oligodendrocyte glycoprotein peptide. J Neuroimmunol 2000; 102:56-66.
198. Comabella M, Balashov K, Issazadeh S et al. Elevated interleukin-12 in progressive multiple sclerosis correlates with disease activity and is normalized by pulse cyclophosphamide therapy. J Clin Invest 1998; 102:671-678.
199. Nicoletti F, Patti F, Cocuzza C et al. Elevated serum levels of interleukin-12 in chronic progressive multiple sclerosis. J Neuroimmunol 1996; 70:87-90.
200. Matusevicius D, Kivisakk P, Navikas V et al. Interleukin-12 and perforin mRNA expression is augmented in blood mononuclear cells in multiple sclerosis. Scand J Immunol 1998; 47:582-590.
201. Balashov KE, Smith DR, Khoury SJ et al. Increased interleukin 12 production in progressive multiple sclerosis: Induction by activated CD4+ T cells via CD40 ligand. Proc Natl Acad Sci USA 1997; 94:599-603.
202. van Boxel-Dezaire AH, Hoff SC, van Oosten BW et al. Decreased interleukin-10 and increased interleukin-12p40 mRNA are associated with disease activity and characterize different disease stages in multiple sclerosis. Ann Neurol 1999; 45:695-703.
203. van Boxel-Dezaire AH, van Trigt-Hoff SC, Killestein J et al. Contrasting responses to interferon beta-1b treatment in relapsing-remitting multiple sclerosis: Does baseline interleukin-12p35 messenger RNA predict the efficacy of treatment? Ann Neurol 2000; 48:313-322.
204. Fassbender K, Ragoschke A, Rossol S et al. Increased release of interleukin-12p40 in MS: Association with intracerebral inflammation. Neurology 1998; 51:753-758.
205. Rohowsky-Kochan C, Molinaro D, Choudhry A et al. Impaired interleukin-12 production in multiple sclerosis patients. Mult Scler 1999; 5:327-334.
206. Gillessen S, Carvajal D, Ling P et al. Mouse interleukin-12 (IL-12) p40 homodimer: A potent IL-12 antagonist. Eur J Immunol 1995; 25:200-206.
207. Ha SJ, Lee CH, Lee SB et al. A novel function of IL-12p40 as a chemotactic molecule for macrophages. J Immunol 1999; 163:2902-2908.
208. Allavena P, Paganin C, Zhou D et al. Interleukin-12 is chemotactic for natural killer cells and stimulates their interaction with vascular endothelium. Blood 1994; 84:2261-2268.
209. Pahan K, Sheikh FG, Liu X et al. Induction of nitric oxide synthase and activation of NF-kappaB by interleukin-12 p40 in microglial cells. J Biol Chem 2000; 7:7.
210. Schmitt EPH, Germann T, Rude E. Differential effects of interleukin-12 on the development of naive mouse CD4+ T cells. Eur J Immunol 1994; 24:343-347.
211. Tarrant TK, Silver PB, Wahlsten JL et al. Interleukin 12 protects from a T helper type 1-mediated autoimmune disease, experimental autoimmune uveitis, through a mechanism involving interferon gamma, nitric oxide, and apoptosis. J Exp Med 1999; 189:219-230.

CHAPTER 8

Chemokines in Experimental Autoimmune Encephalomyelitis and Multiple Sclerosis

Alicia Babcock and Trevor Owens

Introduction

Chemokines are small molecules that direct leukocyte traffic and play a role in cellular activation. Multiple sclerosis (MS) and experimental autoimmune encephalomyelitis (EAE) are inflammatory diseases of the central nervous system (CNS), in which leukocyte infiltration of the CNS accompanies gliosis. This Chapter reviews the roles of chemokines in these diseases.

The term "chemokine" was originally introduced to describe a family of chemotactic cytokines that functioned primarily to recruit cells to sites of inflammation. Since their initial description, the chemokine superfamily has grown to include over 50 members due to advances in molecular cloning techniques and the use of bioinformatics in nucleotide database analyses. The rate at which new chemokine genes are identified is probably nearing its plateau. We now know that chemokine function extends far beyond chemotaxis, having been implicated in leukocyte development and maturation, tumor growth and metastasis, and angiogenesis (see (1) for a recent review). Our review will focus on how the expression of chemokines and their receptors in the central nervous system (CNS) relates to multiple sclerosis (MS) and experimental autoimmune encephalomyelitis (EAE).

Chemokines

Chemokines are small (8 to 10 kD), secreted proteins. Four subfamilies are defined by the position of two highly conserved cysteine residues, which form a defining amino-terminal motif. In the CXC (α) subfamily, this motif consists of two cysteines separated by one nonconserved amino acid residue. No such residue separates the cysteines in the CC (β) subfamily. A single cysteine residue at the amino terminal defines the C (γ) chemokine subfamily, whereas three nonconserved amino acids exist between the pair of cysteine residues in the CX_3C (δ) subgroup. Recently advanced is a nomenclature that describes chemokines as ligands of a particular subfamily (e.g., CC chemokines = CCLs).[2]

Chemokine Families

Most chemokines fall into the CC and CXC subfamilies, as only one member has been described for each of the C and CX3C groups. Lymphotactin, the lone C chemokine, directs T cells that express its receptor, XCR1.[3,4] Fractalkine (neurotactin) is unique, not only because it is the sole member of the CX_3C family, but also because it exists in both secreted and membrane-bound form.[5] This chemokine and its receptor, CX_3CR1, are expressed at highest levels in the CNS,[5] and have been implicated in microglial-neuronal cross-talk.[6] The CXC chemokines may be further subdivided, based on the presence of an ELR (glutamate-leucine-arginine) motif, which confers the ability to recruit neutrophils.[7] ELR-containing CXC chemokines

Cytokines and Chemokines in Autoimmune Disease, edited by Pere Santamaria.

include the prototypic IL-8, KC/Gro-α, and MIP-2. The so-called "nonELR" CXC chemokines include SDF-1, IP-10, and Mig. SDF-1 was originally isolated from bone marrow-derived stromal cells.[8] Through CXCR4 expression, SDF-1 chemoattracts $CD34^+$ stem cells,[9] and may both attract and repel T lymphocytes, depending on its concentration [10]. The IFN-γ-inducible proteins, IP-10 and Mig, function to chemoattract activated T cells.[11] Activated T cells and monocytes/macrophages are usually targeted by CC chemokines. This large subfamily includes MCP-1, MIP-1α, MIP-1β and RANTES. This family of chemokines is the most studied in the context of CNS inflammation and will be discussed extensively, later in our Chapter.

Chemokine Receptors

Chemokine receptors are named based on the classification of the chemokines they bind (e.g., CC chemokines bind CCRs). As members of the seven transmembrane spanning, G-protein coupled receptor superfamily, they are susceptible to inhibition by pertussis toxin. Receptor promiscuity is common within, but not among, chemokine families. The CC chemokines generously share receptors between members. For example, RANTES can interact with CCR1, CCR3, CCR4, and CCR5, whereas MIP-1_ signals through CCR1, CCR5, and CCR9, and MCP-1 through CCR2 and CCR4 (see Table 7.1 and (12-14)). CXC-ELR chemokines bind the receptors CXCR1 and CXCR2, which are primarily expressed on neutrophils.[15] The nonELR CXC chemokines, IP-10 and Mig, share the same receptor, CXCR3, which is expressed on activated T cells.[11] So far, only CXCR3 has been shown to bind chemokines outside of its family. CXCR3 also interacts with CC chemokines eotaxin and MCP4,[16,17] both of which attract eosinophils rather than monocytes and T cells.[18] While some chemokine receptors are expressed constitutively (e.g., CXCR4), many are induced by pro-inflammatory cytokines such as IFN-γ and TNF-α at sites of inflammation, including those in the CNS. Before considering the role of chemokines in CNS inflammatory diseases, we will consider the inflammatory processes that occur in these diseases.

Immunology of Multiple Sclerosis (MS)

MS is the most common inflammatory disease of the brain, affecting 1 person per thousand in Canadian, American, and Northern European populations.[19] The onset of MS usually occurs in young adults. Relapsing-remitting (RR) MS, characterized by episodes of neurological symptoms that lead over time to a more chronic disease with increasing disability, is the most prevalent pattern. Lesions contain activated T lymphocytes, mostly CD4+ but also some CD8+ cells, with evidence of ongoing demyelination. In secondary-progressive (SP) or chronic disease, demyelinated lesions usually show absence of oligodendrocytes and a preponderance of macrophages and activated microglial cells. Macrophages/microglia actively phagocytose myelin, and have been suggested as mediators of tissue damage.[19] Axonal damage also occurs in MS.[20]

The cause of MS is unknown. However, evidence suggests that CD4+ Th1 cells are critical for disease to occur. MS is a collection of heterogeneous pathologies.[21] Importantly, immune cells infiltrate in all cases, and MS continues to be considered an autoimmune disease of the CNS. The fact that immune-directed therapies, such as corticosteroids, IFN-ß and glatiramer acetate (copaxone), are effective[22,23] makes a strong case for immune cell entry to the CNS being a fundamental and necessary aspect of disease. It is important to understand how this is regulated.

Immunology of Experimental Autoimmune Encephalomyelitis (EAE)

The best-studied animal model for MS is EAE. By definition an autoimmune disease of the CNS, EAE reproduces many of the pathological features of MS. EAE can be induced in a wide range of species, but is most often studied in rodents, especially mice. The disease is induced by immunization with myelin protein or peptide in Complete Freund's adjuvant, or by adoptive transfer of $CD4^+$ T cells from an immunized animal. The initiating events are $CD4^+$ Th1 cell

Table 8.1. Chemokines and chemokine receptors implicated in MS and EAE

Chemokine Acronym	Full-Name	Receptor(s)
CXC (α) family		
ELR motif		
Gro-α (KC)	Growth-related oncogene-α	CXCR2 >> CXCR1
MIP-2	Macrophage inflammatory protein-2	CXCR2
NonELR motif		
Mig	Monokine-induced by interferon-γ	CXCR3
IP-10	Interferon-inducible protein-10	CXCR3
CC (α–β) family		
MCP-1	Monocyte chemoattractant protein-1	CCR2
MIP-1α	Macrophage inflammatory protein-1α	CCR1, CCR5
MIP-1β	Macrophage inflammatory protein-1β	CCR1, CCR5
RANTES	Regulation on activation, normal T-cell	CCR1, CCR3, CCR4, expressed and secreted
CCR5		
TCA-3 (I-309)	T-cell activation protein-3	CCR8

activation and clonal expansion in the periphery, and traffic to and migration into the CNS. Macrophages coinfiltrate with T cells, and are essential (though not sufficient) for disease.[24] Both infiltrate the CNS white matter and remain mostly vessel-associated, although some parenchymal dissemination does occur.

A cascade of cellular events is initiated in the CNS that ultimately results in demyelination and axonal damage. Glial cells (astrocytes and microglia) become activated in the inflamed CNS. Both are sources of pro-inflammatory molecules, such as TNF-α, IL-1, IL-6 and reactive nitrogen and oxygen species. IFN-γ, the classic Th1 pro-inflammatory cytokine, is not produced by cells of the adult CNS, but derives from $CD4^+$ and $CD8^+$ T cells and natural killer (NK) cells.

Onset of neurological symptoms in EAE also coincides with T cell entry to the CNS. Many EAE models show a relapsing-remitting progression with increasing severity, and there are also chronic models of disease. The requirement for T cell infiltration has been directly demonstrated in EAE. Nevertheless, in chronic disease, and in transgenic mice that over-express inflammatory cytokines,[25] a preponderance of activated macrophages/microglia is seen. Large numbers of neutrophils infiltrate the CNS during EAE in mice that lack interferon-gamma, as will be discussed later in this chapter. This brief summary underlines that, as is inferred for MS, the regulation of the extent and quality of immune infiltration to the CNS is critical to progression and outcome of EAE.

The Blood Brain Barrier (BBB)

Immune cells reach the CNS the same way they do other organs, via blood circulation. The blood vessels in the CNS have features that distinguish them from vessels elsewhere in the body, notably the presence of tight junctions between endothelial cells, and a perivascular space between the endothelium and the parenchymal basement membrane. The perivascular space contains macrophages that contribute to a general 'Gatekeeper' role for the CNS.[26] For instance, they are implicated in responsiveness to bacterial mediators such as LPS.[27] They may also present antigen to T lymphocytes, although the significance of this ability is not well understood.[26] The basement membrane that separates the perivascular space from the brain parenchyma is "sealed" by the apposition of astrocyte foot-processes. Collectively, these components make up what is termed the Blood-Brain Barrier (BBB).

Concepts that guide our attitudes in the role of chemokines in immune cell entry include the following. The BBB operates to restrict the passage of macromolecules into the CNS. It is now accepted that, although considered relatively immunologically privileged, the CNS is accessible to the immune system. This likely has survival advantage, through immunological surveillance and protection against pathogens and tumors. A variety of studies show that activated or memory T cells can cross the BBB, regardless of their antigen specificity.[28,29] A similar process must underlie the initiation of EAE.

Immune Cell Entry to the CNS

Entry of cells requires active processes of adhesion and chemoattraction (Fig. 8.1). In circumstances such as adoptive transfer without adjuvant, neither the CNS nor the endothelium of the BBB is inflamed, so T cells must direct the initial extravasation. Recently activated T cells may spontaneously release chemokines in the circulation. However, T cell entry to the CNS occurs days after adoptive transfer and it is unlikely that activation-induced chemokine production would persist that long. Whether antigen presentation at the BBB plays a role in chemokine induction is uncertain. Conventional wisdom says that it cannot, because T cells of irrelevant specificities (e.g., anti-ovalbumin) can enter the CNS. Myelin-specific T cells may have an advantage in extravasation to the CNS, but no experiment has yet directly compared them with irrelevant T cells in this regard.

T lymphocytes and the cells of the BBB both respond to chemokines, and are a potential source of them. T cell interaction with endothelial cells or perivascular macrophages may stimulate chemokine release that could promote immune cell entry. Ligands most likely to be involved include adhesion ligands and CD40. CD40 ligation is known to synergize with IFN-γ,[30] thereby conferring an advantage on Th1 cells for initiation of transmigration-promoting chemokines at the BBB. Th1 cells also preferentially express the CCR5 receptor for RANTES and MIP-1α/β[31] and so are favoured for extravasation from blood to tissues in delayed type hypersensitivity reactions, where these chemokines are produced.

Chemokines in EAE

During the induction of EAE, T cells specific for myelin antigens migrate across the BBB and initiate CNS pathology. Selective temporal and spatial chemokine expression provides an attractive explanation for the mechanism by which subsequent leukocytes are recruited to the CNS. In recent years, many studies have documented the presence of a variety of chemokines in the CNS and the stages at which they affect disease.

Initial studies attempted to clarify whether CNS chemokine expression dictated leukocyte infiltration, or simply served to amplify recruitment once the first T cells had crossed the BBB. The expression of MCP-1 and IP-10 correlated with histological inflammation at the onset of proteolipid protein (PLP)-induced EAE in mice.[32] High levels of these chemokines were detected in the liver, prior to the appearance of clinical signs, reflecting systemic immune activation.[32] This suggested that CNS expression of these chemokines did in fact serve to amplify, but not induce, T cell infiltration. Other studies of PLP-induced EAE could not detect elevated levels of MCP-1[33] until late in acute disease, when its expression correlated with the severity of clinical relapse.[34]

CNS production of MIP-1α paralleled both onset and severity of disease.[33-38] In one study, the expression of MIP-1α, MCP-1, IP-10 and a number of other chemokine genes occurred prior to the development of clinical signs.[39] However, it was not determined whether chemokine expression might have correlated with subclinical CNS pathology. The level of MIP-1α expression remained elevated during remission in mice immunized

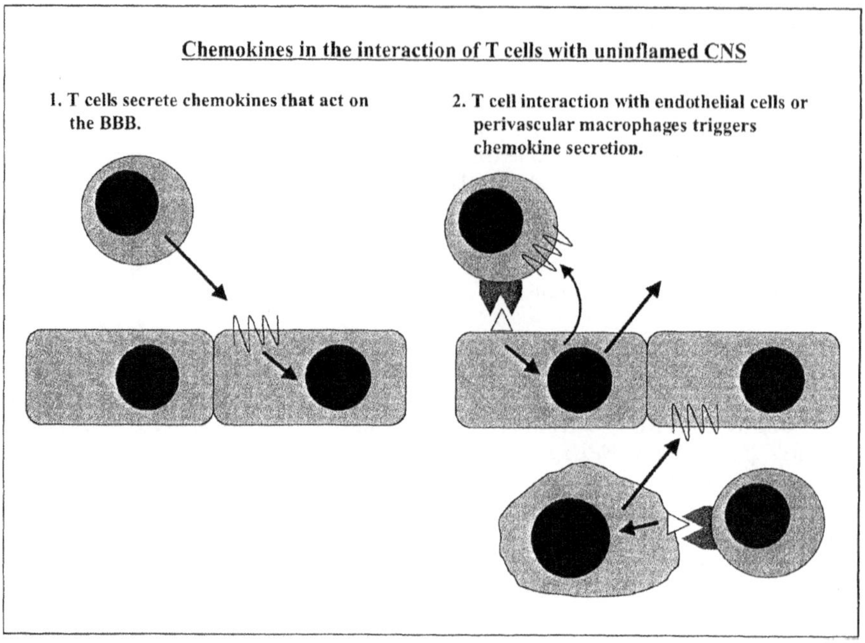

Fig. 8.1. Schematic to illustrate potential modes of chemokine production during entry of activated, memory-effector T cells to the uninflamed CNS. 1). Chemokines produced by recently activated T cells may act through receptors that cells of the BBB constitutively express. This may initiate cascades of chemokine production, which are not exclusive of events described in 2). 2). Interaction of T lymphocytes (via adhesion molecules, antigen recognition) with endothelial cells or perivascular macrophages induces signaling for chemokine production by these cells. Not shown is the possibility of interaction with astrocytes. Chemokines produced by cells of the BBB facilitate T cell extravasation and amplify further T cell recruitment.

with PLP,[33,34,39] but not in myelin basic protein (MBP)-induced EAE in rats.[36,38] MIP-1β shares its pattern of expression with MIP-1β. CNS expression of MIP-1β occurs early in the development of EAE.[36,39] The levels of expression of this chemokine correlated with inflammatory infiltration and disease severity.[36] MIP-1β remained elevated during remission in mice,[39] but decreased in rats.[36,38]

MIP-1α and MIP-1β signal through CCR1 and CCR5, receptors shared with RANTES. While RANTES protein was detected throughout the course of disease, its expression could not be correlated with inflammatory infiltrates at the onset of EAE or during relapse.[34] RANTES mRNA was upregulated with the onset of clinical signs during acute EAE, and this correlated with the intensity and severity of CNS inflammation.[35,36] Levels of RANTES mRNA decreased following acute disease, but remained at a detectable level[35,36] or increased[38] during remission.

RANTES recruits both T cells and macrophages to the site of CNS inflammation. The CC chemokine C10 serves mainly in the recruitment of macrophages during EAE.[40] C10 expression was recently described in the CNS of mice with myelin oligodendrocyte glycoprotein (MOG)-induced EAE,[40] whereas it was not detected in an earlier study.[39]

Neutrophil Versus Macrophage Infiltration

That neutrophils may play a role in EAE has only recently been recognized, and early descriptions of chemokine expression in EAE commented on detection of neutrophil chemoattractants.[33,39] Neutrophils may be recruited to the CNS in response to the chemokine KC/Gro-α, which is expressed early in EAE.[35,39] Overexpression of KC/Gro-α in oligodendrocytes on MBP promoter-driven transgenic mice induced neutrophil infiltration and severe disease.[41] McColl and colleagues showed that neutrophil blockade prevented EAE in SJL/J mice.[42] The presence of neutrophils in mice with EAE has been documented,[43] and recently confirmed by flow cytometric analysis [Brickman and Owens, unpublished]. Expression of neutrophil chemoattracting chemokines in EAE, described below, should be interpreted in light of these still evolving observations.

CNS expression of MIP-2 has rarely been detected.[33,39] MIP-2 is considered the functional homologue of the human chemokine IL-8, both of which chemoattract neutrophils. MIP-2 expression in the CNS during murine EAE has recently been documented in BALB/c mice, which are normally resistant to EAE. This resistance is overcome in IFN-γ -deficient BALB/c mice, which develop a lethal form of EAE that is characterized by pronounced neutrophilia.[44,45] During EAE in mice deficient in IFN-γ, MIP-2 and TCA-3 are strongly expressed, whereas the expression of RANTES and MCP-1 is strikingly absent.[45] The disseminated, nonperivascular distribution of both neutrophils and $CD4^+$ T cells in IFN-γ-deficient mice with EAE contrasts strikingly with discrete perivascular infiltrates in wildtype mice.[45] This speaks to an additional role for chemokines, in regulating parenchymal distribution.

A predominant neutrophil invasion also occurs in BALB/c mice immunized with ultrasound emulsified antigen in adjuvant.[37] In this system, high levels of MIP-2 are accompanied by high levels of MCP-1 and MIP-1α.[37] In BALB/c mice with EAE, astrocytes expressed MIP-2 and MIP-1α proteins, whereas MCP-1 was only expressed by neutrophils.[37]

Cellular Source of Chemokines

Astrocytes and infiltrating leukocytes are the principal cellular sources of chemokines to be described. Astrocytes express KC/Gro-α,[46] MCP-1 and IP-10[32,46,47] in mice with PLP-induced EAE, whereas infiltrating leukocytes elaborated MIP-1α and RANTES.[46] Murine PLP-reactive T cells that could adoptively transfer EAE expressed message for MIP-1α, MIP-1β, RANTES, and TCA-3.[39] In rats with MBP-induced EAE, T cells were identified as the main cells expressing RANTES; astrocytes and macrophages/microglia expressed lower levels of this chemokine.[36] MIP-1α and MIP-1β were expressed predominantly by infiltrating leukocytes in the same study.[36] MIP-1β-producing cells were mainly T cells, but some macrophages and astrocytes also produced this chemokine.[36]

Therapeutic Interventions Directed at Chemokines in EAE

The expression of so many chemokines at various stages of disease and by numerous cell types in the CNS has made these studies difficult to interpret. To assess the relative roles of chemokines, several blocking studies have been performed. Antibodies specific for MIP-1α, but not MCP-1[33] or RANTES,[34] administered at the time of disease induction, prevented onset of EAE without affecting T cell activation.[33] When administered during the remission phase of the disease, antibodies specific for MCP-1, but not RANTES or MIP-1α, reduced macrophage recruitment to the CNS and ameliorated the severity of relapses.[34] DNA vaccination with constructs specific for MIP-1α and MCP-1 prevented EAE, as late as 2 months after vaccine administration.[38,48] MIP-1β-vaccinated rats developed more severe disease, whereas administration of RANTES DNA vaccines had no effect on EAE.[38] These studies highlighted roles for MIP-1α and MCP-1 in the regulation of EAE. In particular, they suggested MIP-1α be involved in initiating EAE and MCP-1 in promoting disease relapse.

Gene knockout mice have revealed that MIP-1α is not required for the induction of EAE. Mice deficient in MIP-1α were fully susceptible to MOG-induced EAE, with similar kinetics and severity to wild-type mice.[49] These mice showed typical Th1 cytokine expression and composition of CNS infiltrates.[49] The chemokine expression profile of MIP-1α knockout mice with EAE included IP-10, RANTES, MCP-1, and lower levels of MIP-1β, MIP-2, lymphotactin and TCA-3.[49] That neutralization studies with anti-MIP-1α antibodies prevented disease, while MIP-1α ablation had no effect on EAE, likely reflects the functional redundancy of chemokines.

MIP-1α interacts with two chemokine receptors: CCR1 and CCR5. Mice deficient in CCR1 developed a less severe form of MOG-induced EAE, with a lower incidence of disease.[50] Levels of IP-10, but not RANTES or MCP-1, mRNA were elevated in CCR1 knockout mice.[50] Protection against EAE was not due to systemic immunosuppression.[50] By contrast, CCR5-deficient mice were susceptible to MOG-induced EAE.[49]

The principal receptor for MCP-1 is CCR2. Deficiency in CCR2 confers resistance to MOG-induced EAE.[51,52] CCR2 knockout mice showed a significant reduction in CNS-infiltrating mononuclear leukocytes.[51,52] It is not clear whether the response of T cells from immunized CCR2 knockout mice is affected.[51,52] Adoptively transferred MOG-specific T cells failed to induce disease in CCR2 knockout mice, whereas CCR2-/- T cells induced EAE in wildtype mice.[51] CCR2 expression on host-derived mononuclear cells is therefore necessary for EAE induction.[51] Levels of RANTES, MCP-1, and IP-10 were not increased in CCR2 knockout mice, nor were the chemokine receptors CCR1 and CCR5.[52]

Speculative Model for Chemokines in EAE

Taken together, these studies implicate different chemokines at various stages of EAE. These are identified in the schematic in Figure 7.2. T cells that express CCR1 on their surface, identified as important to onset of EAE in knockout mice, respond to chemokines induced and/or amplified by T lymphocytes themselves or by the cells of the BBB. The principal ligands for CCR1 include RANTES, MIP-1α, and MIP-1β. Gene knockout and blocking studies have suggested that neither RANTES nor MIP-1α is critical for disease. This implicates MIP-1β in disease onset (see Fig. 8.2). T cell interaction with astrocytes triggers the production of astroglial-derived MCP-1. This chemokine chemoattracts macrophages that express CCR2, which are also necessary for disease to occur. Established disease is more complex. A much-expanded panel of chemokines is produced by a variety of cell types within the CNS. These chemokines promote further immune cell entry and regulate infiltration in the CNS parenchyma. Chemokines may also regulate glial-neuronal interactions and potentially contribute to repair and regeneration.

Chemokines in MS

Although the initial events in MS pathogenesis remain unknown, it has been possible to correlate the expression of chemokines and their receptors with inflammatory infiltrates in demyelinating lesions. Analysis of circulating leukocytes and the cerebral spinal fluid (CSF), whose composition reflects the CNS extracellular space, is less direct. However, such studies support the view that specific chemokine expression amplifies cell recruitment in MS, as in EAE.

Mononuclear cells isolated from the blood of MS patients did not express higher levels of MCP-1 or RANTES than did controls [53]. Rate of migration of T cells from MS blood was increased in Boyden chambers by RANTES and MIP-1α.[54] This migration was partially blocked by anti-CCR5 antibodies.[54] CCR5+ T cells expressed high levels of IFN-gamma[55] and exhibited a Th1/Th0 profile.[54] By contrast, Th2 cells from the blood of healthy individuals migrated more efficiently across an artificial BBB in Boyden chambers, in response to MCP-1 [Biernacki, Prat, and Antel, submitted]. T cells isolated from the blood of MS patients expressed higher levels of CCR5 and CXCR3 than healthy controls.[54,55] The level of expression of both chemokine receptors showed an even greater increase in the CSF of MS patients.[56] Protein levels of MIP-1α were elevated in the CSF during relapse compared to noninflammatory neurological controls, and production correlated with leukocyte infiltration.[57] Levels of IP-10, Mig, and RANTES were elevated during MS attacks.[56] In another study, RANTES and MCP-1 were detected in the CSF of some patients. However, the level of expression was similar to that detected in patients with other neurological disease, indicating that mononuclear cells expressing these chemokines are not specific to MS.[53]

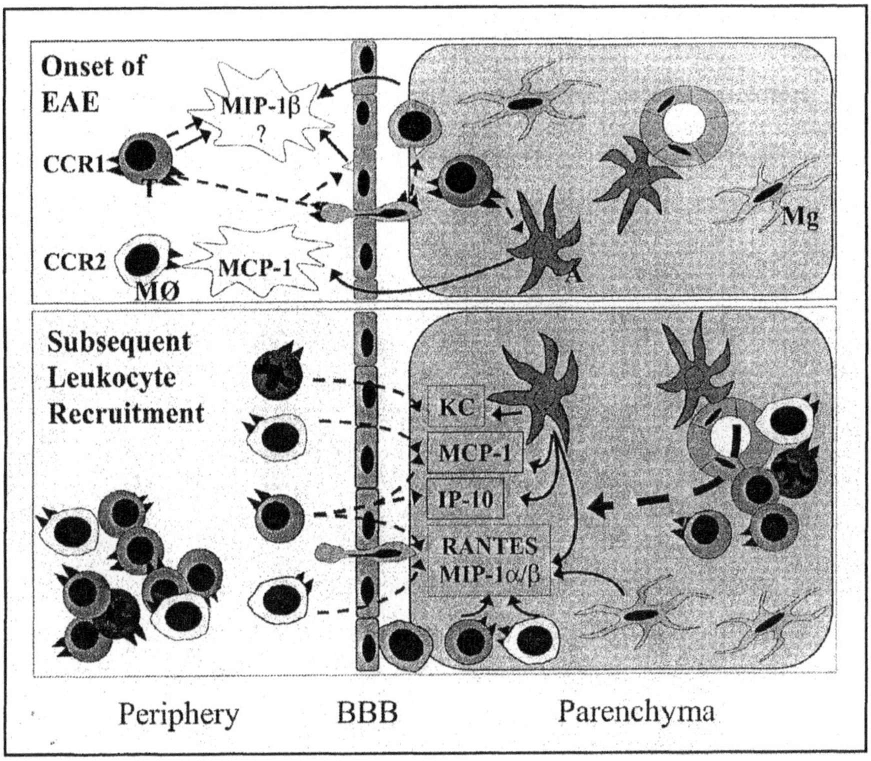

Fig. 8.2. Chemokines involved in onset and progression of CNS inflammatory disease: Speculative model of events in EAE. The top panel identifies chemokines and chemokine receptors that are implicated in EAE through gene knockout and blocking studies, as described in the text. CCR1-expressing T cells (T) respond (dotted arrows) to MIP-1β and extravasate to enter the CNS. Amplification (see Fig. 8.1) through induction of MIP-1β also occurs. CCR2+ macrophages (Mσ) are stimulated by astroglial-derived MCP-1 to enter the CNS. Astrocytes (A) and microglia (Mg) within the CNS are positioned to facilitate perivascular migration. The bottom panel illustrates the complexity of established disease. A much-expanded panel of chemokines is now produced (solid arrows) by a variety of cell types within the CNS. These chemokines promote further entry of cells from the blood and also regulate infiltration by cells in the CNS parenchyma. They may also regulate glial-neuronal interactions and potentially contribute to repair and regeneration.

RANTES, MCP-1 and other chemokines have also been detected in MS lesions. RANTES expression was restricted to perivascular cells and the blood vessel endothelium.[58] MCP-1 was expressed within plaques by astrocytes,[58-61] and by infiltrating lymphocytes,[59,62] in conjunction with MCP-2 and MCP-3.[59] Levels of MCP-1, -2, and -3 expression were reduced in chronic lesions.[59] Although expression of MCP-1 by macrophages has been reported,[58,61] MCP-1 protein was not detected on perivascular or parenchymal "foamy" macrophages,[60] which have phagocytosed myelin. Both astrocytes and macrophages expressed MIP-1α within the plaque,[58] whereas macrophages/microglia were the sole source of IP-10, Mig[62] and MIP-1β.[58] Glial cells surrounding the lesion may also contribute to infiltration and demyelination. Microglia surrounding the plaques expressed MIP-1β.[58] Astrocytes were reactive for RANTES,[58] MCP-1,[58,59] MCP-2, MCP-3,[59] IP-10, and Mig.[62]

Chemokine Receptors in MS

Glial cells and infiltrating leukocytes constitutively express, or can be induced to express, a variety of chemokine receptors. The receptor for IP-10 and Mig, CXCR3, has been implicated in CNS infiltration. MS plaques were infiltrated by CXCR3-expressing T cells[55,56,62] although some astroglial expression of CXCR3 was also described.[62] The chemokines RANTES, Eotaxin, MCP-3 and MCP-4 bind to the CCR3 receptor, which was not detected in the CNS of MS patients.[55] Another study associated the expression of this receptor with foamy macrophages and activated microglia in MS lesions, and to a lesser extent with infiltrating lymphocytes and astrocytes.[61] Foamy macrophages and activated microglia in MS lesions expressed CCR2, the receptor for MCP-1, as did numerous infiltrating lymphocytes.[61] Expression of CCR5, a receptor for MIP-1α, MIP-1β, and RANTES, occurs primarily on infiltrating lymphocytes, macrophages, and microglia in demyelinating lesions.[55,56,61] In some patients, astroglial expression of CCR5 was also noted.[61] Thus, it can be appreciated that chemokines and their receptors are implicated in CNS inflammation in MS, and parallels with EAE are obvious.

Chemokine Genetics and CNS Disease

Polymorphism in chemokines and chemokine receptors contributes to susceptibility to MS and EAE. Individuals who are homozygous for the CCR5 delta 32 deletion do not express this receptor. Because CCR5 is a coreceptor for HIV, these individuals are protected against HIV infection. Three separate studies have shown that CCR5 delta 32 does not confer protection against MS.[63 64 65] However, this mutation correlated with a lower risk for recurrent clinical disease activity [65] and 3-year delay in disease onset.[64] The promoter/enhancer region of the chemokine MCP-3 shows CA/GA repeat polymorphisms.[66] However the frequency of allelic variants was not significantly different in MS and control populations.[66] The MCP-3 A4 allele may protect individuals who are positive for HLA-DRB1*15, which increases their risk for developing MS,[66] whereas MCP-3 A2 seems protective for individuals who do not express MS-susceptibility-associated HLA genes.[66] Polymorphisms in TCA-3, MCP-1, and MCP-3 genes have been implicated in susceptibility to MBP-induced EAE in EAE-susceptible SJL/J mice, but not EAE-resistant B10 or BALB/c mice.[67] These mutations are candidates for *eae7*, a genetic locus which controls susceptibility to and severity of monophasic and relapsing-remitting EAE.[67]

Neural Roles for Chemokines

Chemokines are implicated in developmental regulation in mature animals, as well as in reactive inflammation and immunity. The role of chemokines in normal CNS development and function is well-summarized in a recent review.[13] In mice deficient in CXCR4, the receptor for SDF-1, there is a cerebellar development defect, with abnormal migration of granular cells,[68] pointing to a role for CNS-derived chemokines in neuronal migration in development. This chemokine plays analogous role in lymphoid development, B-cell lymphopoiesis and myelopoiesis.[69] Fractalkine has been implicated in the rapid reaction of CX3CR1 receptor-expressing microglial cells to neuronal injury,[6] and protects microglia from Fas-mediated apoptotic death.[70] Microglial production of chemokines can be induced by ligation of scavenger receptors by β-amyloid protein, as occurs in Alzheimer's Disease.[71] The β-amyloid peptide p25-35 induces microglial production of MCP-1, which is enhanced through synergy with IFN-γ.[72] These findings underline a role for immune cytokines in modulating or promoting endogenous CNS functions, as has been shown in other systems.[73] Phagocytosis of myelin by macrophages and microglia, as occurs in MS and EAE, may also induce chemokine production. MCP-1 was produced in response to sterile head injury in mice, whereas when endotoxin was present, a wide range of both CC and CXC chemokines was detected.[74] Excitotoxic brain injury induced expression of CCR5 on rat microglia and macrophages.[75]

Conclusions

Chemokines are involved as endogenous regulators of CNS glial responses to both inflammation and degeneration. The production of chemokines by microglia or astrocytes at, or proximal to, the BBB may be critical for immune cell entry. The current view is that immune cells have access to the healthy, uninflamed CNS, and it is of interest whether chemokine production by glial cells might contribute to this immune surveillance role. Endogenous CNS programs of response to injury or loss of homeostasis are triggered by autoimmune infiltration, and become part of the complex interplay between immune and nervous systems. It is of interest to determine to what extent the production of chemokines in MS and EAE reflects endogenous CNS programs of glial reactivity. Clearly, there is much more to be learned about this family of important regulatory molecules.

References

1. Rossi D, Zlotnik A. The biology of chemokines and their receptors. Annu Rev Immunol 2000; 18:217-242.
2. Sallusto F, Mackay CR, Lanzavecchia A. The role of chemokine receptors in primary, effector, and memory immune responses. Annu Rev Immunol 2000; 18:593-620.
3. Yoshida T, Imai T, Kakizaki M et al. Identification of single C motif-1/lymphotactin receptor XCR1. J Biol Chem 1998; 273:16551-16554.
4. Kelner GS, Kennedy J, Bacon KB et al. Lymphotactin: A cytokine that represents a new class of chemokine. Science 1994; 266:1395-1399.
5. Schwaeble WJ, Stover CM, Schall TJ et al. Neuronal expression of fractalkine in the presence and absence of inflammation. FEBS Lett 1998; 439:203-207.
6. Harrison JK, Jiang Y, Chen S et al. Role for neuronally derived fractalkine in mediating interactions between neurons and CX3CR1-expressing microglia. Proc Natl Acad Sci USA 1998; 95:10896-10901.
7. Clark-Lewis I, Schumacher C, Baggiolini M et al. Structureactivity relationships of interleukin-8 determined using chemically synthesized analogs. Critical role of NH2-terminal residues and evidence for uncoupling of neutrophil chemotaxis, exocytosis, and receptor binding activities. J Biol Chem 1991; 266:23128-23134.
8. Nagasawa T, Kikutani H, Kishimoto T. Molecular cloning and structure of a preB-cell growth-stimulating factor. Proc Natl Acad Sci USA 1994; 91:2305-2309.
9. Aiuti A, Turchetto L, Cota M et al. Human CD34(+) cells express CXCR4 and its ligand stromal cell-derived factor-1. Implications for infection by T-cell tropic human immunodeficiency virus. Blood 1999; 94:62-73.
10. Poznansky MC, Olszak IT, Foxall R et al. Active movement of T cells away from a chemokine. Nat Med 2000; 6:543-548.
11. Farber JM. Mig and IP-10: CXC chemokines that target lymphocytes. J Leukoc Biol 1997; 61:246-257.
12. Murphy PM, Baggiolini M, Charo IF et al. International union of pharmacology. XXII. Nomenclature for chemokine receptors. Pharmacol Rev 2000; 52:145-176.
13. Mennicken F, Maki R, de Souza EB et al. Chemokines and chemokine receptors in the CNS: A possible role in neuroinflammation and patterning. Trends Pharmacol Sci 1999; 20:73-78.
14. Asensio VC, Campbell IL. Chemokines in the CNS: Plurifunctional mediators in diverse states. Trends Neurosci 1999; 22:504-512.
15. Murphy PM. Neutrophil receptors for interleukin-8 and related CXC chemokines. Semin Hematol 1997; 34:311-318.
16 .Weng Y, Siciliano SJ, Waldburger KE et al. Binding and functional properties of recombinant and endogenous CXCR3 chemokine receptors. J Biol Chem 1998; 273:18288-18291.
17. Lu B, Humbles A, Bota D et al. Structure and function of the murine chemokine receptor CXCR3. Eur J Immunol 1999; 29:3804-3812.
18. Uguccioni M, Loetscher P, Forssmann U et al. Monocyte chemotactic protein 4 (MCP-4), a novel structural and functional analogue of MCP-3 and eotaxin. J Exp Med 1996; 183:2379-2384.
19. Owens T, Sriram S. The immunology of multiple sclerosis and its animal model, experimental allergic encephalomyelitis. Neurol Clin 1995; 13:51-73.
20. Trapp BD, Ransohoff RRudick R. Axonal pathology in multiple sclerosis: Relationship to neurologic disability. Curr Opin Neurol 1999; 12:295-302.

21. Lucchinetti C, Bruck W, Parisi J et al. Heterogeneity of multiple sclerosis lesions: Implications for the pathogenesis of demyelination. Ann Neurol 2000; 47:707-717.
22. Arnason BG. Immunologic therapy of multiple sclerosis. Annu Rev Med 1999; 50:291-302.
23. Genain CP, Zamvil SS. Specific immunotherapy: One size does not fit all. Nat Med 2000; 6:1098-1100.
24. Tran EH, Hoekstra K, van Rooijen N et al. Immune invasion of the central nervous system parenchyma and experimental allergic encephalomyelitis, but not leukocyte extravasation from blood, are prevented in macrophage-depleted mice. J Immunol 1998; 161:3767-3775.
25. Owens T, Wekerle H, Antel J. Genetic models for CNS inflammation. Nature Medicine 2001; 7: 161-166.
26. Owens T, Tran E, Hassan-Zahraee M et al. Immune cell entry to the CNS—A focus for immunoregulation of EAE. Res Immunol 1998; 149:781-789; discussion 844-786, 855-760.
27. Laflamme N, Lacroix S, Rivest S. An essential role of interleukin-1beta in mediating NF-kappaB activity and COX-2 transcription in cells of the blood-brain barrier in response to a systemic and localized inflammation but not during endotoxemia. J Neurosci 1999; 19:10923-10930.
28. Owens T, Renno T, Taupin V et al. Inflammatory cytokines in the brain: Does the CNS shape immune responses? Immunol Today 1994; 15:566-571.
29. Krakowski ML, Owens T. Naive T lymphocytes traffic to inflamed central nervous system, but require antigen recognition for activation. Eur J Immunol 2000; 30:1002-1009.
30. Tan J, Town T, Suo Z et al. Induction of CD40 on human endothelial cells by Alzheimer's beta-amyloid peptides. Brain Res Bull 1999; 50:143-148.
31. Loetscher P, Uguccioni M, Bordoli L et al. CCR5 is characteristic of Th1 lymphocytes. Nature 1998; 391:344-345.
32. Glabinski AR, Tani M, Tuohy VK et al. Central nervous system chemokine mRNA accumulation follows initial leukocyte entry at the onset of acute murine experimental autoimmune encephalomyelitis. Brain Behav Immun 1995; 9:315-330.
33. Karpus WJ, Lukacs NW, McRae BL et al. An important role for the chemokine macrophage inflammatory protein-1 alpha in the pathogenesis of the T cell-mediated autoimmune disease, experimental autoimmune encephalomyelitis. J Immunol 1995; 155:5003-5010.
34. Kennedy KJ, Strieter RM, Kunkel SL et al. Acute and relapsing experimental autoimmune encephalomyelitis are regulated by differential expression of the CC chemokines macrophage inflammatory protein-1alpha and monocyte chemotactic protein-1. J Neuroimmunol 1998; 92:98-108.
35. Glabinski AR, Tuohy VK, Ransohoff RM. Expression of chemokines RANTES, MIP-1alpha and GRO-alpha correlates with inflammation in acute experimental autoimmune encephalomyelitis. Neuroimmunomodulation 1998; 5:166-171.
36. Miyagishi R, Kikuchi S, Takayama C et al. Identification of cell types producing RANTES, MIP-1 alpha and MIP-1 beta in rat experimental autoimmune encephalomyelitis by in situ hybridization. J Neuroimmunol 1997; 77:17-26.
37. Nygardas PT, Maatta JA, Hinkkanen AE. Chemokine expression by central nervous system resident cells and infiltrating neutrophils during experimental autoimmune encephalomyelitis in the BALB/c mouse. Eur J Immunol 2000; 30:1911-1918.
38. Youssef S, Wildbaum G, Maor G et al. Long-lasting protective immunity to experimental autoimmune encephalomyelitis following vaccination with naked DNA encoding C-C chemokines. J Immunol 1998; 161:3870-3879.
39. Godiska R, Chantry D, Dietsch GN et al. Chemokine expression in murine experimental allergic encephalomyelitis. J Neuroimmunol 1995; 58:167-176.
40. Asensio VC, Lassmann S, Pagenstecher A et al. C10 is a novel chemokine expressed in experimental inflammatory demyelinating disorders that promotes recruitment of macrophages to the central nervous system. Am J Pathol 1999; 154:1181-1191.
41. Tani M, Fuentes ME, Peterson JW et al. Neutrophil infiltration, glial reaction, and neurological disease in transgenic mice expressing the chemokine N51/KC in oligodendrocytes. J Clin Invest 1996; 98:529-539.
42. McColl SR, Staykova MA, Wozniak A et al. Treatment with anti-granulocyte antibodies inhibits the effector phase of experimental autoimmune encephalomyelitis. J Immunol 1998; 161:6421-6426.
43. Traugott U, McFarlin DE, Raine CS. Immunopathology of the lesion in chronic relapsing experimental autoimmune encephalomyelitis in the mouse. Cell Immunol 1986; 99:395-410.
44. Krakowski M, Owens T. Interferon-gamma confers resistance to experimental allergic encephalomyelitis. Eur J Immunol 1996; 26:1641-1646.
45. Tran EH, Prince EN, Owens T. IFN-gamma shapes immune invasion of the central nervous system via regulation of chemokines. J Immunol 2000; 164:2759-2768.

46. Glabinski AR, Tani M, Strieter RM et al. Synchronous synthesis of alpha- and beta-chemokines by cells of diverse lineage in the central nervous system of mice with relapses of chronic experimental autoimmune encephalomyelitis. Am J Pathol 1997; 150:617-630.
47. Tani M, Glabinski AR, Tuohy VK et al. In situ hybridization analysis of glial fibrillary acidic protein mRNA reveals evidence of biphasic astrocyte activation during acute experimental autoimmune encephalomyelitis. Am J Pathol 1996; 148:889-896.
48. Youssef S, Wildbaum G, Karin N. Prevention of experimental autoimmune encephalomyelitis by MIP-1alpha and MCP-1 naked DNA vaccines. J Autoimmun 1999; 13:21-29.
49. Tran EH, Kuziel WA, Owens T. Induction of experimental autoimmune encephalomyelitis in C57BL/6 mice deficient in either the chemokine macrophage inflammatory protein-1alpha or its CCR5 receptor. Eur J Immunol 2000; 30:1410-1415.
50. Rottman JB, Slavin AJ, Silva R et al. Leukocyte recruitment during onset of experimental allergic encephalomyelitis is CCR1 dependent. Eur J Immunol 2000; 30:2372-2377.
51. Fife BT, Huffnagle GB, Kuziel WA et al. CC chemokine receptor 2 is critical for induction of experimental autoimmune encephalomyelitis. J Exp Med 2000; 192:899-906.
52. Izikson L, Klein RS, Charo IF et al. Resistance to experimental autoimmune encephalomyelitis in mice lacking the CC chemokine receptor (CCR)2. J Exp Med 2000; 192:1075-1080.
53. Kivisakk P, Teleshova N, Ozenci V et al. No evidence for elevated numbers of mononuclear cells expressing MCP-1 and RANTES mRNA in blood and CSF in multiple sclerosis. J Neuroimmunol 1998; 91:108-112.
54. Zang YC, Samanta AK, Halder JB et al. Aberrant T cell migration toward RANTES and MIP-1 alpha in patients with multiple sclerosis. Overexpression of chemokine receptor CCR5. Brain 2000; 123:1874-1882.
55. Balashov KE, Rottman JB, Weiner HL et al. CCR5(+) and CXCR3(+) T cells are increased in multiple sclerosis and their ligands MIP-1alpha and IP-10 are expressed in demyelinating brain lesions. Proc Natl Acad Sci U S A 1999; 96:6873-6878.
56. Sorensen TL, Tani M, Jensen J et al. Expression of specific chemokines and chemokine receptors in the central nervous system of multiple sclerosis patients. J Clin Invest 1999; 103:807-815.
57. Miyagishi R, Kikuchi S, Fukazawa T et al. Macrophage inflammatory protein-1 alpha in the cerebrospinal fluid of patients with multiple sclerosis and other inflammatory neurological diseases. J Neurol Sci 1995; 129:223-227.
58. Simpson JE, Newcombe J, Cuzner ML et al. Expression of monocyte chemoattractant protein-1 and other beta- chemokines by resident glia and inflammatory cells in multiple sclerosis lesions. J Neuroimmunol 1998; 84:238-249.
59. McManus C, Berman JW, Brett FM et al. MCP-1, MCP-2 and MCP-3 expression in multiple sclerosis lesions: an immunohistochemical and in situ hybridization study. J Neuroimmunol 1998; 86:20-29.
60. Van Der Voorn P, Tekstra J, Beelen RH et al. Expression of MCP-1 by reactive astrocytes in demyelinating multiple sclerosis lesions. Am J Pathol 1999; 154:45-51.
61. Simpson J, Rezaie P, Newcombe J et al. Expression of the beta-chemokine receptors CCR2, CCR3 and CCR5 in multiple sclerosis central nervous system tissue. J Neuroimmunol 2000; 108:192-200.
62. Simpson JE, Newcombe J, Cuzner ML et al. Expression of the interferon-gamma-inducible chemokines IP-10 and Mig and their receptor, CXCR3, in multiple sclerosis lesions. Neuropathol Appl Neurobiol 2000; 26:133-142.
63. Bennetts BH, Teutsch SM, Buhler MM et al. The CCR5 deletion mutation fails to protect against multiple sclerosis. Hum Immunol 1997; 58:52-59.
64. Barcellos LF, Schito AM, Rimmler JB et al. CC-chemokine receptor 5 polymorphism and age of onset in familial multiple sclerosis. Multiple sclerosis Genetics Group. Immunogenetics 2000; 51:281-288.
65. Sellebjerg F, Madsen HO, Jensen CV et al. CCR5 delta32, matrix metalloproteinase-9 and disease activity in multiple sclerosis. J Neuroimmunol 2000; 102:98-106.
66. Fiten P, Vandenbroeck K, Dubois B et al. Microsatellite polymorphisms in the gene promoter of monocyte chemotactic protein-3 and analysis of the association between monocyte chemotactic protein-3 alleles and multiple sclerosis development. J Neuroimmunol 1999; 95:195-201.
67. Teuscher C, Butterfield RJ, Ma RZ et al. Sequence polymorphisms in the chemokines Scya1 (TCA-3), Scya2 (monocyte chemoattractant protein (MCP)-1), and Scya12 (MCP-5) are candidates for eae7, a locus controlling susceptibility to monophasic remitting/nonrelapsing experimental allergic encephalomyelitis. J Immunol 1999; 163:2262-2266.
68. Zou YR, Kottmann AH, Kuroda M et al. Function of the chemokine receptor CXCR4 in haematopoiesis and in cerebellar development [see comments]. Nature 1998; 393:595-599.

69. Cyster JG. Chemokines and cell migration in secondary lymphoid organs. Science 1999; 286:2098-2102.
70. Boehme SA, Lio FM, Maciejewski-Lenoir D et al. The chemokine fractalkine inhibits Fas-mediated cell death of brain microglia. J Immunol 2000; 165:397-403.
71. Paresce DM, Ghosh RN, Maxfield FR. Microglial cells internalize aggregates of the Alzheimer's disease amyloid beta-protein via a scavenger receptor. Neuron 1996; 17:553-565.
72. Meda L, Bernasconi S, Bonaiuto C et al. Beta-amyloid (25-35) peptide and IFN-gamma synergistically induce the production of the chemotactic cytokine MCP-1/JE in monocytes and microglial cells. J Immunol 1996; 157:1213-1218.
73. Jensen MB, Hegelund IV, Lomholt ND et al. IFNgamma enhances microglial reactions to hippocampal axonal degeneration. J Neurosci 2000; 20:3612-3621.
74. Hausmann EH, Berman NE, Wang YY et al. Selective chemokine mRNA expression following brain injury. Brain Res 1998; 788:49-59.
75. Galasso JM, Harrison JK, Silverstein FS. Excitotoxic brain injury stimulates expression of the chemokine receptor CCR5 in neonatal rats. Am J Pathol 1998; 153:1631-1640.

CHAPTER 9

Cytokines and Chemokines in the Pathogenesis of Murine Type 1 Diabetes

C. Meagher, S. Sharif, S. Hussain, M. J. Cameron, G. A. Arreaza and T. L. Delovitch

Introduction

The immune system can be considered as an intricate set of cell-cell interactions initiated by exposure to antigen and regulated by multiple positive and negative signals derived from lymphocytes, antigen presenting cells (APCs), and stromal cells located in primary and secondary lymphoid tissues. These signals are necessary to maintain a balance between tolerance and immunity. When this balance is not maintained, the consequence may be the development of an organ-specific autoimmune disease, such as autoimmune Type I diabetes (T1D). Here we describe how a group of proteins called cytokines/chemokines are involved in mediating tolerance, and also discuss their biological activities as they pertain to the development of insulitis and islet β cell destruction. A more complete understanding of the biological activities of cytokines and chemokines may lead to the development of novel therapeutics aimed at correcting improper cytokine- or chemokine-mediated immune responses, such as those leading to the development of T1D.

Immune Deviation and the NOD Mouse

The NOD mouse spontaneously develops T1D with a similar immunological and pathological profile to the human disease, and is the most widely used animal model for the study of organ-specific autoimmunity. Considerable evidence supports the notion that regulatory cells exist in the NOD mouse, which can suppress the autoimmune response and the development of T1D. In this context, the breakdown of tolerance followed by induction of autoimmunity and islet β cell destruction requires the cooperation of APCs and lymphocytes. Considerable evidence suggests that both $CD4^+$ and $CD8^+$ T cells are required to facilitate the development of T1D and islet β cell death in NOD mice.[1-9] Based on their respective cytokine secretion profiles, activated $CD4^+$ T cells can be categorized into the T helper (Th) 1 and Th2 subsets in mice and humans.[10,11] $CD4^+$ Th1 cells secrete interleukin (IL)-2, interferon (IFN)-γ, and tumor necrosis factor (TNF)-α and -β, whereas $CD4^+$ Th2 cells secrete IL-4, IL-5, IL-6, IL-10 and IL-13.[12-14] Th1 cells are responsible for cell-mediated immunity, promote inflammation, and are believed to be effector cells in the development of T1D and other autoimmune diseases.[3,15-18] Furthermore, a high IFN-γ/IL-4 expression ratio by islet infiltrating T cells is a predictor of destructive insulitis and a high incidence of T1D.[19] Th2 cells are responsible for humoral immunity and the downregulation of inflammatory Th1 cells,[20,21] and as such, may act as regulatory T cells that block the development of T1D. Indeed, many studies have shown

Cytokines and Chemokines in Autoimmune Disease, edited by Pere Santamaria.

that immune deviation to a Th2 phenotype can protect from T1D as well as other autoimmune diseases.[16,22-26] Nonetheless, $CD4^+$ Th2 cells can also result in pathology and the development of T1D.[27,28] As such, a Th2-like environment does not always protect from autoimmunity,[29] and this outcome may be explained by differences in the antigen specificity, kinetics of cytokine production, and/or tissue migration of T cells. Additionally, it is important to emphasize that these studies were performed with Th2 clones generated *in vitro* and that the enrichment of Th2 cells at the site of inflammation has not been demonstrated after in vivo cell transfer. Moreover, it is not clear whether the cells maintain a stable Th2-like phenotype in vivo and whether manipulation in vitro renders these cells more cytotoxic to islet β cells independent of their cytokine secretion profile. Lastly, islet β cell destruction in humans is probably due to a polyclonal rather than monoclonal immune response. Taking into account these considerations, a model illustrating how the balance between Th1 and Th2 cells might mediate the development of or protection from T1D is shown in Figure 9.1.

This chapter focuses upon the ability of cytokines and chemokines to mediate the development of insulitis and T1D. We will present the role of cytokines and chemokines in relation to the breakdown of peripheral tolerance to islet β cell antigens, activation of regulatory T cells that protect against T1D, and cytokine-mediated mechanisms of islet β cell destruction. The contribution of several cytokines and chemokines to the development of T1D is discussed based on their anti-inflammatory or pro-inflammatory properties, recognizing that certain cytokines may be both anti-inflammatory and pro-inflammatory.

Anti-Inflammatory Cytokines and Autoimmune Diabetes

Interleukin-4 (IL-4)

The presence of IL-4 is characteristic of a Th2-like environment, and as such, its role in the development of T1D has been studied extensively. Our lab previously determined that peripheral $CD4^+$ T cells and thymocytes from NOD mice exhibit an in vitro proliferative unresponsiveness beginning at the time of insulitis. This hyporesponsiveness (anergy) is characterized by reduced IL-2 and IL-4 production, and may be reversed by exposure to physiological levels of exogenous IL-4 but not IL-2.[30,31] Interestingly, anergy and decreased IL-4 production by TCR-stimulated human T cells from newly diagnosed T1D patients has also been demonstrated.[32] Analyses of twin/triplet sets discordant for T1D have shown that all $CD4^-CD8^-V\alpha14J\alpha Q^+$ NKT cell clones from diabetic siblings produce only IFN-γ, whereas > 95% of clones derived from at risk nonprogressor siblings secrete both IL-4 and IFN-γ.[33] These results suggest that deficient IL-4 production by $CD4^-CD8^-$ NKT cells may play an important role in the development of T1D.

The in vivo administration of either recombinant IL-4 to prediabetic NOD mice prevents T1D by reversal of $CD4^+$ T cell hyporesponsiveness and stimulation of a Th2-dominant immune response that alters the recruitment of autoreactive T cells to islets and other sites of inflammation in NOD mice.[31,34-36] Protection from T1D is associated with islet infiltrating T cells that secrete elevated levels of IL-4. Splenic T cells isolated from IL-4 treated NOD mice reduce and delay the onset of T1D and also suppress diabetogenic NOD effector T cell function in NOD.*Scid* recipients.[34] These results suggest that IL-4 treatment induces regulatory T cell function in NOD mice.

Anti-CD28 mAb treatment effectively prevents the onset of destructive insulitis and T1D in NOD mice, provided that treatment is administered perinatally at a sufficiently early age (2- to 4-weeks) and prior to the onset of insulitis.[23] Similar treatment of mice after the onset of insulitis (5-7 weeks of age) does not protect from T1D. Examination of the cytokine secretion profiles of stimulated peripheral (spleen) and islet infiltrating T cells and the ex vivo detection of the expression of intra-pancreatic cytokines of anti-CD28 treated (2-4 week-old) mice revealed a significant upregulation of IL-4 production. The levels of IFN-γ expression remain unchanged. Hence, anti-CD28 mediated costimulation and protection from T1D is mediated

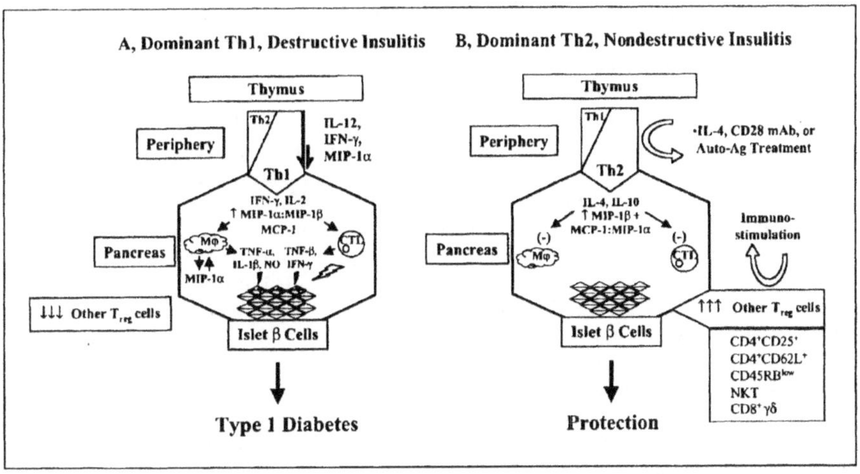

Fig. 9.1. A model illustrating how the balance between Th1 and Th2 cells might mediate the development of or protection against T1D.

by the preferential induction of a Th2-like response in NOD mice. In support of this notion, we found that anti-CD28 mAb treatment in vivo leads to an increased production of IgG1 rather than IgG2a anti-GAD67 autoantibodies in the sera of treated animals.[23] Furthermore, treatment of 2 to 4 week-old NOD mice with anti-IL-4 plus anti-CD28 blocks the anti-CD28 induced protection from T1D. The protective effect of anti-CD28 treatment therefore seems to be mediated by the expansion and survival of IL-4 producing Th2 cells. However, the induction of other regulatory T cells such as $CD4^+CD25^+$ T cells is also possible given that the generation and homeostasis of these cells depends on efficient CD28/B7 costimulation.[37] Taken together, these findings are consistent with the idea that immune dysregulation occurs in NOD mice that results in a predominant Th1-like response at the expense of an impaired Th2-like response, culminating in the development of T1D. However, if a sufficient threshold level of costimulation is provided, the capacity to generate a Th2 and/or other regulatory T cell response is restored and the consequent IL-4 produced mediates protection from destructive insulitis and T1D.

The role of IL-4 in protection against T1D has also been evaluated by generating transgenic NOD mice expressing IL-4 (NOD-IL-4) specifically in islet β cells.[38] These mice are completely protected from insulitis and T1D, and pancreatic expression of IL-4 modulates the effector function of autoreactive lymphocytes as diabetogenic NOD splenocytes are unable to transfer disease to irradiated NOD-IL-4 mice. Splenocytes recognize and proliferate in response to irradiated islet β cells and antigens in vitro, and as such, the NOD-IL-4 mice harbor autoreactive T cells.[26,28] Moreover, islet specific expression of IL-4 shapes the development of Th2-like islet reactive T cells, which is characterized by elevated IL-4 secretion in response to the GAD65 islet autoantigen[39,40] and elevated levels of serum IgG1, IgG2b, and IgG3 specific for GAD65.[26] Elevated levels of IL-4 favors the development of Th2-like islet reactive T cells within islets.[34,36] The importance of the development of a Th2-like immune response in prevention against T1D is also illustrated by the inability of NOD-IL-4 splenocytes to transfer disease to NOD.*Scid* recipients, unless the bioactivity of IL-4 and IL-10 are both neutralized following transfer. Deficient IL-4 and IL-10 is also recently associated with the development of T1D in humans.[41] Furthermore, the ability of NOD-IL-4 splenocytes to delay islet β cell destruction by NOD diabetogenic splenocytes may be reversed by neutralization of IL-4 by anti-IL-4. These studies suggest that IL-4 is required to induce regulatory Th2 cell function that suppresses diabetogenic Th1 cell effector function.[26]

Nonetheless, when NOD-IL-4 and BDC2.5-NOD mice are crossed to generate transgenic mice expressing both IL-4 and a transgenic Vβ4 TCR found on islet autoreactive BDC2.5 T cells, these double transgenic mice rapidly develop T1D.[28] The BDC2.5 Vβ4 TCR recognizes more than 100 peptides with high specificity, and several of the peptides exhibit structural similarity to GAD-65.[42] Given that islet specific expression of IL-4 does not suppress the effector function of transgenic TCR BDC2.5 T cells (limited T cell repertoire) and that onset of T1D is delayed on the NOD.BDC2.5 immune-sufficient genetic background but accelerated on the NOD.*Scid* immunodeficient background,[43] IL-4 might require nonpathogenic T cell populations with expanded specificities to mediate protection.

Plasmid DNA (pDNA) vaccination has also been used to induce Th2 cell effector function in an islet autoantigen specific manner and protect against T1D in NOD mice.[44] One study involved the intramuscular administration of a pDNA vaccine encoding GAD65 linked to IgGFc (GAD65-IgGFc) and IL-4, and prevented the development of T1D if treatment was started either before or after the onset of insulitis. Importantly, pDNA immunization with only GAD65-IgGFc enhanced Th1 cellular activity, indicating that IL-4 is necessary for protection.[44] However, protection was dependent upon GAD65, as pDNA immunization of hen egg lysozyme (HEL)-IgGFc in combination with IL-4 did not prevent T1D. Furthermore, if IL-4 deficient NOD mice are immunized with the pDNA vaccine protection is reversed, demonstrating that both endogenous and exogenous IL-4 production are required for protection. In response to GAD65, splenocyte cultures from nondiabetic NOD mice immunized with the protective pDNA vaccine produce elevated levels of IL-4 and IL-5. The cotransfer of $CD4^+$ T cells from NOD mice immunized with pDNAs encoding GAD65-IgGFc and IL-4 and diabetogenic splenocytes inhibit the transfer of T1D.[44] Thus, regulatory Th2 cells specific for GAD65 require IL-4 to mediate protection. The importance of this study is that it supports the results of Cameron et al[45] and shows that pDNA vaccination appears to be a clinically feasible approach to prevent T1D in humans.[44,45]

The notion that IL-4 suppresses the function of autoreactive T cells and protects from T1D was also suggested by Homann et al,[46] who have used the rat insulin promoter (RIP)-lymphocytic choriomeningitis virus (LCMV) transgenic mouse model for virally induced diabetes to investigate mechanisms of bystander suppression.[46,47] This model involves islet β cell specific expression of LCMV nucleoprotein ($NP^{118\text{-}126}$). After LCMV infection, islet β cells with $NP^{118\text{-}126}$ on their surface are progressively destroyed 1 week after viral clearance by a process involving $NP^{118\text{-}126}$ specific $CD4^+$ and $CD8^+$ T cells and IFN-γ.[14,47,48] The RIP-$NP^{118\text{-}126}$ mice were treated with oral insulin, a diabetes autoantigen, in order to protect from T1D onset by inducing regulatory $CD4^+$ T cells specific for the immunodominant insulin B chain.[46] Transfer of regulatory $CD4^+$ T cell lines derived from the pancreatic draining lymph nodes of insulin fed RIP-$NP^{118\text{-}126}$ mice into prediabetic RIP-$NP^{118\text{-}126}$ mice (5 days after LCMV infection) led to complete protection from T1D. Interestingly, regulatory cells could not be induced to protect from T1D in IL-4 deficient RIP-$NP^{118\text{-}126}$ mice or in Stat6 deficient RIP-$NP^{118\text{-}126}$ mice.[46] Hence, IL-4 was necessary to generate $CD4^+$ regulatory cells by the IL-4 signaling pathway involving Stat6.[49] Furthermore, the regulatory T cells proliferated and $CD4^+$ and $CD8^+$ LCMV-specific T cell responses were downregulated only in pancreatic draining lymph nodes where diabetogenic T cells first encounter islet antigen.[50] In vitro culture of LCMV infected $Stat6^{+/+}$ APCs with Stat6 deficient effector LCMV T cells blocked the expansion and cytotoxic activity of effector LCMV cells specifically when IL-4 was added to cultures.[50] This result shows that APCs play an important role in IL-4 mediated protection from T1D.

Recently, King et al[51] crossed the RIP-NP and NOD-IL-4 mice to generate a transgenic mouse that expresses IL-4 and the LCMV nucleoprotein specifically in islet β cells. These doubly transgenic mice do not develop T1D following LCMV infection. In this model, transgenic IL-4 encourages the development of antigen-specific $CD8^+$ T cells but suppresses the development of effector cytotoxic T lymphocytes (CTL) that recognize LCMV $NP^{118\text{-}126}$ on islet β cells. Effector CTL fail to develop because IL-4 decreases the surface expression of

B7.1 while increasing the surface expression of B7.2 on dendritic cells (DCs). For an efficient CTL response, CD8+ T cells must be activated by two signals, the first being the TCR recognizing a specific antigen presented by MHC class I and the second being costimulatory signals supplied by B7.1 and B7.2.[52] Depending upon the stage of CD8+ T cell development, IL-4-shaped DCs affects the development of CD8+ T cell subsets by changing the expression of B7.1 and B7.2 on their surface. These results suggest that B7.2 potentiates CD8+ T cell expansion, while B7.1 regulates the development of CD8+ T cells into effector CTL. As such, the upregulation of B7.1 on DCs correlates with protection from T1D.

Since IL-4 protects against T1D, one might expect that deficient expression of IL-4 would exacerbate disease onset. However, a lack of IL-4 in NOD.IL-4$^{-/-}$ mice does not accelerate the development of T1D.[53] This result might be explained by the findings that wild-type NOD mice are already quite deficient in IL-4 production.[31,54] Accordingly, the absence of IL-4 maintains the IL-4 deficiency and mediates the development of T1D. In another study, Radu et al[55] examined the outcome of deleting the IL-4Rα gene on the development of T1D in TCR-hemagglutinin (HA) RIP-HA double transgenic mice. This model of T1D requires the transgenic expression of both a class II MHC–restricted TCR (I-E^d) specific for a influenza virus hemagglutinin peptide (TCR-HA) and the hemaglutinin gene under the control of the rat insulin promoter (RIP-HA). Previous analysis of TCR-HA/RIP-HA mice showed that HA-specific lymphocytes play an important role in islet β cell destruction, a process correlated to intensified expression of pro-inflammatory cytokines.[56] Hence, it was unexpected that deletion of the IL-4Rα chain would protect against T1D. Since both IL-4 and IL-13 use IL-4Rα as a signaling chain, deletion of IL-4Rα chain impairs the activity of both IL-4 and IL-13.[55] Thus, the absence of IL-4 and IL-13 activity appears to protect against T1D, a controversial result as both IL-4 and IL-13 have been shown to protect against T1D.[26,34,38,44,57] Further experimentation is required to delineate the mechanism(s) of protection from T1D in IL-4Rα knockout mice.

Overall, IL-4 administration or expression has been shown in various models to inhibit the destructive processes mediated by APCs and T cells in the development of T1D. IL-4 therefore appears to mediate functional tolerance and development of islet destructive CD4+ and CD8+ T cells. It follows that treatments which involve a combination of an autoantigen(s), cytokine(s) and adjuvant-like molecule(s) that shape an anti-inflammatory antigen-specific effector immune response may provide minimally invasive therapies to prevent and/or reverse T1D in humans.

Interleukin-6 (IL-6)

IL-6 possesses both anti-inflammatory and pro-inflammatory properties, binds with high affinity to the IL-6 receptor (IL-6R) and soluble IL-6R,[58,59] and is produced by numerous cell types. The role of IL-6 in inflammation is characterized by the production of acute phase proteins. IL-6 influences the development of activated B cells,[60] stimulates the proliferation of both thymic and peripheral lymphocytes,[61,62] induces the development of effector CTL in the presence of IL-1[63,64] and activates NK cells.[65] Clearly, IL-6 is important for both nonspecific and adaptive immune responses, and may regulate the development of T1D.

In a cyclophosphamide (CY)-induced model of T1D, mice treated with a neutralizing anti-IL-6 antibody were significantly protected from T1D.[66] Transgenic C57BL/6 mice over expressing IL-6 in islet β cells developed insulitis but not T1D. In young mice, the observed structural changes within the islets and surrounding pancreatic tissue changes include islet hyperplasia, neo-ductal formation, and fibrosis.[66] In older mice, islet infiltrating cells were mainly composed of B220+ lymphocytes, but macrophages and T lymphocytes were also observed. These findings suggest an important role for IL-6 in tissue development. By comparison, NOD transgenic mice expressing IL-6 in islet β cells mice develop insulitis with similar kinetics to their nontransgenic littermates, but develop T1D with slower kinetics.[67] Thus, IL-6 appears to have a pathogenic role in the development of insulitis but a protective effect during the conversion from nondestructive to destructive insulitis. The protective mechanism still requires clarification, but could involve interactions of IL-6 with islet β cells and/or islet infiltrating cells.[67] Given that IL-6 influences the development of naïve CD4+ lymphocytes towards an IL-4 producing Th2 phenotype,[68] IL-6 may protect from T1D by inducing the development of IL-4 producing Th2 cells.

Transforming Growth Factor-β (TGF-β)

TGF-β1 functions as a negative regulator of the immune system primarily by inhibition of IL-2 dependent cell proliferation and production of pro-inflammatory cytokines from mononuclear cells, downregulation of MHC expression on APCs and decreased membrane expression of the B cell receptor.[69-73] Due to its immunosuppressive properties, attempts have been made to elucidate the role of TGF-β1 in the pathogenesis of T1D.[74-79]

Transgenic expression of RIP-TGF-β1 in NOD islet β cells reduces the incidence of T1D.[74] This protective effect was not due to a direct effect of TGF-β1 on diabetogenic T cells, as transfer of splenocytes from diabetic NOD mice into irradiated RIP-TGF-β1 mice did not protect against T1D. However, a delay in onset of T1D was observed when splenocytes from RIP-TGF-β1 and diabetic NOD mice were cotransferred into NOD.*Scid* mice. Moreover, the protective effect of RIP-TGF-β1 splenocytes cotransferred with diabetic NOD splenocytes was lost upon administration of a neutralizing anti-IL-4 antibody. This suggests that localized expression of TGF-β1 in the pancreatic islets may shift an IFN-γ producing Th1 phenotype towards an IL-4 producing Th2 phenotype. Furthermore, transgenic expression of RIP-TGF-β1 in the pancreas also shifts the presentation of islet antigen from B cells to macrophages, which may be important for the expansion of autoreactive T cells. Interestingly, transgenic expression of TGF-β1 under the control of the rat glucagon promoter (RGP-TGF-β1) in islet a cells of NOD mice also protects against T1D.[77] The paracrine effect of TGF-β1 is more potent, as RGP-TGF-β1 NOD mice do no develop T1D even after CY administration. Similarly, the RGP-TGF-β1 NOD mice remain diabetes-resistant after adoptive transfer of islet β cell specific CD4+ and CD8+ T cell clones. In addition to the suppressive effect of high concentrations of TGF-β1 on autoreactive T cells, the possibility was raised that TGF-β1 may induce the generation of regulatory T cells that mediate protection against T1D.[77]

Systemic expression of TGF-β1 also protects NOD mice from T1D, and is mediated by deviation of islet infiltrating T cells from a Th1 to a Th2 phenotype.[79] Systemic expression of TGF-β1 in female NOD mice by intramuscular injection of plasmid DNA encoding murine TGF-β1 under the control of the cytomegalovirus promoter (pCMV-TGF-β1) also protects against spontaneous and CY-induced T1D.[78] The protective effect may be the outcome of reduced IL-12 and IFN-γ mRNA expression in pancreatic islets of pCMV-TGF-β1 injected NOD mice, which suggests the development of a Th2 response. Other regulatory activities of TGF-β1, such as the inhibitory effects on T cell proliferation, antigen processing and presentation by APCs, and production of pro-inflammatory cytokines and nitric oxide by islet infiltrating cells may be responsible for protection against the development of T1D.

Induction of oral tolerance by feeding insulin also generates TGF-β1 producing regulatory T cells in gut associated lymphoid tissues, which then migrate to the pancreas to suppress insulitis by a mechanism of bystander suppression.[22] Adoptive transfer of a TGF-β1 secreting CD4+ T cell clone isolated from islet infiltrating lymphocytes of acutely diabetic NOD mice prevents the development of T1D.[80] The protective effect of TGF-β1 may be related to inhibition of the expansion of autoreactive T cells by the blocking of IL-2 dependent cell signaling pathways.[70]

In contrast to its protective role in T1D, transgenic expression of TGF-β1 in pancreatic islet β cells of diabetes-resistant mice results in chronic pancreatitis with fibrosis and accumulation of extracellular matrix.[75] Moreover, double transgenic mice that express LCMV-NP as well as TGF-β1 in their islet β cells fail are not protected against T1D upon LCMV infection.[76] It was reasoned that either local expression of TGF-β1 does not inhibit continued antigen presentation required to perpetuate an immune response to the viral antigen, or continued viral antigen presentation in the islets is not important in the RIP-LCMV-NP model of T1D.[76] In advanced stages of diabetes, hyperglycemia stimulated production of TGF-β1 via activation of protein kinase C by various kidney cells also results in accumulation of extracellular matrix and diabetic nephropathy.[81,82] The over production of active TGF-β1 may be responsible for the pathological changes in the pancreas of these transgenic mice.[77]

Thus, due to its ability to divert an immune response from a Th1 to a Th2 phenotype, TGF-β1 may be a potential candidate for the prevention of T1D. However, achievement of suitable local or systemic threshold levels of TGF-β1 is important as over production may lead to local or systemic pathology.[75,76,82]

Interleukin-10 (IL-10)

Several lines of evidence suggest an immunoregulatory role for IL-10. Autoimmune colitis mediated by Th1 CD45RB(high) $CD4^+$ cells may be prevented by CD45RB(low) cells and the immunoregulatory activity of the latter population is attributable to IL-10.[83] Suppression of experimental allergic encephalomyelitis (EAE) by regulatory cells is IL-10 dependent,[84] and EAE may be prevented by IL-10 gene therapy as well as IL-10 transgene expression.[85,86] Immune privilege in the anterior chamber of eye in the ACAID model is IL-10 dependent.[87] The role of IL-10 in promoting the activity of regulatory cells in transplantation tolerance has also been appreciated.[88] A resident $CD4^+$ population in the skin of patients with nickel allergic contact dermatitis regulates the response of effector cells via secretion of IL-10 and ultimately inhibition of differentiation and maturation of skin DCs.[89,90]

Despite the above findings, the role of IL-10 in the pathogenesis of T1D remains paradoxical. Early studies demonstrated that treatment of NOD mice with recombinant IL-10[91] or by systemic IL-10 gene therapy[92] is protective against T1D even when initiated after the onset of insulitis at 9-10 weeks of age. Similarly, transduction of islet-specific Th1 cells with IL-10 reduces the severity of insulitis and incidence of T1D in adoptively transferred mice.[93] T cell specific expression of an IL-10 transgene is protective against T1D.[94] The function of regulatory T cells that are induced by oral insulin therapy and suppress insulitis in NOD mice is associated with their ability to secrete IL-10.[22] Similarly, IL-10 is produced by diabetes-protective T cells obtained from NOD mice immunized with GAD and insulin via mucosal routes.[95] The importance of IL-10 in islet transplantation emerges from the finding that IL-4/IL-10 combination therapy in diabetic recipients of syngeneic islet grafts effectively prolongs graft survival and prevents the recurrence of T1D.[96] Furthermore, this IL-4/IL-10 combination therapy prevents the primary nonfunction of islet graft islets in NOD mice elicited by CY administration.[97]

IL-10 is also involved in the regulation of susceptibility to T1D in humans. Diabetic patients produce less Th2 cytokines, including IL-10, which precedes an increased production of Th1 cytokines.[98] This finding is supported by discordant twin studies which documented that peripheral blood mononuclear cells of siblings at low risk of disease produce more IL-10 in response to heat shock protein 60 (hsp60).[99]

In contrast, there is a large body of evidence suggesting a pathogenic role for IL-10 in the development of T1D. Syngeneic islet grafts transduced with IL-10 or IL-4 are not protected from the recurrent autoimmune response when transplanted in diabetic NOD mice.[100] Pancreatic-restricted expression of an IL-10 transgene in mice that coexpress LCMV antigens in the pancreas is not protective against LCMV-induced diabetes.[101] Moreover, pancreatic β or α cell specific expression of IL-10 accelerates the development of T1D in an MHC-dependent manner.[76,92,102] IL-10 induced acceleration of T1D is dependent on the presence of $CD8^+$ T cells, while the development of disease is independent of $CD4^+$ T and B cells.[103] Interestingly, administration of CFA and hsp65 prevent T1D in NOD mice but not in NOD.IL-10 transgenic mice.[103] The acceleration of T1D and islet β cell destruction in the latter mice mice does not occur by Fas, perforin, TNFR-1 or TNFR-2 dependent apoptosis pathways.[104] However, acceleration of T1D in these mice is dependent on ICAM-1, which might be present in the immunological synapse and potentiate the formation of islet-specific T cells.[105]

These apparent controversies may be explained by the possibility that the function of IL-10 as an anti-inflammatory or pro-inflammatory cytokine in the pathogenesis of T1D is tissue- and time-dependent. Nonetheless, the consensus view is that the presence of IL-10 at the early stages of diabetes development promotes the generation of effector diabetogenic T cells, whereas expression of IL-10 at the later stages of disease progression is protective.

Interleukin-11 (IL-11)

IL-11 is a member of the gp130 (IL-6) family of cytokines and is produced by a variety of cells in the thymus, bones, connective tissue, central nervous system and lungs.[106] In contrast to IL-6, IL-11 has profound anti-inflammatory effects via the inhibition of the transcription factor, NFkB, which ultimately leads to the decreased production of nitric oxide (NO), IL-1, IL-12 and TNF-α by activated macrophages.[107] The immunoregulatory activity of IL-11 was elucidated in a model of murine model of bone marrow induced graft-*Versus*-Host Disease, in which administration of IL-11 results in reduced expression of Th1 cytokines (IL-12 and IFN-γ) and a significant increase in a Th2 cytokine (IL-4).[108]

T1D may be prevented when female NOD mice are treated with IL-11 starting at 4 weeks of age. However, IL-11 has no protective effect if this treatment is initiated at 18 weeks of age or is withdrawn at 22 weeks of age.[109] Protection is associated with a significant reduction in the systemic production of IL-12, TNF-α and IFN-γ following the injection of anti-CD3 plus LPS.[109]

Interleukin-13 (IL-13)

IL-13 and IL-4 are two closely related Th2 cytokines with overlapping as well as distinct functions.[110] IL-13 plays a critical role in the generation of Th2 responses, as IL-13 deficient mice present with impaired Th2 cell development.[111,112] Similarly, administration of IL-13 promotes an asthma-like phenotype and inactivation of IL-13 ameliorates experimental asthma.[113] This could be partially explained by a significant induction of eotaxin and eosinophil recruitment in the lungs following administration of IL-13. Interestingly, IL-13 was found to be by far the most potent inducer of eotaxin in comparison with other Th2 cytokines.[114] There is also evidence supporting the notion that IL-13 and IL-4 cooperate in an additive manner to initiate Th2 responses.[115] IL-13 exerts its immunregulatory activities by suppression of NF-kB and preservation of IkBα as well as suppression of IL-12, TNF-α and IFN-γ.[116-118] The immunoegulatory effects of IL-13 in promoting the generation of DCs and modulation of monocyte functions have also been described.[117,119]

IL-13 possesses suppressive effects against the development of EAE in the Lewis rat model.[120] Similarly, the suppression of adoptively transferred EAE by a Th2 clone specific for an altered peptide ligand of proteolipid protein was associated with the secretion of IL-13 and other Th2 cytokines.[121] The spontaneous development of T1D is prevented by administration of IL-13 to NOD mice starting at 5 or 14 weeks of age.[57] Protection in this model is associated with the reduced systemic production of TNF-α and IFN-γ and increased production of IL-4.[57] Supportive evidence for a protective role for IL-13 in T1D was obtained in a family study, in which peripheral blood samples from subjects at risk for T1D produced less IL-13 in response to PHA or PHA plus insulin compared to that of healthy controls. In contrast, subjects at low risk of development of T1D produced more IL-13.[122] However, it is noteworthy that whereas IL-4Rα$^{-/-}$ mice are resistant to the development of T1D,[123] IL-4Rα$^{+/+}$ mice are susceptible to the development of T1D in a RIP-influenza hemagglutinin model.[55]

Interleukin-3 (IL-3)

Diabetes may be perceived as a stem cell disease, as bone marrow transplantation from NOD donors into genetically resistant mice renders the recipients susceptible to T1D.[124-126] Transfer of allogeneic bone marrow from T1D-resistant mice into NOD mice prevents T1D.[124,127] NOD mice possess defective responses of their bone marrow myeloid progenitors to IL-3, GM-CSF and IL-5, which may ultimately lead to impaired macrophage development and dysfunction.[128,129] A defect in the differentiation of NOD bone marrow to myeloid DCs was recently identified.[130,131]

IL-3 stimulates the proliferation and differentiation of various hematopoietic progenitor cells. There is however only scant information on its role in autoimmune disease. In the EAE model, nonencephalitogenic T cells produce more IL-3 transcripts than encephalitogenic T cells.[132] This is in contrast to the finding that regulatory T cells in the EAE model downregulate IL-3 production by encephalitogenic T cells.[133] IL-3 is protective against spontaneous T1D when administered to NOD mice starting at 2-4 weeks of age.[134] Moreover, this cytokine or bone marrow cells obtained from IL-3 treated donors can prevent the induction of CY-accelerated disease.[134] It appears that NOD islets are capable of producing IL-3 and that the level of IL-3 production is positively correlated with the degree of mononuclear cell infiltration.[135]

Proinflammatory Cytokines and Autoimmune Diabetes

Interleukin-1 (IL-1)

IL-1 is a pleiotropic multifunctional cytokine implicated in a variety of biological responses, including the inflammatory process that leads to autoimmune disease.[136] IL-1 is a critical effector cytokine for the destructive inflammation of pancreatic islets,[137] and in conjunction with IFN-γ and TNF-α may have a direct cytotoxic effect on islet β cells. This may occur by the induced secretion of other cytotoxic factors and mediators of apoptosis, such as nitric oxide (NO).[138,139] IL-1 stimulates the expression of inducible NO synthase (iNOS) and IFN-γ may increase the sensitivity of islets to such an effect.[140] The main source of IL-1β in rodent and human islets seems to be activated resident macrophages, an important source of NO.[141-144] Thus, the activation of resident islet macrophages and the intra-islet release of IL-1 may mediate the initial dysfunction (cytostatic action) and destruction of β cells (cytocidal action) by inducing the synthesis of iNOS and consequent NO production in islet β cells themselves.[145]

Islet β cell lysis may be mediated by IL-1β induced Fas,[146] and IL-1α and IL-1β together with IFN-γ may sensitize β cells for Fas-dependent destruction by CTL.[147] The stable expression in β cell lines of manganese superoxide dismutase (MnSOD), which interferes with the ability of IL-1β to increase iNOS expression, prevents IL-1β induced cytotoxicity.[148,149] Gene transfer of the IL-1 receptor antagonist protein (IRAP) to cultured human islets can prevent:

1. IL-1β induced β-cell impairment of the dynamic response to a glucose challenge,
2. IL-1β enabled Fas-triggered apoptosis, and
3. Induction of NO production.[150]

However, in vivo data demonstrating the pathogenic effect of IL-1 are scarce. Intraperitoneal injection of IL-1β to normal rats and mice abolishes glucose stimulated insulin secretion from rat pancreas without affecting pancreatic insulin content and islet morphology.[151,152] In diabetes-prone BB rats, a high dose of IL-1β accelerates T1D whereas a low dose reduces the disease incidence.[153] In NOD mice, administration of IL-1α or TNF-α protects from insulitis and T1D.[154] The in vivo neutralization of endogenously produced IL-1 by administration of IRAP or a genetically engineered soluble receptor prevents diabetes onset in NOD mice.[155] IL-1β appears to play a role in the CY-accelerated model of T1D, as treatment of CY-treated NOD mice with an antiIL-1β mAb prevents T1D. These results suggest that specific inhibitors of IL-1β may be attractive targets for therapeutic intervention of T1D.[156] However, it has not yet been possible to create a transgenic mouse expressing and liberating mature IL-1 from islet beta cells.[136] As such, although in vitro data are compelling, more in vivo studies are necessary to confirm the role of IL-1 as critical pathogenic cytokine in the development of autoimmune diabetes.

Interleukin-2 (IL-2)

IL-2 is a pro-inflammatory cytokine that has been linked to a variety of biological responses including B cell, natural killer (NK) cell and T cell activation, T cell development, and Fas induced activation cell death.[157] IL-2 has also been implicated in the development of autoimmunity by breaking self-tolerance.[158] Mice deficient for IL-2 have been generated and are characterized by having normal thymopoiesis and normal numbers of peripheral B and T cells. However, dysregulation of the immune system was characterized by reduced in vitro T cell responses and by variation in serum immunoglobulin isotype levels.[159,160] The IL-2 receptor (IL-2R) is composed of the α and β subunit, and IL-2 binds to the α/β complex with high affinity.[161] This high affinity IL-2R is a specific marker of T cell activation, and is not expressed on memory or resting T cells.[162] Thus, the high affinity IL-2R has for a long time been targeted in models of inflammation, because if successful, it would be possible to kill recently activated T cells and suppress unwanted immune responses, such as the autoimmune destruction of islet β cells.[163] As such, treatment of NOD mice with a cytolytic IL-2/Fcg2a (IL-2/Fc) fusion protein reduced the development of T1D.[164]

Additional evidence that IL-2 plays a pathogenic role in the development of T1D was provided by the report that treatment with an anti-IL-12R mAb prevents the development of insulitis in NOD mice.[165] These results suggested that cells expressing the IL-2R were required for islet β cell destruction and that targeting of the IL-2R may be an efficacious treatment of T1D. In addition to targeting IL-2R expressing cells, it was shown that transgenic RIP-IL-2 mice expressing low levels of IL-2 in islet β cells develop insulitis but not T1D.[166] Importantly, these mice expressed a single copy of the transgene, but when these RIP-IL-2 mice were made homozygous for the transgene, diabetes did develop. In comparing islet infiltrates of the single copy and homozygous RIP-IL-2 mice, the homozygous mice had a high proportion of memory $CD4^+$ T cells at a young age. Expression of the IL-2 transgene in NOD.*Scid* mice resulted in inflammation and T1D in some cases.[167] Hence, it appears that IL-2 can mediate islet β cell destruction in the absence of antigen-specific T or B cells, possibly by modulation of APC function. In a low incidence strain of NOD mice expressing a single copy of the IL-2 transgene, T1D developed at an accelerated rate and required a suitable genetic background that included the diabetes susceptibility loci *Idd*1, *Idd*3, and *Idd*10. Moreover, $CD8^+$ T cells seem to play a key role in accelerating disease onset in this strain.[168] A similar result was also observed in double transgenic mice expressing the LCMV $NP^{118-126}$ and IL-2 in islet β cells. Following challenge with LCMV, IL-2 enhances the development of T1D.[169]

Congenic mapping has localized the *Idd*3 locus to a 145 kb interval that encompasses the IL-2 gene on mouse chromosome 3.[170] As such, the IL-2 gene is a strong candidate for the *Idd*3 locus. Moreover, a serine to proline substitution at position six of IL-2 is associated with both increased glycosylation of IL-2 and diabetes susceptibility.[171-172] These results suggest that there is a good chance that *Idd*3 is an allelic variant of the IL-2 gene, and that functionally distinct variants of IL-2 may be important for diabetes development. However, current data do not demonstrate a functional difference between the allotypes of IL-2.[170]

Thus, IL-2 appears to play an important role in accelerating the development of T1D by modulating APC function or by promoting a toxic cytokine milieu in the pancreas. The concept of targeting T cells expressing the IL-2R following activation has been known for some time, but the technique continues to be developed and may in the future prove effective in preventing human diabetes.

Interleukin-12 (IL-12)

IL-12 is primarily secreted by professional APCs in the form of a bioactive heterodimer comprised of covalently linked 40 kDa and 35 kDa subunits.[171-173] Biologically active IL-12 binds to a high affinity IL-12R consisting of at least two β-type subunits and signals through the IL-12R-β2 subunit.[173] The role of IL-12 in the pathogenesis of autoimmune disease has been extensively investigated.[173-175] In NOD mice, the levels of IL-12p40 mRNA expression in pancreatic islets progressively increase from 5 weeks of age until onset of T1D at 13 weeks of age and correlate with islet β cell destruction.[176] The protective effect of complete Freund's adjuvant (CFA) against T1D is also related to the reduced expression of IL-12p40, IFN-γ and

IL-2 mRNA.[176] CY treatment of NOD mice accelerates diabetes onset by increasing IL-12p40 mRNA expression in pancreatic islets and shifting the T cell response to a Th1-like phenotype.[177]

Systemic administration of recombinant mouse IL-12 to prediabetic female NOD mice induces insulitis and rapid onset of T1D.[175] T cells isolated from insulitis lesions produce elevated amounts of IFN-γ and reduced amounts of IL-4 upon TCR stimulation.[175] This finding suggests that IL-12 promotes T1D in NOD mice by polarizing the T cell response towards a Th1 phenotype. In contrast, another study showed protection against T1D in IL-12 injected female NOD mice.[178] However, IL-12 treatment was ineffective in irradiated male NOD mice that received diabetogenic spleen cells from female NOD mice, suggesting that IL-12 may be important in preventing the development of diabetogenic T cells.[178] The discrepancy between these two studies may be due to the use of different doses and timing of administration of IL-12. Neutralization of endogenous IL-12 by treatment of female NOD mice with an anti-IL-12 antibody at an early age protects against T1D, but this treatment is ineffective if anti-IL-12 treatment is administered after insulitis is established.[179] In contrast, daily i.p. injection of anti-IL-12 accelerates the onset of T1D in female NOD mice.[180] However, the same antibody provided full protection when used twice weekly from 5 to 25 weeks of age. The short term IL-12 neutralization resulted in an increase of Th2 producing $CD4^+CD25^+CD44^{high}$ splenic T cells, suggesting the accumulation of activated memory T cells that might include autoreactive diabetogenic T cells. Neutralization of IL-12 in female NOD mice at an early age also accelerates diabetes onset.[180] Thus, short-term treatment at an early age with anti-IL-12 antibody may inhibit IL-2 production and enhance diabetes onset by accumulation of progenitors of effector T cells. Neutralization of endogenous IL-12 production by a natural IL-12 antagonist has also identified a pathogenic role of IL-12 in T1D.[181] Systemic administration of an IL-12 antagonist, a homodimer of the IL-12p40 subunit called $(p40)_2$, reduces both spontaneous and CY-induced diabetes onset.[181,182] However, when NOD mice are injected with $(p40)_2$ after insulitis is established, only minimal protection against T1D ensues.[182] The IL-12 antagonist protects against T1D by deviating pancreas infiltrating $CD4^+$ T cells from a Th1 to a Th2 phenotype. The reason for reduced protection by IL-12 antagonists in NOD mice with advanced prediabetes is not yet known. However, other cytokines that appear late during the insulitis process, such as IL-18, may compensate for the IL-12 mediated deficiency in IFN-γ production.[181,183] The local effects of IL-12p40 in the pancreas were examined by transfecting NOD islets with an adenovirus vector containing the mouse IL-12p40 gene (Ad.IL-12p40), and transplanting the transfected NOD islets under the renal capsule of female diabetic NOD mice.[184] Local production of IL-12p40 prolonged islet survival and recipient mice remained normoglycemic for more than 4 weeks following transplantation. Ad.IL-12p40 transfected islets produced increased amounts of TGF-β1 and reduced amounts of IFN-γ suggesting that localized expression of IL-12p40 in islets generates TGF-β1 secreting regulatory cells, which protect β cells from destruction.[184]

IL-12 deficient NOD mice still develop T1D.[185] These mice possess normal numbers of $CD4^+$ T cells in pancreatic islets and lack IL-4 and IL-10 producing T cells. The $CD4^+$ T cells in pancreatic islets may be residual Th1 cells that infiltrated the islets due to their increased surface expression of the P-selectin ligand. Hence, impairment of a Th1 response may not be sufficient to prevent T1D, but induction of a regulatory pathway may be necessary for protection against Th1 mediated autoimmunity.[185]

Interferon (IFN)-γ

A direct correlation exists between elevated levels of IFN-γ mRNA expression in islet infiltrating mononuclear cells and onset of T1D in NOD mice.[186] As a pro-inflammatory cytokine, IFN-γ may contribute to the pathogenesis of T1D by several mechanisms, including the activation of autoreactive CTL, upregulation of Fas and MHC class I antigen on islet β cells, and enhanced expression of MHC class II and other costimulatory molecules on APCs.[14,147,187,188] IFN-γ may also exert a direct cytotoxic effect on islet β cells,[189] and elevated concentrations of IFN-γ can inhibit glucose stimulated insulin secretion by islet β cells.[190] The in vivo effect of IFN-γ on pancreatic islet β cells has been examined in NOD.RIP-ΔγR transgenic mice,[191] which express a dominant negative IFN-γR a chain on their β cells rendering these β cells unresponsive to IFN-γ. Both wild type and RIP-ΔγR NOD mice develop T1D at the same rate, suggesting that direct action of IFN-γ may not be required for the development of T1D in NOD mice. Transgenic expression of IFN-γ in islet β cells of diabetes-resistant mice results in insulitis, progressive islet β cell destruction and T1D.[192] Islet β cell destruction in these mice may be due to the generation of autoreactive T lymphocytes as a result of upregulation of costimulatory molecules on APCs.[192]

IFN-γ deficient mice expressing LCMV-NP or GP on their islet β cells show resistance to diabetes upon LCMV infection.[14] Islets of these diabetes-resistant mice lack MHC class II positive cells, suggesting that IFN-γ is required for APCs to infiltrate islets. Reduced insulitis and diabetes incidence was also observed by neutralizing endogenous IFN-γ with anti-IFN-γ antibodies and soluble IFN-γ receptor (sIFN-γR) in both NOD mice and DP-BB rats.[187,193,194] The protective effect of anti-IFN-γ antibody treatment may be due to reduced expression of MHC and other costimulatory molecules on APCs required for the generation of CTL.

The role of IFN-γ in the pathogenesis of T1D has also been examined in IFN-γR deficient NOD mice, and interestingly, complete protection against T1D was observed.[195] Back crossing of IFN-γRα deficient mice onto the NOD genetic background showed that the protective effect was due to transfer of a diabetes resistance gene(s) linked to the IFN-γRα locus in the 129 mouse strain.[196] These results were further supported by congenic transfer of a functionally inactive gene for the IFN-γR β chain from a 129 donor to the NOD background. These NOD.IFN-γRb^{null} mice were unable to provide protection against T1D.[197]

The destructive role of IFN-γ in the pathogenesis of T1D is not conclusive, as some studies also describe a protective role for IFN-γ against T1D.[198,199] IFN-γ deficiency in NOD mice only delays the onset of T1D but does not reduce the severity of disease.[198] Reduced expression of IL-4 and IL-10 in islets of these mice was observed, and suggests that IFN-γ is required for the development of a Th2 response. This finding is supported by the recent demonstration that CFA- and BCG-mediated resistance to T1D in NOD mice is lost by disruption of the IFN-γ gene.[199]

Interleukin-18 (IL-18)

IL-18 was originally identified as an IFN-γ inducing factor expressed by the liver Kupffer cells and activated macrophages.[200] The IFN-γ inducing effects of IL-18 exceed those of IL-12, but they can operate synergistically.[201] Thus, one of the prime effects of IL-18 on the adaptive immune response, in concert with IL-12, is to generate a Th1 environment. However, this cytokine has potent pro-inflammatory effects by inducing the production of TNF-α, and the innate response is also influenced by this cytokine through the activation of NK cells and secretion of IL-8, prostaglandin E_2 and iNOS.[201]

The involvement of IL-18 in the pathogenesis of T1D was indicated by the finding of elevated levels of IL-18 in the NOD pancreata with early insulitis lesions.[202] IL-18 mRNA transcripts in the pancreas of NOD mice are significantly increased following the administration of CY and this precedes the elevation of pancreatic IFN-γ transcripts.[203] In BDC2.5 TCR transgenic mice, IL-18 in concert with IL-12 and IFN-γ plays an instrumental role in the onset of aggressive autoimmune response in the pancreas.[204] IL-18 lacks direct cytotoxic effects on islet β cells and only possesses minor stimulatory effects.[205] It is important to note that although islet β cells express IL-18, they fail to express the IL-18R.[206] Thus, the role of IL-18 in islet β cell destruction is indirect. Interestingly, systemic administration of IL-18 in NOD mice starting at 10 weeks of age delays and partially protects against the onset of T1D.[207] Protection is associated with lower IFN-γ/IL-10 and IFN-γ/IL-4 ratios in the pancreas of IL-18 treated mice, and these mice present less severe insulitis lesions.[207] This raises the possibility that under different conditions IL-18 may act differently. For example, IL-18 seems to prevent the progression of nondestructive insulitis to destructive insulitis. Indeed, there is accumulating evidence suggesting a dual role for IL-18 in regulation of the immune response. IL-18 can drive the differentiation of Th2 cells in the absence of IL-4, and this IL-18 activity may be blocked by IL-12.[208] However, IL-18 can induce the production of IL-4 in an IL-12 independent manner, and this activity seems to be dependent on the presence of NKT cells.[209]

Tumor Necrosis Factor-α (TNF-α)

Another important cytokine involved in the development of insulitis and T1D is TNF-α,[210] an inflammatory Th1 cytokine that is secreted mainly by activated macrophages and $CD4^+$ T cells.[211] TNF-α upregulates adhesion molecules such as ICAM1 and VCAM1 on endothelial cells, and as such, TNF-α might play a role in recruiting lymphocytes to islets.[212,213] TNF-α may also directly induce apoptosis of islet β cells.[214]

The first evidence suggesting a role for TNF-α in islet β cell destruction was the induced upregulation of surface MHC class II on islet β cells in vitro.[215] TNF-α was also shown to upregulate MHC class I on islet cells.[216] Subsequently, TNF-α mRNA was detected in islet infiltrating cells, mainly $CD4^+$ lymphocytes, during the development of T1D.[217] These initial studies correlating TNF-α with the development of T1D were substantiated in follow up studies supporting a key role for TNF-α in the development of T1D.[218] TNF-α was shown to increase T cell autoreactivity to islet β cells and exacerbate T1D when administered in low doses to neonatal NOD mice, while injection of neutralizing anti-TNF-α antibodies during this same neonatal period completely prevents the development of T1D.[218] In contrast, injection of TNF-α to adult NOD mice > 6 weeks of age blocked the development of T1D, whereas injection of anti-TNF-α antibodies exacerbates T1D in age-matched NOD mice.[154,219] Moreover, islet β cell specific expression of TNF-α in adult NOD mice led to the development of insulitis yet prevented the development of T1D,[220-222] while neonatal expression of TNF-α resulted in T1D by 9-12 weeks of age in male and female NOD mice.[211] Importantly, these studies suggested that age-related differences exist in the susceptibility and resistance to T1D by TNF-α treatment.

Hypotheses describing the role of TNF-α in autoimmunity were proposed based upon results showing that chronic TNF-α exposure diminishes T cell effector function characterized by decreased proliferation and reduces Th1 and Th2 cytokine production.[223] In contrast, chronic anti-TNF-α exposure up-regulates antigen-specific T cell responses and effector function.[223] Interpretation of these results led to speculation that TNF-α in neonatal mice acts as a growth factor for thymic T cells that are specific for both self and foreign antigens, augments peripheral T cell effector function, and intensifies the recruitment of activated T cells to the pancreas.[224] Anti-TNF-α blocks these effects and prevents primary follicle and germinal formation in the lymph nodes. Thus, anti-TNF-α treatment may decrease autoreactive B cell formation and B cell APC function, and interfere with the development and migration of autoreactive T cells to the pancreas.

Chronic exposure to anti-TNF-α and TNF-α increases and reduces TCR signaling, respectively.[223] Based on this result, it was proposed that TNF-α, which is constitutively expressed in the neonatal thymus, may hinder TCR signaling in the thymus and negative selection, thus increasing the number of autoreactive T cells escaping to the periphery and migrating to the pancreas.[224] In contrast, anti-TNF-α treatment would enhance negative selection and lower the number of autoreactive T cells escaping to the periphery culminating with the prevention of T1D. In adult NOD mice, treatment with TNF-α might hinder TCR signaling and autoreactive T cell effector function, leading to protection from T1D, while anti-TNF-α treatment might intensify TCR signaling and T cell effector function leading to T1D in adult NOD mice.[16] Thus, it was hypothesized that TNF-α modulates the immune system in an age-dependant manner.

Support for this hypothesis was provided by the report that neonatal TNF-α expression exacerbates the development of T1D, which was associated with the ability of APCs, primarily islet-infiltrating DCs, to present islet autoantigens to both $CD4^+$ and $CD8^+$ T cells.[225] Further studies showed that antigen presentation by DCs to effector $CD8^+$ T cells was critical for the progression to T1D, while $CD4^+$ T cells played a smaller role.[211,226] These data posed the question of whether TNF-α alters interactions between CD154 on $CD4^+$ T cells and CD40 on APCs. This interaction activates APCs, which then activate $CD8^+$ T cells. Previously, CD154-CD40 interactions were shown to be necessary for the development of insulitis and T1D.[227] Using several transgenic and knockout mouse models, it was determined that $CD4^+$ T cell help was not required for the development of naïve $CD8^+$ T cells into islet-specific effector $CD8^+$ T cells. Rather, the results suggested TNF-α can obviate the need for CD154 signals necessary for APC activation, and act as a substitute for $CD4^+$ T cell priming of $CD8^+$ T cells towards islet autoantigens.[226] Accordingly, TNF-α production at a site of inflammation may stimulate an environment conducive to autoimmunity by negating $CD4^+$ T cell-dependent CD154 immunoregulatory mechanisms.[226]

To further clarify the role of TNF-α in the development of T1D, the activities of TNF-α transduced by its two receptors TNFR1 (p55) and TNFR2 (p75) were investigated. The ability of TNF-α to induce cell death appears to be regulated mainly by TNFR1, while TNFR2 plays a minor role.[228] TNFR1-deficient NOD mice mice develop insulitis but not T1D.[229] Furthermore, the adoptive transfer of diabetic NOD splenocytes into sublethally irradiated TNFR1-deficient NOD mice delays the onset of T1D. TNFR1-deficient NOD mice are also resistant to CY induced T1D. These results show that surface expression of TNFR1 on islet β cells mediates islet β cell death.[229] In another well-designed study,[230] NOD.*Scid* mice were treated with streptozotocin to induce T1D and then engrafted with TNFR1 deficient-islets under the kidney capsule resulting in normoglycemia. These NOD chimeric mice were then transferred with diabetogenic BDC2.5/NOD.*Scid* $CD4^+$ T cells. Interestingly, the engrafted TNFR1-deficient islets were only mildly infiltrated (peri-insulitis), were not apoptotic, and remained functional. However, TNFR1-deficient islets were destroyed when engrafted together with TNFR1-sufficient islets.[230] These studies indicated that a nonapoptotic islet response to TNF-α is required to activate $CD4^+$ T cells and develop a destructive intra-islet infiltrate. Essentially, these results imply that islet β cells dictate their own destruction.[230]

Recently, a novel transgenic model (Tet-TNF-α) has been developed where islet β cell specific expression of TNF-α is controlled by the tetracycline-regulated gene transcription system.[231] This model allows for TNF-α gene expression to be turned on or off depending on the absence or presence of tetracycline, respectively. Initiation of TNF-α expression at birth results in insulitis but not T1D. The Tet-TNF-α mice were then crossed to transgenic C57BL/6 (RIP-B7-1) mice expressing the costimulatory molecule B7-1 to evaluate whether the age of the animal or the duration of TNF-α mediated inflammation is key to breaking T cell peripheral tolerance to islet antigens.[231] Significantly, this model removes unknown genetic variables by including a genetic background not partial to diabetes development. Previously, islet specific expression of both TNF-α and B7-1 in C57BL/6 mice resulted in T1D,[232,233] whereas transgenic expression of B7-1 rarely caused insulitis and T1D.[234] In the Tet-TNF-α model, constitutive TNF-α expression beginning at birth or at 6 weeks of age results in T1D.[231] Hence, TNF-α can break peripheral tolerance to islet antigens in an age-independent manner. Subsequently, by downregulating TNF-α expression at critical timepoints leading to T1D (i.e., prior to insulitis, during insulitis but before islet β cell death, or after islet β cell death), it was determined that TNF-α needs to be expressed up to 25 days of age (early stages of insulitis) in order to break peripheral tolerance to islet autoantigens.[231] Furthermore, the results indicate that the duration of TNF-α expression, not the age of the mice, is a critical factor in shaping an environment favoring islet β cell destruction. Based upon this result, it is interesting to consider whether the duration of expression of other cytokines during inflammation, such as IL-10 or IFN-γ, plays a role in breaking peripheral tolerance to islet autoantigens.

Chemokines and Autoimmune Diabetes

Chemokines are a superfamily of low molecular weight (8-14 kDa) chemotactic cytokines that mediate leukocyte migration through interactions with seven-transmembrane, rhodopsin-like G protein-coupled receptors.[235] Depending on the position of the first two cysteine residues in the primary structure of these molecules, chemokines can be divided into four families. The CXC (one amino acid lies between the first two cysteines) family of chemokines includes at least 15 ligands, which mediate mainly neutrophil chemotaxis and binds at least five receptors (CXCR). The CC (no intervening amino acid) chemokine family includes at least 27 members that bind at least ten receptors (CCR). CC chemokine targets include monocytes, T cells, DCs and NK cells. Recently, it has become more apparent that in addition to their role in the recruitment of leukocytes to sites of inflammation, chemokines play a fundamental role in mediating innate and adaptive immune responses by their ability to recruit, activate, and costimulate cells of the immune system.[236,237] Moreover, Th1 and Th2 cells can be differentiated by their responsiveness to specific chemokines and their distinctive expression of chemokine receptors.[237] Currently, there is limited information describing the role of chemokines in the pathogenesis of diabetes. Previously, we determined that the control of T cell hyporesponsiveness in NOD mice, a phenotype shown to be involved in diabetogenesis,[31] is linked to a central region of chromosome 11 that encompasses the *Idd4* diabetogenic locus and the CC chemokine gene family.[238] Interestingly, the *eae7* genetic locus, which controls susceptibility to monophasic remitting/relapsing EAE, is also linked to a region of chromosome 11 encompassing the CC chemokine gene family.[239] Moreover, sequence polymorphisms in the TCA-3, MCP-1, and MCP-5 CC chemokine genes are considered possible candidates for *eae7*.[240] Together with the well-established role of chemokines in inflammation and the effects of IL-4 and CD28 signaling on chemokine expression,[241,242] the possibility that different chemokines might be associated with T cell differentiation as well as diabetes susceptibility in NOD mice was plausible. In a previous study from our group,[243] candidate CC chemokines that mediate the establishment of insulitis were identified. Macrophage inflammatory protein-1α (MIP-1α) and MCP-1 were shown to play an early effector role in the establishment of insulitis in NOD mice. Interestingly, it was suggested that MCP-1 resident in the pancreas may contribute to early islet infiltration by attracting lymphocytes, the outcome of which depends on the presence or subsequent expression of other chemokines.[243] Transgenic expression of MCP-1 under the control of the rat insulin promoter can establish a monocytic infiltrate in the pancreas.[244] Moreover, the ratio of MIP-1α/MIP-1β appears to be important during the initial stages of islet mononuclear cell infiltration in determining the nature of insulitis progression in NOD mice. Additional evidence regarding the role of MIP-1α in the development of T1D was provided by monitoring the spontaneous incidence of diabetes in MIP-1α deficient NOD mice (NOD.MIP-1$\alpha^{-/-}$). The incidence of T1D is significantly reduced and delayed in NOD.MIP-1$\alpha^{-/-}$ mice as compared to NOD.MIP-1$\alpha^{+/+}$ mice, indicating that MIP-1α is an important effector chemokine in the pathogenesis of T1D.[243] Similarly, an effector role of MIP-1α in EAE pathogenesis has also been determined,[245] and is consistent with its association with Th1-like immune responses.[246-248]

In addition to identifying MIP-1α as an effector molecule in the development of T1D, we found that a close correlation exists between Th2 cytokine responses and a high MIP-1β + MCP-1/MIP-1α-chemokine ratio elicited by IL-4 treatment in the pancreas of mice protected from T1D.[243] In accordance with this result, CCR5 was the only CC chemokine receptor whose expression was modulated in the pancreas upon IL-4 treatment. As CCR5 is linked with Th1 responses,[249] this may reflect a diminished function of Th1-like cells in the pancreas. These results are supported by previous studies that have linked certain chemokine expression to a Th1/Th2 paradigm,[237,246,247,249] and by results showing that islet specific Th1 and Th2 cells can be differentiated by their respective chemokine expression patterns.[250] Additionally, Th1 cell-mediated destruction of islet β cells correlates with a specific chemokine expression pattern.[250]

Overall, the interrelationship of cytokines, chemokines, and chemokine receptors in mediating the distinctive recruitment of effector Th1 cells versus regulatory Th2 cells to the pancreas, and in modulating effector Th1 versus regulatory Th2 cell function, plays a significant role in the development of T1D. Currently, knowledge of the roles that chemokines and chemokine receptors play in the development of T1D is limited, but as chemokines represent an attractive therapeutic target in preventing diabetes, our knowledge is likely to increase rapidly.

Conclusions

The roles of cytokines in the development of T1D have been extensively investigated during the past 18 years, and the roles of chemokines in T1D are rapidly being dissected. The results presented in this chapter are derived from studies that either determined relative cytokine and chemokine levels in the islet β cell environment, deleted the biological activities of a given cytokine, or examined the effects of increasing cytokine and chemokine levels by transgenic expression or systemic administration. One must consider the fact that these studies generally investigate one or at most a few cytokines or chemokines, but it is the cooperation of many cytokines and chemokines that mediates the development of insulitis and T1D. The common theme of the results presented is that if a dominant anti-inflammatory cytokine/chemokine milieu can be generated in the periphery, and locally in the pancreas, as for IL-4, the ability of regulatory cells to prevent insulitis and diabetes can be elicited. Figure 9.1 attempts to summarize this concept. It is noteworthy that several cytokines appear to have both anti- and pro-inflammatory properties in different models of diabetes, such as IL-10, and support the notion that it is the entire cytokine/chemokine milieu that modulates whether or not T1D will develop. Thus, manipulation of the cytokine and chemokine system and/or the cells that create a dominant anti-inflammatory immune response offers great potential for redirecting an autoimmune response towards functional tolerance, and thus preventing organ specific autoimmunity.

References

1. Wong FS, Visintin I, Wen L et al. CD8 T cell clones from young nonobese diabetic (NOD) islets can transfer rapid onset of diabetes in NOD mice in the absence of CD4 cells. J Exp Med 1996; 183:67-76.
2. Nagata M, Santamaria P, Kawamura T et al. Evidence for the role of CD8+ cytotoxic T cells in the destruction of pancreatic beta-cells in nonobese diabetic mice. J Immunol 1994; 152:2042-2050.
3. Wong FS, Dittel BN, Janeway CA. Transgenes and knockout mutations in animal models of type 1 diabetes and multiple sclerosis. Immunol Rev 1999; 169:93-104.
4. Wang B, Gonzalez A, Benoist C et al. The role of CD8+ T cells in the initiation of insulin-dependent diabetes mellitus. Eur J Immunol 1996; 26:1762-1769.
5. Christianson SW, Shultz LD, Leiter EH. Adoptive transfer of diabetes into immunodeficient NOD-scid/scid mice. Relative contributions of CD4+ and CD8+ T-cells from diabetic versus prediabetic NOD.NONThy-1a donors. Diabetes 1993; 42:44-55.
6. Haskins K, McDuffie M. Acceleration of diabetes in young NOD mice with a CD4+ islet-specific T cell clone. Science 1990; 249:1433-1436.
7. Rohane PW, Shimada A, Kim DT et al. Islet-infiltrating lymphocytes from prediabetic NOD mice rapidly transfer diabetes to NOD-scid/scid mice. Diabetes 1995; 44:550-554.
8. Kay TW, Parker JL, Stephens LA et al. RIP-beta 2-microglobulin transgene expression restores insulitis, but not diabetes, in beta 2-microglobulin null nonobese diabetic mice. J Immunol 1996; 157:3688-3693.
9. Bendelac A, Carnaud C, Boitard C et al. Syngeneic transfer of autoimmune diabetes from diabetic NOD mice to healthy neonates. Requirement for both L3T4+ and Lyt-2+ T cells. J Exp Med 1987; 166:823-832.
10. Paul WE, Seder RA. Lymphocyte responses and cytokines. Cell 1994; 76:241-251.
11. Seder RA, Paul WE. Acquisition of lymphokine-producing phenotype by CD4+ T cells. Annu Rev Immunol 1994; 12:635-673.
12. Rabinovitch A. Immunoregulatory and cytokine imbalances in the pathogenesis of IDDM. Therapeutic intervention by immunostimulation? Diabetes 1994; 43:613-621.
13. De Carli M, D'Elios MM, Zancuoghi G et al. Human Th1 and Th2 cells: functional properties, regulation of development and role in autoimmunity. Autoimmunity 1994; 18:301-308.

14. von Herrath MG, Oldstone MB. Interferon-gamma is essential for destruction of beta cells and development of insulin-dependent diabetes mellitus. J Exp Med 1997; 185:531-539.
15. Liblau RS, Singer SM, McDevitt HO. Th1 and Th2 CD4+ T cells in the pathogenesis of organ-specific autoimmune diseases. Immunol Today 1995; 16:34-38.
16. Delovitch TL, Singh B. The nonobese diabetic mouse as a model of autoimmune diabetes: Immune dysregulation gets the NOD. Immunity 1997; 7:727-738.
17. Zamvil SS, Steinman L. The T lymphocyte in experimental allergic encephalomyelitis. Annu Rev Immunol 1990; 8:579-621.
18. Kuchroo VK, Martin CA, Greer JM et al. Cytokines and adhesion molecules contribute to the ability of myelin proteolipid protein-specific T cell clones to mediate experimental allergic encephalomyelitis. J Immunol 1993; 151:4371-4382.
19. Fox CJ, Danska JS. IL-4 expression at the onset of islet inflammation predicts nondestructive insulitis in nonobese diabetic mice. J Immunol 1997; 158:2414-2424.
20. Fiorentino DF, Bond MW, Mosmann TR. Two types of mouse T helper cell. IV. Th2 clones secrete a factor that inhibits cytokine production by Th1 clones. J Exp Med 1989; 170:2081-2095.
21. Rocken M, Urban J, Shevach EM. Antigen-specific activation, tolerization, and reactivation of the interleukin 4 pathway in vivo. J Exp Med 1994; 179:1885-1893.
22. Hancock WW, Polanski M, Zhang J et al. Suppression of insulitis in nonobese diabetic (NOD) mice by oral insulin administration is associated with selective expression of interleukin-4 and -10, transforming growth factor-beta, and prostaglandin-E. Am J Pathol 1995; 147:1193-1199.
23. Arreaza GA, Cameron MJ, Jaramillo A et al. Neonatal activation of CD28 signaling overcomes T cell anergy and prevents autoimmune diabetes by an IL-4-dependent mechanism. J Clin Invest 1997; 100:2243-2253.
24. Chen Y, Kuchroo VK, Inobe J et al. Regulatory T cell clones induced by oral tolerance: Suppression of autoimmune encephalomyelitis. Science 1994; 265:1237-1240.
25. Falcone M, Bloom BR. A T helper cell 2 (Th2) immune response against nonself antigens modifies the cytokine profile of autoimmune T cells and protects against experimental allergic encephalomyelitis. J Exp Med 1997; 185:901-907.
26. Gallichan WS, Balasa B, Davies JD et al. Pancreatic IL-4 expression results in islet-reactive Th2 cells that inhibit diabetogenic lymphocytes in the nonobese diabetic mouse. J Immunol 1999; 163:1696-1703.
27. Pakala SV, Kurrer MO, Katz JD. T helper 2 (Th2) T cells induce acute pancreatitis and diabetes in immune-compromised nonobese diabetic (NOD) mice. J Exp Med 1997; 186:299-306.
28. Mueller R, Bradley LM, Krahl T et al. Mechanism underlying counterregulation of autoimmune diabetes by IL-4. Immunity 1997; 7:411-418.
29. Lafaille JJ, Keere FV, Hsu AL et al. Myelin basic protein-specific T helper 2 (Th2) cells cause experimental autoimmune encephalomyelitis in immunodeficient hosts rather than protect them from the disease. J Exp Med 1997; 186:307-312.
30. Zipris D, Lazarus AH, Crow AR et al. Defective thymic T cell activation by concanavalin A and anti-CD3 in autoimmune nonobese diabetic mice. Evidence for thymic T cell anergy that correlates with the onset of insulitis. J Immunol 1991; 146:3763-3771.
31. Rapoport MJ, Jaramillo A, Zipris D et al. Interleukin 4 reverses T cell proliferative unresponsiveness and prevents the onset of diabetes in nonobese diabetic mice. J Exp Med 1993; 178:87-99.
32. Berman MA, Sandborg CI, Wang Z et al. Decreased IL-4 production in new onset type I insulin-dependent diabetes mellitus. J Immunol 1996; 157:4690-4696.
33. Wilson SB, Kent SC, Patton KT et al. Extreme Th1 bias of invariant Valpha24JalphaQ T cells in type 1 diabetes. Nature 1998; 391:177-181.
34. Cameron MJ, Arreaza GA, Zucker P et al. IL-4 prevents insulitis and insulin-dependent diabetes mellitus in nonobese diabetic mice by potentiation of regulatory T helper-2 cell function. J Immunol 1997; 159:4686-4692.
35. Dosch H, Cheung RK, Karges W et al. Persistent T cell anergy in human type 1 diabetes. J Immunol 1999; 163:6933-6940.
36. Tominaga Y, Nagata M, Yasuda H et al. Administration of IL-4 prevents autoimmune diabetes but enhances pancreatic insulitis in NOD mice. Clin Immunol Immunopathol 1998; 86:209-218.
37. Salomon B, Lenschow DJ, Rhee L et al. B7/CD28 costimulation is essential for the homeostasis of the CD4+CD25+ immunoregulatory T cells that control autoimmune diabetes. Immunity 2000; 12:431-440.
38. Mueller R, Krahl T, Sarvetnick N. Pancreatic expression of interleukin-4 abrogates insulitis and autoimmune diabetes in nonobese diabetic (NOD) mice. J Exp Med 1996; 184:1093-1099.

39. Tian J, Atkinson MA, ClareSalzler M et al. Nasal administration of glutamate decarboxylase (GAD65) peptides induces Th2 responses and prevents murine insulin-dependent diabetes. J Exp Med 1996; 183:1561-1567.
40. Tian J, ClareSalzler M, Herschenfeld A et al. Modulating autoimmune responses to GAD inhibits disease progression and prolongs islet graft survival in diabetes-prone mice. Nat Med 1996; 2:1348-1353.
41. Farilla L, Dotta F, Di Mario U et al. Presence of interleukin 4 or interleukin 10, but not both cytokines, in pancreatic tissue of two patients with recently diagnosed diabetes mellitus type I. Autoimmunity 2000; 32:161-166.
42. Judkowski V, Pinilla C, Schroder K et al. Identification of MHC class II-restricted peptide ligands, including a glutamic acid decarboxylase 65 sequence, that stimulate diabetogenic T cells from transgenic BDC2.5 nonobese diabetic mice. J Immunol 2001; 166:908-917.
43. Asano M, Toda M, Sakaguchi N et al. Autoimmune disease as a consequence of developmental abnormality of a T cell subpopulation. J Exp Med 1996; 184:387-396.
44. Tisch R, Wang B, Weaver DJ et al. Antigen-specific mediated suppression of beta cell autoimmunity by plasmid DNA vaccination. J Immunol 2001; 166:2122-2132.
45. Cameron MJ, Strathdee CA, Holmes KD et al. Biolistic-mediated interleukin 4 gene transfer prevents the onset of type 1 diabetes. Hum Gene Ther 2000; 11:1647-1656.
46. Homann D, Holz A, Bot A et al. Autoreactive CD4+ T cells protect from autoimmune diabetes via bystander suppression using the IL-4/Stat6 pathway. Immunity 1999; 11:463-472.
47. von Herrath MG, Dockter J, Oldstone MB. How virus induces a rapid or slow onset insulin-dependent diabetes mellitus in a transgenic model. Immunity 1994; 1:231-242.
48. Oldstone MB, Nerenberg M, Southern P et al. Virus infection triggers insulin-dependent diabetes mellitus in a transgenic model: Role of anti-self (virus) immune response. Cell 1991; 65:319-331.
49. Kaplan MH, Schindler U, Smiley ST et al. Stat6 is required for mediating responses to IL-4 and for development of Th2 cells. Immunity 1996; 4:313-319.
50. Hoglund P, Mintern J, Waltzinger C et al. Initiation of autoimmune diabetes by developmentally regulated presentation of islet cell antigens in the pancreatic lymph nodes. J Exp Med 1999; 189:331-339.
51. King C, Hoenger RM, Cleary MM, Murali-Krishna K, Ahmed R et al. Interleukin-4 acts at the locus of the antigen-presenting dendritic cell to counter-regulate cytotoxic $CD8^+$ T-cell responses. Nat Med 2001; 7:206-214.
52. Lenschow DJ, Walunas TL, Bluestone JA. CD28/B7 system of T cell costimulation. Annu Rev Immunol 1996; 14:233-258.
53. Wang B, Gonzalez A, Hoglund P et al. Interleukin-4 deficiency does not exacerbate disease in NOD mice. Diabetes 1998; 47:1207-1211.
54. Cameron MJ, Arreaza GA, Delovitch TL. Cytokine- and costimulation-mediated therapy of IDDM. Crit Rev Immunol 1997; 17:537-544.
55. Radu DL, Noben-Trauth N, Hu-Li J et al. A targeted mutation in the IL-4Ralpha gene protects mice against autoimmune diabetes. Proc Natl Acad Sci USA 2000; 97:12700-12704.
56. Sarukhan A, Lanoue A, Franzke A et al. Changes in function of antigen-specific lymphocytes correlating with progression towards diabetes in a transgenic model. EMBO J 1998; 17:71-80.
57. Zaccone P, Phillips J, Conget I et al. Interleukin-13 prevents autoimmune diabetes in NOD mice. Diabetes 1999; 48:1522-1528.
58. Novick D, Engelmann H, Wallach D et al. Soluble cytokine receptors are present in normal human urine. J Exp Med 1989; 170:1409-1414.
59. Ward LD, Howlett GJ, Discolo G et al. High affinity interleukin-6 receptor is a hexameric complex consisting of two molecules each of interleukin-6, interleukin-6 receptor, and gp- 130. J Biol Chem 1994; 269:23286-23289.
60. Muraguchi A, Hirano T, Tang B et al. The essential role of B cell stimulatory factor 2 (BSF-2/IL-6) for the terminal differentiation of B cells. J Exp Med 1988; 167:332-344.
61. Lotz M, Jirik F, Kabouridis P et al. B cell stimulating factor 2/interleukin 6 is a costimulant for human thymocytes and T lymphocytes. J Exp Med 1988; 167:1253-1258.
62. Uyttenhove C, Coulie PG, Van Snick J. T cell growth and differentiation induced by interleukin-HP1/IL-6, the murine hybridoma/plasmacytoma growth factor. J Exp Med 1988; 167:1417-1427.
63. Okada M, Kitahara M, Kishimoto S et al. IL-6/BSF-2 functions as a killer helper factor in the in vitro induction of cytotoxic T cells. J Immunol 1988; 141 :1543-1549.
64. Renauld JC, Vink A, Van Snick J. Accessory signals in murine cytolytic T cell responses. Dual requirement for IL-1 and IL-6. J Immunol 1989; 143:1894-1898.
65. Luger TA, Krutmann J, Kirnbauer R et al. IFN-beta 2/IL-6 augments the activity of human natural killer cells. J Immunol 1989; 143:1206-1209.

66. Campbell IL, Kay TW, Oxbrow L et al. Essential role for interferon-gamma and interleukin-6 in autoimmune insulin-dependent diabetes in NOD/Wehi mice. J Clin Invest 1991; 87:739-742.
67. DiCosmo BF, Picarella D, Flavell RA. Local production of human IL-6 promotes insulitis but retards the onset of insulin-dependent diabetes mellitus in nonobese diabetic mice. Int Immunol 1994; 6:1829-1837.
68. Rincon M, Anguita J, Nakamura T et al. Interleukin (IL)-6 directs the differentiation of IL-4-producing CD4+ T cells. J Exp Med 1997; 185:461-469.
69. Gorelik L, Flavell RA. Abrogation of TGFbeta signaling in T cells leads to spontaneous T cell differentiation and autoimmune disease. Immunity 2000; 12:171-181.
70. Han HS, Jun HS, Utsugi T et al. Molecular role of TGF-beta, secreted from a new type of CD4+ suppressor T cell, NY4.2, in the prevention of autoimmune IDDM in NOD mice. J Autoimmun 1997; 10:299-307.
71. Strober W, Kelsall B, Fuss I et al. Reciprocal IFN-gamma and TGF-beta responses regulate the occurrence of mucosal inflammation. Immunol Today 1997; 18:61-64.
72. Czarniecki CW, Chiu HH, Wong GH et al. Transforming growth factor-beta 1 modulates the expression of class II histocompatibility antigens on human cells. J Immunol 1988; 140:4217-4223.
73. Kehrl JH, Taylor A, Kim SJ et al. Transforming growth factor-beta is a potent negative regulator of human lymphocytes. Ann N Y Acad Sci 1991; 628:345-353.
74. King C, Davies J, Mueller R et al. TGF-beta1 alters APC preference, polarizing islet antigen responses toward a Th2 phenotype. Immunity 1998; 8:601-613.
75. Sanvito F, Nichols A, Herrera PL et al. TGF-beta 1 overexpression in murine pancreas induces chronic pancreatitis and, together with TNF-alpha, triggers insulin-dependent diabetes. Biochem Biophys Res Commun 1995; 217:1279-1286.
76. Lee MS, Mueller R, Wicker LS et al. IL-10 is necessary and sufficient for autoimmune diabetes in conjunction with NOD MHC homozygosity. J Exp Med 1996; 183:2663-2668.
77. Moritani M, Yoshimoto K, Wong SF et al. Abrogation of autoimmune diabetes in nonobese diabetic mice and protection against effector lymphocytes by transgenic paracrine TGF- beta1. J Clin Invest 1998; 102:499-506.
78. Piccirillo CA, Chang Y, Prud'homme GJ. TGF-beta1 somatic gene therapy prevents autoimmune disease in nonobese diabetic mice. J Immunol 1998; 161:3950-3956.
79. Suarez-Pinzon W, Korbutt GS, Power R et al. Testicular sertoli cells protect islet beta-cells from autoimmune destruction in NOD mice by a transforming growth factor-beta1-dependent mechanism. Diabetes 2000; 49:1810-1818.
80. Han HS, Jun HS, Utsugi T et al. A new type of CD4+ suppressor T cell completely prevents spontaneous autoimmune diabetes and recurrent diabetes in syngeneic islet-transplanted NOD mice. J Autoimmun 1996; 9:331-339.
81. Ziyadeh FN. Role of transforming growth factor beta in diabetic nephropathy. Exp Nephrol 1994; 2:137.
82. Reeves WB, Andreoli TE. Transforming growth factor beta contributes to progressive diabetic nephropathy. Proc Natl Acad Sci USA 2000; 97:7667-7669.
83. Asseman C, Powrie F. Interleukin 10 is a growth factor for a population of regulatory T cells. Gut 1998; 42:157-158.
84. Stohlman SA, Pei L, Cua DJ et al. Activation of regulatory cells suppresses experimental allergic encephalomyelitis via secretion of IL-10. J Immunol 1999; 163:6338-6344.
85. Cua DJ, Groux H, Hinton DR et al. Transgenic interleukin 10 prevents induction of experimental autoimmune encephalomyelitis. J Exp Med 1999; 189:1005-1010.
86. Cua DJ, Hutchins B, LaFace DM et al. Central nervous system expression of IL-10 inhibits autoimmune encephalomyelitis. J Immunol 2001; 166:602-608.
87. Sonoda KH, Faunce DE, Taniguchi M et al. NK T cell-derived IL-10 is essential for the differentiation of antigen-specific T regulatory cells in systemic tolerance. J Immunol 2001; 166:42-50.
88. Zhai Y, Kupiec-Weglinski JW. What is the role of regulatory T cells in transplantation tolerance? Curr Opin Immunol 1999; 11:497-503.
89. Cavani A, Mei D, Guerra E et al. Patients with allergic contact dermatitis to nickel and nonallergic individuals display different nickel-specific T cell responses. Evidence for the presence of effector CD8+ and regulatory CD4+ T cells. J Invest Dermatol 1998; 111:621-628.
90. Cavani A, Nasorri F, Prezzi C et al. Human CD4+ T lymphocytes with remarkable regulatory functions on dendritic cells and nickel-specific Th1 immune responses. J Invest Dermatol 2000; 114:295-302.
91. Pennline KJ, Roque-Gaffney E, Monahan M. Recombinant human IL-10 prevents the onset of diabetes in the nonobese diabetic mouse. Clin Immunol Immunopathol 1994; 71:169-175.

92. Nitta Y, Tashiro F, Tokui M et al. Systemic delivery of interleukin 10 by intramuscular injection of expression plasmid DNA prevents autoimmune diabetes in nonobese diabetic mice. Hum Gene Ther 1998; 9:1701-1707.
93. Moritani M, Yoshimoto K, Ii S et al. Prevention of adoptively transferred diabetes in nonobese diabetic mice with IL-10-transduced islet-specific Th1 lymphocytes. A gene therapy model for autoimmune diabetes. J Clin Invest 1996; 98:1851-1859.
94. Pauza ME, Neal H, Hagenbaugh A et al. T-cell production of an inducible interleukin-10 transgene provides limited protection from autoimmune diabetes. Diabetes 1999; 48:1948-1953.
95. Maron R, Melican NS, Weiner HL. Regulatory Th2-type T cell lines against insulin and GAD peptides derived from orally- and nasally-treated NOD mice suppress diabetes. J Autoimmun 1999; 12:251-258.
96. Rabinovitch A, Suarez-Pinzon WL, Sorensen O et al. Combined therapy with interleukin-4 and interleukin-10 inhibits autoimmune diabetes recurrence in syngeneic islet-transplanted nonobese diabetic mice. Analysis of cytokine mRNA expression in the graft. Transplantation 1995; 60:368-374.
97. Faust A, Rothe H, Schade U et al. Primary nonfunction of islet grafts in autoimmune diabetic nonobese diabetic mice is prevented by treatment with interleukin-4 and interleukin-10. Transplantation 1996; 62:648-652.
98. Rapoport MJ, Mor A, Vardi P et al. Decreased secretion of Th2 cytokines precedes up-regulated and delayed secretion of Th1 cytokines in activated peripheral blood mononuclear cells from patients with insulin-dependent diabetes mellitus. J Autoimmun 1998; 11:635-642.
99. Kallmann BA, Lampeter EF, Hanifi-Moghaddam P et al. Cytokine secretion patterns in twins discordant for Type I diabetes. Diabetologia 1999; 42:1080-1085.
100. Smith DK, Korbutt GS, Suarez-Pinzon WL et al. Interleukin-4 or interleukin-10 expressed from adenovirus-transduced syngeneic islet grafts fails to prevent beta cell destruction in diabetic NOD mice. Transplantation 1997; 64:1040-1049.
101. Lee MS, Wogensen L, Shizuru J et al. Pancreatic islet production of murine interleukin-10 does not inhibit immune-mediated tissue destruction. J Clin Invest 1994; 93:1332-1338.
102. Moritani M, Yoshimoto K, Tashiro F et al. Transgenic expression of IL-10 in pancreatic islet A cells accelerates autoimmune insulitis and diabetes in nonobese diabetic mice. Int Immunol 1994; 6:1927-1936.
103. Balasa B, Davies JD, Lee J et al. IL-10 impacts autoimmune diabetes via a CD8+ T cell pathway circumventing the requirement for CD4+ T and B lymphocytes. J Immunol 1998; 161:4420-4427.
104. Balasa B, Van Gunst K, Jung N et al. Islet-specific expression of IL-10 promotes diabetes in nonobese diabetic mice independent of Fas, perforin, TNF receptor-1, and TNF receptor-2 molecules. J Immunol 2000; 165:2841-2849.
105. Balasa B, La Cava A, Van Gunst K et al. A mechanism for IL-10-mediated diabetes in the nonobese diabetic (NOD) mouse: ICAM-1 deficiency blocks accelerated diabetes. J Immunol 2000; 165:7330-7337.
106. Trepicchio WL, Wang L, Bozza M et al. IL-11 regulates macrophage effector function through the inhibition of nuclear factor-kappaB. J Immunol 1997; 159:5661-5670.
107. Trepicchio WL, Bozza M, Pedneault G et al. Recombinant human IL-11 attenuates the inflammatory response through down-regulation of proinflammatory cytokine release and nitric oxide production. J Immunol 1996; 157:3627-3634.
108. Hill GR, Cooke KR, Teshima T et al. Interleukin-11 promotes T cell polarization and prevents acute graft- versus-host disease after allogeneic bone marrow transplantation. J Clin Invest 1998; 102:115-123.
109. Nicoletti F, Zaccone P, Conget I et al. Early prophylaxis with recombinant human interleukin-11 prevents spontaneous diabetes in NOD mice. Diabetes 1999; 48:2333-2339.
110. McKenzie AN. Regulation of T helper type 2 cell immunity by interleukin-4 and interleukin-13. Pharmacol Ther 2000; 88:143-151.
111. McKenzie GJ, Emson CL, Bell SE et al. Impaired development of Th2 cells in IL-13-deficient mice. Immunity 1998; 9:423-432.
112. McKenzie GJ, Bancroft A, Grencis RK et al. A distinct role for interleukin-13 in Th2-cell-mediated immune responses. Curr Biol 1998; 8:339-342.
113. Grunig G, Warnock M, Wakil AE et al. Requirement for IL-13 independently of IL-4 in experimental asthma. Science 1998; 282:2261-2263.
114. Li L, Xia Y, Nguyen A et al. Effects of Th2 cytokines on chemokine expression in the lung: IL-13 potently induces eotaxin expression by airway epithelial cells. J Immunol 1999; 162:2477-2487.
115. McKenzie GJ, Fallon PG, Emson CL et al. Simultaneous disruption of interleukin (IL)-4 and IL-13 defines individual roles in T helper cell type 2-mediated responses. J Exp Med 1999; 189:1565-1572.

116. Lentsch AB, Shanley TP, Sarma V et al. In vivo suppression of NF-kappa B and preservation of I kappa B alpha by interleukin-10 and interleukin-13. J Clin Invest 1997; 100:2443-2448.
117. Cosentino G, Soprana E, Thienes CP et al. IL-13 down-regulates CD14 expression and TNF-alpha secretion in normal human monocytes. J Immunol 1995; 155:3145-3151.
118. Muchamuel T, Menon S, Pisacane P et al. IL-13 protects mice from lipopolysaccharide-induced lethal endotoxemia: Correlation with down-modulation of TNF-alpha, IFN-gamma, and IL-12 production. J Immunol 1997; 158:2898-2903.
119. Morse MA, Lyerly HK, Li Y. The role of IL-13 in the generation of dendritic cells in vitro. J Immunother 1999; 22:506-513.
120. Cash E, Minty A, Ferrara P et al. Macrophage-inactivating IL-13 suppresses experimental autoimmune encephalomyelitis in rats. J Immunol 1994; 153:4258-4267.
121. Young DA, Lowe LD, Booth SS et al. IL-4, IL-10, IL-13, and TGF-beta from an altered peptide ligand-specific Th2 cell clone down-regulate adoptive transfer of experimental autoimmune encephalomyelitis . J Immunol 2000; 164:3563-3572.
122. Kretowski A, Mysliwiec J, Kinalska I. In vitro inerleukin-13 production by peripheral blood in patients with newly diagnosed insulin-dependent diabetes mellitus and their first degree relatives. Scand J Immunol 2000; 51:321-325.
123. Mohrs M, Ledermann B, Kohler G et al. Differences between IL-4 and IL-4 receptor alpha-deficient mice in chronic leishmaniasis reveal a protective role for IL-13 receptor signaling. J Immunol 1999; 162:7302-7308.
124. LaFace DM, Peck AB. Reciprocal allogeneic bone marrow transplantation between NOD mice and diabetes-nonsusceptible mice associated with transfer and prevention of autoimmune diabetes. Diabetes 1989; 38:894-901.
125. Wicker LS, Miller BJ, Chai A et al. Expression of genetically determined diabetes and insulitis in the nonobese diabetic (NOD) mouse at the level of bone marrow-derived cells. Transfer of diabetes and insulitis to nondiabetic (NOD X B10) F1 mice with bone marrow cells from NOD mice. J Exp Med 1988; 167:1801-1810.
126. Ikehara S, Kawamura M, Takao F et al. Organ-specific and systemic autoimmune diseases originate from defects in hematopoietic stem cells. Proc Natl Acad Sci USA 1990; 87:8341-8344.
127. Yasumizu R, Sugiura K, Iwai H et al. Treatment of type 1 diabetes mellitus in nonobese diabetic mice by transplantation of allogeneic bone marrow and pancreatic tissue. Proc Natl Acad Sci USA 1987; 84:6555-6557.
128. Langmuir PB, Bridgett MM, Bothwell AL et al. Bone marrow abnormalities in the nonobese diabetic mouse. Int Immunol 1993; 5:169-177.
129. Serreze DV, Gaedeke JW, Leiter EH. Hematopoietic stem-cell defects underlying abnormal macrophage development and maturation in NOD/Lt mice: defective regulation of cytokine receptors and protein kinase C. Proc Natl Acad Sci USA 1993; 90:9625-9629.
130. Morel PA, Vasquez AC, Feili-Hariri M. Immunobiology of DC in NOD mice. J Leukoc Biol 1999; 66:276-280.
131. Lee M, Kim AY, Kang Y. Defects in the differentiation and function of bone marrow-derived dendritic cells in nonobese diabetic mice. J Korean Med Sci 2000; 15:217-223.
132. Jeong MC, Izikson L, Uccelli A, Brocke S, Oksenberg JR. Differential display analysis of murine encephalitogenic mRNA. Int Immunol 1998; 10:1819-1823.
133. Offner H, Vainiene M, Celnik B et al. Coculture of TCR peptide-specific T cells with basic protein-specific T cells inhibits proliferation, IL-3 mRNA, and transfer of experimental autoimmune encephalomyelitis. J Immunol 1994; 153:4988-4996.
134. Ito A, Aoyanagi N, Maki T. Regulation of autoimmune diabetes by interleukin 3-dependent bone marrow-derived cells in NOD mice. J Autoimmun 1997; 10:331-338.
135. Burlinson EL, Drakes ML, Wood PJ. Differential patterns of production of granulocyte macrophage colony stimulating factor, IL-2, IL-3 and IL-4 by cultured islets of Langerhans from nonobese diabetic and nondiabetic strains of mice. Int Immunol 1995; 7:79-87.
136. Dinarello CA. Biologic basis for interleukin-1 in disease. Blood 1996; 87:2095-2147.
137. Mandrup-Poulsen T. The role of interleukin-1 in the pathogenesis of IDDM. Diabetologia 1996; 39:1005-1029.
138. Lacy PE, Finke EH. Activation of intraislet lymphoid cells causes destruction of islet cells. Am J Pathol 1991; 138:1183-1190.
139. Lacy PE. The intraislet macrophage and type I diabetes. Mt Sinai J Med 1994; 61:170-174.
140. Heitmeier MR, Scarim AL, Corbett JA. Interferon-gamma increases the sensitivity of islets of Langerhans for inducible nitric-oxide synthase expression induced by interleukin 1. J Biol Chem 1997; 272:13697-13704.

141. Corbett JA, McDaniel ML. Intraislet release of interleukin 1 inhibits beta cell function by inducing beta cell expression of inducible nitric oxide synthase. J Exp Med 1995; 181:559-568.
142. Arnush M, Scarim AL, Heitmeier MR et al. Potential role of resident islet macrophage activation in the initiation of autoimmune diabetes. J Immunol 1998; 160:2684-2691.
143. Arnush M, Heitmeier MR, Scarim AL et al. IL-1 produced and released endogenously within human islets inhibits beta cell function. J Clin Invest 1998; 102:516-526.
144. McDaniel ML, Kwon G, Hill JR et al. Cytokines and nitric oxide in islet inflammation and diabetes. Proc Soc Exp Biol Med 1996; 211:24-32.
145. Corbett JA, Wang JL, Sweetland MA et al. Interleukin 1 beta induces the formation of nitric oxide by beta-cells purified from rodent islets of Langerhans. Evidence for the beta-cell as a source and site of action of nitric oxide. J Clin Invest 1992; 90:2384-2391.
146. Yamada K, Takane-Gyotoku N, Yuan X et al. Mouse islet cell lysis mediated by interleukin-1-induced Fas. Diabetologia 1996; 39:1306-1312.
147. Amrani A, Verdaguer J, Thiessen S et al. IL-1alpha, IL-1beta, and IFN-gamma mark beta cells for Fas-dependent destruction by diabetogenic CD4(+) T lymphocytes. J Clin Invest 2000; 105:459-468.
148. Hohmeier HE, Thigpen A, Tran VV et al. Stable expression of manganese superoxide dismutase (MnSOD) in insulinoma cells prevents IL-1beta- induced cytotoxicity and reduces nitric oxide production. J Clin Invest 1998; 101:1811-1820.
149. Stassi G, De Maria R, Trucco G et al. Nitric oxide primes pancreatic beta cells for Fas-mediated destruction in insulin-dependent diabetes mellitus. J Exp Med 1997; 186:1193-1200.
150. Giannoukakis N, Rudert WA, Ghivizzani SC et al. Adenoviral gene transfer of the interleukin-1 receptor antagonist protein to human islets prevents IL-1beta-induced beta-cell impairment and activation of islet cell apoptosis in vitro. Diabetes 1999; 48:1730-1736.
151. Wogensen L, Helqvist S, Pociot F et al. Intra-peritoneal administration of interleukin-1 beta induces impaired insulin release from the perfused rat pancreas. Autoimmunity 1990; 7:1-12.
152. Wang Y, Goodman M, Lumerman J et al. In vivo administration of interleukin-1 inhibits glucose-stimulated insulin release. Diabetes Res Clin Pract 1989; 7:205-211.
153. Wilson CA, Jacobs C, Baker P et al. IL-1 beta modulation of spontaneous autoimmune diabetes and thyroiditis in the BB rat. J Immunol 1990; 144:3784-3788.
154. Jacob CO, Also S, Michie SA et al. Prevention of diabetes in nonobese diabetic mice by tumor necrosis factor (TNF): Similarities between TNF-alpha and interleukin-1. Proc Natl Acad Sci USA 1990; 87:968-972.
155. Nicoletti F, Di Marco R, Barcellini W et al. Protection from experimental autoimmune diabetes in the nonobese diabetic mouse with soluble interleukin-1 receptor. Eur J Immunol 1994; 24:1843-1847.
156. Cailleau C, Diu-Hercend A, Ruuth E et al. Treatment with neutralizing antibodies specific for IL-1beta prevents cyclophosphamide-induced diabetes in nonobese diabetic mice. Diabetes 1997; 46:937-940.
157. Refaeli Y, Van Parijs L, London CA et al. Biochemical mechanisms of IL-2-regulated Fas-mediated T cell apoptosis. Immunity 1998; 8:615-623.
158. Kroemer G, Wick G. The role of interleukin 2 in autoimmunity. Immunol Today 1989; 10:246-251.
159. Schorle H, Holtschke T, Hunig T et al. Development and function of T cells in mice rendered interleukin-2 deficient by gene targeting. Nature 1991; 352:621-624.
160. Smith KA. Interleukin-2. Curr Opin Immunol 1992; 4:271-276.
161. Miyajima A, Kitamura T, Harada N et al. Cytokine receptors and signal transduction. Annu Rev Immunol 1992; 10:295-331.
162. Strom TB, Kelley VR, Murphy JR et al. Interleukin-2 receptor-directed therapies: Antibody-or cytokine-based targeting molecules. Annu Rev Med 1993; 44:343-353.
163. Waldmann TA. The IL-2/IL-2 receptor system: A target for rational immune intervention. Immunol Today 1993; 14:264-270.
164. Zheng XX, Steele AW, Hancock WW et al. IL-2 receptor-targeted cytolytic IL-2/Fc fusion protein treatment blocks diabetogenic autoimmunity in nonobese diabetic mice. J Immunol 1999; 163:4041-4048.
165. Kelley VE, Gaulton GN, Hattori M et al. Anti-interleukin 2 receptor antibody suppresses murine diabetic insulitis and lupus nephritis. J Immunol 1988; 140:59-61.
166. Allison J, Malcolm L, Chosich N et al. Inflammation but not autoimmunity occurs in transgenic mice expressing constitutive levels of interleukin-2 in islet beta cells. Eur J Immunol 1992; 22:1115-1121.
167. Allison J, Oxbrow L, Miller JF. Consequences of in situ production of IL-2 for islet cell death. Int Immunol 1994; 6:541-549.

168. Allison J, McClive P, Oxbrow L et al. Genetic requirements for acceleration of diabetes in nonobese diabetic mice expressing interleukin-2 in islet beta-cells. Eur J Immunol 1994; 24:2535-2541.
169. von Herrath MG, Allison J, Miller JF et al. Focal expression of interleukin-2 does not break unresponsiveness to "self" (viral) antigen expressed in beta cells but enhances development of autoimmune disease (diabetes) after initiation of an anti-self immune response. J Clin Invest 1995; 95:477-485.
170. Lyons PA, Armitage N, Argentina F et al. Congenic mapping of the type 1 diabetes locus, Idd3, to a 780-kb region of mouse chromosome 3: Identification of a candidate segment of ancestral DNA by haplotype mapping. Genome Res 2000; 10:446-453.
171. Denny P, Lord CJ, Hill NJ et al. Mapping of the IDDM locus Idd3 to a 0.35-cM interval containing the interleukin-2 gene. Diabetes 1997; 46:695-700.
172. Podolin PL. Wilusz MB, Cubbon RM et al. Differential glycosylation of interleukin-2, the molecular basis for the NOD Idd3 type 1 diabetes gene? Cytokine 2000; 12:477-482.
173. Caspi RR. IL-12 in autoimmunity. Clin Immunol Immunopathol 1998; 88:4-13.
174. Adorini L. Interleukin 12 and autoimmune diabetes. Nat Genet 2001; 27:131-132.
175. Trembleau S, Germann T, Gately MK et al. The role of IL-12 in the induction of organ-specific autoimmune diseases. Immunol Today 1995; 16:383-386.
176. Rabinovitch A, Suarez-Pinzon WL, Sorensen O. Interleukin 12 mRNA expression in islets correlates with beta-cell destruction in NOD mice. J Autoimmun 1996; 9:645-651.
177. Rothe H, Burkart V, Faust A et al. Interleukin-12 gene expression mediates the accelerating effect of cyclophosphamide in autoimmune disease. Ann NY Acad Sci 1996; 9:645-651.
178. O'Hara RM, Henderson SL, Nagelin A. Prevention of a Th1 disease by a Th1 cytokine: IL-12 and diabetes in NOD mice. Ann NY Acad Sci 1996; 795:241-249.
179. Nicoletti F, DiMarco R, Zaccone P et al. Endogenous interleukin-12 only plays a key pathogenetic role in nonobese diabetic mouse diabetes during the very early stages of the disease. Immunology 1999; 97:367-370.
180. Fujihira K, Nagata M, Moriyama H et al. Suppression and acceleration of autoimmune diabetes by neutralization of endogenous interleukin-12 in NOD mice. Diabetes 2000; 49:1998-2006.
181. Rothe H, O'Hara RM, Martin S et al. Suppression of cyclophosphamide induced diabetes development and pancreatic Th1 reactivity in NOD mice treated with the interleukin (IL)-12 antagonist IL-12(p40)2. Diabetologia 1997; 40:641-646.
182. Trembleau S, Penna G, Gregori S et al. Deviation of pancreas-infiltrating cells to Th2 by interleukin-12 antagonist administration inhibits autoimmune diabetes. Eur J Immunol 1997; 27:2330-2339.
183. Gracie JA, Forsey RJ, Chan WL et al. A proinflammatory role for IL-18 in rheumatoid arthritis. J Clin Invest 1999; 104:641-646.
184. Yasuda H, Nagata M, Arisawa K et al. Local expression of immunoregulatory IL-12p40 gene prolonged syngeneic islet graft survival in diabetic NOD mice. J Clin Invest 1998; 102:1807-1814.
185. Trembleau S, Penna G, Gregori S et al. Pancreas-infiltrating Th1 cells and diabetes develop in IL-12-deficient nonobese diabetic mice. J Immunol 1999; 163:2960-2968.
186. Rabinovitch A, Suarez-Pinzon WL, Sorensen O et al. IFN-gamma gene expression in pancreatic islet-infiltrating mononuclear cells correlates with autoimmune diabetes in nonobese diabetic mice. J Immunol 1995; 154:4874-4882.
187. Debray-Sachs M, Carnaud C, Boitard C et al. Prevention of diabetes in NOD mice treated with antibody to murine IFN gamma. J Autoimmun 1991; 4:237-248.
188. Kay TW, Campbell IL, Oxbrow L et al. Overexpression of class I major histocompatibility complex accompanies insulitis in the nonobese diabetic mouse and is prevented by anti-interferon-gamma antibody. Diabetologia 1991; 34:779-785.
189. Campbell IL, Iscaro A, Harrison LC. IFN-gamma and tumor necrosis factor-alpha. Cytotoxicity to murine islets of Langerhans. J Immunol 1988; 141:2325-2329.
190. Dunger A, Cunningham JM, Delaney CA et al. Tumor necrosis factor-alpha and interferon-gamma inhibit insulin secretion and cause DNA damage in unweaned-rat islets. Extent of nitric oxide involvement. Diabetes 1996; 45:183-189.
191. Thomas HE, Parker JL, Schreiber RD et al. IFN-gamma action on pancreatic beta cells causes class I MHC upregulation but not diabetes. J Clin Invest 1998; 102:1249-1257.
192. Sarvetnick N, Liggitt D, Pitts SL et al. Insulin-dependent diabetes mellitus induced in transgenic mice by ectopic expression of class II MHC and interferon-gamma. Cell 1988; 52:773-782.
193. Nicoletti F, Zaccone P, Di Marco R et al. Prevention of spontaneous autoimmune diabetes in diabetes-prone BB rats by prophylactic treatment with antirat interferon-gamma antibody. Endocrinology 1997; 138:281-288.

194. Nicoletti F, Zaccone P, Di Marco R et al. The effects of a nonimmunogenic form of murine soluble interferon-gamma receptor on the development of autoimmune diabetes in the NOD mouse. Endocrinology 1996; 137:5567-5575.
195. Wang B, Andre I, Gonzalez A et al. Interferon-gamma impacts at multiple points during the progression of autoimmune diabetes. Proc Natl Acad Sci USA 1997; 94:13844-13849.
196. Kanagawa O, Xu G, Tevaarwerk A et al. Protection of nonobese diabetic mice from diabetes by gene(s) closely linked to IFN-gamma receptor loci. J Immunol 2000; 164:3919-3923.
197. Serreze DV, Post CM, Chapman HD et al. Interferon-gamma receptor signaling is dispensable in the development of autoimmune type 1 diabetes in NOD mice. Diabetes 2000; 49:2007-2011.
198. Hultgren B, Huang X, Dybdal N et al. Genetic absence of gamma-interferon delays but does not prevent diabetes in NOD mice. Diabetes 1996; 45:812-817.
199. Serreze DV, Chapman HD, Post CM et al. Th1 to Th2 cytokine shifts in nonobese diabetic mice: Sometimes an outcome, rather than the cause, of diabetes resistance elicited by immunostimulation. J Immunol 2001; 166:1352-1359.
200. Okamura H, Tsutsi H, Komatsu T et al. Cloning of a new cytokine that induces IFN-gamma production by T cells. Nature 1995; 378:88-91.
201. Dayer JM. Interleukin-18, rheumatoid arthritis, and tissue destruction. J Clin Invest 1999; 104:1337-1339.
202. Rothe H, Jenkins NA, Copeland NG et al. Active stage of autoimmune diabetes is associated with the expression of a novel cytokine, IGIF, which is located near Idd2. J Clin Invest 1997; 99:469-474.
203. Rothe H, Hibino T, Itoh Y et al. Systemic production of interferon-gamma inducing factor (IGIF) versus local IFN-gamma expression involved in the development of Th1 insulitis in NOD mice. J Autoimmun 1997; 10:251-256.
204. Andre-Schmutz I, Hindelang C, Benoist C et al. Cellular and molecular changes accompanying the progression from insulitis to diabetes. Eur J Immunol 1999; 29:245-255.
205. Krook H, Wallstrom J, Sandler S. Function of rat pancreatic islets exposed to interleukin-18 in vitro. Autoimmunity 1999; 29:263-267.
206. Hong TP, Andersen NA, Nielsen K et al. Interleukin-18 mRNA, but not interleukin-18 receptor mRNA, is constitutively expressed in islet beta-cells and up-regulated by interferon-gamma. Eur Cytokine Netw 2000; 11:193-205.
207. Rothe H, Hausmann A, Casteels K et al. IL-18 inhibits diabetes development in nonobese diabetic mice by counterregulation of Th1-dependent destructive insulitis. J Immunol 1999; 163:1230-1236.
208. Xu D, Trajkovic V, Hunter D et al. IL-18 induces the differentiation of Th1 or Th2 cells depending upon cytokine milieu and genetic background. Eur J Immunol 2000; 30:3147-3156.
209. Leite-De-Moraes MC, Hameg A, Pacilio M et al. IL-18 enhances IL-4 production by ligand-activated NKT lymphocytes: A pro-Th2 effect of IL-18 exerted through NKT cells. J Immunol 2001; 166:945-951.
210. Green EA, Flavell RA. Tumor necrosis factor-alpha and the progression of diabetes in nonobese diabetic mice. Immunol Rev 1999; 169:11-22.
211. Green EA, Eynon EE, Flavell RA. Local expression of TNFalpha in neonatal NOD mice promotes diabetes by enhancing presentation of islet antigens. Immunity 1998; 9:733-743.
212. Campbell IL, Cutri A, Wilkinson D et al. Intercellular adhesion molecule 1 is induced on isolated endocrine islet cells by cytokines but not by reovirus infection. Proc Natl Acad Sci USA 1989; 86:4282-4286.
213. Yagi N, Yokono K, Amano K et al. Expression of intercellular adhesion molecule 1 on pancreatic beta- cells accelerates beta-cell destruction by cytotoxic T-cells in murine autoimmune diabetes. Diabetes 1995; 44:744-752.
214. Stephens LA, Thomas HE, Ming L et al. Tumor necrosis factor-alpha-activated cell death pathways in NIT-1 insulinoma cells and primary pancreatic beta cells. Endocrinology 1999; 140:3219-3227.
215. Pujol-Borrell R, Todd I, Doshi M et al. HLA class II induction in human islet cells by interferon-gamma plus tumour necrosis factor or lymphotoxin. Nature 1987; 326:304-306.
216. Campbell IL, Oxbrow L, West J et al. Regulation of MHC protein expression in pancreatic beta-cells by interferon-gamma and tumor necrosis factor-alpha. Mol Endocrinol 1988; 2:101-107.
217. Held W, MacDonald HR, Weissman IL et al. Genes encoding tumor necrosis factor alpha and granzyme A are expressed during development of autoimmune diabetes. Proc Natl Acad Sci USA 1990; 87:2239-2243.
218. Yang XD, Tisch R, Singer SM et al. Effect of tumor necrosis factor alpha on insulin-dependent diabetes mellitus in NOD mice. I. The early development of autoimmunity and the diabetogenic process. J Exp Med 1994; 180:995-1004.

219. Jacob CO, Aiso S, Schreiber RD et al. Monoclonal anti-tumor necrosis factor antibody renders nonobese diabetic mice hypersensitive to irradiation and enhances insulitis development. Int Immunol 1992; 4:611-614.
220. Picarella DE, Kratz A, Li CB et al. Transgenic tumor necrosis factor (TNF)-alpha production in pancreatic islets leads to insulitis, not diabetes. Distinct patterns of inflammation in TNF-alpha and TNF-beta transgenic mice. J Immunol 1993; 150:4136-4150.
221. Higuchi Y, Herrera P, Muniesa P et al. Expression of a tumor necrosis factor alpha transgene in murine pancreatic beta cells results in severe and permanent insulitis without evolution towards diabetes. J Exp Med 1992; 176:1719-1731.
222. Grewal IS, Grewal KD, Wong FS et al. Local expression of transgene encoded TNF alpha in islets prevents autoimmune diabetes in nonobese diabetic (NOD) mice by preventing the development of auto-reactive islet-specific T cells. J Exp Med 1996; 184:1963-1974.
223. Cope AP, Liblau RS, Yang XD et al. Chronic tumor necrosis factor alters T cell responses by attenuating T cell receptor signaling. J Exp Med 1997; 185:1573-1584.
224. Cope A, Ettinger R, McDevitt H. The role of TNF alpha and related cytokines in the development and function of the autoreactive T-cell repertoire. Res Immunol 1997; 148:307-312.
225. Rothe H, Kolb H. The APC1 concept of type I diabetes. Autoimmunity 1998.
226. Green EA, Wong FS, Eshima K et al. Neonatal tumor necrosis factor alpha promotes diabetes in nonobese diabetic mice by CD154-independent antigen presentation to CD8(+) T cells. J Exp Med 2000; 191:225-238.
227. Balasa B, Krahl T, Patstone G et al. CD40 ligand-CD40 interactions are necessary for the initiation of insulitis and diabetes in nonobese diabetic mice. J Immunol 1997; 159:4620-4627.
228. Aggarwal BB, Natarajan K. Tumor necrosis factors: developments during the last decade. Eur Cytokine Netw 1996; 7:93-124.
229. Kagi D, Ho A, Odermatt B et al. TNF receptor 1-dependent beta cell toxicity as an effector pathway in autoimmune diabetes. J Immunol 1999; 162:4598-4605.
230. Pakala SV, Chivetta M, Kelly CB et al. In autoimmune diabetes the transition from benigh to pernicious insulitis requires an islet cell response to tumor necrosis factor alpha. J Exp Med 1999; 189:1053-1062.
231. Green EA, Flavell RA. The temporal importance of TNFalpha expression in the development of diabetes. Immunity 2000; 12:459-469.
232. Guerder S, Meyerhoff J, Flavell R. The role of the T cell costimulator B7-1 in autoimmunity and the induction and maintenance of tolerance to peripheral antigen. Immunity 1994; 1:155-166.
233. Herrera PL, Harlan DM, Vassalli P. A mouse CD8 T cell-mediated acute autoimmune diabetes independent of the perforin and Fas cytotoxic pathways: Possible role of membrane TNF. Proc Natl Acad Sci USA 2000; 97:279-284.
234. Herrera PL, Harlan DM, Fossati L et al. A CD8+ T-lymphocyte-mediated and CD4+ T-lymphocyte-independent autoimmune diabetes of early onset in transgenic mice. Diabetologia 1994; 37:1277-1279.
235. Zlotnik A, Yoshie O. Chemokines: A new classification system and their role in immunity. Immunity 2000; 12:121-127.
236. Ward SG, Bacon K, Westwick J. Chemokines and T lymphocytes: More than an attraction. Immunity 1998; 9:1-11.
237. Sallusto F, Mackay CR, Lanzavecchia A. The role of chemokine receptors in primary, effector, and memory immune responses. Annu Rev Immunol 2000; 18:593-620.
238. Gill BM, Jaramillo A, Ma L et al. Genetic linkage of thymic T-cell proliferative unresponsiveness to mouse chromosome 11 in NOD mice. A possible role for chemokine genes. Diabetes 1995; 44:614-619.
239. Butterfield RJ, Sudweeks JD, Blankenhorn EP et al. New genetic loci that control susceptibility and symptoms of experimental allergic encephalomyelitis in inbred mice. J Immunol 1998; 161:1860-1867.
240. Teuscher C, Butterfield RJ, Ma RZ et al. Sequence polymorphisms in the chemokines Scya1 (TCA-3), Scya2 (monocyte chemoattractant protein (MCP)-1), and Scya12 (MCP-5) are candidates for eae7, a locus controlling susceptibility to monophasic remitting/nonrelapsing experimental allergic encephalomyelitis. J Immunol 1999; 163:2262-2266.
241. Standiford TJ, Kunkel SL, Liebler JM et al. Gene expression of macrophage inflammatory protein-1 alpha from human blood monocytes and alveolar macrophages is inhibited by interleukin-4. Am J Respir Cell Mol Biol 1993; 9:192-198.
242. Herold KC, Lu J, Rulifson I et al. Regulation of C-C chemokine production by murine T cells by CD28/B7 costimulation. J Immunol 1997; 159:4150-4153.

243. Cameron MJ, Arreaza GA, Grattan M et al. Differential expression of CC chemokines and the CCR5 receptor in the pancreas is associated with progression to type I diabetes. J Immunol 2000; 165:1102-1110.
244. Grewal IS, Rutledge BJ, Fiorillo JA et al. Transgenic monocyte chemoattractant protein-1 (MCP-1) in pancreatic islets produces monocyte-rich insulitis without diabetes: Abrogation by a second transgene expressing systemic MCP-1. J Immunol 1997; 159:401-408.
245. Karpus WJ, Lukacs NW, McRae BL et al. An important role for the chemokine macrophage inflammatory protein-1 alpha in the pathogenesis of the T cell-mediated autoimmune disease, experimental autoimmune encephalomyelitis. J Immunol 1995; 155:5003-5010.
246. Karpus WJ, Kennedy KJ. MIP-1alpha and MCP-1 differentially regulate acute and relapsing autoimmune encephalomyelitis as well as Th1/Th2 lymphocyte differentiation. J Leukoc Biol 1997; 62:681-687.
247. Kunkel SL. Th1- and Th2-type cytokines regulate chemokine expression. Biol Signals 1996; 5:197-202.
248. Loetscher P, Uguccioni M, Bordoli L et al. CCR5 is characteristic of Th1 lymphocytes. Nature 1998; 391:344-345.
249. Bonecchi R, Bianchi G, Bordignon PP et al. Differential expression of chemokine receptors and chemotactic responsiveness of type 1 T helper cells (Th1s) and Th2s. J Exp Med 1998; 187:129-134.
250. Bradley LM, Asensio VC, Schioetz LK et al. Islet-specific Th1, but not Th2, cells secrete multiple chemokines and promote rapid induction of autoimmune diabetes. J Immunol 1999; 162:2511-2520.

CHAPTER 10

Immunoregulation by Cytokines in Autoimmune Diabetes

Alex Rabinovitch

Introduction

In the previous Chapter, Meagher and colleagues discuss the role of a number of cytokines and chemokines in the pathogenesis of murine type 1 (insulin dependent) diabetes mellitus. Here I provide an integrated view of type 1 diabetes as a disorder of immunoregulation. T cells specific for pancreatic islet β cell constituents (autoantigens) exist normally but are restrained by regulatory mechanisms (self-tolerant state). When regulation fails, β cell-specific autoreactive T cells become activated and expand clonally. Current evidence indicates that islet β cell-specific autoreactive T cells belong to a T helper 1 (Th1) subset, and these Th1 cells and their characteristic cytokine products, IFNγ and IL-2, are believed to cause islet inflammation (insulitis) and β cell destruction. Immune-mediated destruction of β cells precedes hyperglycemia and clinical symptoms by many years because these become apparent only when most of the insulin-secreting β cells have been destroyed. Therefore, several approaches are being tested or are under consideration for clinical trials to prevent or arrest complete autoimmune destruction of islet β cells and insulin-dependent diabetes. Approaches that attempt to correct underlying immunoregulatory defects in autoimmune diabetes include interventions aimed at i) deleting β cell autoreactive Th1 cells and cytokines (IFNγ and IL-2) and/or ii) increasing regulatory Th2 cells and/or Th3 cells and their cytokine products (IL-4, IL-10 and TGFβ1).

Type 1 Diabetes Viewed as a Disorder of Immunoregulation

Type 1 diabetes mellitus results from selective destruction of the insulin-producing β cells in the pancreatic islets of Langerhans. The current concept is that pancreatic islet β cells are destroyed by an autoimmune response mediated by T lymphocytes (T cells) that react specifically to one or more β cell proteins (autoantigens).[1] Although it has not been excluded that a primary β cell lesion, intrinsic or acquired (possibly viral or chemical), might be involved in initiating an autoimmune response,[2] it is clear that, once established, an immune response is the cause of β cell destruction. For example, diabetes transfer studies have demonstrated that bone marrow-derived cells from hosts with autoimmune diabetes can transfer β cell destructive insulitis to nondiabetes-prone human, mouse, or rat pancreas, thereby indicating that an underlying abnormality in type 1 diabetes resides in the immune system.[3-8]

The autoimmune response to islet β cells is thought to occur in persons who possess certain susceptibility alleles and who lack other protective alleles of the major histocompatibility (MHC) gene complex, which regulates immune responses. In addition, non-MHC genes may contribute to the autoimmune response. The traditional concept is that environmental factors (e.g., microbial, chemical, dietary) may trigger an autoimmune response against β cells in a genetically diabetes-prone individual. Studies in animal models with spontaneous autoimmune diabetes, however, have revealed that environmental factors (particularly microbial agents) may

Cytokines and Chemokines in Autoimmune Disease, edited by Pere Santamaria.

either promote or protect against diabetes development.[9] Therefore, the current concept being explored is that both genetic and environmental inputs may be either pathogenic (i.e., diabetes-promoting) or protective against type 1 diabetes, and that disease appearance is influenced by the net effects of genetic and environmental factors on immune responses. According to this concept, type 1 diabetes, like other organ-specific autoimmune diseases, results from a disorder of immunoregulation.[1] This posits that T cells specific for islet β cell molecules (i.e., autoantigens) exist normally but are restrained by immunoregulatory mechanisms (the self-tolerant state), and that type 1 diabetes develops when one or another immunoregulatory mechanism (e.g., regulatory T cells) fails, allowing β cell-autoreactive T cells to become activated, expand clonally, and entrain a cascade of immune and inflammatory processes in the islets, culminating in β cell destruction (Fig. 10.1).

Although it is not known what may trigger loss of self-tolerance to islet antigens in type 1 diabetes, it appears that defective immunoregulatory (suppressor) mechanisms allow the autoimmune state to progress to a pathological level and cause β cell destruction. There is now abundant evidence that suppressor cell defects may contribute to diabetes development in rodent models of type 1 diabetes. In the nonobese diabetic (NOD) mouse, diabetes onset is accelerated by thymectomy performed at 3 weeks of age[10] and by administration of cyclophosphamide,[11,12] a drug known for its selective effects on suppressor T cells. Diabetes transfer is obtained only in immunodeficient recipients, that is, neonates[13] and adults that have been sublethally irradiated[14] or thymectomized and treated with a monoclonal antibody to $CD4^+$ T cells.[15] One can prevent diabetes transfer by spleen cells from diabetic mice by preinfusion of $CD4^+$ spleen cells from nondiabetic syngeneic mice.[16] $CD4^+$ and $CD8^+$ suppressor clones have been reported,[17-19] as has the production of a suppressor factor.[19] Treatment of young NOD mice with an anti-MHC class II monoclonal antibody protects them from diabetes, and this protection is transferable to nonantibody-treated mice by infusion of $CD4^+$ T cells from protected mice.[20] In the Biobreeding (BB) rat, diabetes is accelerated by the administration of a monoclonal antibody to $RT6.1^+$ T cells[21] and prevented by transfusion of lymphoid cells from diabetes-resistant BB rats.[22] Finally, the mechanisms by which islet autoreactive T cells may be suppressed are unknown; however recent studies have pointed to cytokines as important immunoregulatory molecules.

Immune Responses: Roles of Cytokines

Characteristics of Cytokines

Cytokines are peptide molecules synthesized and secreted by activated lymphocytes (lymphokines), macrophages/monocytes (monokines) and cells outside the immune system (e.g., endothelial cells, bone marrow stromal cells, and fibroblasts). Cytokines are used mainly by immune system cells to communicate with each other and to control local and systemic events of immune and inflammatory responses. More than 30 immunologically active cytokines exist and are generally grouped as interleukins (ILs), interferons (IFNs), tumor necrosis factors (TNFs), and colony-stimulation factors (CSFs).[23] Both the production of cytokines by cells and the actions of cytokines on cells are complex: A single cell can produce several different cytokines, a given cytokine can be produced by several different cell types, and a given cytokine can act on one or more cell types. Also, cytokine actions are usually local: It can act i) between two cells that are conjugated to one another, ii) on neighboring cells (paracrine), and iii) on the cell that secretes the cytokine (autocrine). In some cases (notably the macrophage-derived inflammatory cytokines, such as IL-1, IL-6, and TNFα) cytokines exert actions on distant organs (endocrine).

Interpretation of the actions of cytokines in general is complicated by the very nature of cytokine biology. First, large amounts of a cytokine are often produced when a cell is stimulated by an antigen, mitogen, or other cytokines (e.g., up to 2% of cell protein synthesis can be devoted to a single cytokine). Second, cytokine receptors have high affinities for their specific

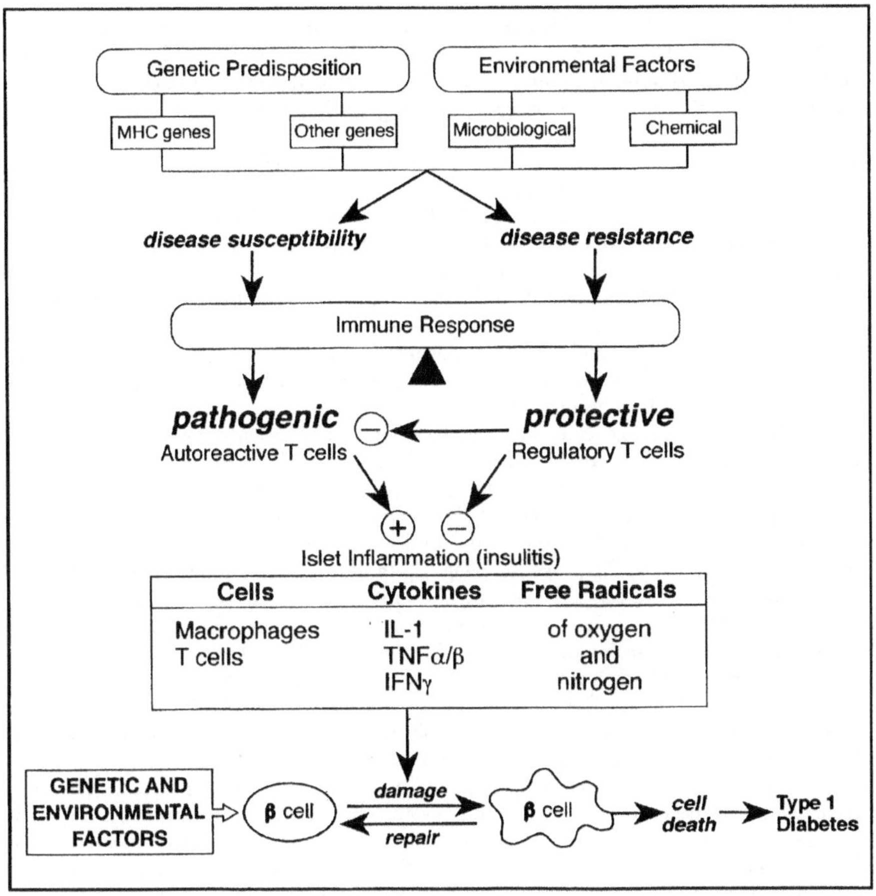

Fig. 10.1. A current formulation of the pathogenesis of type 1 diabetes. Genetic and environmental factors interact and confer either susceptibility or resistance to disease, depending on the gene/allele possessed by the individual and the environmental agent to which that individual is exposed. Disease susceptibility leads to a pathogenic immune response whereas disease resistance leads to a protective immune response. The pathogenic immune response is believed to be mediated by T lymphocytes (T cells) that are reactive to islet β cell self-antigen(s) (autoreactive T cells), whereas a protective immune response may be mediated by T cells that suppress the autoreactive T cells (regulatory T cells). Dominance of the pathogenic immune response would lead to islet inflammation (insulitis). This is characterized by infiltration of the islet by macrophages and T cells that are cytotoxic, both directly and indirectly by producing cytokines (e.g., IL-1, TNFα, TNFβ, and IFNγ) and free radicals that damage β cells. Genetic and environmental factors may also directly increase or decrease the ability of β cells to repair damage and prevent irreversible β cell death, insulinopenia, and diabetes. (Reproduced from Rabinovitch A. and Skyler JS. Prevention of type 1 diabetes 1998;82:739, with permission of W.B. Saunders Co.)

cytokine ligands, so most cytokines have very high specific activity. The consequences of these properties of cytokines and cytokine receptors is that one activated cell can produce enough cytokine to activate 1,000-10,000 other cells (i.e., a very small number of antigen-reactive cells can have widespread effects). Third, cytokine synthesis is regulated by the differentiation of cells into the various cytokine-secreting phenotypes and by the selective activation of different cell types to produce some or all of their characteristic set of cytokines.

T Cell Subsets, Cytokine Profiles and Immune Response Regulation

Antigen-activated T cells are termed T helper (Th) cells because they help to mediate both cellular and humoral (antibody) immune responses. In 1986, Mosmann and colleagues,[24] started a conceptual revolution in immunology by dividing T helper (Th) cells into two populations with contrasting and crossregulating cytokine profiles. The Th1 and Th2 patterns of cytokine production were originally described among mouse CD4+ T cell clones[24,25] and later among human T cells.[26] Mouse Th1 cells produce IL-2, IFNγ, and TNFβ (also termed lymphotoxin), whereas Th2 cells produce IL-4, IL-5, IL-6, IL-9, IL-10 and IL-13. Cytokine production by human Th1 and Th2 cells follows similar patterns, although the synthesis of IL-2, IL-6, IL-10 and IL-13 is not as tightly restricted to a single subset as in mouse T cells. Several other proteins are secreted both by Th1 and Th2 cells, including IL-3, TNFα, granulocyte-macrophage colony-stimulating factor (GM-CSF) and members of the chemokine families.[27] Th1 and Th2 responses are not the only cytokine patterns possible: T cells expressing cytokines of both patterns have been called Th0 cells[28] and those producing high amounts of transforming growth factor β (TGFβ) have been termed Th3.[29]

The functional significance of Th1 and Th2 cell subsets is that their distinct patterns of cytokine secretion lead to strikingly different T cell actions.[27,28,30-32] Th1 cells and their cytokine products (IL-2, IFNγ and TNFβ) are the mediators in cell-mediated immunity (formerly termed delayed-type hypersensitivity). IFNγ and TNFβ activate vascular endothelial cells to recruit circulating leukocytes into the tissues at the local site of antigen challenge, and they activate macrophages to eliminate the antigen-bearing cell. In addition, IL-2 and IFNγ activate i) cytotoxic T cells to destroy target cells expressing the appropriate MHC-associated antigen, and ii) natural killer (NK) cells to destroy target cells in an MHC-independent fashion. Thus, Th1 cytokines activate cellular immune responses. In contrast, Th2 cytokines are much more effective stimulators of humoral immune responses, i.e., immunoglobulin (antibody) production, especially immunoglobulin E, by B cells. Furthermore, responses of Th1 and Th2 cells are mutually inhibitory. Thus, the Th1 cytokine IFNγ inhibits the production of the Th2 cytokines IL-4 and IL-10; these, in turn, inhibit Th1 cytokine production.

Protective responses to pathogens depend on activation of the appropriate Th subset accompanied by its characteristic set of immune effector functions. For example, human Th1 cells develop in response to intracellular bacteria and viruses, whereas Th2 cells develop in response to allergens and helminth components.[30] Th1 and Th2 cells play different roles not only in protection against exogenous offending agents, but also in immunopathology. Th1 cells are involved in contact dermatitis, organ-specific autoimmunity, and allograft rejection, whereas Th2 cells are responsible for initiation of the allergic cascade.[30]

Among signals that may orient the immune response in the direction of either a Th1 or a Th2 cell response, the macrophage-derived cytokines, IL-10[33] and IL-12[34] have been discovered to play important roles. IL-12 is a potent stimulant of Th1 cells and cytokines, notably IFNγ. Thus, IL-12 can initiate cell-mediated immunity. In contrast, IL-10 (derived from macrophages and Th2 cells) exerts anti-inflammatory effects by inhibiting production of IL-12 and other pro-inflammatory macrophage cytokines (e.g., IL-1, IL-6, IL-8, TNFα), by increasing macrophage production of IL-1 receptor antagonist, and by inhibiting the generation of oxygen and nitrogen free radicals by macrophages. In addition, IL-10 may favor Th2 over Th1 cell differentiation and function by inhibiting expression of MHC class II molecules and the B7 accessory molecule on macrophages, a major costimulator of T cells.[35] The combination of IL-4 and IL-10 is particularly effective in inhibiting Th1 effector function (i.e., cell-mediated immunity) in vivo.[36]

Th1-like and Th2-like polarized cytokine secretion patterns have now been described for many different cell types: CD4, CD8 and γδ T cells, NK cells, B cells, dendritic cells, macrophages, mast cells and eosinophils.[37] In recognition of the fact that cytokine secretion patterns are not restricted to certain cell types, they are often described as type 1 and type 2 rather than Th1 and Th2. Thus a cytokine can be classified on the basis of the response it evokes rather than on the cell type that produces it.[38] Type 1 cytokines (IFNγ, IL-2, TNFβ and IL-12) primarily stimulate cell-mediated immunity; type 2 cytokines (IL-4, IL-5, IL-6, IL-10 and IL-13) primarily induce humoral immunity and diminish cellular immunity; and the type 3 cytokine (TGFβ) also diminishes cellular immunity. This functional classification is used in this chapter, and the primarily (but not exclusively) macrophage-derived cytokines, IL-1 (both α and β isoforms), TNFα and IFNα are referred to as proinflammatory cytokines (see Tables 8.1-8.3).

Approaches Used to Study Roles of Cytokines in Type 1 Diabetes

Studies over the last 15 years have examined the possible involvement of cytokines in the pathogenesis of type 1 diabetes through a variety of approaches. These can be classified as: i) correlation studies of cytokines expressed in islets in relation to diabetes development; ii) cytokine augmentation studies; and iii) cytokine deficiency studies.

Cytokine augmentations have been created by:

1. Adding cytokines to islets in vitro,
2. Expressing cytokine genes transgenically in β cells, and
3. Administering cytokines and cytokine-producing cells.

Cytokine deficiencies have been created by:

1. Disrupting genes encoding cytokines or their receptors,
2. Neutralizing cytokines by anti-cytokine antibodies or soluble cytokine receptors,
3. Blocking cytokine receptors by receptor antagonists or antibodies, and
4. Deleting cytokine receptor-positive cells.

Information obtained from these studies is summarized in Tables 8.1-8.3.

Cytokines Expressed in the Insulitis Lesion

A variety of cytokines have been found to be expressed at the gene or protein level, or both, in the insulitis lesion of autoimmune diabetes-prone NOD mice and BB rats, as well as in the pancreata of humans with type 1 diabetes. The simple presence of a cytokine in islets, however, does not identify its role in the pathogenesis of diabetes. Thus, a given cytokine might promote autoimmunity and β cell destruction (β cell destructive insulitis) or, alternatively, may regulate (i.e., suppress) the autoimmune and/or inflammatory processes that would otherwise result in β cell destruction (benign insulitis). The term "benign insulitis" has been used to describe the accumulation of mononuclear leukocytes (macrophages, monocytes, T and B lymphocytes) around and within pancreatic islets, without progression to significant β cell destruction.[39]

Correlations have been observed between β cell destructive insulitis and expression of proinflammatory cytokines (IL-1, TNFα and IFNγ) and type 1 cytokines (IFNγ, IL-2, TNFβ and IL-12) in NOD mice and BB rats, whereas expression of type 2 cytokines (IL-4 and IL-10) and the type 3 cytokine (TGFβ) tended to correlate with benign insulitis in these animals (Table 10.1).[40-62] At present, only IFNα[63-65] and IFNγ[66,67] have been associated with β cell destructive insulitis in human type 1 diabetes (Table 10.1). The specific cell sources of the cytokines expressed in the insulitis lesions of NOD mice, BB rats and humans have not been identified, except in one study where IFNγ-producing cells in NOD islets with β cell destructive insulitis were identified to be equally divided into $CD4^+$ (Th1) cells and $CD8^+$ (Tc1) cells.[54]

Cytokine Studies in Isolated Islets

It is now well documented that certain cytokines are cytotoxic to pancreatic islets in vitro.[68,69] IL-1, TNFα, TNFβ, and IFNγ (in piconanomolar concentrations) are cytostatic to β cells, in that they inhibit insulin synthesis and secretion; however, these functions may recover after the cytokine is removed. In addition, these cytokines can be cytocidal: usually when added in combination, they destroy β cells in both rodent and human islets. Because the cytodestructive effects of cytokines on islet β cells in vitro are not specific to β cells (e.g., α cells in the islets are also damaged),[69] cytokines may not qualify as mediators of β cell destruction in type 1 diabetes, which is β cell specific. Even agents with known β cell specificity in vivo (alloxan and

Table 10.1. Correlations of cytokines expressed in islets with β cell destructive or benign insulitis

	Proinflammatory Cytokines			Type 1 Cytokines				Type 2 Cytokines			Type 3 Cytokine
	IL-1	TNFα	IFNα	IL-12	IFNγ	TNFβ	IL-2	IL-4	IL-6	IL-10	TGFβ
NOD mice	+	+	0	+	+	+	+	S	+	S	S
BB rats	+	+	+	+	+	?	+	0	?	S	S
Humans	0	nd	+	?	+	?	nd	0	0	?	?

+, cytokine presence correlates with β cell destructive insulitis; S, cytokine presence correlates with benign insulitis; 0, cytokine presence does not correlate with either destructive or benign insulitis; nd, not detected; ? not reported. (Reproduced from Rabinovitch A. An update on cytokines in the pathogenesis of insulin-dependent diabetes mellitus. Diabetes/Metabolism Reviews 1998;14:129, with permission of John Wiley & Sons, Ltd.)

streptozocin), however, can damage other islet endocrine cells in vitro,[70] possibly because of nonspecific damage to non-β cells adjacent to damaged β cells. For example, β cells separated from non-β cells in islets are destroyed by streptozocin and alloxan, and non-β cells are not.[71] Similarly, IL-1 is cytotoxic to both α and β cells in isolated rat islets, but selectively inhibits β cell secretion of insulin and not α cell secretion of glucagon in separated purified preparations of these islet endocrine cells.[72]

Cytokine applications to islets in vitro may not mimic the molecular pathology of the pancreatic insulitis lesion in vivo. Polar release of cytokines by Th cells conjugated to B cells has been reported,[73] and membrane forms of IL-1[74] and TNFα[75] may contribute to macrophage-mediated cytotoxicity. Similarly, cytokine products of islet-infiltrating macrophages and T cells could be delivered in a targeted fashion into the microenvironment of the β cell or even directly into the β cell by contiguous cytotoxic T cells. Highly localized and directed delivery of cytokines from T cells and macrophages to β cells might explain why rejection of islet allografts in rats was found not to destroy syngeneic islets mixed in with the allogeneic islets (whole islets, not single cell preparations, were admixed).[76] Also, syngeneic islets were not destroyed after cotransplantation with allogeneic or xenogeneic islets in mice; however, insulin secretory responses from the syngeneic islets cotransplanted with xenogeneic islets were severely impaired, suggesting inhibitory effects of xenogeneic macrophage-derived products (e.g., IL-1, TNFα, nitric oxide) on islet β cell function.[77]

From the aforementioned studies, we may conclude that IL-1, TNFα, TNFβ and IFNγ impair insulin secretion and, usually when added in combinations of two or more, these cytokines are destructive to rodent and human β cells in whole islet preparations in vitro (Table 10.2). Although IL-1, TNFα, TNFβ, and IFNγ are produced by islet-infiltrating macrophages and T cells in the insulitis lesion of type 1 diabetes (Table 10.1), they have not been proven directly cytotoxic to β cells in vivo. A recent study has challenged the relevance of IFNγ effects on β cells in vitro to the pathogenesis of β cell destruction in autoimmune diabetes. Transgenic NOD mice expressing dominant negative mutant IFNγ receptors on pancreatic islet β cells developed diabetes at a rate similar to that of wild-type animals.[78] This suggests that β cells are not immediate targets of IFNγ in autoimmune diabetes. Nevertheless, IFNγ may lead to β cell destruction indirectly, likely by activating macrophages or cytotoxic T cells. For example, activation of resident macrophages in rat[79] and human[80] islets by treatment of the islets with TNFα, IFNγ, and lipopolysaccharide in vitro resulted in inhibition of insulin secretion that was mediated by intra-islet release of IL-1, followed by expression of inducible nitric oxide synthase (iNOS) in the β cells. Current evidence for mechanisms of cytokine-induced impaired insulin secretion and β cell destruction points to nitric oxide and/or oxygen free radicals produced in β cells exposed to the cytokines.[81-84]

Transgenic Expression of Cytokines by β Cells

Transgene technology has been used to examine the possible roles of different immunoregulatory molecules (MHC proteins, costimulatory molecules and cytokines) in the immunopathogenesis of the insulitis lesion and β cell destruction in autoimmune diabetes. Selective expression of gene products in islet β cells has been achieved by fusing the regulatory elements of the rat insulin gene (rat insulin promoter, RIP) with the structural gene of interest, microinjecting the hybrid gene (DNA construct) into the pronucleus of a fertilized mouse egg, and screening the resultant mice for phenotypic expression of the integrated transgene. This technique has provided abundant information on the consequences of local intra-islet production of a variety of cytokines in normally nondiabetes-prone mice and in autoimmune diabetes-prone NOD mice.[85-106] Transgenic expression of IFNα, IFNγ and IL-2 by β cells in nondiabetes-prone mice induced β cell destructive insulitis and autoimmune diabetes, whereas expression of TNFα, TNFβ and IL-6 induced insulitis that did not progress to β cell destruction and diabetes, and IL-10 and TGFβ induced only peri-islet inflammatory responses (Table 10.2). Transgenic expression of TNFα, IL-4, IL-6 and TGFβ by β cells in autoimmune

Table 10.2. Effects of cytokine additions to islets in vitro, to β cells transgenically, and to NOD mice and BB rats by systemic administration

	Proinflammatory Cytokines			Type 1 Cytokines				Type 2 Cytokines			Type 3 Cytokine
Cytokine Additions	**IL-1**	**TNFα**	**IFNα**	**IL-12**	**IFNγ**	**TNFβ**	**IL-2**	**IL-4**	**IL-6**	**IL-10**	**TGFβ**
To islets/β cells in vitro (rodent and human)	Toxic	Toxic	0	?	Toxic	Toxic	0	0	0	0	0
To β cells, transgenically in:											
Nondiabetes-prone mice	?	Insulitis	+	?	+	Insulitis	+	?	Insulitis	Peri-islet inflamm.	Peri-islet inflamm.
NOD mice	?	S	?	?	?	?	+	S	S	+	S
By systemic (parenteral) administration to:											
NOD mice	S	+(<3wks) S(>4wks)	S	S/+	S/Insulitis	S	S	S	?	S	S
BB rats	S/+	S	S/+	?	S	S	S/+	?	?	?	?

0, no effect; ? not reported; +, diabetes produced, accelerated, or incidence increased; S, diabetes delayed or incidence decreased. (Modified from Rabinovitch A. An update on cytokines in the pathogenesis of insulin-dependent diabetes mellitus. Diabetes/Metabolism Reviews 1998;14:129, with permission of John Wiley & Sons, Ltd.).

diabetes-prone NOD mice protected against diabetes development, whereas expression of IL-2 and IL-10 accelerated it (Table 10.2). These results suggest pathogenic roles for IFNα, IFNγ, IL-2 and IL-10 in type 1 diabetes development, and protective roles for TNFα, IL-4, IL-6 and TGFβ. However, because these cytokines were expressed in islets constitutively during ontogeny, they may have affected maturation of the immune system and the insulitis process in ways that do not mimic the roles of the cytokines during diabetes pathogenesis.

Effects of Systemic Administration of Cytokines

Systemic administration of a wide variety of cytokines has been shown to prevent diabetes development in NOD mice and/or BB rats (Table 10.2).[107-116] Because deficiencies in the endogenous production of IL-1,[118] IL-2,[111,118] IL-4,[111] TNFα,[108,113,119] and TNFβ[116] have been reported in diabetes-prone NOD mice and/or BB rats, chronic administration of these cytokines may prevent diabetes by correcting deficits in cytokine production in these animals. This appears to be the case with IL-4 administration, because deficient IL-4 production has been identified as an underlying cause of the emergence of autoreactive Th1 cells in NOD mice.[111] Moreover, systemic treatment of NOD mice with IL-4 induces a Th2 cell-enriched environment in the pancreatic islets of these mice.[120] This suggests that systemic cytokine delivery can target the local β cell-directed autoimmune process.

Systemic administration of a cytokine, however, produces a gradient for the cytokine which is higher outside than inside the islet, and this may result in immunologic effects different from those induced by the same cytokine secreted in the islet. For example, IFNα is proinflammatory and induces autoimmune diabetes when expressed transgenically by islet β cells in nondiabetes-prone mice,[87,88] whereas systemic administration of IFNα inhibits insulitis and diabetes in NOD mice[121] and BB rats.[122] Similarly, all evidence points to IFNγ acting as a proinflammatory cytokine when expressed in islets (Tables 8.1 and 8.2); however, when administered systemically with TNFα, IFNγ decreased insulitis in NOD mice,[123] and when administered alone, IFNγ significantly decreased the incidence of diabetes in BB rats.[124] IFNγ has both pro- and anti-inflammatory actions:[125,126] the former may be manifested when IFNγ is produced locally in islets, whereas the latter may result from systemic administration of the cytokine. In addition to acting on immunologic circuits outside the islet, systemically-administered cytokines may act on targets outside the immune system. For example, IL-1 and TNFα can stimulate the hypothalamic-pituitary axis, leading to secretion of adrenocorticotropic hormone and consequently adrenal glucocorticosteroids that suppress immune and inflammatory responses.[127] Recent studies have shown that glucocorticosteroids suppress production of type 1 cytokines such as IL-2 and IFNγ, while type 2 cytokines such as IL-4 and IL-10 may be increased.[128,129] Therefore, systemic cytokine delivery may prevent autoimmune diabetes by acting directly or indirectly on the islet β cell immunopathogenic process.

In addition to route of administration, dose and frequency of administration may influence the effects of a cytokine on diabetes development. For example, a low dose of IL-1β decreased diabetes incidence in diabetes-prone BB rats, whereas a high dose of IL-1β accelerated diabetes in these animals.[107] Similarly, systemic administration of large daily doses of IL-12, a cytokine that induces Th1 cell differentiation,[135] accelerated β cell-destructive insulitis and diabetes onset in NOD mice.[130] Surprisingly, at a lower dose and injections once a week, IL-12 administration suppressed diabetes development in NOD mice.[131] Another explanation for the dichotomous effects of a given cytokine may relate to the timing of its participation in the disease process. For example, TNFα administered before age 3 weeks accelerated diabetes development in NOD mice, whereas TNFα administration after age 4 weeks decreased diabetes incidence.[132] These findings suggest that TNFα may function in some way as a growth factor for T cells during development, whereas chronic TNFα exposure may suppress the function of mature T cells in adult mice. Indeed, chronic repeated systemic injections of TNFα have been found to suppress a broad range of T cell responses in mice, including proliferation and cytokine production by both Th1 and Th2 cells, and this has been attributed to attenuation of T cell receptor signaling.[133] Interestingly, chronic TNFα administration protected β cells in syngeneic islet grafts from autoimmune destruction after transplantation into diabetic NOD mice, and protection was associated with selective decreases in expression of Th1-type cytokines (IFNγ, IL-2, TNFβ) in spleens and islet grafts.[134] Similarly, β cell transgenic expression of IL-10 (before autoimmune disease development) has been reported to favor the generation of diabetogenic $CD8^+$ T cells,[135] whereas systemic administration of IL-10 to adult NOD mice (from age 5 to 25 weeks) prevented diabetes development through the induction of $CD4^+$ Th2 cells.[48] In addition, systemic delivery of the immunosuppressive cytokine, TGFβ1, by a somatic gene therapy approach (intramuscular injection of a DNA expression vector encoding TGFβ1) protected NOD mice from β cell destructive insulitis and diabetes, whereas NOD mice injected with a DNA vector encoding the proinflammatory cytokine, IFNγ, developed diabetes earlier.[136] Interestingly, testicular Sertoli cells prolong survival of syngeneic islet grafts transplanted into diabetic NOD mice, and protection against autoimmune destruction of islet β cells was found to be due to TGFβ1 production and systemic release by the implanted Sertoli cells, resulting in decreased IFNγ production in the islet grafts.[137]

In summary, the effects of cytokines on autoimmune diabetes development depend to a large extent on dose, frequency and route of administration, as well as time of administration in relation to disease development. Therefore, systemic delivery of cytokines may not mimic their roles in the pathogenesis of autoimmune diabetes. Nevertheless, elucidation of the mechanisms by which systemic cytokine delivery prevents diabetes development may point to immunotherapies that target the β cell-directed autoimmune response more specifically than does systemic cytokine delivery.

Effects of Cytokine Deletions

Studies in which cytokines are deleted from expression in autoimmune diabetes-prone animals have the potential of revealing whether the cytokine plays an essential (necessary) role in type 1 diabetes development. Cytokine deficiencies have been created in diabetes-prone animals by disrupting genes encoding cytokines or their receptors (gene knockout), neutralizing cytokines by anti-cytokine antibodies or soluble cytokine receptors, blocking cytokine receptors by receptor antagonists or antibodies, and deleting cytokine receptor-positive cells (Table 10.3).[78,138-155] NOD mice with deletions of IL-12 and IFNγ genes have been created to study the consequences of genetic absence of these cytokines on autoimmune diabetes. In a preliminary report, cyclophosphamide-accelerated diabetes was found to be decreased but not prevented in NOD mice with disruption of the IL-12 gene.[149] In another study, IFNγ gene disruption was found to delay but not prevent diabetes in NOD mice.[150] Although IFNγ receptor knockout NOD mice were reported to be protected from diabetes,[153] more recent work has demonstrated such an effect only in cyclophosphamide-induced acceleration of diabetes; a second gene, linked to the IFNγ receptor, plays a role in the resistance originally noted.[154,155]

Although the aforementioned gene knockout studies demonstrate that autoimmune diabetes in NOD mice is not prevented by the genetic absence of the type 1 cytokines, IL-12 and IFNγ, deletions of these cytokines after birth can prevent diabetes development. Thus, the homodimeric IL-12p40 subunit, an antagonist of the bioactive IL-12p35/p40 heterodimer (IL-12), suppressed diabetes development in cyclophosphamide-injected NOD mice.[151] In addition, NOD mouse islets that hyperexpressed IL-12p40 (antagonist of IL-12), after transfection with an adenoviral IL-12p40 gene construct, survived and corrected hyperglycemia after transplantation into diabetic NOD mice.[152] Interestingly, IFNγ mRNA was decreased and TGFβ mRNA was increased in the IL-12p40-expressing (protected) islet grafts. In other studies, neutralization of IFNγ with antibodies in NOD mice[41,141] and BB rats,[142] and with a soluble receptor in NOD mice,[143] significantly decreased diabetes incidence in these diabetes-prone animals. Disruptions of IL-12 and IFNγ genes may not be as effective in preventing autoimmune diabetes development as deleting these cytokines after maturation of the immune system (e.g., by administrations of anti-cytokine antibodies, soluble cytokine receptors, etc.)

Table 10.3. Effects of cytokine deletions in NOD mice and BB rats

	Proinflammatory Cytokines			Type 1 Cytokines					Type 2 Cytokines		Type 3 Cytokine
Cytokine Additions	**IL-1**	**TNFα**	**IFNα**	**IL-12**	**IFNγ**	**TNFβ**	**IL-2**	**IL-4**	**IL-6**	**IL-10**	**TGFβ**
By knockout of gene for cytokine/or its receptor in NOD mice	?	?	?	0	0	?	?	0	?	0	?
By neutralization of cytokine, blockade of receptor, or deletion of receptor-positive cells in:											
NOD mice	S	S (< 3wks) + (> 4wks)	?	S	S	?	S	0	S	S Insulitis (<3wks) 0 (>10wks)	?
BB rats	S	?	?	?	S	?	?	?	?	?	?

?, not reported; 0, no effect; S, diabetes delayed or incidence decreased; +, diabetes accelerated or incidence increased. (Modified from Rabinovitch A. An update on cytokines in the pathogenesis of insulin-dependent diabetes mellitus. Diabetes/Metabolism Reviews 1998;14:129, with permission of John Wiley & Sons, Ltd.)

because genetic absences of IL-12 and IFNγ may allow the development of compensatory immunological mechanisms that would not be available to NOD mice in which the cytokines are deleted after maturation of the immune system.

In summary, deletions of a wide variety of cytokines (IL-1, TNFα, IL-12, IFNγ, IL-2 and IL-6), by one or more of the approaches listed above, have been reported to delay or decrease diabetes incidence, or both, in NOD mice, and deletion of IL-1 and IFNγ has decreased diabetes incidence in BB rats (Table 10.3). These findings reveal that multiple cytokines likely participate in the autoimmune response that leads to β cell destruction and that deletion of a single pathogenic cytokine may not be sufficient to prevent diabetes development completely. The findings are not surprising, given the overlap of functions that different cytokines perform. Therefore, therapy of autoimmune diabetes might require neutralizing or blocking more than one cytokine. Alternatively, a pathogenic mechanism common to the diabetogenic cytokines may be identified.

Cytokines in Human Type 1 Diabetes

It is evident from the aforementioned studies that most of our current information on cytokines implicated in the pathogenesis of type 1 diabetes comes from studies using NOD mouse and BB rat models of the human disease. The possible roles of cytokines in the pathogenesis of the human disease are less well characterized. Histological studies of the pancreas of humans with type 1 diabetes have been limited by necessity to patients in whom clinical diabetes has already developed, and in these patients the insulitis lesion is likely near or at an end stage. In this situation, IFNα[63-65] and IFNγ,[66,67] but not other cytokines, have been detected in human islets (Table 10.1). IFNα expression by human β cells may result from viral or other β cell stresses, and IFNα, in turn could activate autoreactive T cells.[87,88] This remains an attractive but unproven hypothesis for the cause of the β cell-directed autoimmune response in human type 1 diabetes. IFNγ, produced by T cells that infiltrate human islets,[66] and possibly macrophage-derived IL-1 and TNFα, may be directly cytotoxic to human islet β cells in vivo, as demonstrated for these cytokines in vitro.[68,69] In addition, cytokines may sensitize human islet β cells to T cell-mediated cytotoxicity in vivo by upregulating MHC class I protein expression on β cells (an action of IFNγ), and inducing Fas (CD95) protein expression on β cells (an action of IL-1β). Indeed, increased β cell expression of MHC class I protein[156,157] and Fas protein[158] has been reported in the pancreas of patients with recent-onset type 1 diabetes.

Studies of serum levels of different cytokines, as well as secretion of cytokines by peripheral blood mononuclear cells (PBMC) from patients with type 1 diabetes, have not yielded consistent results. One study reported that cells in whole blood from patients with type 1 diabetes produced significantly higher amounts of Th1 cytokines (IFNγ and TNFα) than cells from normal control subjects, while production of Th2 cytokines (IL-4 and IL-10) was similar in diabetic and control subjects.[159] Another study found that secretion of Th2 cytokines was decreased and Th1 cytokines increased in activated PBMC from diabetic subjects.[160] Yet another study reported decreased IL-4 secretion from stimulated PBMC and T cells of diabetic subjects and normal IFNγ expression.[161] It is unclear from these studies, however, whether changes in serum levels or production of cytokines by cells from patients with type 1 diabetes preceded or resulted from diabetes. In one study, circulating levels of IL-1α, TNFα, IL-2 and IFNγ were found to be elevated at the time of diagnosis of diabetes and in the prediabetic period.[162] Similarly, circulating levels of TNFα and soluble IL-2 receptor were reported to be elevated in nondiabetic first degree relatives of patients with type 1 diabetes; also, IL-1α and TNFα production by mitogen-stimulated PBMC was increased in both diabetic and healthy family members.[163] In another study, the ratio of IFNγ/IL-4 production by PBMC was significantly increased in high risk first degree relatives of type 1 diabetic children.[164]

Recently, a subset of cells that express surface markers for both T cells and NK cells, NK1.1$^+$ T cells (TCRαβ$^+$CD4$^-$CD8$^-$) was isolated from the blood of type 1 diabetic patients and their nondiabetic twin/triplet siblings positive for islet cell antibodies (at-risk nonprogressors).[165]

All the NK1.1⁺ T cell clones from the diabetic twin/triplets secreted only IFNγ on stimulation with a monoclonal antibody to $CD3^+$ T cells, whereas the clones from the at-risk nonprogressors and normal subjects secreted both IL-4 and IFNγ. Also, half (7 of 14) of the at-risk nonprogressors had high serum levels of IL-4 (and IFNγ), whereas significantly fewer diabetic patients had elevated serum IL-4 levels. These findings suggest that Th1 cell-mediated damage of islet β cells is initially regulated by $NK1.1^+$ T cells (with a particular Vα24JαQ T cell receptor) that produce both IFNγ and IL-4, and that loss of their capacity to secrete IL-4 correlates with type 1 diabetes.[165] It remains to be determined, however, whether the loss of IL-4 secretion precedes or follows β cell destruction and diabetes appearance. Further studies of cytokine production by peripheral blood mononuclear cells from prediabetic subjects, possibly in response to putative islet autoantigens, may improve prediction of type 1 diabetes development.

Autoimmune Diabetes: A Dominance of Th1 Over Th2 Cells?

There is now abundant evidence that autoreactive T cells are present in the normal immune system but are prevented from expressing their autoreactive potential by other regulatory (suppressor) T cells. For example, reconstitution of lymphopenic, prediabetic BB rats with the IL-4-producing $CD4^+$ $CD45RC^{low}$ subset of Th cells but not with the IL-2-producing $CD4^+$ $CD45RC^{high}$ Th subset protects against autoimmune diabetes.[166] In a different model, adult thymectomy combined with sublethal irradiation causes diabetes in a nonautoimmune diabetes-prone rat strain, and insulitis and autoimmune diabetes are completely prevented by injection of $CD4^+$ $CD45RC^{low}$ T cells that secrete IL-2 and IL-4, not IFNγ.[166,167] Diabetes can be adoptively transferred into neonatal NOD mice or immunocompromised NOD-*scid* by splenic cells from diabetic NOD mice, whereas splenic cells from young nondiabetic NOD mice can prevent diabetic splenic cells from adoptively transferring disease. Interestingly, both the pathogenic and protective functions of $CD4^+$ cells in the diabetic and nondiabetic NOD donor spleens were found to reside in a $CD45RB^{low}$ subset of $CD4^+$ T cells; however, the pathogenic cells had a significantly higher IFNγ/IL-4 production ratio than did the protective ones.[168] These findings support the concept that Th1 cells (IFNγ-producing) are pathogenic and Th2 cells (IL-4-producing) prevent diabetes development; however, diabetes transfer and prevention were observed using polyclonal populations of T cells, and the autoimmune response in type 1 diabetes is believed to be dependent on T cells specifically reactive to islet β-cell autoantigens.

A variety of islet-reactive T cell lines and clones that either adoptively transfer diabetes or prevent against its development in NOD mice have been described, and some of these T cell lines/clones have been characterized in terms of their cytokine production profiles. In one study, $CD4^+$ T cells reactive to the islet autoantigen, glutamic acid decarboxylase (GAD), were reported to secrete IFNγ, TNFα, and TNFβ, but not IL-4 in response to GAD antigen, and these cells adoptively transferred diabetes into NOD-*scid* mice.[169] Interestingly, several diabetes-preventive $CD4^+$ T cell clones were found to produce a variety of cytokines, including type 1 cytokines (IFNγ and TNFβ), a type 2 cytokine (IL-10), and a type 3 cytokine (TGFβ).[170-172] TGFβ was implicated as the mediator of the diabetes-preventive effects of these islet-reactive $CD4^+$ T cell clones.[171,172] In another study, $CD4^+$ T cell lines that react to rat insulinoma cells and secrete either IFNγ or IL-4 were developed from spleens of diabetic NOD mice.[173] The IFNγ-secreting $CD4^+$ T cells (Th1-type) adoptively transferred β cell destructive insulitis and diabetes into neonatal NOD mice, whereas the IL-4-secreting $CD4^+$ T cells (Th2-type) induced a nondestructive peri-islet insulitis.[173] Similarly, Th1 cells expressing a diabetogenic T cell receptor adoptively transferred β cell destructive insulitis and diabetes in neonatal NOD mice, whereas Th2 cells expressing the same T cell receptor did not; however, the Th2 cells did not prevent the Th1 cells from transferring diabetes.[174] This suggests that Th2 cells cannot downregulate Th1 cells whose effector functions (e.g., type 1 cytokine production) are fully differentiated.

In contrast, a subset of natural killer thymocytes (NKT), TCRαβ^{+}CD4^{-}CD8^{-}, has recently been reported to prevent adoptive transfer of diabetes by diabetogenic NOD splenocytes, and protection was related to IL-4 and/or IL-10 production.[175] The protection provided by the NKT cells is believed to represent diabetes prevention by correction of an underlying deficiency of NKT cells[176,177] and IL-4 production[111] in NOD mice. In another recent study, a subset of TCRαβ^{+}CD4^{+}CD62L^{+} thymocytes was reported to prevent adoptive transfer of diabetes by diabetogenic NOD splenocytes,[178] however, the cytokine-producing phenotype of these CD4^{+} regulatory T cells was not determined. Collectively, these studies have given rise to the concept that the autoimmune response in type 1 diabetes involves disturbances in immunoregulatory circuits manifested as a dominance of Th1 over Th2 cell function and cytokine production (Fig. 10.2).

According to the scheme depicted in Figure 8.2, certain β cell protein(s) act as autoantigens after being processed by antigen-presenting cells (APCs), such as macrophages, dendritic cells, and B cells. APCs appear to play an important role in the initiation of insulitis. Thus, many studies indicate that macrophages and dendritic cells are the first cells to infiltrate pancreatic islets,[179-181] and inactivation of macrophages results in the near-complete prevention of insulitis and diabetes in both NOD mice and BB rats.[182,183] Recent studies have found that macrophages play an essential role in diabetes development in NOD mice by activating, largely through IL-12 secretion, Th1 cells and CD8^{+} cytotoxic T cells.[184,185] Also, recent studies have revealed that B cells clearly influence diabetes development in a manner that probably relates to their APC function, and lack of B cells prevents diabetes development.[186-188] The immunogenicity of a β cell protein may depend upon the peptide fragment derived from processing by the APC,[189] the amino acid sequences of the MHC class II molecules that bind and present the β cell peptide (antigen), and the precursor frequency of autoreactive T cells with T cell receptors to match the β cell antigen-MHC complex.[190] Interestingly, both nonMHC genes[191] and MHC class II genes[192] have been reported to determine the polarity of the Th1/Th2 immune response in NOD mice.

In addition to the MHC-antigen complex interaction with T cell receptors, T cell activation by APCs involves costimulation through multiple ligand/receptor pairs, e.g., B7/CD28, CD40L/CD40, and ICAM-1/LFA-1.[193,194] There is evidence that APC-T cell interactions via these costimulatory molecules are involved in diabetes pathogenesis. For example, transgenic expression of the costimulator molecule, B7-1 (CD80) in islet β cells has been shown to accelerate diabetes in NOD mice.[195] Also, NOD female mice did not develop diabetes when treated, at the onset of insulitis (2-4 weeks of age), with CTLA4 immunoglobulin (a soluble antagonist to CD28, the T cell receptor for the B7 ligand on APCs) or a monoclonal antibody specific for B7-2 (CD86).[196] In addition, anti-CD40L monoclonal antibody treatment of NOD female mice (3-4 weeks of age, but not greater than 9 weeks of age) completely prevented insulitis and diabetes.[197] Blockade of ICAM-1 and LFA-1 by injection of monoclonal antibodies[198,199] or soluble forms of ICAM-1,[200] reduced insulitis and diabetes incidence in NOD mice, and treatments with the soluble forms of ICAM-1 were found to decrease IFNγ mRNA expression in the pancreas.[200]

The direction taken by the T cell response, in terms of Th phenotype, is largely regulated by cytokines. Thus, naive T cells are not precommitted to any particular Th phenotype; the Th phenotype varies with the cytokines in the microenvironment. The presence of IL-12, a macrophage and B cell product, favors Th1 cell differentiation, and anti-IL-12 antiserum blocks expression of the Th1 phenotype.[129] Indeed, administration of IL-12 to prediabetic NOD female mice was found to accelerate diabetes onset, and this was associated with i) enhanced IFNγ and decreased IL-4 production by islet-infiltrating lymphocytes, and ii) selective β cell destruction.[130] IL-4, a Th2 and possibly a mast cell product,[201] favors Th2 cell differentiation, and anti-IL-4 monoclonal antibody promotes expression of a Th1 phenotype.[201,202] The results of Th1 cell activation are induction of IL-2 and IFNγ production, inhibition of Th2 cytokine production, and activation of macrophages, cytotoxic T cells, and natural killer cells.

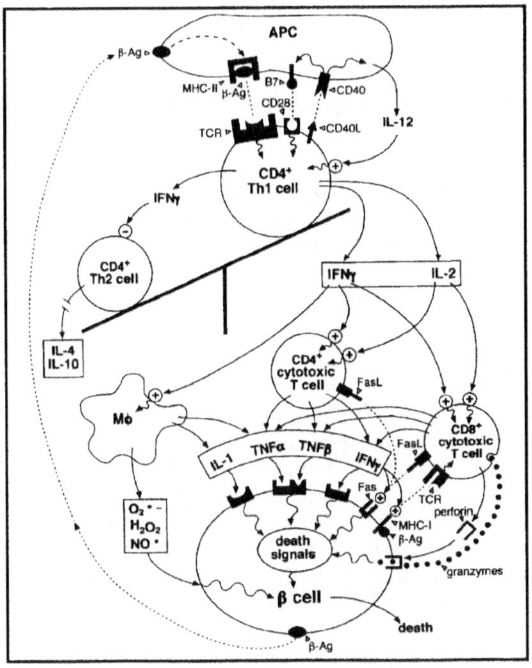

Fig. 10.2. A scheme of the immune system cells and cytokines believed to mediate destruction of pancreatic islet β cells in type 1 diabetes. The concept illustrated posits that certain β cell protein(s) are processed by antigen-presenting cells (APC), such as macrophages and dendritic cells, then presented as antigen(s) (β-Ag) in a complex with MHC class II molecules on the surface of the APC. APC and CD4⁺ T cells interact via i) the binding of a β-Ag-MHC II complex on the APC surface to a T cell receptor (TCR) specific for β-Ag, ii) the binding of costimulator molecules (e.g., B7, CD40, ICAM-1) on the APC surface to their corresponding receptors or ligands (e.g., CD28, CD40L, LFA-1) on the T cell, and iii) the production by the APC of cytokines such as IL-12 that promote differentiation of CD4⁺ T cells into Th1-type cells. Collectively, these interactions, and perhaps others, activate CD4⁺ Th1 cells to produce their characteristic cytokines (IFNγ, IL-2). IFNγ i) inhibits CD4⁺ Th2 cell production of IL-4 and IL-10, and ii) activates macrophages (Mϕ) and cytotoxic T cells; also, IL-2 activates cytotoxic T cells. CD8⁺ T cells are cytotoxic to β cells following specific recognition of β-Ag on the β cell. This necessitates direct contact of CD8⁺ T cells with β cells via the binding of a CD8⁺ TCR specific for β-Ag to the β-Ag-MHC I complex on the β cell surface. This T cell-β cell interaction activates CD8⁺ T cells, and these cells may then destroy β cells via i) the binding of Fas ligand (FasL) on the CD8⁺ T cell to a Fas receptor on the β cell, and ii) the secretion of cytotoxic molecules, such as perforin and granzymes. In addition, T cells and Mϕ may destroy β cells indirectly, that is, the immunologic cells are not in direct contact with β cells and there is no requirement for specific recognition of β-Ag on β cells. Rather, activated Mϕ may destroy β cells by producing free radicals, such as superoxide ($O_2^{\bullet -}$), hydrogen peroxide (H_2O_2), and nitric oxide ($NO^{\bullet}$), and cytokines (IL-1, TNFα) that are cytotoxic to β cells. Also, activated CD4⁺ T cells and CD8⁺ T cells may destroy β cells by producing cytokines (TNFα, TNFβ, IFNγ) that are cytotoxic to β cells. In addition, cytokines (IL-1, TNFα, TNFβ, IFNγ) may i) induce Fas receptors on β cells and so allow CD4⁺ and CD8⁺ T cells to destroy the β cells via FasL/Fas-mediated mechanisms, and ii) increase expression of MHC-I molecules on β cells and so increase interactions of CD8⁺ T cells and β cells. Finally, β cell death may result from direct toxic effects of free radicals (death by necrosis), and from actions of cytokines (IL-1, TNFα, TNFβ, IFNγ), FasL/Fas, perforin and granzymes that activate death signals (e.g., caspase enzymes) in β cells and lead to β cell self-destruction (death by apoptosis and sometimes necrosis). (Reproduced from Rabinovitch A. Roles of cell-mediated immunity and cytokines in the pathogenesis of Type 1 diabetes mellitus: In Diabetes mellitus: A Fundamental and Clinical Text. 2nd edition, 2000, Eds. LeRoith, Taylor, Olefsky, with permission of Lippincott Williams & Wilkins.)

These activated effector cells may be cytotoxic to islet β cells through a variety of antigen-specific and nonspecific mechanisms.

Antigen-Specific and Nonspecific Mechanisms of Islet β Cell Destruction

Both antigen-specific and nonspecific immune or inflammatory responses appear to be involved in mediating islet β cell destruction in type 1 diabetes.[203] Antigen-specific β cell destruction involves binding of $CD8^+$ cytotoxic T cells, through T cell receptors specific for β cell antigen(s), to the β cell antigen-MHC class I complex on β cells (Fig. 10.2). This leads to activation of the $CD8^+$ T cells (cytotoxic T cells) which, then, may destroy islet β cells by i) the binding of Fas ligand (FasL or CD95L) on the $CD8^+$ T cells to Fas receptors (Fas or CD95) on β cells, and ii) the secretion of cytotoxic molecules (granzymes and perforin) by the $CD8^+$ T cells.[204] Antigen-specific $CD8^+$ T cell- mediated cytotoxicity as a mechanism for islet β cell destruction is supported by several lines of evidence. First, diabetes can be transferred to young NOD mice by $CD8^+$ T cell clones[205,206] even in the absence of $CD4^+$ T cells.[206] Second, $CD8^+$ T cells are necessary to transfer diabetes to fully immunoincompetent irradiated or neonatal NOD mice[13,207,208] and BB rats.[209] Third, NOD mice backcrossed with $CD8^+$ T cell-deprived mice whose MHC class I genes have been inactivated by homologous recombination do not develop diabetes.[210] Fourth, $CD8^+$ T cells expressing the cytolytic mediator perforin are found in the NOD mouse insulitis lesion,[211] and diabetes incidence is reduced and onset is delayed in perforin-deficient NOD mice.[212] There is some evidence that $CD8^+$ T cells from diabetic patients and animals lyse β cells,[205,213] but these results have been difficult to reproduce. $CD8^+$ T cells have also been shown to inhibit insulin release by islet cells cultured in vitro,[214] but the interpretation is complicated by the absence of MHC restriction in this model.

Antigen-nonspecific β cell destruction could result from free radicals ($O_2^{\bullet-}$, H_2O_2, $NO^{\bullet}$), cytokines (IL-1, TNFα, TNFβ, IFNγ), and other inflammatory products of activated macrophages and T cells, both $CD4^+$ and $CD8^+$ cytotoxic T cells (Fig. 10.2). Antigen-nonspecific mechanisms for islet β cell destruction are supported by several lines of evidence. First, diabetes can be transferred to young NOD mice by $CD4^+$ T cells and T cell clones,[14,207] even after administration of an anti-CD8 monoclonal antibody to rule out any involvement of host $CD8^+$ T cells.[208,215] This observation is at variance with previously mentioned evidence that $CD8^+$ T cells are necessary for diabetes transfer. Perhaps young NOD mice (3-4 weeks) used for T cell clone transfer have some $CD8^+$ T cells (even after anti-CD8 antibody treatment) that cooperate with the $CD4^+$ T cell clones. For example, the addition of polyclonal $CD8^+$ T cells from diabetic mice accelerates diabetes transfer by $CD4^+$ T cell clones in irradiated recipients.[205] Second, it appears that β cell destruction is not MHC-restricted because diabetes recurs after transplantation of MHC-incompatible islet grafts in NOD mice[216] or BB rats[217] under conditions excluding allogeneic rejection (prior islet culture in vitro). Third, anti-CD4 monoclonal antibodies prevent recurrence of diabetes in islets grafted in NOD mice, whereas anti-CD8 monoclonals do not.[216]

$CD4^+$ T cells could mediate antigen-nonspecific β cell destruction by secreting various cytokines (IFNγ, TNFα, TNFβ) that can be directly toxic to β cells or can attract into the islets and activate other cell types such as monocytes and macrophages. These cells could, in turn, produce β cell toxic mediators such as the proinflammatory cytokines, IL-1 and TNFα, and free radicals such as superoxide ($O_2^{\bullet-}$), hydrogen peroxide (H_2O_2), and nitric oxide ($NO^{\bullet}$). Selective destruction of β cells in islets might occur if these inflammatory mediators are more toxic to β cells than to other islet cell types; however, data on this question are inconclusive.[69,72] Nevertheless, there is abundant evidence in vitro that β cells are sensitive to oxygen and nitrogen-based free radicals.[68,69] In addition, cytokines (IL-1, TNFα, TNFβ, IFNγ) are cytotoxic to β cells via mechanisms that appear to involve the production of free radicals in β cells themselves.[81-84] Importantly, $NO^{\bullet}$ production has been demonstrated in pancreatic islets in situ in conjunction with autoimmune diabetes development in BB rats[218] and NOD mice.[42]

Both macrophages and β cells have been reported to produce NO• in islets of NOD mice.[47] Also, peroxynitrite (ONOO⁻), the highly reactive oxidant produced by the combination of O_2^- and NO•, has been detected in β cells of NOD mice in conjunction with diabetes development.[83] Prevention of diabetes in rodent models by treatment with antioxidants and nicotinamide[219,220] fits with the hypothesis that oxygen free radicals and nitric oxide contribute to autoimmune destruction of islet β cells in type 1 diabetes.[81-84]

In addition to direct cytotoxic actions of cytokines on islet β cells, cytokines may render β cells susceptible to destruction by islet-infiltrating T cells (e.g., MHC class I-restricted CD8⁺ T cells) (Fig. 10.2). Thus, IFNγ upregulates MHC class I expression on rodent and human β cells,[156] and increased expression of MHC class I proteins on islet β cells (and other endocrine and nonendocrine cells) is consistently observed in the insulitis lesion of NOD mice and BB rats[156,221] and recently diagnosed type 1 diabetic patients.[157,158] A recent study, however, has demonstrated that increased MHC class I expression on β cells was not required for diabetes development in NOD mice.[78]

Another mechanism whereby cytokines may render β cells susceptible to T cell-mediated killing is via induction of Fas (CD95) receptors on β cells (Fig. 10.2). Ligation of Fas receptors on β cells by FasL (CD95L) on CD4⁺ and/or CD8⁺ T cells has been postulated to be a mechanism of β cell death by apoptosis in type 1 diabetes. IL-1β induces Fas on mouse[222] and human[223] β cells in vitro, and IL-1-sensitized, Fas-expressing islet cells are killed by addition of anti-Fas monoclonal antibody.[222] In a recent study, IL-1β-induced Fas expression on human islet β cells was reported to be β cell selective.[158] In the same report, Fas expression was detected only on β cells in pancreatic sections from two children with recent-onset type 1 diabetes and not on β cells in normal human pancreas; also, apoptosis was detected in the Fas-positive β cells located close to FasL-positive T cells infiltrating the islets.[158] In another study, however, Fas was not detected on β cells (or any other cells) in islets of NOD mice (without or with destructive insulitis), whereas FasL was present, but only on islet α cells.[224] In contrast, both Fas and FasL were found to be expressed on β cells in syngeneic islet grafts undergoing autoimmune destruction in NOD mice, and Fas expression correlated with expression of IL-1α, TNFα and IFNγ in the islet grafts.[225] Also, constitutive expression of FasL on human islet cells has been reported.[226] Taken together, these studies suggest that cytokine-induced Fas expression on islet β cells could target the β cells for destruction by FasL-expressing T cells (CD4⁺ and CD8⁺) and, possibly, by FasL-expressing β cells themselves.[225,226] Reports that NOD mice lacking Fas (NOD-*lpr*/*lpr* mice created by crossing NOD mice with MRL-*lpr*/*lpr* mice that have an incapacitating mutation in the *fas* gene) do not develop diabetes and are resistant to adoptive transfer of diabetes[227,228] suggested that Fas expression by β cells may be a limiting factor for β cell destruction during the course of the insulitis process. Subsequent studies, however, suggested that failure of diabetes development in Fas-deficient NOD-*lpr*/*lpr* mice could be due to immune defects in *lpr* mice other than Fas deletion.[229-231] This possibility was circumvented in two newly derived NOD mouse strains in which either FasL or Fas gene expression was deleted, and diabetes still was prevented.[232] In addition, diabetes was prevented when Fas expression was abrogated in transgenic NOD mice with CD4⁺ T cells bearing highly diabetogenic β cell-specific T cell receptors.[233] Also, activation of β cell cytotoxic CD8⁺ T cells in T cell receptor transgenic NOD mice was associated with expression of FasL on the diabetogenic T cells.[234] Further evidence for a role for FasL in autoimmune β cell destruction was reported in a study in which an anti-FasL antibody prolonged survival of syngeneic islet grafts transplanted into diabetic NOD mice.[235]

In summary, both CD4⁺ and CD8⁺ T cell subsets are needed for diabetes development because elimination of either subset can prevent diabetes in NOD mice and BB rats. Interestingly, depletion of either CD4⁺ or CD8⁺ T cells in diabetes-prone BB rats, by administration of monoclonal antibodies to these T cell subsets, completely prevented IFNγ mRNA expression by islet-infiltrating leukocytes, β cells were preserved, and diabetes did not develop.[236] This finding is concordant with reports that CD4⁺ and CD8⁺ T cells are interdependent for IFNγ production.[237-239] Therefore, prevention of IFNγ production in islets might explain why deletion of either T cell subset prevents autoimmune β destruction and diabetes development. It is still not clear, however, which cell(s) are the final effector(s) of islet β cell destruction, and exactly how each cell type regulates the other. A study in NOD mice transgenically expressing islet β cell-reactive $CD4^+$ or $CD8^+$ monoclonal T cells suggested that β cell destruction may be initiated by $CD4^+$ T cells which then recruit β cell-reactive $CD8^+$ T cells.[240] Also, it is not known whether $CD4^+$ and $CD8^+$ T cells recognize the same, or different, autoantigens. Nevertheless, it appears that both β cell antigen-specific $CD8^+$ T cells and antigen-nonspecific cytotoxic mechanisms induced by β cell antigen-specific $CD4^+$ T cells contribute to β cell destruction in type 1 diabetes.

In addition to T cells and macrophages, other cellular elements in and around the islet (not shown in Fig. 10.2) are likely participants in the insulitis lesion. For example, vascular endothelial cells may contribute cytokines (IL-1 and IL-6) and may respond to inflammatory cytokines (IL-1, TNF, and IFNγ) by expressing adhesion molecules to circulating leukocytes.[241] This response would permit migration of macrophages and lymphocytes from the circulation into the islet. Also, endothelial cells may respond to inflammatory cytokines (IL-1, TNF, and IFNγ) by expressing MHC class II molecules,[241] which could allow endothelial cells to act as APCs and possibly present β cell autoantigen(s) to T cells. Thus, intra- and peri-islet vascular endothelial cells could participate actively in amplifying the β cell-directed autoimmune process.[242] Indeed, immunohistochemical studies of the pancreas in subjects with recent-onset type 1 diabetes,[157,158] as well as in patients with disease recurrence after pancreas transplantation,[243] have revealed expression of intercellular adhesion molecule (ICAM-1) and MHC class II molecules on vascular endothelium of islets and small vessels near the islets. MHC class II molecules also were expressed on islet-infiltrating macrophages and T cells. Therefore, by increasing expressions of adhesion molecules and MHC class II molecules on macrophages and endothelial cells (collectively APCs) the inflammatory cytokines, IL-1, TNF, and IFNγ provide a positive feedback loop to the autoimmune response depicted in Figure 8.2.

Immunostimulatory Procedures to Prevent Type 1 Diabetes

The concept has been presented above that the autoimmune response in type 1 diabetes involves disturbances in immunoregulatory circuits that may be manifested as dominance of Th1 over Th2 cell function and cytokine production (Fig. 10.2). A corollary of this proposition is that measures leading to reversal of this Th subset balance, with Th2 cells/cytokines dominating over Th1 cells/cytokines, should block the autoimmune response and prevent diabetes development. There is evidence to support this hypothesis. Thus, administrations of a variety of immunostimulants—microbial agents, immune adjuvants, and T cell mitogens—have been discovered to prevent the development of insulitis, β cell destruction, and autoimmune diabetes in genetically diabetes-prone NOD mice and BB rats.[244-265] Importantly, these immunostimulatory procedures prevented diabetes development without structural changes or complete remodelling of the immune system, unlike procedures that involve bone marrow, thymic, or lymphoid cell replacement or deletion (e.g., anti-lymphocyte serum, cyclosporine, monoclonal antibodies to T cells, silica, and anti-macrophage antibodies).[8] Rather, the diabetes-preventive effects of immune adjuvants have been attributed to stimulation of T regulatory (suppressor) cells and cytokines whose effects were to suppress[260-263] or render dormant[259] autoreactive T cells. Taken together, these studies suggest that certain immunostimulatory procedures may reset the Th subset balance so that Th2 cells/cytokines dominate over Th1 cells/cytokines (Fig. 10.3).

The hypothesis that immunostimulatory procedures may prevent diabetes development in autoimmune diabetes-prone rodents by upregulating Th2 cells/cytokines is supported by several lines of evidence. Complete Freund's adjuvant (CFA)-induced protection of NOD mice from β cell-destructive insulitis and diabetes was found to be associated with a relative increase in IL-4-producing cells and a decrease in IFNγ-producing cells recovered from "sentinel"

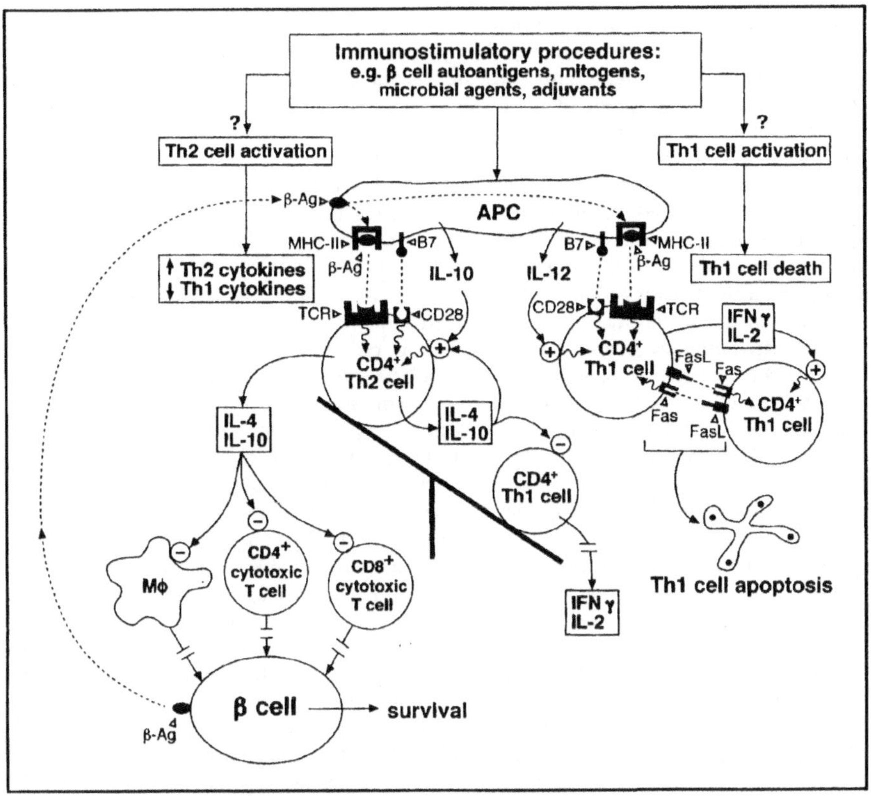

Fig. 10.3. Two distinct mechanisms by which immunostimulatory procedures (e.g., β cell autoantigens, mitogens, microbial agents, adjuvants), possibly acting via APC stimulation, may prevent or block the autoimmune response leading to β cell destruction in type 1 diabetes. One mechanism may be by Th2 cell activation. Thus, strong B7-CD28 costimulation during APC-CD4+ T cell interactions is thought to favor differentiation of Th2 over Th1 cells.[268,269] Also, IL-4 and IL-10 induce Th2 over Th1 cell differentiation. Th2 cells produce IL-4 and IL-10 which downregulate Th1 cells that produce IFNγ and IL-2. The combination of increased IL-4 and IL-10 production and decreased IFNγ and IL-2 production inhibits cytotoxic Mϕ and T cell activities, thereby preventing β cell damage and diabetes development. A second mechanism by which immunostimulatory procedures may prevent autoimmune β cell destruction may be by activating β cell-autoreactive Th1 cells along pathways leading to their self-destruction (apoptosis) by IFNγ and IL-2-dependent, FasL/Fas-mediated mechanisms, while Th2 cells that are relatively resistant to activation-induced cell death would survive.[271-273] (Reproduced from Rabinovitch A. Roles of cell-mediated immunity and cytokines in the pathogenesis of Type 1 diabetes mellitus: In Diabetes mellitus: A Fundamental and Clinical Text. 2nd edition, 2000. Eds. LeRoith, Taylor, Olefsky, with permission of Lippincott Williams & Wilkins).

syngeneic islet grafts placed under the renal capsule.[40] However, in a subsequent study, it was found that diabetes suppression following CFA administration to diabetes-prone NOD mice may be mediated only in part by Th2-type cytokines because combined anti-IL-4 and anti-IL-10 antibody treatment induced a state of glucose intolerance but did not abrogate diabetes prevention by CFA.[265] In another study, treatment of already diabetic NOD mice with CFA at the time of syngeneic islet transplantation prevented destruction of β cells in the islet graft and diabetes did not recur.[258] Lymphocytes and monocytes/macrophages still accumulated around the transplanted islets (peri-islet insulitis) in the CFA-treated NOD mice, but these mononuclear cells did not invade the islets and β cells remained intact.[258] In yet another study, IL-10 mRNA expression was significantly increased and IL-2 and IFNγ mRNA levels were significantly decreased in syngeneic islet grafts of CFA-injected NOD mice compared with saline-injected NOD mice.[53] This suggested that CFA treatment upregulated IL-10 production in the islet graft, resulting in decreased production of Th1 cytokines (IL-2 and IFNγ) and conversion of a β cell-destructive islet infiltrate into a nondestructive peri-insulitis lesion. This interpretation was supported in a subsequent study, in which the combined administration of IL-10 plus IL-4 (Th2 cytokines) was found to produce significantly prolonged survival of syngeneic islet grafts in diabetic NOD mice.[266] The diabetes-preventive effects of IL-4 and IL-10 are in accord with the known actions of these cytokines to downregulate inflammatory responses mediated by monocytes/macrophages and their cytokine products, as well as to downregulate cell-mediated immune responses triggered by Th1 cells and their cytokine products.[34,37] Indeed, IL-4 is consistently diabetes-preventive in NOD mice, either expressed transgenically by β cells,[98] or administered systemically.[111] Transgenic studies suggest a proinflammatory and diabetogenic role for IL-10 when this cytokine is expressed locally in islets;[102-104] however, systemic administrations of IL-10[48,112] and islet-specific T cells that hyperexpress IL-10 (by gene transfection)[267] have been reported to prevent diabetes development in NOD mice.

Interestingly, the ability of immunostimulatory procedures, such as microbial agents and immune adjuvants to promote Th2 over Th1 immune responses is concordant with the concept that the intensity of T cell signalling can dramatically affect the balance of Th1/Th2 subsets. According to this "strength of signal" hypothesis, any reagent or situation that results in strong costimulation of CD28 receptors on T cells by B7 costimulatory molecules on APCs will promote Th2 immune responses, whereas lower intensities of B7/CD28 costimulation will promote Th1 responses.[268] In support of this hypothesis, diabetes in NOD mice is exacerbated when the mice are bred onto the CD28 knockout background as a direct result of a reduction in the protective Th2 response and concomitant enhancement of the Th1 response.[269] Also, activation of CD28 signalling in T cells by anti-CD28 monoclonal antibody treatment of NOD mice at 2 weeks of age (but not at 5-6 weeks) was recently reported to increase IL-4 production by islet-infiltrating T cells and prevent diabetes development.[60] These findings suggest that immunostimulatory procedures may promote Th2 immune responses and prevent diabetes by upregulating B7/CD28 costimulation (Fig. 10.3). Recently, it was reported that B7-1 and B7-2 expression is decreased on dendritic cells in peripheral blood of humans at high risk for type 1 diabetes, and this was accompanied by reduced stimulation of autologous $CD4^+$ T cells.[270] Therefore, according to the strength of signal hypothesis, low levels of B7/CD28 costimulation in individuals at risk for type 1 diabetes would favor a Th1 cell-mediated immune response that destroys islet β cells at the expense of a protective Th2 response.

Recent studies suggest a novel mechanism for differential regulation of Th1 and Th2 subsets, namely a differential ability of Th1 and Th2 cells to undergo activation-induced cell death (AICD), also termed apoptosis. Thus, Th1, but not Th2, cells have been reported to undergo rapid FasL/Fas-mediated apoptosis after antigen stimulation.[271-273] Therefore, it is tempting to speculate that immunostimulatory procedures, such as microbial agents, adjuvants, mitogens, and β cell autoantigens, might prevent autoimmune diabetes development by preferentially inducing apoptosis of autoreactive Th1-type cells (Fig. 10.3). According to this scenario, prevention of autoimmune destruction of β cells would be associated with a decrease in the ratio of Th1/Th2 cells as a consequence of decreases in Th1 cells without any increase in Th2 cells, rather their selective survival.

Indeed, this has recently been reported to be the mechanism of the protective effect of immune adjuvants against diabetes development in NOD mice. It was found that BCG and CFA-induced diabetes prevention in NOD mice persisted in NOD mice genetically deficient in either IL-4 or IL-10, whereas IFNγ-deficient NOD mice were not protected from diabetes by BCG or CFA.[274] Thus, immune adjuvants protected against diabetes by mechanisms independent of Th2-type cytokines (IL-4 and IL-10); rather, the Th1-type cytokine, IFNγ was required, unexpectedly, for immune adjuvant-induced diabetes prevention. The dependency on IFNγ for immune adjuvant-induced diabetes prevention was due, presumably, to deletion of autoreactive Th1 cells by IFNγ, because NOD Th1 splenic cells were more sensitive to activation-induced cell death than NOD Th2 splenic cells.[274] Similarly, IFNγ, induced by BCG infection of nondiabetes-prone mice, has been reported to act as regulator of the immune response by inducing apoptosis of CD4 T cells initially activated by BCG.[275]

Anti-T cell antibodies have been found to induce tolerance and prevent β cell destruction in NOD mice, even when the antibodies are administered after insulitis has started and effector T cells have been activated.[276] The mechanism of nondepleting anti-CD4 monoclonal antibody to induce tolerance in a primed immune system has been reported to be by activation of $CD4^+$ T regulatory cells,[277] and recently by direct prevention of effector cell function, presumably by deletion of activated autoreactive T cells.[278] Other studies have revealed that induction of tolerance to cardiac and pancreatic islet allografts in mice is critically dependent upon IFNγ[279,280] and IL-2[281,282] production. This supports the concept that Th1 cell activation can lead to self-deletion via apoptosis and, consequently, specific T cell tolerance to the stimulating antigen.

In addition, the protective effect of peripheral NKT cells against autoimmunity in NOD mice, originally proposed to be due to shifting the profile of autoreactive T cells toward a protective Th2 type,[165] was recently reported to be related, instead, to IL-12-induced activation and IFNγ secretion by NKT cells, and these Th1 type immunoregulatory responses were deficient in NOD mice.[283] Further evidence that Th1 cell activation is required to prevent autoimmune diabetes development was provided by a recent study that reported acceleration of diabetes in NOD mice in which endogenous IL-12 was neutralized by anti-IL-12 antibody administered to young NOD mice (2 weeks of age) for 6 days only.[284] By contrast, when anti-IL-12 antibody was administered to older NOD mice (from age 5 to 30 weeks), insulitis and diabetes were suppressed.[284] These findings reveal the dual role of Th1 cytokines (IL-12 and IFNγ): i) they act as early regulators of immune responses, by deleting autoreactive Th1 cells and, if this regulatory action is inadequate and islet β cell autoreactive Th1 cells persist, then ii) they act as effectors of β cell destruction.

The aforementioned studies support the general consensus that Th1 cells/cytokines are the major disease effectors in autoimmune diabetes,[285-288] and that deletion of Th1 cells or blockade of Th1 cell/cytokine actions can prevent diabetes development. There is conflicting evidence, however, on whether Th2 cells/cytokines have a protective effect. For example, cotransfer of polarized Th1 and Th2 cells did not inhibit the ability of the Th1 population to provoke diabetes.[174] Also, NOD mice with an IL-4 gene knockout mutation did not manifest intensified insulitis or accelerated diabetes.[289] These findings do not support the concept that Th2 cells provide dominant protection against β cell destruction in the insulitis lesion. This conclusion must be tempered, though, by the fact that IL-4 knockout mice still produce other Th2 cell-derived cytokines (e.g., IL-5, IL-10),[290,291] and possibly the Th3 cell-derived cytokine, TGFβ, any or all of which could still downregulate Th1 cytokine production in IL-4-deficient NOD

Future Prospects: Clinical Considerations

The clinical hope from observations that certain immunostimulatory procedures prevent autoimmune diabetes development in genetically diabetes-prone animals is that clinically safe means of immune stimulation may be similarly effective in preventing type 1 diabetes in human subjects at risk for this disease. Immunostimulatory agents that have a broad spectrum of immune stimulation affecting macrophages and T cells (e.g., the immune adjuvant, bacille Calmette-Guérin [BCG] vaccine) and polyclonal T cell activators (e.g., microbial superantigens, lectins) may not be optimal for clinical trials because of possible undesirable side effects from generalized immunostimulation.

Recent findings, however, demonstrate that more selective immunostimulation may be at hand. Thus, administration of the peptide GAD65, an islet β cell autoantigen, can prevent autoimmune diabetes development in NOD mice, and this prevention is associated with the induction of specific tolerance to this peptide.[292,293] Moreover, GAD-responsive T cells from diabetes-prone NOD mice were characterized as Th1, IFNγ-producing.[292] In contrast, IFNγ production was reduced in antigen-stimulated spleen cell cultures from GAD65-tolerant (and diabetes-protected) NOD mice, indicating that tolerance may result from suppression of GAD65-responsive Th1 cells.[293] Because this effect was not accompanied by a corresponding reduction of the humoral (antibody) response to GAD and other β cell autoantigens, a GAD65 induction of Th2 cells with suppression of Th1 cells was suggested.[293] Importantly, GAD65 administration to NOD mice was reported to suppress an ongoing diabetogenic response (late insulitis, prehyperglycemic stage of type 1 diabetes), and this protection was mediated through the induction of regulatory $CD4^+$ T cells with a Th2 phenotype.[294] Furthermore, induction of GAD65-specific Th2 cells and suppression of diabetes in NOD mice is IL-4 dependent, because NOD mice genetically deficient in IL-4 production (IL-4 gene knockout NOD mice) were not protected from diabetes development after immunization with GAD65-specific peptides,[295] or a novel plasmid DNA construct encoding both a GAD65 peptide linked to IgG Fc and IL-4.[296] These findings are directly relevant to reports that there is an inverse relation between humoral (Th2 cell-mediated) and cellular (Th1 cell-mediated) autoimmunity to GAD in human subjects at risk for type 1 diabetes[297] and that a strong humoral (serum antibody) response to GAD correlates with a slow progression to diabetes.[297,298]

Administration of β cell candidate autoantigens other than GAD may also induce self-tolerance and prevent diabetes development. For example, insulin (and insulin B chain) can prevent diabetes in NOD mice and BB rats, and possibly in human subjects at high risk for type 1 diabetes.[299] Recently, a T cell response to a particular epitope of the insulin B chain, $B_{(9-23)}$ was described in peripheral blood lymphocytes obtained from human subjects with recent-onset type 1 diabetes and from prediabetic subjects at high risk for disease; also these insulin peptide-reactive T cells produced IFNγ.[300] The significance of these findings is that therapies that are directed at this autoantigenic response might be of benefit in controlling human type 1 diabetes, as was achieved by administration of the B chain or $B_{(10-24)}$ peptide of insulin in NOD mice.[301,302] In addition, reports that NOD mice can be protected from diabetes development by administering the β cell autoantigens, GAD[303,304] and insulin[46,301,302,305-308] by oral, intranasal or aerosol inhalation routes may be of practical importance for clinical application. The mechanisms of the protective effects of these treatments in NOD mice have been ascribed to activation of $CD4^+$ αβ T cells or $CD8^+$ γδ T cells that produced one or more suppressor cytokines (IL-4, IL-10 and TGFβ).

Immune-mediated destruction of insulin-secreting β cells precedes the overt expression of clinical symptoms by many years because these become apparent only when a majority of the β cells have been destroyed. Interrupting this pathogenetic sequence by immune intervention offers the opportunity to alter the natural history of type 1 diabetes. Several approaches are currently being explored in clinical trials or are under consideration for such trials. These include the following therapeutic approaches used singly or in combination:

1. administration of β cell autoantigens (e.g., insulin) via parenteral, oral, nasal or aerosol inhalation routes;
2. manipulation of expression of costimulatory molecules (e.g., B7/CD28, CD40/CD40L) on antigen-presenting cells and T cells in attempts to delete autoreactive Th1 cells or direct T cell signalling pathways from Th1 to Th2 cell dominance; and
3. administration of cytokine-based therapies (e.g., cytokines, antibodies to cytokines and cytokine receptors, soluble cytokine receptors and receptor antagonists, cytokine receptor-targeted cytotoxic drugs) to block the production and/or action of proinflammatory cytokines (IL-1 and TNFα) and type 1 cytokines (IFNγ, IL-2, TNFβ and IL-12), while maintaining or increasing the production and/or action of regulatory cytokines (IL-4, IL-10, TGFβ).

Acknowledgments

The author thanks Wilma Suarez-Pinzon, Chris Bleackley, Tim Mosmann, Robert Power, Jonathan Lakey, Ray Rajotte and David Serreze, who have contributed to the research from his laboratory cited in this review. This work was supported by the Alberta Heritage Foundation for Medical Research, the Canadian Institutes for Health Research, the Juvenile Diabetes Foundation International, the Canadian Diabetes Association, the Muttart Diabetes Research and Training Centre at the University of Alberta, and the MacLachlan Fund of the University of Alberta Hospitals.

References

1. Bach JF. Insulin-dependent diabetes mellitus as an autoimmune disease. Endocrine Rev 1994; 15:516-542.
2. Wilkin TJ. The primary lesion theory of autoimmunity: A speculative hypothesis. Autoimmunity 1990; 7:225-235.
3. Lampeter EF, Homberg M, Quabeck K et al. Transfer of insulin-dependent diabetes between HLA-identical siblings by bone marrow transplantation. Lancet 1993; 341:1243-1244.
4. Vialettes B, Maraninchi D, San Marco MP et al. Autoimmune polyendocrine failureCType 1 (insulin-dependent) diabetes mellitus and hypothyroidismCafter allogeneic bone marrow transplantation in a patient with lymphoblastic leukaemia. Diabetologia 1993; 36:541-546.
5. Lampeter EB. Discussion remark to Session 24: BMT in autoimmune diseases. Exp Hematol 1993; 21:1155.
6. Calcinaro F, Hao L, Chase HP et al. Detection of cell-mediated immunity in type I diabetes mellitus. J Autoimmun 1992; 5:137-147.
7. Petersen JS, Marshall MO, Bækkeskov S et al. Transfer of Type 1 (insulin-dependent) diabetes mellitus associated autoimmunity to mice with severe combined immunodeficiency (SCID). Diabetologia 1993; 36:510-515.
8. Rossini AA, Greiner DL, Friedman HP et al. Immunopathogenesis of diabetes mellitus. Diabetes Rev 1993; 1:43-75.
9. Singh B, Rabinovitch A. Influence of microbial agents on the development and prevention of autoimmune diabetes. Autoimmunity 1993; 15:209-213.
10. Dardenne M, Lepault F, Bendelac A et al. Acceleration of the onset of diabetes in NOD mice by thymectomy at weaning. Eur J Immunol 1989; 19:889-895.
11. Harada M, Makino S. Promotion of spontaneous diabetes in nonobese diabetes-prone mice by cyclophosphamide. Diabetologia 1984; 27:604-606.
12. Yasunami R, Bach JF. Anti-suppressor effect of cyclophosphamide on the development of spontaneous diabetes in NOD mice. Eur J Immunol 1988; 18:481-484.
13. Bendelac A, Carnaud C, Boitard C et al. Syngeneic transfer of autoimmune diabetes from diabetic NOD mice to healthy neonates. Requirement for both L3T4+ and Lyt-2+ T cells. J Exp Med 1987; 166:823-832.
14. Haskins K, McDuffie M. Acceleration of diabetes in young NOD mice with a CD4+ islet-specific T cell clone. Science 1990; 249:1433-1436.
15. Sempé P, Richard MF, Bach JF et al. Evidence of CD4+ regulatory T cells in the nonobese diabetic male mouse. Diabetologia 1994; 37:337-343.
16. Boitard C, Yasunami R, Dardenne M et al. T cell-mediated inhibition of the transfer of autoimmune diabetes in NOD mice. J Exp Med 1989; 169:1669-1680.
17. Pankewycz OG, Guan JX, Benedict JF. A protective NOD islet-infiltrating CD8+ T cell clone, I.W. 2.15, has *in vitro* immunosuppressive properties. Eur J Immunol 1992; 22:2017-2023.
18. Pankewycz O, Strom TB, Rubin-Kelley VE. Islet-infiltrating T cell clones from nonobese diabetic mice that promote or prevent accelerated onset diabetes. Eur J Immunol 1991; 21:873-879.
19. Diaz-Gallo C, Moscovitch-Lopatin M, Strom TB et al. An anergic, islet-infiltrating T-cell clone that suppresses murine diabetes secretes a factor that blocks interleukin 2/interleukin 4-dependent proliferation. Proc Natl Acad Sci USA 1992; 89:8656-8660.
20. Boitard C, Bendelac A, Richard MF et al. Prevention of diabetes in nonobese diabetic mice by anti-I-A monoclonal antibodies: Transfer of protection by splenic T cells. Proc Natl Acad Sci USA 1988; 85:9719-9723.
21. Greiner DL, Mordes JP, Handler ES et al. Depletion of RT6.1+ T lymphocytes induces diabetes in resistant biobreeding/Worcester (BB/W) rats. J Exp Med 1987; 166:461-475.
22. Rossini AA, Faustman D, Woda BA et al. Lymphocyte transfusions prevent diabetes in the Bio-Breeding/Worcester rat. J Clin Invest 1984; 74:39-46.

23. Thorpe R, Wadhwa M, Bird CR et al. Detection and measurement of cytokines. Blood Rev 1992; 6:133-148.
24. Mosmann TR, Cherwinski H, Bond MW et al. Two types of murine helper T cell clone: I. Definition according to profiles of lymphokine activities and secreted proteins. J Immunol 1986; 136:2348-2357.
25. Cherwinski HC, Schumacher JH, Brown KD et al. Two types of mouse helper T cell clone. III. Further differences in lymphokine synthesis between Th1 and Th2 clones revealed by RNA hybridization, functionally monospecific bioassays and monoclonal antibodies. J Exp Med 1987; 166:1229-1244.
26. Del Prete GF, De Carli M, Mastromauro C et al. Purified protein derivative (PPD) of *Mycobacterium tuberculosis* and excretory-secretory antigen(s) (TES) of *Toxocara canis* expand in vitro human T cells with stable and opposite (type 1 T helper or type 2 T helper) profiles of cytokine production. J Clin Invest 1991; 88:346-350.
27. Mosmann TR, Sad S. The expanding universe of T-cell subsets: Th1, Th2 and more. Immunol Today 1996; 17:138-146.
28. Mosmann TR, Coffman RL. TH1 and TH2 cells: Different patterns of lymphokine secretion lead to different functional properties. Annu Rev Immunol 1989; 7:145-173.
29 Chen Y, Kuchroo VK, Inobe J et al. Regulatory T cell clones induced by oral tolerance: suppression of autoimmune encephalomyelitis. Science 1994; 265:1237-1240.
30. Romagnani S. Lymphokine production by human T cells in disease states. Annu Rev Immunol 1994; 12:227-257.
31. Fitch FW, McKisic MD, Lancki DW et al. Differential regulation of murine T lymphocyte subsets. Annu Rev Immunol 1993; 11:29-48.
32. Seder RA, Paul WE. Acquisition of lymphokine-producing phenotype by CD4+ T-cells. Annu Rev Immunol 1994; 12:635-673.
33. Moore KW, O'Garra A, de Waal Malefyt R et al. Interleukin 10. Annu Rev Immunol 1993; 11:165-190.
34. Trinchieri G. Interleukin-12: a proinflammatory cytokine with immunoregulatory functions that bridge innate resistance and antigen-specific adaptive immunity. Annu Rev Immunol 1995; 13:251-276.
35. Ding L, Linsley PS, Huang L-Y et al. IL-10 inhibits macrophage costimulatory activity by selectively inhibiting the upregulation of B7 expression. J Immunol 1993; 151:1224-1234.
36. Powrie F, Menon S, Coffman RL. Interleukin-4 and interleukin-10 synergize to inhibit cell-mediated immunity in vivo. Eur J Immunol 1993; 23:3043-3049.
37. Mosmann T. Complexity or coherence? Cytokine secretion by B cells. Nature Immunol 2000; 1:465-466.
38. Clerici M, Shearer GM. The Th1-Th2 hypothesis of HIV infection: new insights. Immunol Today 1994; 15:575-581.
39. Kolb H. Benign versus destructive insulitis. Diabetes Metab Rev 1997; 13:139-146.
40. Shehadeh NN, LaRosa F, Lafferty KJ. Altered cytokine activity in adjuvant inhibition of autoimmune diabetes. J Autoimmun 1993; 6:291-300.
41. Campbell IL, Kay TWH, Oxbrow L et al. Essential role for interferon-gamma and interleukin-6 in autoimmune insulin-dependent diabetes in NOD/Wehi mice. J Clin Invest 1991; 87:739-742.
42. Rothe H, Faust A, Schade U et al. Cyclophosphamide treatment of female nonobese diabetic mice causes enhanced expression of inducible nitric oxide synthase and interferon-gamma, but not of interleukin-4. Diabetologia 1994; 37:1154-1158.
43. Pilstrom B, Bjork L, Bohme J. Demonstration of a TH1 cytokine profile in the late phase of NOD insulitis. Cytokine 1995; 7:806-814.
44. Rabinovitch A, Suarez-Pinzon WL, Sorensen O et al. IFN-gamma gene expression in pancreatic islet-infiltrating mononuclear cells correlates with autoimmune diabetes in nonobese diabetic mice. J Immunol 1995; 154:4874-4882.
45. Muir A, Peck A, ClareSalzler M et al. Insulin immunization of nonobese diabetic mice induces a protective insulitis characterized by diminished intraislet interferon-γ transcription. J Clin Invest 1995; 95:628-634.
46. Hancock WW, Polanski M, Zhang J et al. Suppression of insulitis in nonobese diabetic (NOD) mice by oral insulin administration is associated with selective expression of interleukin-4 and -10, transforming growth factor-β, and prostaglandin-E. Am J Pathol 1995; 147:1193-1199.
47. Rabinovitch A, Suarez-Pinzon WL, Sorensen O et al. Inducible nitric oxide synthase (iNOS) in pancreatic islets of nonobese diabetic mice: Identification of iNOS-expressing cells and relationships to cytokines in the islets. Endocrinology 1996; 137:2093-2099.

48. Zheng XX, Steele AW, Hancock WW et al. A noncytolytic IL-10/Fc fusion protein prevents diabetes, blocks autoimmunity, and promotes suppressor phenomena in NOD mice. J Immunol 1997; 158:4507-4513.
49. Kolb H, Wörz-Pagenstert U, Kleemann R et al. Cytokine gene expression in the BB rat pancreas: Natural course and impact of bacterial vaccines. Diabetologia 1996; 39:1448-1454.
50. Zipris D, Greiner DL, Malkani S et al. Cytokine gene expression in islets and thyroids of BB rats. IFN-gamma and IL-12p40 mRNA increase with age in both diabetic and insulin-treated nondiabetic BB rats. J Immunol 1996; 156:1315-1321.
51. Rabinovitch A, Suarez-Pinzon W, El-Sheikh A et al. Cytokine gene expression in pancreatic islet-infiltrating leukocytes of BB rats: Expression of Th1 cytokines correlates with beta-cell destructive insulitis and IDDM. Diabetes 1996; 45:749-754.
52. Chung Y-H, Jun H-S, Kang Y et al. Role of macrophages and macrophage-derived cytokines in the pathogenesis of Kilham rat virus-induced autoimmune diabetes in diabetes-resistant BioBreeding rats. J Immunol 1997; 159:466-471.
53. Rabinovitch A, Sorensen O, Suarez-Pinzon WL et al. Analysis of cytokine mRNA expression in syngeneic islet grafts of NOD mice: interleukin 2 and interferon gamma mRNA expression correlate with graft rejection and interleukin 10 with graft survival. Diabetologia 1994; 37:833-837.
54. Suarez-Pinzon W, Rajotte RV, Mosmann TR et al. Both CD4+ and CD8+ T-cells in syngeneic islet grafts in NOD mice produce interferon-gamma during beta-cell destruction. Diabetes 1996; 45:1350-1357.
55. Rothe H, Burkart V, Faust A et al. Interleukin-12 gene expression is associated with rapid development of diabetes mellitus in nonobese diabetic mice. Diabetologia 1996; 39:119-122.
56. Rabinovitch A, Suarez-Pinzon WL, Sorensen O. Interleukin 12 mRNA expression in islets correlates with β-cell destruction in NOD mice. J Autoimmun 1996; 9:645-651.
57. Rothe H, Jenkins NA, Copeland NG et al. Active stage of autoimmune diabetes is associated with the expression of a novel cytokine, IGIF, which is located near Idd2. J Clin Invest 1997; 99:469-474.
58. Rothe H, Hibino T, Itoh Y et al. Systemic production of interferon-gamma inducing factor (IGIF) versus local IFN-γ expression involved in the development of Th1 insulitis in NOD mice. J Autoimmun 1997; 10:251-256.
59. Fox CJ, Danska JS. IL-4 expression at the onset of islet inflammation predicts nondestructive insulitis in nonobese diabetic mice. J Immunol 1997; 158:2414-2424.
60. Arreaza GA, Cameron MJ, Jaramillo A et al. Neonatal activation of CD28 signaling overcomes T cell anergy and prevents autoimmune diabetes by an IL-4 dependent mechanism. J Clin Invest 1997; 100:2243-2253.
61. Cetkovic-Cvrlje M, Gerling IC, Muir A et al. Retardation or acceleration of diabetes in NOD/Lt mice mediated by intrathymic administration of candidate β-cell antigens. Diabetes 1997; 46:1975-1982.
62. Huang X, Hultgren B, Dybdal N et al. Islet expression of interferon-γ precedes diabetes in both the BB rat and streptozotocin-treated mice. Immunity 1994; 1:469-478.
63. Foulis AK, Farquharson MA, Meager A. Immunoreactive alpha-interferon in insulin-secreting beta cells in type 1 diabetes mellitus. Lancet 1987; 2:1423-1427.
64. Huang X, Yuan J, Goddard A et al. Interferon expression in the pancreases of patients with type 1 diabetes. Diabetes 1995; 44:658-664.
65. Somoza N, Vargas F, Roura-Mir C et al. Pancreas in recent onset insulin-dependent diabetes mellitus: changes in HLA, adhesion molecules and autoantigens, restricted TCR Vβ usage, and cytokine profile. J Immunol 1994; 153:1360-1377.
66. Foulis AK, McGill M, Farquharson MA. Insulitis in type 1 (insulin-dependent) diabetes mellitus in man: Macrophages, lymphocytes, and interferon-γ containing cells. J Pathol 1991; 165:97-103.
67. Yamagata K, Nakajima H, Tomita K et al. Dominant TCR α-chain clonotypes and interferon-_ are expressed in the pancreas of patients with recent-onset insulin-dependent diabetes mellitus. Diabetes Res Clin Pract 1996; 34:37-46.
68. Mandrup-Poulsen T, Helqvist S, Wogensen LD et al. Cytokines and free radicals as effector molecules in the destruction of pancreatic β-cells. Curr Top Microbiol Immunol 1990; 164:169-193.
69. Rabinovitch A. Roles of cytokines in IDDM pathogenesis and islet β-cell destruction. Diabetes Rev 1993; 1:215-240.
70. Bolaffi JL, Nowlain RE, Cruz L et al. Progressive damage of cultured pancreatic islets after single early exposure to streptozotocin. Diabetes 1986; 35:1027-1033.
71. Pipeleers D, Van de Winkel M. Pancreatic β cells possess defense mechanisms against cell-specific toxicity. Proc Natl Acad Sci USA 1986; 83:5267-5271.

72. Ling Z, In't Veld PA, Pipeleers DG. Interaction of interleukin-1 with islet β-cells. Distinction between indirect, aspecific cytotoxicity and direct, specific functional suppression. Diabetes 1993; 42:56-65.
73. Poo W-J, Conrad L, Janeway Jr CA. Receptor-directed focusing of lymphokine release by helper T cells. Nature 1988; 332:378-380.
74. Kurt-Jones EA, Beller DI, Mizel SB et al. Identification of a membrane associated interleukin-1 in macrophages. Proc Natl Acad Sci USA 1985; 82:1204-1208.
75. Kriegler M, Perez C, DeFay K et al. A novel form of TNF/cachetin is a cell surface cytotoxic transmembrane protein: ramifications for the complex physiology of TNF. Cell 1988; 53:45-53.
76. Sutton R, Gray DWR, McShane P et al. The specificity of rejection and the absence of susceptibility of pancreatic islet β-cells to nonspecific immune destruction in mixed strain islets grafted beneath the renal capsule in the rat. J Exp Med 1989; 170:751-762.
77. Korsgren O, Jansson L. Characterization of mixed syngeneic-allogeneic and syngeneic-xenogeneic islet-graft rejections in mice. Evidence of functional impairment of the remaining syngeneic islets in xenograft rejections. J Clin Invest 1994; 93:1113-1119.
78. Thomas HE, Parker JL, Schreiber RD et al. IFN-γ action on pancreatic beta cells causes class I MHC upregulation but not diabetes. J Clin Invest 1998; 102:1249-1257.
79. Corbett JA, McDaniel ML. Intraislet release of IL-1 inhibits β-cell function by inducing β-cell expression of iNOS. J Exp Med 1995; 181:559-568.
80. Arnush M, Heitmeier MR, Scarim AL et al. IL-1 produced and released endogenously within human islets inhibits β cell function. J Clin Invest 1998; 102:516-526.
81. Mandrup-Poulsen T. The role of interleukin-1 in the pathogenesis of IDDM. Diabetologia 1996; 39:1005-1029.
82. Eizirik DL, Pavlovic D. Is there a role for nitric oxide in β-cell dysfunction and damage in IDDM? Diabetes Metab Rev 1997; 13:293-307.
83. Suarez-Pinzon WL, Szabó C, Rabinovitch A. Development of autoimmune diabetes in NOD mice is associated with the formation of peroxynitrite in pancreatic islet β-cells. Diabetes 1997; 46:907-911.
84. Rabinovitch A, Suarez-Pinzon WL. Cytokines and their roles in pancreatic islet β-cell destruction and insulin-dependent diabetes mellitus. Biochem Pharmacol 1998; 55:1139-1149.
85. Sarvetnick N, Liggitt D, Pitts SL et al. Insulin-dependent diabetes mellitus induced in transgenic mice by ectopic expression of class II MHC and interferon-gamma. Cell 1988; 52:773-782.
86. Sarvetnick N, Shizuru J, Liggitt D et al. Loss of pancreatic islet tolerance induced by β-cell expression of interferon-γ. Nature 1990; 346:844-847.
87. Stewart TA, Hultgren B, Huang X et al. Induction of type 1 diabetes by interferon-α in transgenic mice. Science 1993; 260:1942-1946.
88. Chakrabarti D, Hultgren B, Stewart TA. IFN-α induces autoimmune T cells through the induction of intracellular adhesion molecule-1 and B7.2. J Immunol 1996; 157:522-528.
89. Allison J, Malcolm L, Chosich N et al. Inflammation but not autoimmunity occurs in transgenic mice expressing constitutive levels of interleukin-2 in islet β cells. Eur J Immunol 1992; 22:1115-1121.
90. Allison J, Oxbrow L, Miller JFA. Consequences of *in situ* production of IL-2 for islet cell death. Int Immunol 1994; 6:541-549.
91. Allison J, McClive P, Oxbrow L et al. Genetic requirements for acceleration of diabetes in nonobese diabetic mice expressing interleukin-2 in islet β-cells. Eur J Immunol 1994; 24:2535-2541.
92. Higuchi Y, Herrera P, Muniesa P et al. Expression of a tumour necrosis factor α transgene in murine pancreatic β cells results in severe and permanent insulitis without evolution towards diabetes. J Exp Med 1992; 176:1719-1731.
93. Picarella DE, Kratz A, Li C et al. Transgenic tumor necrosis factor (TNF)-α production in pancreatic islets leads to insulitis, not diabetes. Distinct patterns of inflammation in TNF-α and TNF-β transgenic mice. J Immunol 1993; 150:4136-4150.
94. Guerder S, Meyerhoff J, Flavell R. The role of the T cell costimulator B7-1 in autoimmunity and the induction and maintenance of tolerance to peripheral antigen. Immunity 1994; 1:155-166.
95. Grewal IS, Grewal KD, Wong FS et al. Local expression of transgene encoded TNF-α in islets prevents autoimmune diabetes in NOD mice by preventing the development of autoreactive islet specific T cells. J Exp Med 1996; 184:1963-1974.
96. Sarvetnick N. Mechanisms of cytokine-mediated localized immunoprotection. J Exp Med 1996; 184:1597-1600.
97. McSorley SJ, Soldera S, Malherbe L et al. Immunological tolerance to a pancreatic antigen as a result of local expression of TNFα by islet β cells. Immunity 1997; 7:401-409.

98. Mueller R, Krahl T, Sarvetnick N. Pancreatic expression of interleukin-4 abrogates insulitis and autoimmune diabetes in nonobese diabetic (NOD) mice. J Exp Med 1996; 184:1093-1099.
99. Mueller R, Bradley LM, Krahl T et al. Mechanism underlying counterregulation of autoimmune diabetes by IL-4. Immunity 1997; 7:411-418.
100. Campbell IL, Hobbs MV, Dockter J et al. Islet inflammation and hyperplasia induced by the pancreatic islet-specific overexpression of interleukin-6 in transgenic mice. Am J Pathol 1994; 145:157-166.
101. Di Cosmo BF, Picarella D, Flavell RA. Local production of human IL-6 promotes insulitis but retards the onset of insulin-dependent diabetes mellitus in nonobese diabetic mice. Int Immunol 1994; 6:1829-1837.
102. Wogensen L, Huang X, Sarvetnick N. Leukocyte extravasation into the pancreatic tissue in transgenic mice expressing interleukin 10 in the islets of Langerhans. J Exp Med 1993; 178:175-185.
103. Wogensen L, Lee M-S, Sarvetnick N. Production of interleukin 10 by islet cells accelerates immune-mediated destruction of β cells in nonobese diabetic mice. J Exp Med 1994; 179:1379-1384.
104. Moritani M, Yoshimoto K, Tashiro F et al. Transgenic expression of IL-10 in pancreatic islet A cells accelerates autoimmune insulitis and diabetes in nonobese diabetic mice. Int Immunol 1994; 6:1927-1936.
105. King C, Davies J, Mueller R et al. TGF-β1 alters APC preference, polarizing islet antigen responses toward a Th2 phenotype. Immunity 1998; 8:601-613.
106. Moritani M, Yoshimoto K, Wong SF et al. Abrogation of autoimmune diabetes in nonobese diabetic mice and protection against effector lymphocytes by transgenic paracrine TGF-β1. J Clin Invest 1998; 102:499-506.
107. Wilson CA, Jacobs C, Baker P et al. IL-1β modulation of spontaneous autoimmune diabetes and thyroiditis in the BB rat. J Immunol 1990; 144:3784-3788.
108. Jacob CO, Asiso S, Michie SA et al. Prevention of diabetes in nonobese diabetic mice by tumor necrosis factor (TNF): Similarities between TNF-α and interleukin-1. Proc Natl Acad Sci USA 1990; 87:968-972.
109. Serreze DV, Hamaguchi K, Leiter EH. Immunostimulation circumvents diabetes in NOD/Lt mice. J Autoimmun 1989; 2:759-776.
110. Zielasek J, Burkart V, Naylor P et al. Interleukin-2-dependent control of disease development in spontaneously diabetic BB rats. Immunology 1990; 69:209-214.
111. Rapoport MJ, Jaramillo A, Zipris D et al. Interleukin-4 reverses T cell proliferative unresponsiveness and prevents the onset of diabetes in nonobese diabetic mice. J Exp Med 1993; 178:87-99.
112. Pennline KJ, Roque-Gaffney E, Monahan M. Recombinant human IL-10 (rHU IL-10) prevents the onset of diabetes in the nonobese diabetic (NOD) mouse. Clin Immunol Immunopathol 1994; 71:169-175.
113. Satoh J, Seino H, Abo T et al. Recombinant human tumor necrosis factor-α suppresses autoimmune diabetes in nonobese diabetic mice. J Clin Invest 1989; 84:1345-1348.
114. Satoh J, Seino H, Shintani S et al. Inhibition of type I diabetes in BB rats with recombinant human tumor necrosis factor-α. J Immunol 1990; 145:1395-1399.
115. Seino H, Takahashi K, Satoh J et al. Prevention of autoimmune diabetes with lymphotoxin in NOD mice. Diabetes 1993; 42:398-404.
116. Takahashi K, Satoh J, Seino H et al. Prevention of type I diabetes with lymphotoxin in BB rats. Clin Immunol Immunopathol 1993; 69:318-323.
117. Zaccone P, Phillips J, Conget I et al. Interleukin-13 prevents autoimmune diabetes in NOD mice. Diabetes 1999; 48:1522-1528.
118. Serreze DV, Leiter EH. Defective activation of T suppressor cell function in nonobese diabetic mice. Potential relation to cytokine deficiencies. J Immunol 1988; 140:3801-3807.
119. Lapchak PH, Guilbert LJ, Rabinovitch A. Tumor necrosis factor production is deficient in diabetes-prone BB rats and can be corrected by complete Freund's adjuvant: A possible immunoregulatory role of tumor necrosis factor in the prevention of diabetes. Clin Immunol Immunopathol 1992; 65:129-134.
120. Cameron MJ, Arreaza GA, Zucker P et al. IL-4 prevents insulitis and insulin-dependent diabetes mellitus in nonobese diabetic mice by potentiation of regulatory T helper-2 cell function. J Immunol 1997; 159:4686-4692
121. Sobel DO, Ahvazi B. α-interferon inhibits the development of diabetes in NOD mice. Diabetes 1998; 47:1867-1872.
122. Sobel DO, Creswell K, Yoon J-W et al. Alpha interferon administration paradoxically inhibits the development of diabetes in BB rats. Life Sciences 1998; 62:1293-1302.

123. Campbell IL, Oxbrow L, Harrison LC. Reduction in insulitis following administration of IFN-gamma and TNF-alpha in the NOD mouse. J Autoimmun 1991; 4:249-262.
124. Nicoletti F, Zaccone P, Di Marco R et al. Paradoxical antidiabetogenic effect of γ-interferon in DP-BB rats. Diabetes 1998; 47:32-38.
125. Balashov KE, Khoury SJ, Hafler DA et al. Inhibition of T cell responses by activated human CD8+ T cells is mediated by interferon-gamma and is defective in chronic progressive multiple sclerosis. J Clin Invest 1995; 95:2711-2719.
126. Willenborg DO, Fordham S, Bernard CCA et al. IFN-γ plays a critical down-regulatory role in the induction and effector phase of myelin oligodendrocyte glycoprotein-induced autoimmune encephalomyelitis. J Immunol 1996; 157:3223-3227.
127. Mandrup-Poulsen T, Nerup J, Reimers JI et al. Cytokines and the endocrine system. I. The immunoendocrine network. Eur J Endocrinol 1995; 133:660-671.
128. Ramierz F, Fowell DJ, Puklavec M et al. Glucocorticoids promote a Th2 cytokine response by CD4+ T cells in vitro. J Immunol 1996; 156:2406-2412.
129. Blotta MH, DeKruyff RH, Umetsu DT. Corticosteroids inhibit IL-12 production in human monocytes and enhance their capacity to induce IL-4 synthesis in CD4+ lymphocytes. J Immunol 1997; 158:5589-5595.
130. Trembleau S, Penna G, Bosi E et al. Interleukin 12 administration induces T helper type 1 cells and accelerates autoimmune diabetes in NOD mice. J Exp Med 1995; 181:817-821.
131. O'Hara Jr RM, Henderson SL, Nagelin AM. Prevention of a Th1 disease by a Th1 cytokine: IL-12 and diabetes in NOD mice. Ann NY Acad Sci 1996; 795:241-249.
132. Yang X-D, Tisch R, Singer SM et al. Effect of tumor necrosis factor α on insulin-dependent diabetes mellitus in NOD mice. I. The early development of autoimmunity and the diabetogenic process. J Exp Med 1994; 180:995-1004.
133. Cope AP, Liblau RS, Yang X-D et al. Chronic tumor necrosis factor alters T cell responses by attenuating T cell receptor signaling. J Exp Med 1997; 185:1573-1584.
134. Rabinovitch A, Suarez-Pinzon WL, Sorensen O et al. TNF-α down-regulates type 1 cytokines and prolongs survival of syngeneic islet grafts in nonobese diabetic mice. J Immunol 1997; 159:6298 6303.
135. Balasa B, Davies DJ, Lee J et al. Interleukin-10 impacts autoimmune diabetes via a CD8+ T cell pathway circumventing the requirement for CD4+ T- and B-lymphocytes. J Immunol 1998; 161:4420-4427.
136. Piccirillo CA, Chang Y, Prud=homme GJ. TGF-β1 somatic gene therapy prevents autoimmune disease in nonobese diabetic mice. J Immunol 1998; 161:3950-3956.
137. Suarez-Pinzon W, Korbutt GS, Power R et al. Testicular Sertoli cells protect islet β-cells from autoimmune destruction in NOD mice by a transforming growth factor-β1-dependent mechanism. Diabetes 2000; 49:1810-1818.
138. Nicoletti F, Dimarco R, Barcellini W et al. Protection from experimental autoimmune diabetes in the nonobese diabetic mouse with soluble interleukin-1 receptor. Eur J Immunol 1994; 24:1843-1847.
139. Cailleau C, Diu-Hercend A, Ruuth E et al. Treatment with neutralizing antibodies specific for IL-1β prevents cyclophosphamide-induced diabetes in nonobese diabetic mice. Diabetes 1997; 46:937-940.
140. Sandberg J-O, Eizirik DL, Sandler S. IL-1 receptor antagonist inhibits recurrence of disease after syngeneic pancreatic islet transplantation to spontaneously diabetic nonobese diabetic (NOD) mice. Clin Exp Immunol 1997; 108:314-317.
141. Debray-Sachs M, Carnaud C, Boitard C et al. Prevention of diabetes in NOD mice treated with antibody to murine IFN-γ. J Autoimmun 1991; 4:237-248.
142. Nicoletti F, Zaccone P, Di Marco R et al. Prevention of spontaneous autoimmune diabetes in diabetes-prone BB rats by prophylactic treatment with antirat interferon-γ antibody. Endocrinology 1997; 138:281-288.
143. Nicoletti F, Zaccone P, Di Marco R et al. The effects of a nonimmunogenic form of murine soluble interferon-γ receptor on the development of autoimmune diabetes in the NOD mouse. Endocrinology 1996; 137:5567-5575.
144. Kurschner C, Ozmen L, Garotta G et al. IFN-gamma receptor-Ig fusion proteins. Half-life, immunogenicity and in vivo activity. J Immunol 1992; 149:4096-4100.
145. Kelley VE, Gaulton GN, Hattori M et al. Anti-interleukin-2 receptor antibody suppresses murine diabetic insulitis and lupus nephritis. J Immunol 1988; 140:59-61.
146. PachecoSilva A, Bastos MG, Muggia RA et al. Interleukin-2 receptor targeted fusion toxin (DAB486-IL-2) treatment blocks diabetogenic autoimmunity in nonobese diabetic mice. Eur J Immunol 1992; 22:697-702.

147. Hunger RE, Carnaud C, Garcia I et al. Prevention of autoimmune diabetes mellitus in NOD mice by transgenic expression of soluble tumor necrosis factor receptor p55. Eur J Immunol 1997; 27:255-261.
148. Lee M-S, Mueller R, Wicker LS et al. IL-10 is necessary and sufficient for autoimmune diabetes in conjunction with NOD MHC homozygosity. J Exp Med 1996; 183:2663-2668.
149. Trembleau S, Penna G, Gregori S et al. The role of endogenous IL-12 in the development of spontaneous diabetes in NOD mice. Autoimmunity 1996; 21:A087.
150. Hultgren B, Huang X, Dybdal N et al. Genetic absence of γ-interferon delays but does not prevent diabetes in NOD mice. Diabetes 1996; 45:812-817.
151. Rothe H, O'Hara Jr RM, Martin S et al. Suppression of cyclophosphamide induced diabetes development and pancreatic Th1 reactivity in NOD mice treated with the interleukin (IL)-12 antagonist IL-12(p40)$_2$. Diabetologia 1997; 40:641-646.
152. Yasuda H, Nagata M, Arisawa K et al. Local expression of immunoregulatory IL-12p40 gene prolonged syngeneic islet graft survival in diabetic NOD mice. J Clin Invest 1998; 102:1807-1814.
153. Wang B, André I, Gonzalez A et al. Interferon-γ impacts at multiple points during the progression of autoimmune diabetes. Proc Natl Acad Sci USA 1997; 94:13844-13849.
154. Kanagawa O, Xu G, Tevaarwerk A et al. Protection of nonobese diabetic mice from diabetes by gene(s) closely linked to IFN-γ receptor loci. J Immunol 2000; 164:3919-3923.
155. Serreze DV, Post CM, Chapman HD et al. Interferon-γ receptor signalling is dispensable in the development of autoimmune type 1 diabetes in NOD mice. Diabetes 2000; 49:2007-2011.
156. Hänninen A, Jalkanen S, Salmi M et al. Macrophages, T-cell receptor usage, and endothelial cell activation in the pancreas at the onset of insulin-dependent diabetes mellitus. J Clin Invest 1992; 90:1901-1910.
157. Itoh N, Hanafusa T, Miyazaki A et al. Mononuclear cell infiltration and its relation to the expression of major histocompatibility complex antigens and adhesion molecules in pancreas biopsy specimens from newly diagnosed insulin-dependent diabetes mellitus patients. J Clin Invest 1993; 92:2313-2322.
158. Stassi G, De Maria R, Trucco G et al. Nitric oxide primes pancreatic β cells for Fas-mediated destruction in insulin-dependent diabetes mellitus. J Exp Med 1997; 186:1193-1200.
159. Kallmann BA, Hüther M, Tubes M et al. Systemic bias of cytokine production toward cell-mediated immune regulation in IDDM and toward humoral immunity in Graves' disease. Diabetes 1997; 46:237-243.
160. Rapoport MJ, Mor A, Vardi P et al. Decreased secretion of Th2 cytokines precedes up-regulated and delayed secretion of Th1 cytokines in activated peripheral blood mononuclear cells from patients with insulin-dependent diabetes mellitus. J Autoimmun 1998; 11:635-642.
161. Berman MA, Sandborg CI, Wang Z et al. Decreased IL-4 production in new onset Type I insulin-dependent diabetes mellitus. J Immunol 1996; 157:4690-4696.
162. Hussain MJ, Peakman M, Gallati H et al. Elevated serum levels of macrophage-derived cytokines precede and accompany the onset of IDDM. Diabetologia 1996; 39:60-69.
163. Hussain MJ, Maher J, Warnock T et al. Cytokine overproduction in healthy first degree relatives of patients with IDDM. Diabetologia 1998; 41:343-349.
164. Karlsson MGE, Sederholm Lawesson S, Ludvigsson J. Th1-like dominance in high-risk first-degree relatives of Type 1 diabetic patients. Diabetologia 2000; 43:742-749.
165. Wilson SB, Kent SC, Patton KT et al. Extreme Th1 bias of regulatory Vα24JαQ T cells in type 1 diabetes. Nature 1998; 391:177-181.
166. Fowell D, McKnight AJ, Powrie F et al. Subsets of CD4+ T-cells and their roles in the induction and prevention of autoimmunity. Immunol Rev 1991; 123:37-64.
167. Fowell D, Mason D. Evidence that the T cell repertoire of normal rats contains cells with the potential to cause diabetes. Characterization of the CD4+ T cell subset that inhibits this autoimmune potential. J Exp Med 1993; 177:627-636.
168. Shimada A, Rohane P, Fathman CG et al. Pathogenic and protective roles of $CD45RB^{low}$ $CD4^{+}$ cells correlate with cytokine profiles in the spontaneously autoimmune diabetic mouse. Diabetes 1996; 45:71-78.
169. Zekzer D, Wong FS, Ayalon O et al. GAD-reactive CD4+ Th1 cells induce diabetes in NOD/SCID mice. J Clin Invest 1998; 101:68-73.
170. Akhtar I, Gold JP, Pan L-Y et al. $CD4^{+}$ β islet cell-reactive T cell clones that suppress autoimmune diabetes in nonobese diabetic mice. J Exp Med 1995; 182:87-97.
171. Han H-S, Jun H-S, Utsugi T et al. A new type of CD4+ suppressor T cell completely prevents spontaneous autoimmune diabetes and recurrent diabetes in syngeneic islet-transplanted NOD mice. J Autoimmun 1996; 9:331-339.

172. Zekzer D, Wong FS, Wen L et al. Inhibition of diabetes by an insulin-reactive CD4 T-cell clone in the nonobese diabetic mouse. Diabetes 1997; 46:1124-1132.
173. Healey D, Ozegbe P, Arden S et al. *In vivo* activity and *in vitro* specificity of CD4+ Th1 and Th2 T cells derived from the spleens of diabetic NOD mice. J Clin Invest 1995; 95:2979-2985.
174. Katz JD, Benoist C, Mathis D. T helper cell subsets in insulin-dependent diabetes. Science 1995; 268:1185-1188.
175. Hammond KJL, Poulton LD, Palmisano LJ et al. α/β-T cell receptor $(TCR)^+CD4^SCD8^S$ (NKT) thymocytes prevent insulin-dependent diabetes mellitus in nonobese diabetic (NOD)/Lt mice by the influence of interleukin (IL)-4 and/or IL-10. J Exp Med 1998; 187:1047-1056.
176. Gombert J-M, Herbelin A, Tancrede-Bohin E et al. Early quantitative and functional deficiency of $NK1^+$-like thymocytes in the NOD mouse. Eur J Immunol 1996; 26:2989-2998.
177. Godfrey DI, Kinder SJ, Silveira P et al. Flow cytometric study of T cell development in NOD mice reveals a deficiency in $\alpha/\beta TCR^+CD4^SCD8^S$ thymocytes. J Autoimmun 1997; 10:279-285.
178. Herbelin A, Gombert J-M, Lepault F et al. Mature mainstream $TCR\alpha\beta^+CD4^+$ thymocytes expressing L-selectin mediate Aactive tolerance@ in the nonobese diabetic mouse. J Immunol 1998; 161:2620-2628.
179. Lee K, Kim M, Amano K et al. Preferential infiltration of macrophages during early stages of insulitis in diabetes-prone BB rats. Diabetes 1988; 37:1053-1058.
180. Voorbij HA, Jeucken PH, Kabel PJ et al. Dendritic cells and scavenger macrophages in pancreatic islets of prediabetic BB rats. Diabetes 1989; 38:1623-1629.
181. Jansen A, Homo-Delarche F, Hooijkaas H et al. Immunohistochemical characterization of monocyte-macrophages and dendritic cells involved in the initiation of insulitis and beta-cell destruction in NOD mice. Diabetes 1994; 43:667-675.
182. Oschilewski U, Kiesel U, Kolb H. Administration of silica prevents diabetes in BB rats. Diabetes 1985; 34:197-199.
183. Lee K, Amano K, Yoon JW. Evidence for initial involvement of macrophages in development of insulitis in NOD mice. Diabetes 1988; 37:989-991.
184. Jun H-S, Yoon C-S, Zbytnuik L et al. The role of macrophages in T cell-mediated autoimmune diabetes in nonobese diabetic mice. J Exp Med 1999; 189:347-358.
185. Jun H-S, Santamaria P, Lim H-W et al. Absolute requirement of macrophages for the development and activation of β-cell cytotoxic $CD8^+$ T-cells in T-cell receptor transgenic NOD mice. Diabetes 1999; 48:34-42.
186. Serreze DV, Chapman HD, Varnum DS et al. B lymphocytes are essential for the initiation of T cell-mediated autoimmune diabetes: analysis of a new Aspeed congenic@ stock of NOD.Ig mu null mice. J Exp Med 1996; 184:2049-2053.
187. Noorchashm H, Noorchashm N, Kern J et al. B-cells are required for the initiation of insulitis and sialitis in nonobese diabetic mice. Diabetes 1997; 46:941-946.
188. Akashi T, Seiho N, Keizo A et al. Direct evidence for the contribution of B cells to the progression of insulitis and the development of diabetes in nonobese diabetic mice. Int Immunol 1997; 9:1159-1164.
189. Shimizu J, Kanagawa O, Unanue ER. Presentation of β-cell antigens to CD4+ and CD8+ T-cells of nonobese diabetic mice. J Immunol 1993; 151:1723-1730.
190. Nepom GT, Erlich H. MHC class-II molecules and autoimmunity. Annu Rev Immunol 1991; 9:493-525.
191. Scott B, Liblau R, Degermann S et al. A role for nonMHC genetic polymorphism in susceptibility to spontaneous autoimmunity. Immunity 1994; 1:73-83.
192. Singer SM, Tisch R, Yang X-D et al. Prevention of diabetes in NOD mice by a mutated I-Ab transgene. Diabetes 1998; 47:1570-1577.
193. van Seventer GA, Shimizu Y, Shaw S. Roles of multiple accessory molecules in T-cell activation. Curr Opin Immunol 1991; 3:294-303.
194. Grewal IS, Flavell RA. CD40 and CD154 in cell-mediated immunity. Annu Rev Immunol 1998; 16:111-135.
195. Wong S, Guerder S, Visintin I et al. Expression of the costimulator molecule B7-1 in pancreatic β-cells accelerates diabetes in the NOD mouse. Diabetes 1995; 44:326-329.
196. Lenschow DJ, Ho SC, Sattar H et al. Differential effects of anti-B7-1 and anti-B7-2 monoclonal antibody treatment on the development of diabetes in the nonobese diabetic mouse. J Exp Med 1995; 181:1145-1155.
197. Balasa B, Krahl T, Patstone G et al. CD40 ligand-CD40 interactions are necessary for the initiation of insulitis and diabetes in nonobese diabetic mice. J Immunol 1997; 159:4620-4627.

198. Yagi N, Yokono K, Amano K et al. Expression of intercellular adhesion molecule 1 on pancreatic beta-cells accelerates beta-cell destruction by cytotoxic T-cells in murine autoimmune diabetes. Diabetes 1995; 44:744-752.
199. Moriyama H, Yokono K, Amano K et al. Induction of tolerance in murine autoimmune diabetes by transient blockade of leukocyte function-associated antigen-/intercellular adhesion molecule-1 pathway. J Immunol 1996; 157:3737-3743.
200. Martin S, Heidenthal E, Schulte B et al. Soluble forms of intercellular adhesion molecule-1 inhibit insulitis and onset of autoimmune diabetes. Diabetologia 1998; 41:1298-1303.
201. Hsieh C-S, Heimberger AB, Gold JS et al. Differential regulation of T helper phenotype development by interleukins 4 and 10 in an αβ T-cell-receptor transgenic system. Proc Natl Acad Sci USA 1992; 89:6065-6069.
202. Seder RA, Paul WE, Davis MM et al. The presence of interleukin 4 during in vitro priming determines the lymphokine-producing potential of CD4+ T-cells from T-cell receptor transgenic mice. J Exp Med 1992; 176:1091-1098.
203. Kolb H, Kolb-Bachofen V, Roep BO. Autoimmune *versus* inflammatory type I diabetes: a controversy? Immunol Today 1995; 16:170-172.
204. Benoist C, Mathis D. Cell death mediators in autoimmune diabetes C no shortage of suspects. Cell 1997; 89:1-3.
205. Utsugi T, Yoon J-W, Park B-J et al. Major histocompatibility complex class I-restricted infiltration and destruction of pancreatic islets by NOD mouse-derived β-cell cytotoxic $CD8^+$ T-cell clones in vivo. Diabetes 1996; 45:1121-1131.
206. Wong FS, Visintin I, Wen L et al. CD8 T cell clones from young nonobese diabetic (NOD) islets can transfer rapid onset of diabetes in NOD mice in the absence of CD4 cells. J Exp Med 1996; 183:67-76.
207. Miller BJ, Appel MC, O'neil JJ et al. Both the Lyt-2+ and L3T4+ T cell subsets are required for the transfer of diabetes in nonobese diabetic mice. J Immunol 1988; 140:52-58.
208. Christianson SW, Shultz LD, Leiter EH. Adoptive transfer of diabetes into immunodeficient NOD-scid/scid mice. Relative contributions of CD4+ and CD8+ T cells from diabetic *versus* prediabetic NOD.NONThy-1a donors. Diabetes 1993; 42:44-55.
209. Edouard P, Hiserodt JC, Plamondon C et al. CD8+ T-cells are required for adoptive transfer of the BB rat diabetic syndrome. Diabetes 1993; 42:390-397.
210. Katz J, Benoist C, Mathis D. Major histocompatibility complex class I molecules are required for the development of insulitis in nonobese diabetic mice. Eur J Immunol 1993; 23:3358-3360.
211. Young LH, Peterson LB, Wicker LS et al. *In vivo* expression of perforin by CD8+ lymphocytes in autoimmune disease. Studies on spontaneous and adoptively transferred diabetes in nonobese diabetic mice. J Immunol 1989; 143:3994-3999.
212. Kägi D, Odermatt B, Seiler P et al. Reduced incidence and delayed onset of diabetes in perforin-deficient nonobese diabetic mice. J Exp Med 1997; 186:989-997.
213. Nagata M, Yokono K, Hayakawa M et al. Destruction of pancreatic islet cells by cytotoxic T lymphocytes in nonobese diabetic mice. J Immunol 1989; 143:1155-1162.
214. Boitard C, Chatenoud L, Debray-Sachs M. *In vitro* inhibition of pancreatic B cell function by lymphocytes from diabetics with associated autoimmune diseases: A T cell phenomenon. J Immunol 1982; 129:2529-2531.
215. Bradley BJ, Haskins K, La Rosa FG et al. CD8 T cells are not required for islet destruction induced by a CD4+ islet-specific T-cell clone. Diabetes 1992; 41:1603-1608.
216. Wang Y, Pontesilli O, Gill RG et al. The role of CD4+ and CD8+ T cells in the destruction of islet grafts by spontaneously diabetic mice. Proc Natl Acad Sci USA 1991; 88:527-531.
217. Weringer EJ, Like AA. Immune attack on pancreatic islet transplants in the spontaneously diabetic BioBreeding/Worcester (BB/W) rat is not MHC restricted. J Immunol 1985; 134:2383-2386.
218. Kleemann R, Rothe H, Kolb-Bachofen V et al. Transcription and translation of inducible nitric oxide synthase in the pancreas of prediabetic BB rats. FEBS Lett 1993; 328:9-12.
219. Rabinovitch A, Suarez WL, Power RF. Lazaroid antioxidant reduces incidence of diabetes and insulitis in nonobese diabetic mice. J Lab Clin Med 1993; 121:603-607.
220. Mandrup-Poulsen T, Reimers JI, Andersen HU et al. Nicotinamide treatment in the prevention of insulin-dependent diabetes mellitus. Diabetes Metab Rev 1993; 9:295-309.
221. Kay TWH, Campbell IL, Oxbrow L et al. Overexpression of class I major histocompatibility complex accompanies insulitis in the nonobese diabetic mouse and is prevented by anti-interferon-γ antibody. Diabetologia 1991; 34:779-785.
222. Yamada K, Takane-Gyotoku N, Yuan X et al. Mouse islet lysis mediated by interleukin-1-induced Fas. Diabetologia 1996; 39:1306-1312.

223. Stassi G, Todaro M, Richiusa P et al. Expression of apoptosis-inducing CD95 (Fas/Apo-1) on human β-cells sorted by flow-cytometry and cultured in vitro. Transplant Proc 1995; 27:3271-3275.
224. Signore A, Annovazzi A, Procaccini E et al. CD95 and CD95-ligand expression in endocrine pancreas of NOD, NOR and BALB/c mice. Diabetologia 1997; 40:1476-1479.
225. Suarez-Pinzon WL, Sorensen O, Bleackley RC et al. β-cell destruction in NOD mice correlates with Fas (CD95) expression on β-cells and proinflammatory cytokine expression in islets. Diabetes 1999; 48:21-28.
226. Loweth AC, Williams GT, James RFL et al. Human islets of Langerhans express Fas ligand and undergo apoptosis in response to interleukin-1β and Fas ligation. Diabetes 1998; 47:727-732.
227. Chervonsky AV, Wang Y, Wong FS et al. The role of Fas in autoimmune diabetes. Cell 1997; 89:17-24.
228. Itoh N, Imagawa A, Hanafusa T et al. Requirement of Fas for the development of autoimmune diabetes in nonobese diabetic mice. J Exp Med 1997; 186:613-618.
229. Allison J, Strasser A. Mechanisms of β cell death in diabetes: A minor role for CD95. Proc Natl Acad Sci USA 1998; 95:13818-13822.
230. Kim Y-H, Kim S, Kim K-A et al. Apoptosis of pancreatic β-cells detected in accelerated diabetes of NOD mice: no role of Fas-Fas ligand interaction in autoimmune diabetes. Eur J Immunol 1999; 29:455-465.
231. Kim S, Kim KA, Hwang DY et al. Inhibition of autoimmune diabetes by Fas ligand: The paradox is solved. J Immunol 2000; 164:2931-2936.
232. Su X, Hu Q, Kristan JM. Significant role for Fas in the pathogenesis of autoimmune diabetes. J Immunol 2000; 164:2523-2532.
233. Amrani A, Verdaguer J, Thiessen S et al. IL-1α, IL-1β, and IFN-γ mark β cells for Fas-dependent destruction by diabetogenic CD4+ T lymphocytes. J Clin Invest 2000; 105:459-468.
234. Jun H-S, Santamaria P, Lim H-W et al. Absolute requirement of macrophages for the development and activation of β cell cytotoxic T cells in T cell receptor transgenic NOD mice. Diabetes 1999; 48:34-42.
235. Suarez-Pinzon WL, Power RF, Rabinovitch A. Fas ligand-mediated mechanisms are involved in autoimmune destruction of islet beta-cells in nonobese diabetic mice. Diabetologia 2000; 43;1149-1156.
236. El-Sheikh A, Suarez-Pinzon WL, Power RF et al. Both CD4+ and CD8+ T cells are required for IFN-γ gene expression in pancreatic islets and autoimmune diabetes development in biobreeding rats. J Autoimmun 1999; 12:109-119.
237. Kemeny DM, Noble A, Holmes BJ et al. Immune regulation: a new role for the CD8+ T cell. Immunol Today 1994; 15:107-110.
238. Rus V, Svetic A, Nguyen P et al. Kinetics of Th1 and Th2 cytokine production during the early course of acute and chronic murine graft-*Versus*-Host Disease. J Immunol 1995; 155:2396-2406.
239. Williams NS, Engelhard VH. Perforin-dependent cytotoxic activity and lymphokine secretion by CD4+ T cells are regulated by CD8+ T cells. J Immunol 1997; 159:2091-2099.
240. Verdaguer J, Schmidt D, Amrani A et al. Spontaneous autoimmune diabetes in monoclonal T cell nonobese diabetic mice. J Exp Med 1997; 186:1663-1676.
241. Bevilacqua MP. Endothelial-leukocyte adhesion molecules. Annu Rev Immunol 1993; 11:767-804.
242. Doukas J, Mordes JP. T lymphocytes capable of activating endothelial cells in vitro are present in rats with autoimmune diabetes. J Immunol 1993; 150:1036-1046.
243. Santamaria P, Nakhleh RE, Sutherland DE et al. Characterization of T lymphocytes infiltrating human pancreas allograft affected by isletitis and recurrent diabetes. Diabetes 1992; 41:53-61.
244. Oldstone MB. Prevention of type I diabetes in nonobese diabetic mice by virus infection. Science 1988; 239:500-502.
245. Dyrberg T, Schwimmbeck PL, Oldstone MB. Inhibition of diabetes in BB rats by virus infection. J Clin Invest 1988; 81:928-931.
246. Wilberz S, Partke HJ, Dagnaes-Hansen F et al. Persistent MHV (mouse hepatitis virus) infection reduces the incidence of diabetes mellitus in nonobese diabetic mice. Diabetologia 1991; 34:2-5.
247. Hermitte L, Vialettes B, Naquet P et al. Paradoxical lessening of autoimmune processes in nonobese diabetic mice after infection with the diabetogenic variant of encephalomyocarditis virus. Eur J Immunol 1990; 20:1297-1303.
248. Takei I, Asaba Y, Kasatani T et al. Suppression of development of diabetes in NOD mice by lactate dehydrogenase virus infection. J Autoimmun 1992; 5:665-673.
249. Toyota T, Satoh J, Oya K et al. Streptococcal preparation (OK-432) inhibits development of type I diabetes in NOD mice. Diabetes 1986; 35:496-499.
250. Satoh J, Shintani S, Oya K et al. Treatment with streptococcal preparation (OK-432) suppresses anti-islet autoimmunity and prevents diabetes in BB rats. Diabetes 1988; 37:1188-1194.

251. Kawamura T, Nagata M, Utsugi T et al. Prevention of autoimmune type I diabetes by CD4+ suppressor T-cells in superantigen-treated nonobese diabetic mice. J Immunol 1993; 151:4362-4370.
252. Kino K, Mizumoto K, Sone T et al. An immunomodulating protein Ling Zhi-8 (LZ-8) prevents insulitis in nonobese diabetic mice. Diabetologia 1990; 33:713-718.
253. Elias D, Markovits D, Reshef T et al. Induction and therapy of autoimmune diabetes in the nonobese diabetic (NOD/Lt) mouse by a 65-kDa heat shock protein. Proc Natl Acad Sci USA 1990; 87:1576-1580.
254. Sadelain MWJ, Qin H-Y, Lauzon J et al. Prevention of type I diabetes in NOD mice by adjuvant immunotherapy. Diabetes 1990; 39:583-589.
255. Sadelain MWJ, Qin H-Y, Sumoski W et al. Prevention of diabetes in the BB rat by early immunotherapy using Freund's adjuvant. J Autoimmun 1990; 3:671-680.
256. McInerney MF, Pek SB, Thomas DW. Prevention of insulitis and diabetes onset by treatment with complete Freund's adjuvant in NOD mice. Diabetes 1991; 40:715-725.
257. Pearce RB, Peterson CM. Studies of concanavalin A in nonobese diabetic mice. I. Prevention of insulin-dependent diabetes. J Pharmacol Exp Ther 1991; 258:710-715.
258. Wang T, Singh B, Warnock GL et al. Prevention of recurrence of IDDM in islet-transplanted diabetic NOD mice by adjuvant immunotherapy. Diabetes 1992; 41:114-117.
259. Ulaeto D, Lacy PE, Kipnis DM et al. A T-cell dormant state in the autoimmune process of nonobese diabetic mice treated with complete Freund's adjuvant. Proc Natl Acad Sci USA 1992; 89:3927-3931.
260. Qin H-Y, Suarez WL, Parfrey N et al. Mechanisms of complete Freund's adjuvant protection against diabetes in BB rats: induction of nonspecific suppressor cells. Autoimmunity 1992; 12:193-199.
261. Qin H-Y, Sadelain MWY, Hitchon C et al. Complete Freund's adjuvant-induced T-cells prevent the development and adoptive transfer of diabetes in nonobese diabetic mice. J Immunol 1993; 150:2072-2080.
262. Yagi H, Matsumoto M, Suzuki S et al. Possible mechanism of the preventive effect of BCG against diabetes mellitus in NOD mouse. I. Generation of suppressor macrophages in spleen cells of BCG-vaccinated mice. Cell Immunol 1991; 138:130-141.
263. Yagi H, Matsumoto M, Kishimoto Y et al. Possible mechanism of the preventive effect of BCG against diabetes mellitus in NOD mice. II. Suppression of pathogenesis by macrophage transfer from BCG-vaccinated mice. Cell Immunol 1991; 138:142-149.
264. Lakey JRT, Singh B, Warnock GL et al. BCG immunotherapy prevents recurrence of diabetes in islet grafts transplanted into spontaneously diabetic NOD mice. Transplantation 1994; 57:1213-1217.
265. Calcinaro F, Gambelunghe G, Lafferty KJ. Protection from autoimmune diabetes by adjuvant therapy in the nonobese diabetic mouse: The role of interleukin-4 and interleukin-10. Immunology and Cell Biology 1997; 75:467-471
266. Rabinovitch A, Suarez-Pinzon WL, Sorensen O et al. Combined therapy with interleukin-4 and interleukin-10 inhibits autoimmune diabetes recurrence in syngeneic islet-transplanted nonobese diabetic mice: analysis of cytokine mRNA expression in the graft. Transplantation 1995; 60:368-374.
267. Moritani M, Yoshimoto K, Ii S et al. Prevention of adoptively transferred diabetes in nonobese diabetic mice with IL-10-transduced islet-specific Th1 lymphocytes. A gene therapy model for autoimmune diabetes. J Clin Invest 1996; 98:1851-1859.
268. Rulifson IC, Sperling AI, Fields PE et al. CD28 costimulation promotes the production of Th2 cytokines. J Immunol 1997; 158:658-665.
269. Lenschow DJ, Rhee L, Patel B et al. CD28/B7 regulation of Th1 and Th2 subsets in the development and progression of autoimmune diabetes. Immunity 1996; 5:285-293.
270. Takahashi K, Honeyman MC, Harrison LC. Impaired yield, phenotype, and function of monocyte-derived dendritic cells in humans at risk for insulin-dependent diabetes. J Immunol 1998; 161:2629-2635.
271. Ramsdell F, Seaman MS, Miller RE et al. Differential ability of Th1 and Th2 T cells to express Fas ligand and to undergo activation-induced cell death. Int Immunol 1994; 6:1545-1553.
272. Varadhachary AS, Perdow SM, Hu C et al. Differential ability of T cell subsets to undergo activation-induced cell death. Proc Natl Acad Sci USA 1997; 94:5778-5783.
273. Zhang X, Brunner T, Carter L et al. Unequal death in T helper (Th)1 and Th2 effectors: Th1, but not Th2, effectors undergo rapid Fas/FasL-mediated apoptosis. J Exp Med 1997; 185:1837-1849.
274. Serreze DV, Chapman HD, Post CM et al. Th1 to Th2 cytokine shifts in NOD mice: sometimes an outcome, rather than the cause of diabetes resistance elicited by immunostimulation. J Immunol 2001; 166:1352-1359.
275. Dalton DK, Haynes L, Chu C-Q et al. Interferon γ eliminates responding CD4 T cells during mycobacterial infection by inducing apoptosis of activated CD4 T cells. J Exp Med 2000; 192:117-122.

276. Parish NM, Hutchings PR, O'Reilly L et al. Tolerance induction as a therapeutic strategy for the control of autoimmune endocrine disease in mouse models. Immunol Rev 1995; 144:269-300.
277. Waldmann H, Cobbold S. How do monoclonal antibodies induce tolerance? A role for infectious tolerance? Annu Rev Immunol 1998; 16:619-644.
278. Phillips JM, Harach SZ, Parish NM et al. Nondepleting anti-CD4 has an immediate action on diabetogenic effector cells, halting their destruction of pancreatic β cells. J Immunol 2000; 165:1949-1955.
279. Konieczny BT, Dai Z, Elwood ET et al. IFN-γ is critical for long-term allograft survival induced by blocking the CD28 and CD40L T cell costimulation pathways. J Immunol 1998; 160:2059-2064.
280. Diamond A, Gill RG. Biphasic roles for IFNγ in islet allograft immunity and tolerance. Transplantation 1999; 67:S23.
281. Dai Z, Konieczny BT, Baddoura FK et al. Impaired alloantigen-mediated T cell apoptosis and failure to induce long-term allograft survival in IL-2-deficient mice. J Immunol 1998; 161:1659-1663.
282. Steiger B, Nickerson PW, Steurer W et al. IL-2 knockout recipient mice reject islet cell allografts. J Immunol 1995; 155:489-498.
283. Falcone M, Yeung B, Tucker L et al. A defect in interleukin 12-induced activation and interferon γ secretion of peripheral natural killer T cells in nonobese diabetic mice suggests new pathogenic mechanisms for insulin-dependent diabetes mellitus. J Exp Med 1999; 190:963-972.
284. Fujihira K, Nagata M, Moriyama H et al. Suppression and acceleration of autoimmune diabetes by neutralization of endogenous interleukin-12 in NOD mice. Diabetes 2000; 49:1998-2006.
285. Rabinovitch A. Immunoregulatory and cytokine imbalances in the pathogenesis of IDDM. Therapeutic intervention by immunostimulation? Diabetes 1994; 43:613-621.
286. Liblau RS, Singer SM, McDevitt HO. Th1 and Th2 $CD4^+$ cells in the pathogenesis of organ-specific autoimmune diseases. Immunology Today 1995; 16:34-38.
287. Charlton B, Lafferty KJ. The Th1/Th2 balance in autoimmunity. Curr Opin Immunol 1995; 7:793-798.
288. Delovitch T, Singh B. The nonobese diabetic mouse as a model of autoimmune diabetes: immune dysregulation gets the NOD. Immunity 1997; 7:727-738.
289. Wang B, Gonzalez A, Höglund P et al. Interleukin-4 deficiency does not exacerbate disease in NOD mice. Diabetes 1998; 47:1207-1211.
290. Pearce EJ, Cheever A, Leonard S et al. *Schistosoma mansoni* in IL-4-deficient mice. Int Immunol 1996; 8:435-444.
291. Noben-Trauth N, Kropf P, Muller I. Susceptibility to *Leishmani major* infection in interleukin-4-deficient mice. Science 1996; 271:987-990.
292. Kaufman DL, ClareSalzler M, Tian J et al. Spontaneous loss of T-cell tolerance to glutamic acid decarboxylase in murine insulin-dependent diabetes. Nature 1993; 366:69-72.
293. Tisch R, Yang X-D, Singer SM et al. Immune response to glutamic acid decarboxylase correlates with insulitis in nonobese diabetic mice. Nature 1993; 366:72-75.
294. Tisch R, Liblau RS, Yang X-D et al. Induction of GAD65-specific regulatory T-cells inhibits ongoing autoimmune diabetes in nonobese diabetic mice. Diabetes 1998; 47:894-899.
295. Tisch R, Wang B, Serreze DV. Induction of GAD65-specific Th2 cells and suppression of autoimmune diabetes at late stages of disease is epitope-dependent. J Immunol 1999 163; 1178-1187.
296. Tisch R, Wang B, Weaver DJ, et al. Antigen-specific mediated suppression of β cell autoimmunity by plasmid DNA vaccination. J Immunol 2001 166; 2122-2132.
297. Harrison LC, Honeyman MC, DeAizpurua HJ et al. Inverse relation between humoral and cellular immunity to glutamic acid decarboxylase in subjects at risk of insulin-dependent diabetes. Lancet 1993; 341:1365-1369.
298. Yu L, Gianani R, Eisenbarth GS. Quantitation of glutamic acid decarboxylase autoantibody levels in prospectively evaluated relatives of patients with type I diabetes. Diabetes 1994; 43:1229-1233.
299. Ramiya V, Muir A, Maclaren N. Insulin prophylaxis in insulin-dependent diabetes mellitus. Immunological rationale and therapeutic use. Clin Immunother 1995; 3:177-183.
300. Alleva DG, Crowe PD, Jin L et al. (2001). A disease-associated cellular immune response in type 1 diabetics to an immunodominant epitope of insulin. J Clin Invest 2001; 107:173-180.
301. Daniel D, Wegmann DR. Protection of nonobese diabetic mice from diabetes by intranasal or subcutaneous administration of insulin peptide B-chain (9-23). Proc Natl Acad Sci USA 1996; 93:956-960.
302. Polanski M, Melican NS, Zhang J et al. Oral administration of the immunodominant B-chain of insulin reduces diabetes in a cotransfer model of diabetes in the NOD mouse and is associated with a switch from Th1 to Th2 cytokines. J Autoimmun 1997; 10:339-346.

303. Tian J, Atkinson MA, ClareSalzler MC et al. Nasal administration of glutamate decarboxylase (GAD65) peptides induces Th2 responses and prevents murine insulin-dependent diabetes. J Exp Med 1996; 183:1561-1567.
304. Ma S-W, Zhao D-L, Yin Z-Q et al. Transgenic plants expressing autoantigens fed to mice to induce oral immune tolerance. Nature Medicine 1997; 3:793-796.
305. Zhang ZJ, Davidson LE, Eisenbarth G et al. Suppression of diabetes in NOD mice by oral administration of porcine insulin. Proc Natl Acad Sci USA 1991; 88:10252-10256.
306. Bergerot I, Fabien N, Maguer V et al. Oral administration of human insulin to NOD mice generates $CD4^+$ T cells that suppress adoptive transfer of diabetes. J Autoimmun 1994; 7:655-663.
307. Ploix C, Bergerot I, Fabien N et al. Protection against autoimmune diabetes with oral insulin is associated with the presence of IL-4 type 2 T-cells in the pancreas and pancreatic lymph nodes. Diabetes 1998; 47:39-44.
308. Harrison LC, Dempsey-Collier M, Kramer DR et al. Aerosol insulin induces regulatory CD8 γδ T cells that prevent murine insulin-dependent diabetes. J Exp Med 1996; 184:2167-2174.

CHAPTER 11

Cytokines in the Pathogenesis of Rheumatoid Arthritis and Collagen-Induced Arthritis

Erik Lubberts and Wim B. van den Berg

Introduction

The cytokine network in rheumatoid arthritis (RA) is a complex field, with a lot of cytokines showing pleiotropic actions and many different targets. To keep it simple, the network can be divided in two groups, the pro-inflammatory and anti-inflammatory cytokines. Controling the balance between these two groups is considered as an important therapeutic goal.

Two key pro-inflammatory cytokines in RA are IL-1 and TNFα. Regulation of these cytokines is of crucial importance in the RA disease. First data of clinical trials showed efficacy, however, revealed also that blockade of these cytokines did not fully control the arthritis in all patients. Recent discoveries of novel cytokines in the pathology of arthritis, such as IL-17, IL-18 and RANK ligand (RANKL) will help us to get a better understanding of the pathogenesis of chronic arthritis and may contribute to improvement of current therapies. IL-4 and IL-10 are pleiotropic cytokines, and are considered as promising modulators in the control of RA.

Rheumatoid arthritis (RA) is a chronic systemic disorder of unknown etiology. This disease affects about 1% of the population worldwide, most commonly middle-aged women. It is characterized by chronic inflammation of the synovium, particularly of small joints, which often leads to destruction of articular cartilage and juxtaarticular bone.[1] The clinical and laboratory features are suggestive of an autoimmune disease. However, the autoantigen is still unknown, hampering specific immunomodulation as a straightforward therapeutic approach.

The pathogenesis of RA is not identified and seems to be multifactorial. A major research goal in the field of arthritis is to unravel the pathogenesis of chronic arthritis and the concomitant joint destruction. During the last 20 years, the understanding of the basic biology of RA has increased enormously. This will help to define targeted therapies, selectively inhibiting the progression of destructive arthritis, yet leaving host defence mechanisms virtually intact. Targeting the cytokine disbalance might represent a solid way to control this disease.[2]

Pathways in the Pathogenesis of RA

The current concept is that inflammation and tissue destruction results from complex cell-cell interactions in the rheumatoid synovium.[3, 4] The major cytokines and cellular pathways currently implicated in the pathogenesis of RA are presented in Figure 11.1. These events can be amplified or initiated by an interaction between antigen presenting cells (APC) and CD4+ T cels; APC display complexes of class II major histocompatibility complex (MHC) molecules and peptide antigen(s) that bind to specific receptors on the T cells. Macrophage activation occurs, with abundant secretion of proinflammatory cytokines such as IL-1 and TNFα. These cytokines stimulate synovial fibroblasts and chondrocytes in the nearby articular cartilage to secrete enzymes that degrade proteoglycans and collagen, leading to tissue destruction.

Cytokines and Chemokines in Autoimmune Disease, edited by Pere Santamaria.

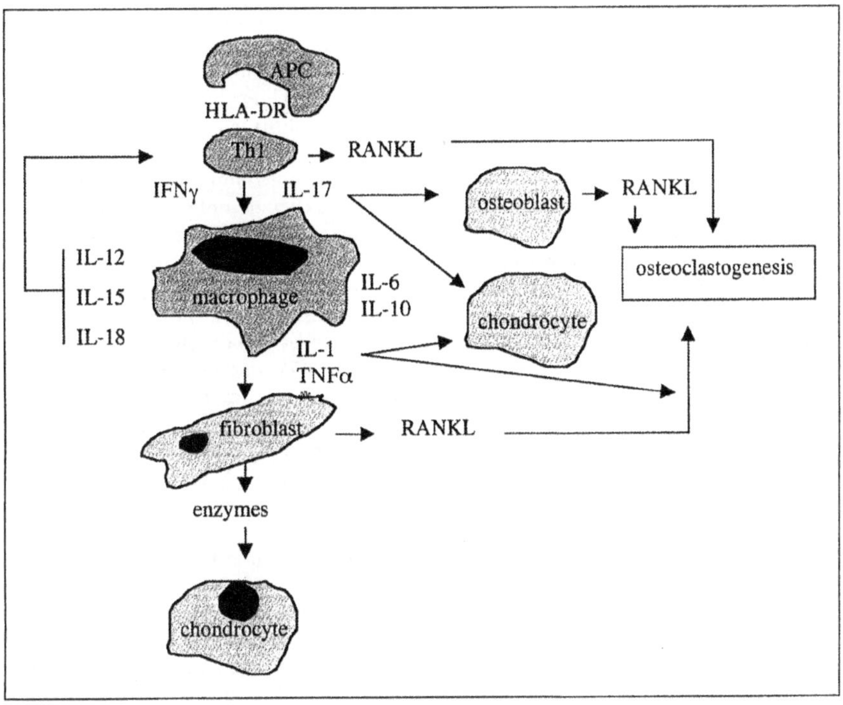

Fig. 11.1. Schematic overview of cytokines in RA.

Whether this process of destruction is driven by T cells or reflects mainly macrophage and synovial fibroblast activation is still a matter of debate. It has been shown that RA synovial fibroblasts are capable of mediating progressive joint destruction in the absence of T cells or other inflammatory cells,[5] suggesting T cell independent pathways in joint destruction.[6] Detailed analysis of mediators production in the inflamed synovial tissue reveals a relative lack of T cell factors and an abundance of cytokines and growth factors, produced by macrophages and synovial fibroblasts.[7]

Proinflammatory Cytokines IL-1 and TNF

It is well established that TNF and IL-1 are key cytokines in the process of chronic joint inflammation and the concomitant erosive changes in cartilage and bone. Animal model studies have greatly contributed to this identification. The initial studies analysed the arthritogenic potential of recombinant cytokines when directly injected into the knee joints of rabbits and rodents. This provided the first suggestive evidence that TNFα was an inflammatory mediator, whereas IL-1 was a crucial cytokine in both arthritis and cartilage destruction. TNFα alone was hardly destructive, but it could enhance in a synergistic way the destructive behaviour of IL-1.[8, 9] Follow-up studies in TNFα transgenics further underlined the fact that TNFα overexpression, in the absence of functional T and B cells, was arthritogenic.[10] Recent observations clarified that there is no requirement for soluble TNFα but that the full expression of arthritis can occur even with a membrane-bound form of TNFα (mTNFα).[11] The consequences of this is that therapies focused on TNF blockade should preferably make use of antibodies or scavenging soluble receptors that have excellent access to cell surfaces. The development of arthritis in TNF transgenic mice could be prevented with antibodies to TNFα, which seems obvious. More interestingly, pathology could also be fully blocked with antibodies against the IL-1 receptor.[12] This strongly indicates that 1) IL-1 is the secondary mediator responsible for the arthritic changes, and 2) TNFα alone is neither arthritogenic nor destructive towards joints.

Meanwhile, studies with neutralizing antibodies against TNFα have been instrumental in the elucidation of TNF as a major target in more natural arthritis models, with a T cell-driven pathogenetic pathway, compared with the plain over-expression of a single mediator.

Ample studies have been performed in the generally accepted murine autoimmune model of collagen-induced arthritis (CIA). CIA is based on T cell and antibody mediated autoimmune reactivity against collagen type II, the major component of cartilage. The model is characterized by severe and rapid cartilage and bone erosion. Suppression of collagen arthritis was achieved both with neutralizing antibodies against TNFα and with soluble TNF receptors.[13,14] Intriguingy, it was found that TNFα was crucial at the onset of the arthritis but appeared less dominant in the later stages.[15] In fact, studies in TNF receptor knockout mice demonstrated that the incidence and severity of arthritis were less in such mice; once the joints became affected, however, full progression to erosive damage was noted in an apparently TNF-independent fashion.[16]

As state above, IL-1 is a potent cytokine in the induction of cartilage destruction[8,9] and a pivotal secondary mediator in arthritis and tissue destruction in TNF transgenic over-expression models.[12] In addition, it has been found that IL-1 is not necessarily a dominant cytokine in the acute, inflammatory stages of most arthritis models, but plays a crucial role in the propagation of joint inflammation and concomitant cartilage and bone erosion in collagen arthritis. Transgenic over-expression of IL-1 produced erosive arthritis.[17,18]

In CIA, it was shown that treatment with a set of neutralizing antibodies against both IL-1α and IL-1β was still highly effective in established arthritis, reducing both inflammation and the progression of cartilage destruction. Studies with antibodies to seperate IL-1 isoforms revealed that IL-1β is more crucial.[15,19] This is in line with the clear efficacy in this model of ICE (IL-1β-converting enzyme) inhibitors and the observation of reduced CIA in ICE-deficient mice.[20] Similarly, the local over-expression of IL-1ra by retroviral gene transfer in inflamed knee joints was effective at the site.[21] In line with the identification of TNFα and IL-1β as separate targets in animal models of arthritis, it has been convincingly demonstrated that combination therapy with both TNF and IL-1 blockers provides optimal protection.[22]

Role of T Cell Cytokines in Pathology of RA

Rheumatoid arthritis is considered as an Th1-associated disease.[23] However, the factors that initiate and sustain Th1 responses in RA synovium are still not identified. The discovery of new cytokines such as IL-15, IL-17 and RANKL have reconsidered the importance of T cells in the pathology of RA.

IL-15

IL-15 shares many biologic activities with the T cell cytokine IL-2. IL-15 is produced in substantial amounts by macrophages and fibroblasts in the rheumatoid synovial membrane.[24] It may recruit and activate synovial T cells in the relative absence of IL-2.[25] IL-15 induces T cell proliferation, B cell maturation and isotype switching, and may protect T cells from apoptosis.[25,26] In addition, IL-15 has novel activity to stimulate the differentiation of osteoclast progenitors into preosteoclasts.[27] Blocking endogenous IL-15 by a soluble IL-15 receptor α-chain prevents murine collagen-induced arthritis, indicating a role of IL-15 in development of antigen-induced immunopathology.[28] IL-15 recruits and activates CD45RO+ memory T cell subset in the synovial membrane and induces TNFα production in RA.[25,26] Interestingly, these T cell subsets are IL-17 producer cells after stimulation and it has been shown that IL-15 triggers IL-17 production in vitro.[29]

Fig. 11.2. Local IL-17 gene transfer promotes collagen arthrits.

IL-17

IL-17 is a recently discovered cytokine that is secreted by a restricted set of cells, whereas its receptor is ubiquitously expressed on many cell types.[30-32] IL-17 production has been demonstrated in RA synovial tissue[33] and it enhances IL-1 mediated IL-6 production in vitro.[34] The CD4+CD45RO are the major source of IL-17. Th1/Th0, but not Th2 subsets of CD4+ T cell clones isolated from rheumatoid synovium produced IL-17.[35] It is not clear whether IL-17 operates downstream of IL-15 and whether IL-17 has a direct role in T cell activation. The contribution of IL-17 in destructive arthritis was suggested by the fact that the cellular responses induced by IL-17 look similar to that of IL-1. Synergistic effects together with IL-1 and TNFα have been shown.[36] Recently, adenoviral vector mediated overexpression of IL-17 in the knee joint of type II collagen immunized mice was shown to promote destructive collagen arthritis (Fig. 11.2). It induces relatively high levels of IL-1β. Of extreme interest, part of the destructive effect of local overexpression of IL-17 in the knee joints of mice with collagen induce arthritis seems independent of IL-1.[37] Furthermore, amelioration of destructive collagen arthritis was noted after blocking endogenous IL-17 using soluble IL-17 receptor.[37] IL-17 could therefore be a novel target for the treatment of destructive arthritis and this may have implications for tissue destruction in other autoimmune diseases as well.

RANKL

T cell IL-17 may be a crucial cytokine for osteoclastic bone resorption in vitro via RANKL expression.[38-40] Osteoclasts are potent bone resorbing cells and RANKL has been shown to be a key regulator of osteoclastogenesis.[39] RANKL binds to its receptor, RANK (receptor activator of nuclear factor κB) inducing NFkB activation via TRAF 6.[41] The decoy receptor OPG binds with the soluble and cell-bound forms of RANKL and thus prevents their interaction with, and stimulation of, RANK (Fig. 11.3).[42-45] The RANKL/RANK/OPG balance seems of crucial importance in osteoclastogenesis and the bone erosion process during RA.[46] Immunohistochemical and in situ hybridization studies have localized RANKL expression to T cells within lymphoid aggregates of inflamed synovial tissues in patients with RA.[47-49] RANKL mRNA and protein were also detected in synovial fibroblasts from RA patients, [48, 49] and these fibroblasts promoted osteoclastogenesis when stimulated with 1,25-dihydroxyvitamin D_3. This was mediated by increase in RANKL and a decrease in OPG production and could be abrogated by administration of OPG.[49] In CIA, RANKL expression was found in synovial infiltrating mononuclear cells, fibroblast-like cells and chondrocytes.[50,51] In vivo it was demonstrated that neutralization of RANKL by daily injections of recombinant OPG completely prevents bone and joint abnormalities in rat adjuvant arthritis, without interfering with the inflammatory process.[39] However, recently a RANKL-independent role of TNF in osteoclastogenesis in vitro has

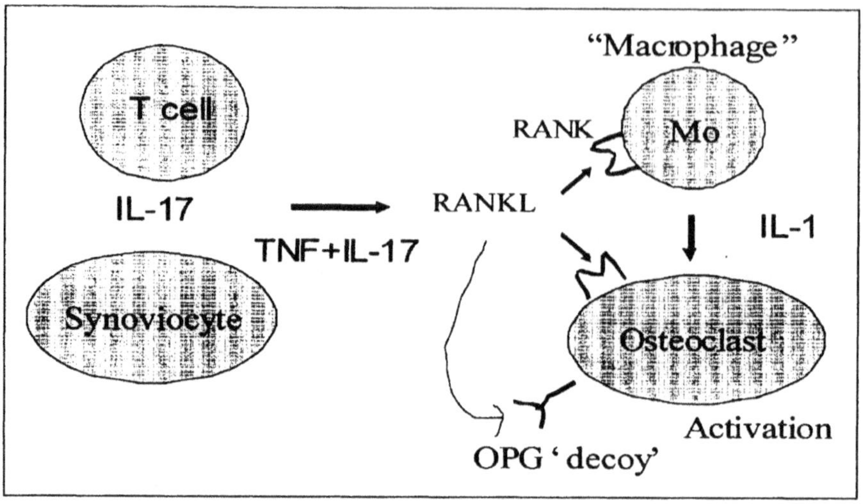

Fig. 11.3. Schematic overview of mediators involved in osteoclastogenesis and bone erosion.

been reported. Further studies in vivo are needed to evaluate the relation between the proinflammatory cytokines IL-17, IL-1, TNF and the RANKL/RANK/OPG pathway.

IL-12/IL-18

The production of the pro-inflammatory cytokines IL-1 and TNF is influenced by other cytokines. Disease promoting mediators can on the one hand induce or sustain direct production of IL-1 and TNF or on the other hand propagate arthritis via Th1 immune-stimulatory activity. IL-12 and the novel cytokine IL-18 (and IL-15 see above) have been shown to be potent Th1-driving cytokines, but can also induce the production of TNF and IL-1 in a T cell independent way. Administration of IL-12 during the early onset of collagen-induced arthritis accelerated onset and enhanced severity.[52] Blocking endogenous IL-12 during onset using specific antibodies inhibited the onset of CIA, indicating that IL-12 is a pivotal mediator in the expression of CIA. However, continued treatment did not suppress established arthritis. Instead, these mice showed marked exacerbation of arthritis shortly after cessation of anti-IL-12 treatment, implying impairment of endogenous control. Enhanced expression of IL-1β and TNFα was noted in the synovium. Treating established CIA with recombinant mIL-12 suppresses the arthritis. Elevated levels of IL-10 seems responsible for this effect, since the anti-inflammatory effects of IL-12 is reversed by anti-IL-10 treatment. This dual role of IL-12 in early and late stages of CIA needs subtle tuning of IL-12-directed therapy in human arthritis.

Another pivotal cytokine for the development of Th1 responses is the recently discovered proinflammatory cytokine IL-18.[53] IL-18 is a member of the IL-1 family of proteins and has been demonstrated in RA synovium.[53] Synergistic activity was noted with IL-12 and IL-15 in sustaining both Th1 responses (IFNγ) and monokine production in RA.[53] Both articular chondrocytes and osteoblasts express IL-18. Mice lacking IL-18 revealed reduced incidence and severity of collagen-induced arthritis.[54] This was accompanied by reduced Ag-specific proliferation and pro-inflammatory cytokine (IFNγ, TNFα, IL-6 and IL-12) production by spleen and lymph node cells in response to bovine type II collagen in vitro, paralleled in vivo by a significant reduction in serum anti-CII IgG2a Ab level. Interestingly, blockade of endogenous IL-18 in murine streptococcal cell wall-induced arthritis revealed an IFNγ-independent role of IL-18.[55] Significant suppression of local TNFI and IL-1 was found under these conditions,

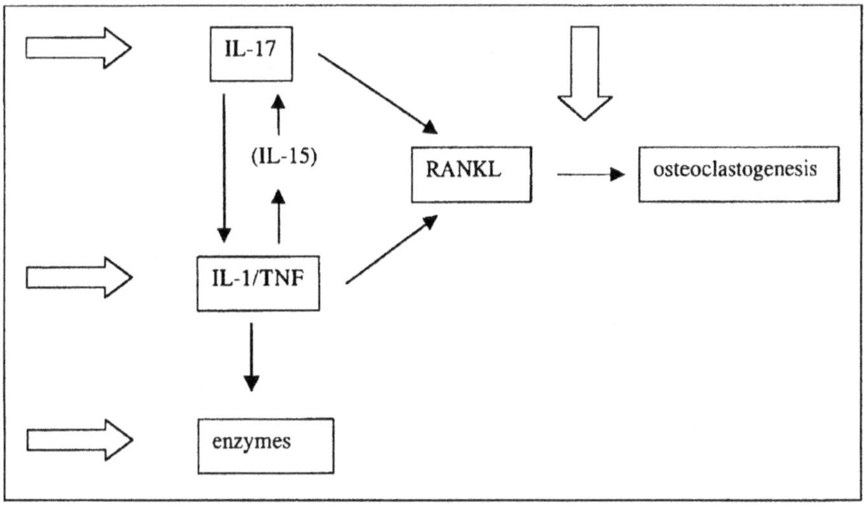

Fig. 11.4. Potential targets of IL-4.

indicating regulation of these proinflammatory cytokines by IL-18. Blocking IL-18 could therefore represent a new therapeutic approach that warrants further testing in the clinic.

Regulation by IL-4/IL-10

Apart from direct interference with TNF and IL-1, regulation of arthritis can also be exerted at the level of modulatory cytokines, such as interleukin-4 (IL-4) and interleukin-10 (IL-10). These regulatory mediators can inhibit Th1 cell activity by suppressing IFNγ expression. In addition, they may have a direct inhibitory effect on the macrophage activity in the synovium. Both actions will lead to less IL-1 and TNFα production in the synovium. Moreover, IL-4 and IL-10 may upregulate natural inhibitors of IL-1 and TNFα, such as IL-1 receptor antagonist (IL-1Ra), soluble TNFα receptor (sTNFαR), and tissue inhibitor of metalloproteinase (TIMP), suggesting surplus value to anti-IL-1/TNFα treatment.

Elevated levels of IL-10 has been shown in the synovial fluid of RA patients. No IL-4 has been found in the synovial fluid of RA patients. In vitro studies have shown that IL-4 and IL-10 regulated the production of IL-1 and TNFα by RA synovial tissue.[56-61]

IL-10 is a dominant suppressive cytokine in the CIA model.[56, 62-68] Blocking both IL-4 and IL-10, however, resulted in the best acceleration of CIA onset. Treatment with IL-10 was only marginally effective, with variation probably linked to variable involvement of endogenous IL-10. Low dose of IL-4 alone did not provoke any effect. Pronounced protection against cartilage destruction was only achieved with combination treatment of IL-4 and IL-10. This cooperatieve effect was noted after early treatment but also occurred when treatment was started during full blown arthritis. The mechanism of protection is linked to suppressed generation of TNFα and IL-1 and upregulation of the IL-1Ra/IL-1β balance in the synovium and, in particular, in the arthritic cartilage.[62] Initial trials with IL-10 were disappointing, and it is expected that in the treatment of RA patients too, IL-10 and IL-4 have to be combined.

IL-4 could not be detected in synovial fluid, synovial supernatants, or synovium of RA patients.[23] This lack of IL-4 is likely to contribute to the uneven Th1/Th2 balance and to the chronic nature of RA. Local IL-4 overexpression in the knee joint of type II collagen immunized mice has been shown to enhance the onset and aggravated the synovial inflammation. However, impressive prevention of chondrocyte death and cartilage erosion was noted.[69]

Chondrocyte proteoglycan synthesis was enhanced in the articular cartilage by local IL-4. Reduction of cartilage erosion was substantiated by lack of expression of the MMP-dependent cartilage proteoglycan breakdown neoepitope VDIPEN in the local IL-4 treated knee joints. The protective effect was associated with a reduction of PMN's in the synovial joint space, decreased NO synthesis, down-regulation of IL-1β and a reduction of the MMP-3/TIMP disbalance in the synovium. Furthermore, IL-4 gene therapy reduced IL-17 and RANKL expression in the synovium and prevents bone erosion.[51] This protective effect was associated with decreased formation of osteoclast-like cells and reduced mRNA levels of cysteine proteinase cathepsin K. Interestingly, IL-4 prevented collagen type I breakdown, but enhanced the formation of type I procollagen in bone samples from RA patients, suggesting promotion of tissue repair. This data suggest that therapeutic strategies that enhance local IL-4 production may protect against cartilage and bone destruction in RA (Fig. 11.4).

References

1. Harris ED, Jr. Rheumatoid arthritis: Pathophysiology and implications for therapy. N Engl J Med 1990; 322:1277-1289.
2. Miossec P, Chomarat P, Dechanet J. Bypassing the antigen to control rheumatoid arthritis. Immunol Today 1996; 17:170-173.
3. Arend WP. The pathophysiology and treatment of rheumatoid arthritis. Arthritis Rheum 1997; 40:595-597.
4. Moreland LW, Heck LW, Jr., Koopman WJ. Biologic agents for treating rheumatoid arthritis. Arthritis Rheum 1997; 40:397-409.
5. Müller-Ladner U, Kriegsmann J, Franklin BN et al. Synovial fibroblast of patients with rheumatoid arthritis attach to and invade normal human cartilage when engrafted into SCID mice. Am J Pathol 1996; 49:1607-1615.
6. Franz JK, Pap T, Müller-Ladner U et al. T cell-independent joint destruction. In: Miossec P, Van den Berg WB, Firestein GS, eds. T Cells in Arthritis. Basel: Birkhäuser Verlag, 1998.
7. Firestein GS, Alvaro-Garcia JS, Maki R. Quantitative analysis of cytokine gene expression in rheumatoid arthritis. J Immunol 1990; 144:3347-3353.
8. Van de Loo AAJ, Van den Berg WB. Effects of murine recombinant IL-1 on synovial joints in mice: Measurements of patellar cartilage metabolism and joint inflammation. Ann Rheum Dis 1990; 49:238-245.
9. Henderson B, Pettipher ER. Arthritogenic actions of recombinant IL-1 and TNF in the rabbit: Evidence for synergistic interactions between cytokines in vivo. Clin Exp Immunol 1989; 75:306-310.
10. Keffer J, Probert L, Cazlaris H et al. Transgenic mice expressing human tumor necrosis factor: A predictive genetic model of arthritis. EMBO Journal 1991; 4025-4031.
11. Georgopoulos S, Plows D, Kollias G. Transmembrane TNF is sufficient to induce localized tissue toxicity and chronic inflammatory arthritis in transgenic mice. J Inflamm 1996; 46:86-97.
12. Probert L, Plows D, Kontogeorgos G et al. The type I IL-1 receptor acts in serie with TNFα to induce arthritis in TNFα transgenic mice. Eur J Immunol 1995; 25:1794-1797.
13. Williams RO, Feldmann M, Maini RN. Anti-tumor necrosis factor ameliorates joint disease in murine collagen-induced arthritis. Proc Natl Acad Sci USA 1992; 89:9784-9788.
14. Wooley PH, Dutcher J, Widmer MB et al. Influence of a recombinant human soluble tumor necrosis factor receptor FC fusion protein on type II collagen-induced arthritis in mice. J Immunol 1993; 151:6602-6607.
15. Joosten LAB, Helsen MMA, Van de Loo FAJ et al. Anticytokine treatment of established type II collagen-induced arthritis in DBA/1 mice: A comparative study using anti-TNFα, anti-IL-1α/β, and IL-1Ra. Arthritis Rheum 1996; 39:797-809.
16. Mori L, Iselin S, de Libero G et al. Attenuation of collagen-induced arthritis in 55-kDa TNF receptor type I (TNFRI)-IgGI-treated and TNFRI-deficient mice. J Immunol 1996; 157:3178-3182.
17. Ghivizzani SC, Kang R, Georgescu HI et al. Constitutive intra-articular expression of human IL-1β following gene transfer to rabbit synovium produces all major pathologies of human rheumatoid arthritis. J Immunol 1997; 159:3604-3612.
18. Niki Y, Yadmada H, Kikuchi T et al. Membrane associated IL-1 contributes to chronic synovitis in human IL-1α transgenic mice. Arthritis Rheum 1998; 41:S212.
19. Van den Berg WB, Joosten LAB, Helsen MMA et al. Amelioration of established murine collagen induced arthritis with anti-IL-1 treatment. Clin Exp Immunol 1994; 95:237-243.

20. Ku G, Faust T, Laufer LL et al. IL-1β converting enzyme inhibition blocks progression of type II collagen induced arthritis in mice. Cytokine 1996; 8:377-386.
21. Bakker AC, Joosten LAB, Arntz OJ et al. Prevention of murine collagen-induced arthritis in the knee and ipsilateral paw by local expression of human IL-1Ra protein in the knee. Arthritis Rheum 1997; 40:893-900.
22. Van den Berg WB. Anti-cytokine therapy in chronic destructive arthritis. Arthritis Res 2001; 3:18-26.
23. Miossec P, van den Berg WB. Th1/Th2 cytokine balance in arthritis. Arthritis Rheum 1997; 40:2105-2115.
24. McInnes IB, Leung BP, Stuock RD et al. Interleukin-15 mediates T cell-dependent regulation of tumor necrosis factor-α production in rheumatoid arthritis. Nature Med 1997; 3:189-195.
25. McInnes IB, Al-Mughales J, Field M et al. The role of interleukin-15 in T-cell migration and activation in rheumatoid arthritis. Nat Med 1996; 2:175-182.
26. McInnes IB, Liew FY. Interleukin-15: A proinflammatory role in rheumatoid arthritis synovitis. Immunol Today 1998; 19:75-79.
27. Ogata Y, Kukita A, Kukita T et al. A novel role of IL-15 in the development of osteoclasts: Inability to replace its activity with IL-2. J Immunol 1999; 162:2754-2760.
28. Ruchatz H, Leung BP, Wei X et al. Soluble IL-15 receptor α-chain administration prevents murine collagen-induced arthritis: A role for IL-15 in development of antigen-induced immunopathology. J Immunol 1998; 160:5654-5660.
29. Ziolkowska M, Koch A, Luszczykiewics G et al. High levels of IL-17 in rheumatoid arthritis patients: IL-15 triggers in vitro IL-17 production via cyclosporin A-sensitive mechanism. J Immunol 2000; 164:2832.
30. Fossiez F, Djossou O, Chomarat P et al. T-cell IL-17 induces stromal cells to produce proinflammatory and hematopoietic cytokines. J Exp Med 1996; 183:2593.
31. Yao Z, Painter SL, Fanslow WC et al. Human IL-17: A novel cytokine derived from T cells. J Immunol 1995; 155:5483.
32. Yao Z, Fanslow WC, Seldin MF et al. Herpesvirus Saimiri encodes a new cytokine IL-17, which binds to a novel cytokine receptor. Immunity 1995; 3:811.
33. Chabaud M, Durand JM, Buchs N et al. Human interleukin-17. A T cell-derived proinflammatory cytokine produced by the rheumatoid synovium. Arthritis Rheum 1999; 42:963.
34. Chabaud M, Fossiez F, Taupin JL et al. Enhancing effect of IL-17 on IL-1-induced IL-6 and leukemia inhibitory factor production by rheumatoid arthritis synoviocytes and its regulation by Th2 cytokines. J Immunol 1998; 161:409.
35. Aarvak T, Chabaud M, Miossec P et al. IL-17 is produced by some proinflammatory Th1/Th0 cells but not by Th2 cells. J Immunol 1999; 162:1246.
36. Chabaud M, Lubberts E, Joosten L et al. IL-17 derived from juxta-articular bone and synovium ontributes to joint degradation in rheumatoid arthritis. Arthritis Res 2001; 3:168-177.
37. Lubberts E, Joosten LAB, Oppers B et al. IL-1 independent role of IL-17 in synovial inflammation and joint destruction during collagen induced arthritis. J Immunol 2001; in press.
38. Kotake S, Udagawa N, Takahashi N et al. IL-17 in synovial fluids from patients with rheumatoid arthritis is a potent stimulator of osteoclastogenesis. J Clin Invest 1999; 103:1345.
39. Kong YY, Feige U, Sarosi I et al. Activated T cells regulate bone loss and joint destruction in adjuvant arthritis through osteoprotegerin ligand. Nature 1999; 402:304.
40. Van Bezooijen RL, Farih-Sips HCM, Papapoulos SE et al. Interleukin-17: A new bone acting cytokine in vitro. J Bone Min Res 1999; 14:1513.
41. Schwandner R, Yamaguchi K, Cao Z. Requirement of tumor necrosis factor receptor-associated factor (TRAF)6 in interleukin-17 signal transduction. J Exp Med 2000; 191:1233.
42. Quinn JMW, Elliott J, Gillespie MT et al. A combination of osteoclast differentiation factor and macrophage-colony stimulating factor is sufficient for both human and mouse osteoclat formation in vitro. Endocrinology 1998; 139:4424-4427.
43. Fuller K, Wong B, Fox S et al. TRANCE is necessary and sufficient for osteoblast-mediated activation of bone resorption in osteoclasts. J Exp Med 1998; 188:997-1001.
44. Burgess TL, Qian Y, Kaufman S et al. The ligand for osteoprotegerin (OPGL) directly activates mature osteoclasts. J Cell Biol 1999; 145:527-538.
45. Jimi E, Akiyama S, Tsurukai T et al. Osteoclast differentiation factor acts as a multifunctional regulator in murine osteoclast differentiation and function. J Immunol 1999; 163:434-442.
46. Hofbauer LC, Heufelder AE. The role of osteoprotegerin and receptor activator of nuclear factor kB ligand in the pathogenesis and treatment of rheumatoid arthritis. Arthritis Rheum 2001; 44:253-259.
47. Horwood NJ, Kartsogiannis V, Quinn JMW et al. Activated T cells support osteoclast formation in vitro. Biochem Biophys Res Commun 1999; 265:144-150.

48. Gravallese EM, Manning C, Tsay A et al. Synovial tissue in rheumatoid arthritis is a source of osteoclast differentiation factor. Arthritis Rheum 2000; 43:250-258.
49. Takayanagi H, Iizuka H, Juji T et al. Involvement of receptor activator of nuclear factor kB ligand/ osteoclast differentiation factor in osteoclastogenesis from synoviocytes in rheumatoid arthritis. Arthritis Rheum 2000; 43:259-269.
50. Romas E, Bakharevski O, Hards DK et al. Expression of osteoclast differentiation factor at sites of bone erosion in collagen-induced arthritis. Arthritis Rheum 2000; 43:821-826.
51. Lubberts E, Joosten LAB, Chabaud M et al. IL-4 gene therapy for collagen arthritis suppresses synovial IL-17 and osteoprotegerin ligand and prevents bone erosion. J Clin Invest 2000; 105:1697-1710.
52. Joosten LAB, Lubberts E, Helsen MMA et al. Dual role of IL-12 in early and late stages of murine collagen type II arthritis. J Immunol 1997; 159:4094-4102.
53. Gracie AJ, Forsey RJ, Chan WL et al. A proinflammatory role for IL-18 in rheumatoid arthritis. J Clin Invest 1999; 104:1393.
54. Wei XQ, Leung BP, Arthur HML et al. Reduced incidence and severity of collagen-induced arthritis in mice lacking IL-18. J Immunol 2001; 166:517-521.
55. Joosten LAB, Van de Loo FAJ, Lubberts E et al. An IFN-γ-independent proinflammatory role of IL-18 in murine streptococcal cell wall arthritis. J Immunol 2000; 165:6553-6558.
56. Katsikis PD, Chu C-Q, Brennan FM et al. Immuno regulatory role of interleukin-10 (IL-10) in rheumatoid arthritis. J Exp Med 1994; 179:1517-1527.
57. Llorente L, Richaud-Patin Y, Kior R et al. In vivo production of interleukin-10 by non-T cells in rheumatoid arthritis, Sjögren's syndrome, and systemic lupus erythematosus. A potential mechanism of B lymphocyte hyperactivity and autoimmunity. Arthritis Rheum 1994; 37:1647-1655.
58. Cush JJ, Splawski JB, Thomas R et al. Elevated interleukin-10 levels in patient with rheumatoid arthritis. Arthritis Rheum 1995; 38:96-104.
59. Cohen SBA, Katsikis PD, Chu C-Q et al. High level of interleukin-10 production by the activated T cell population within the rheumatoid synovial membrane. Arthritis Rheum 1995; 38:946-952.
60. Chomarat P, Vannier E, Dechanet J et al. Balance of IL-1 receptor antagonist/IL-1J in rheumatoid synovium and its regulation by IL-4 and IL-10. J Immunol 1995; 154:1432-1439.
61. Isomäki P, Luukkainen R, Saario R et al. Interleukin-10 functions as an antiinflammatory cytokine in rheumatoid synovium. Arthritis Rheum 1996; 39:386-395.
62. Joosten LAB, Lubberts E, Durez P et al. Role of interleukin-4 and interleukin-10 in murine collagen-induced arthritis. Protective effect of interleukin-4 and interleukin-10 treatment on cartilage destruction. Arthritis Rheum 1997; 40:249-260.
63. Lubberts E, Joosten LAB, Helsen MMA et al Regulatory role of interleukin-10 in joint inflammation and cartilage dstruction in murine streptococcal cell wall arthritis. More therapeutic benefit with IL-4/IL-10 combination therapy than with IL-10 treatment alone. Cytokine 1998; 10:361-369.
64. Lubberts E, Joosten LAB, Van den Bersselaar L et al. Intra-articular IL-10 gene transfer regulates the expression of collagen-induced arthritis in the knee and ipsilateral paw. Clin Exp Immunol 2000; 120:375-383.
65. Walmsley M, Katsikis PD, Abney E et al. Interleukin-10 inhibition of the progression of established collagen-induced arthritis. Arthritis Rheum 1996; 39:495-503.
66. Apparaily F, Verwaerde C, Jacquet C et al. Adenovirus-mediated transfer of viral IL-10 gene inhibits murine collagen-induced arthritis. J Immunol 1998; 160:5213-5220.
67. Ma Y, Thornton S, Duwel LE et al. Inhibition of collagen-induced arthritis in mice by viral IL-10 gene transfer. J Immunol 1998; 16:1516-1524.
68. Whalen JD, Lechman El, Carlos CA et al. Adenoviral transfer of the viral IL-10 gene periarticularly to mouse paws suppresses development of collagen-induced arthritis in both injected and uninjected paws. J Immunol 1999; 162:3625-3632.
69. Lubberts E, Joosten LAB, Van den Bersselaar L et al. Adenoviral vector-mediated overexpression of IL-4 in the knee joints of mice with collagen-induced arthritis prevents cartilage destruction. J Immunol 1999; 163:4546-4556.

CHAPTER 12

Cytokines and Chemokines in Virus-Induced Autoimmunity

Urs Christen and Matthias G. von Herrath

Introduction

Virus infections usually elicit a massive inflammatory reaction characterized by release of chemokines and cytokines that attract and activate cells of the host's immune system with the goal to eliminate the foreign pathogen from the organism. In addition to the load and presentation of viral antigens, the distinct profile of chemokines and cytokines is characteristic for an individual virus infection and therefore determines the pattern of cells that infiltrate into the infected tissue or organ and the magnitude and type of the anti-viral immune response. Because viral infections induce strong cellular and humoral immune responses, their association with autoimmune diseases including type 1 diabetes, keratitis, and multiple sclerosis has been proposed and forms the basis for some animal models of autoimmune disease, such as the RIP-LCMV model for type 1 diabetes. In these experimental systems, their ability to induce diabetes,[1,2] keratitis[3] or allergic encephalomyelitis[4] either through direct T cell cross-reactivity or cytokine/chemokine mediated 'bystander' activation of autoreactive processes has essentially been demonstrated. However, their association with human disorders has never been conclusively proven. A more recent concept proposes that the association of viral infections with autoimmunity is complex in so far as viruses can likely enhance as well as ameliorate an ongoing (for example, genetically determined) autoimmune process rather than initiating and causing all organ damage by themselves. An interesting and central question is whether and how the type of cytokine and/or chemokine profile induced by a viral infection can influence its ability to enhance or abrogate autoimmunity. In this chapter we will discuss, focused on experimental scenarios in type 1 diabetes, how this can occur, for example, by over-expression of a single cytokine, such as TNFα, or by superimposing a second viral infection on an already established auto-aggressive process. These insights should allow in the long run a better understanding of the possible pathways involved in the immunopathogenesis of human type 1 diabetes and identification of viral infections that enhance the auto-aggressive response.

Cytokines and Chemokines as 'Conductors' of the Immune Response

Cytokines and chemokines play an important role in orchestrating inflammatory reactions. In response to external stimuli such as viral infections, these effector molecules act synergistically as 'conductors' and coordinate both timing and location of effector functions by the individual cell populations of the immune defense system. On the one hand there are chemokines; these chemoattractant cytokines are released by resident macrophages or endothelial cells after they are activated, for example by cytokines such as TNFα or IFNγ. The pattern of chemokine secretion then determines the composition of immune competent cells (macrophages, dendritic cells, CD8 and CD4 T-cells, and B-cells) that infiltrate an infected tissue or organ (for review see 5-8). On the other hand there are pro-inflammatory cytokines (TNFα,

Cytokines and Chemokines in Autoimmune Disease, edited by Pere Santamaria.

IL-1 or IFNγ), which constitute a major factor that influences activation, differentiation, and proliferation of an inflammatory cell population. The immune response to a particular virus or other pathogen is often characterized by a distinct pattern of cytokine production. For example, the so-called Th1-type cytokines (IFNγ, TNFα, IL-2 and IL-12) produced by CD4 T cell subsets drive the immune system toward a predominantly cell-mediated response targeting the clearance of intracellular organisms. In contrast, Th2-type cytokines (IL-4, IL-5, IL-10 and IL-13) favor a humoral/allergic type response by stimulating the differentiation of B-lymphocytes, mast cells, and eosinophils.[9,10]

Cytokines have been implicated in a variety of autoimmune diseases and were demonstrated to have both beneficial as well as exacerbating effects. In insulin-dependent diabetes mellitus (IDDM, type 1 diabetes) cytokines are one of the key factors that locally influence the islet-inflammatory reaction. It has been proposed that Th1-type cytokines have mainly prodiabetic effects and enhance the autoimmune process.[11,12] In contrast, Th2-type cytokines are thought to have a more regulatory function.[12,13] A third group of cytokines termed Th3 (TGF-β) might have regulatory functions as well,[14] especially after oral administration of islet (self)-antigens.[15,16] However, experiments have shown that this is not a 'black and white' situation and Th2 cytokines might actually enhance or maintain inflammatory processes in certain situations[17,18] and Th1 cytokines might conversely lead to the termination of an inflammation[19,20] possibly through induction of activation-dependent cell death.[21] Therefore, more than the cytokine profile will characterize a truly diabetogenic or regulatory autoreactive lymphocyte, a consideration that has to be taken into account when designing immune interventions based on cytokine secretion and/or administration.

Cytokines and Chemokines in Autoimmune Type 1 Diabetes

Spontaneous animal models are widely used for studying the etiology of type 1 diabetes and provide a very important tool to gain insight into critical mechanisms that lead to autoimmunity and to develop possible treatments for human IDDM. The most commonly used spontaneous model for IDDM is the nonobese diabetic (NOD) mouse that was discovered in 1980 by researchers at the Shinogi Company.[22] The NOD mouse model shares many clinical and immunological features with human IDDM, including the appearance of autoantibodies against similar islet (self)-antigens, disease susceptibility genes (MHC alleles), intra-islet infiltration of mixed lymphocyte populations (insulitis), and the dependence on autoreactive T lymphocytes (for reviews see 23,24). In the NOD model an increased production of proinflammatory cytokines (TNFα, IL-1 IFNγ) and Th1 cytokines (IFNγ, TNFβ, IL-2 and IL-12) was associated with β-cell destructive (malignant) insulitis, whereas enhanced expression of Th2 cytokines (IL-4 and IL-10) and Th3 cytokine TGFβ was associated with a nondestructive (benign) insulitis.[12,25] These findings led to the generation of several models where a broad variety of cytokines were either administered systemically to regular NOD mice or overexpressed in an islet-specific manner in rat insulin promoter transgenic NOD mice. Islet-specific expression turned out to be the most reliable form of cytokine delivery since systemic administration or expression may result in additional sometimes opposing effects on the overall lymphocyte repertoire and its effector functions and development that are difficult to sort out. In addition, IDDM is an organ-specific autoimmune disease that is mostly restricted to the destruction of the pancreatic islets of Langerhans, and therefore it is important to examine the influence of cytokines that are present locally in the islets rather than observing 'diluted' systemic effects. As expected islet-specific expression of some proinflammatory and Th1 cytokines resulted in an increase in diabetes incidence and/or in an acceleration of diabetes onset in nondiabetes prone mice (IFNγ;[26,27] IL-2;[28,29] TNFα[30]) or in NOD mice (TNFα;[31,32] IL-2[33]). In addition, islet-specific expression of Th2 and Th3 cytokines was demonstrated to induce only peri-insulitis that did not progress to diabetes in nondiabetes prone mice (TGFβ[34]) or to slow down or prevent diabetes in NOD mice (IL-4;[35-37] TGFβ[38]). In contrast, some cytokines did not behave as expected or had opposing effects on diabetes development in different models. For example, TNFα was found to prevent diabetes when expressed in adult NOD mice[39] (the influence of TNFα on type 1 diabetes will be discussed below in detail) and IL-10 turned out to rather enhance than abrogate diabetes when expressed in the islets of Langerhans.[40-42]

Mouse models in which transgene-encoded 'target-antigens' are expressed in the pancreatic β-cells, such as the RIP-LCMV[1,2] and the INS-HA[43,44] mouse, have demonstrated that the presence of autoaggressive T cells alone is not enough to cause disease. For example, when RIP-LCMV mice were crossed with mice expressing an inactive mutated form of IFNγR the diabetes incidence was drastically reduced.[45] The role of chemokines and cytokines in the RIP-LCMV model will be discussed below. Mice expressing the hemagglutinin (HA) of the influenza virus under control of the insulin promotor (INS-HA) as a target antigen and a transgenic anti-HA TcR suffer either from insulitis only or from insulitis as well as diabetes.[46,47] Interestingly, IFNγ transcription levels were significantly higher in islets from diabetic mice, while TNFα was expressed at a higher level in nondiabetic mice.[48] These results indicate that IFNγ might be particularly important to drive the autoaggressive response (β-cell destruction) in 'antigen-specific' models for type 1 diabetes.

The Role of Cytokines and Chemokines in the RIP-LCMV Transgenic Mouse Model for Autoimmune Diabetes

It is important to note here that investigations analyzing the role of messenger molecules, such as chemokines and cytokines, that are extremely focused in terms of both location and time of action require a clearly defined model system that allows the precise dissection of changes in chemokine and/or cytokine expression relative to the current stage of immunopathogenesis. Spontaneous models have the major disadvantage that the starting point of the autoimmune process is poorly defined. In general, the NOD mouse develops insulitis at 3-4 weeks of age and progresses to destruction of insulin-producing β-cells and subsequent clinically overt diabetes by 4-6 months of age. However, it is very difficult if not impossible to predict the onset of diabetes in a given individual NOD mouse that is under observation. Furthermore, the incidence of diabetes ranges from 10%-40% in males up to 80%-90% in females.[23] The RIP-LCMV transgenic mouse model for type 1 diabetes offers an attractive alternative to the NOD mouse. The initiation of immunopathogenesis can be precisely set by infection of these mice with the lymphocytic choriomeningitis virus (LCMV) that induces a strong anti-viral and simultaneously anti-islet transgene (autoreactive) CD4 and CD8 response modeling the breaking of self-tolerance to a transgenically expressed viral protein in the pancreatic β-cells of the host. In addition, the target antigen (transgene) is clearly defined allowing the specific tracking of autoaggressive T lymphocytes with experimental tools, such as specific MHC-peptide-tetramers and the autoimmune attack is focused to the β-cells in the islets of Langerhans. In contrast, additional generalized autoimmune effects are present in the NOD mouse system (sialitis, orchitis and inducibility of EAE) indicating that the NOD mouse more accurately reflects human diabetes cases that suffer from polyendocrinopathies.[49] Further, the tracking of autoreactive T cell responses in the NOD mouse has incurred a significant amount of variability comparing different laboratories and investigators, an issue that is only now being resolved by joint T cell workshops and has been making comparison of experimental findings sometimes difficult.[50]

Since viruses can cause massive inflammation in infected tissues, they have the potential to break tolerance against localized auto-antigens and are therefore thought of as good candidates for initiating or enhancing autoimmunity. For example, virally induced activation of antigen presenting cells (APCs) leads to subsequent presentation of CNS auto-antigens and autoimmunity in Theiler's virus infected mice.[51] Viruses have been implicated in the initiation and/or precipitation of a broad variety of autoimmune diseases such as autoimmune (type 1) diabetes,[52-54] Herpes stromal keratitis[3] and multiple sclerosis.[4] There are several lines of evidence that link viruses or other microbes to the development of autoimmune diabetes: First, entero, rubella and mumps viruses could be detected and isolated from the pancreatic islets of Langerhans[55] and some strains of these viruses are known to infect and replicate in islets in vitro (reviewed in 53,56). It was demonstrated that β-cell destruction can result from direct viral lysis[57,58] or from indirect bystander killing likely mediated by antiviral cytokines. Alternatively, antiviral T lymphocytes that cross-react with islet-cell antigens (molecular mimicry[59]) or become bystander activated[60] have been proposed to exist.[61,62] In addition, nonlytic viruses could possibly persist and replicate in β-cells and alter their function without killing them.[54,63] Further, inoculation of mice with virus lead to autoimmune diabetes in multiple experimental models.[52,56,64] However, viruses have been also implicated in preventing autoimmune disease.[65,66]

Based on that extensive body of literature the following scenario for the pathogenesis of IDDM was hypothesized and formed the basis for establishing a virus-induced animal model for type 1 diabetes as an alternative to study the etiology and pathogenesis of human diabetes. First, restricted but low level expression of self, environmental, or viral antigen occurs in β-cells of the islets of Langerhans. This event by itself does not cause disease, since the host is hyporesponsive or tolerant to the antigen. Tolerance can be achieved by thymic expression of the self or viral antigen or through peripheral tolerance mechanisms.[67-70] Later in life, a triggering event occurs, which is exposure to the same environmental factor, infection with the same virus or pathogen or with cross-reacting antigenic determinants.[71] The result is an immune response to the virus that eventually localizes to the β-cells and progresses to IDDM after a lag period. This scenario was experimentally reconstructed in the late 1980s by the laboratories of Michael Oldstone[2] and Rolf Zinkernagel and colleagues.[1] Both groups used a rat insulin promoter (RIP) to create separate lines of transgenic mice whose pancreatic β-cells expressed either the nucleoprotein (NP) or the glycoprotein (GP) of the lymphocyte choriomeningitis virus (LCMV) as defined target antigen. The expression of the target (self)-antigen does not lead to β-cell dysfunction, islet cell infiltration, hyperglycemia, or spontaneous activation of autoreactive (anti-LCMV) lymphocytes.[71] However, infection with LCMV results in autoimmune diabetes in >95% of RIP-LCMV mice. In contrast, nontransgenic littermates never develop diabetes or insulitis after LCMV challenge.[71] The concept of the RIP-LCMV mouse model is displayed in Figure 12.1. Hence the RIP-LCMV model has become a very useful tool to further delineate events leading to type 1 diabetes and, in particular, understand the possible role of viral infections in its etiology.

Just as proposed for human type 1 diabetes, the onset of diabetes in RIP-LCMV mice depends on the action of both autoreactive CD4 and CD8 lymphocytes and correlates with the numbers of auto-aggressive lymphocytes generated. In accordance, the incidence of disease varied between the individual transgenic lines ranging from 2 weeks (RIP-GP lines) to 1-6 months (RIP-NP lines). Further studies revealed the mechanism involved is the rapid compared to the slow progressive diabetes. Transgenic lines expressing the LCMV-GP transgene exclusively in the β-cells of the islets manifested rapid-onset IDDM, usually 10-14 days after viral challenge.[71] T lymphocytes developed normally and had equivalent cytotoxic T lymphocyte (CTL) activity to splenic lymphocytes from nontransgenic age- and sex-matched littermates. In these lines the high systemic numbers of auto-aggressive CD8 lymphocytes were sufficient to induce diabetes and did not require help from CD4 cells. In contrast, in lines expressing the LCMV-NP transgene in both the b-cells and in the thymus, IDDM took longer to occur after subsequent LCMV challenge. Several lines of evidence indicated that the anti-self (viral) CTL were of lower affinity and that CD4 lymphocytes were essential to generate anti-self (viral) CD8 lymphocyte-mediated IDDM of adult transgenic mice.[71]

Chemokines and Cytokines Are Expressed Early After LCMV Infection and Are Important in Breaking Tolerance

Uninfected RIP-LCMV mice are perfectly healthy and show no signs of diabetes, such as elevated blood glucose values or insulitis. Thus, these mice seem to immunologically view the transgenic viral proteins (NP or GP) as a true self-component. Tolerance (unresponsiveness) to these viral (self) molecules is due to the fact that under normal circumstances LCMV proteins

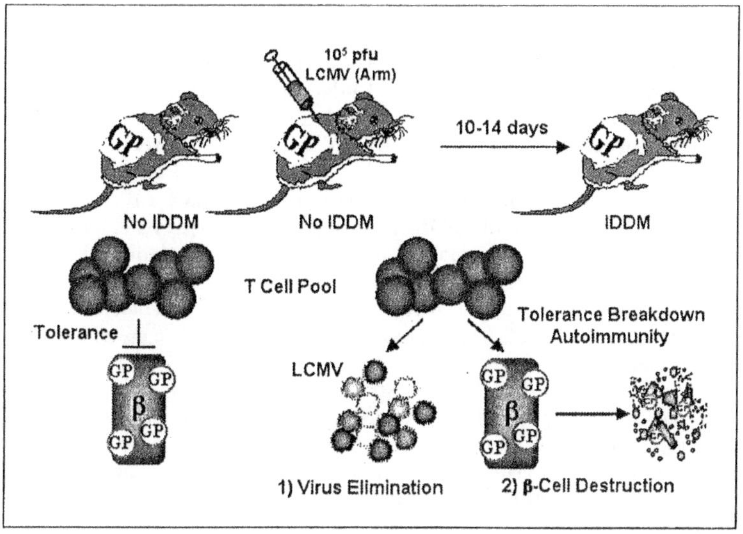

Fig. 12.1. The RIP-LCMV transgenic mouse model for type 1 diabetes. Transgenic mice express the glycoprotein (GP) or the nucleoprotein (NP) of the lymphocytic choriomeningitis virus (LCMV) in the β-cells of the pancreatic islets of Langerhans. At that stage the mice do not develop disease since they tolerate the viral proteins as self components. However, self (viral) tolerance can be broken by intraperitoneal injection of 10^5 pfu LCMV strain Armstrong. RIP-LCMV mice are mounting an anti-LCMV immune response and develop at a stage distinctly after viral clearance (as displayed for RIP-GP mice 10-14 days post-infection) diabetes as a result of autoimmune destruction of β-cells.

are not expressed on and presented by professional activated APCs. It can only be broken after infection with LCMV, which directly leads to the presentation of viral (self) antigens on APCs such as dendritic cells. The anti-viral immune response is also accompanied by a massive inflammation that occurs independently from the presence of the transgene and is intended to eliminate the virus. Only in the transgenic RIP-LCMV mice the anti-LCMV response continues in a kinetically distinct step to eradicate an additional target resembling virally infected APCs which are the β-cells expressing the transgenic LCMV proteins. Therefore it is important to clearly distinguish between direct viral effects and immunopathogenic events that ultimately lead to β-cell destruction and diabetes.

Analysis of pancreatic chemokine and cytokine mRNA expression levels by RNase protection assay (RPA) revealed a burst of chemokine expression immediately after LCMV infection. Crg-2, the mouse homologue to human IP-10, could be detected as early as day 1 post-infection. The magnitude of its expression already decreased at day 2 post-infection and was reduced to preinfection levels after 4-7 days. Similarly, Mig, a close relative to Crg-2, is expressed very early (days 2-4 post-infection) and decreased to background levels after 7 days (Christen and von Herrath, unpublished observations). It is around that time (days 5-7 post-infection) that most of the virus is already eliminated from the pancreas and cannot easily be detected in standard screening analyses, such as virus plaque assays. It is important to point out that LCMV infects the pancreas but only rarely the islets themselves.[72] Interestingly, pancreatic mRNA expression of a variety of chemokines and cytokines peaked around the same time when increased numbers of infiltrating lymphocytes into the pancreatic tissue are present (day 7 post infection). In contrast, early production of chemokines is more likely due to the presence of virally-infected and consequently activated pancreatic resident macrophages, dendritic cells and endothelial cells. Peak expression levels of the chemokines RANTES, MIP-1a, Eotaxin, and MCP-1 and the cytokines TNFα, IFNγ, TGFβ, LTβ, and IL-1 were detected in RIP-LCMV and nontransgenic C57BL/6 mice around this time, which is 7 days post-infection (Christen and von Herrath, unpublished observations). Expression of all chemokines and most cytokines returned to preinfection levels 10-14 days after LCMV infection in nontransgenic C57BL/6 as well as RIP-LCMV transgenic mice, with the exception of IFNγ and to a certain extent TNFα that remained high in RIP-LCMV mice until diabetes developed. Especially IFNγ expression was still found to be elevated in RIP-LCMV (GP) mice after 21 days, a time at which diabetes is already ongoing, indicating an important role for IFNγ in β-cell destruction and immunopathogenesis of IDDM.[45] In contrast to IFN-γ, TNFα levels did not differ very much comparing transgenic and nontransgenic LCMV infected mice. However, the presence of elevated TNFa levels early (days 5-8) post infection is crucial for breaking of tolerance to the self (viral) proteins expressed in b-cells since blockade of TNFα bioactivity via treatment with neutralizing TNFR55-IgG1 fusion protein resulted in complete prevention of IDDM in the RIP-GP line.[73]

No difference was found in mRNA expression patterns for any chemokine comparing nontransgenic C57BL/6 mice and RIP-LCMV mice. Similar to the situation with TNFα, this does not mean that chemokines are not important for the initiation of the auto-aggressive process in RIP-LCMV transgenics. It however suggests that the role of chemokines in the RIP-LCMV model might be restricted to the initial attraction of immune competent cells to the location of infection and, simultaneously to the area of the later-developing auto-aggressive process. Once the self-tolerance to the transgenically expressed LCMV proteins is broken, chemokines might not be as important as during the initiation stage of disease. The effect of an individual chemokine is unequivocally closely linked to the attracted cell that expresses the corresponding receptor. Crg-2, expressed very early and highly after LCMV infection, was demonstrated to attract predominantly T lymphocytes and monocytes.[74] However, CXCR3, the cell surface receptor for Crg-2 and Mig, is expressed predominantly on Th1 type lymphocytes and NK cells.[75] There is now growing evidence for an association of specific chemokines with Th1 and Th2-type immune responses, and the differential expression of chemokine receptors on Th1 and Th2 cells is well established (for example CXCR3 and CCR5 on Th1 cells and CCR3, CCR4, and CCR8 on Th2 cells).[7,8] Therefore, considering the early expression of the Th1 chemokine Crg-2 after LCMV infection as an isolated event, blocking of its bioactivity should lead to a reduced attraction and activation of Th1 cells during the anti-viral response. However, when RIP-LCMV are treated with a neutralizing anti-Crg-2 antibody immediately after LCMV infection, the onset of diabetes and its frequency are not significantly changed indicating that the network of chemokines and cytokines involved in the initial immune response against the viral infection is complex and highly regulated (Christen and von Herrath, unpublished observations).

Thus, during the initiation stage of diabetes in RIP-LCMV mice, chemokines and cytokines are very important to attract and activate immune competent cells to the site of infection in order to ensure efficient virus elimination. Chemokines are one of the major factors that maintain a continuous supply of leukocytes infiltrating the infected organ or tissue, and their importance in anti-viral defense is further underlined by the fact that some viruses even express chemokine and chemokine receptor homologues in order to neutralize or even exploit the chemokine mediated cell attraction system.[76] Chemokines may be also important for initiation autoimmunity in the BDC T cell receptor transgenic NOD mouse model. Adoptive transfer of islet specific BDC2.5 TcR transgenic Th1- or Th2-type CD4 lymphocytes from BDC mice to immunodeficient NOD.*scid* recipients resulted in infiltration of the pancreas by both cell types but only Th1-type cells caused diabetes. Interestingly, only Th1-type but not Th2-type CD4 lymphocytes secreted multiple chemokines including lymphotactin, MIP-1α and β, MCP-1, Crg-2/IP-10, and RANTES indicating that a polarized chemokine expression in the pancreas can elicit a destructive inflammatory infiltration that initiates disease.[77]

Table 12.1. Diabetogenic events after LCMV-infection of RIP-GP mice

Stage	Time	Effects	Dominating Factors
Initiation	Days 0-7	Breakdown of	Chemokines (IP-10, RANTES)
		self tolerance *No diabetes*	Inflammatory cytokines (TNFα, IFNγ)
Expansion	Days 4-10	Initial inactivation and destruction of some β-cells *No diabetes*	Autoaggressive CD8 cells Perforin (IFNγ)
Termination	Days 10-14	Inactivation and destruction of the majority of β-cells *Diabetes*	Autoaggressive CD8 cells IFNγ (Perforin)

Regulation of β-Cell Destruction by Cytokines in RIP-LCMV Mice

After extensive studies over the past 10 years with the RIP-LCMV mouse model, the following scenario can be proposed for the pathogenesis of viral-induced type 1 diabetes (Table 12.1 and Fig.12.2): During a first stage (initiation), antigen presenting cells (APCs) in the islets are activated and later on will be required for locally driving expansion and activation of autoreactive lymphocytes. APC activation is achieved through local presence of the viral (LCMV) infection in the pancreas but not necessarily the islets. Virally induced chemokines attract the first auto-aggressive lymphocytes that reach the islets at day 4-7-post infection where they are further propagated by activated APCs and maybe cytokines such as TNFα and IFNγ. In a second stage (expansion), autoreactive CD8 cells kill some β-cells by perforin-mediated cytolysis resulting in additional presentation of islet antigens by APCs. At that time one would think that complete β-cell destruction and IDDM are unavoidable and will develop immediately concurrent with systemic elimination of the LCMV infection. However, it is clear from several studies discussed in the following that destruction of the majority of β-cells occurs in a third stage distinctly AFTER viral (LCMV) clearance, thus representing a true 'hit and run' event with respect to the autoimmune process in the islets. That third stage (islet-destructive terminal stage) can take from 1 week (RIP-GP line) up to 2 months or more (RIP-NP line). Importantly, rather than perforin, inflammatory cytokines such as IFN-γ that can act directly on β-cells and, in conjunction with TNF and maybe IL-1β can induce nitric oxide, are required for β-cell destruction.[45]

Importantly, the degree of APC activation and inflammation can be regulated even after b-cell destruction has already begun. In RIP-NP mice with slow-onset IDDM the majority of lymphocytes found in or around the islets produce IL-4 prior to complete islet destruction.[16] This profile shifts in favor of IFN-γ around the time when clinical IDDM develops;[16] similar findings were made in NOD mice[13,23,78] and humans.[79] Consequently, it is possible to differentiate between 'benign' (maybe Th2-like profile) and 'malignant' (maybe Th1-like) insulitis.

Typical Th2-type cytokines, such as IL-4 and IL-10, were investigated in the past for their immune-regulatory potential to prevent and/or control autoimmune diseases. Whereas IL-4 prevented autoimmune diseases in various animal models, IL-10 was successfully used to prevent experimental allergic encephalomyelitis (EAE) and collagen induced arthritis (CIA), but not type 1 diabetes.[80] In mouse models of type 1 diabetes, such as the NOD and RIP-LCMV, IL-10 had paradoxical effects. IL-10 was found to be beneficial in young NOD mice treated

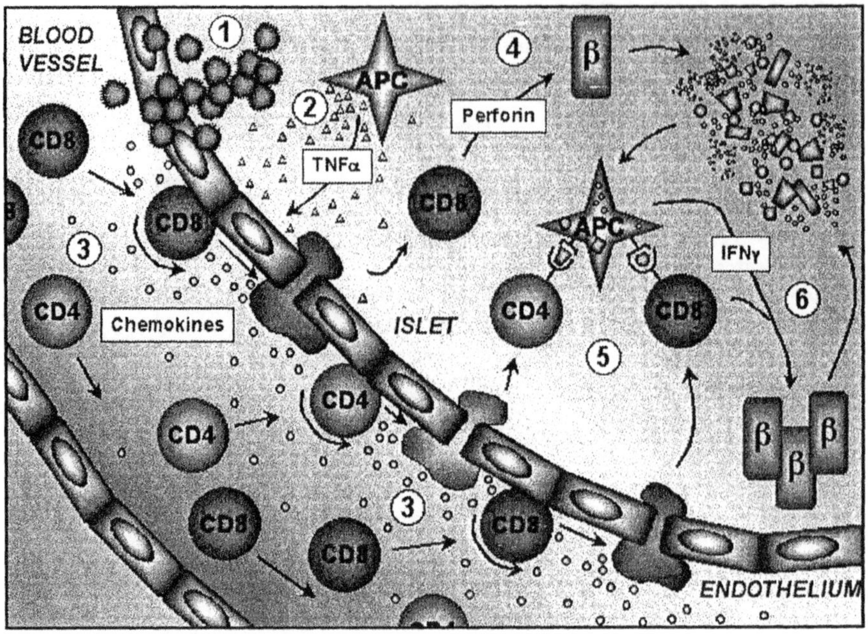

Fig. 12.2 Immunopathogenesis associated with virus-induced diabetes. After LCMV infection active virus is found in the pancreas but not necessarily in the islets (1). In an initial stage local resident macrophages or dendritic cells are activated and release pro-inflammatory cytokines, such as TNFα, IFNγ, and IL-1 (2). Chemokines, including Crg-2/IP-10, Mig, and RANTES, are released by macrophages and by endothelial cells that were activated by TNFα and/or IFNγ and attract a mixed population of lymphocytes that are able to roll along and migrate through the activated endothelium (3). The first self (viral) specific CD8 cells destroy some β-cells by perforin dependent cytolysis resulting in release of β-cell antigens (4). These antigens include transgenic viral proteins and additional nonviral components and are processed and presented to infiltrated CD8 and CD4 cells by antigen presenting cells (APC) (5). In the terminal stage of immunopathogenesis the majority of β-cells are being destroyed by autoreactive CD8 cells in a IFNγ dependent manner (6). It is only in this final stage where clinically overt diabetes is apparent and can be assessed by blood glucose measurements.

with recombinant IL-10,[81] with an IL-10/Fc fusion protein[82] and adoptive transfer of IL-10 transduced islet-specific Th1 clones was found to be protective as well.[83] In contrast, treatment with anti-IL-10 mAb prevented insulitis in young NOD mice[84] and transgenic expression of IL-10 in either pancreatic α- or β-cells accelerated diabetes.[40-42, 85-87] In the RIP-LCMV mouse model, it was demonstrated very recently that regulatory CD4 cells induced by oral treatment with porcine insulin were protective upon adoptive transfer and produced IFNγ, IL-4 and IL-10, whereas nonprotective cells secreted IFNγ only.[88]

IFNγ seems to play an important role in the final stages of IDDM pathogenesis. In the RIP-GP mice as well as in wildtype C57BL/6 mice, infection with LCMV leads to a production of TNFα and IFNγ that reach their highest levels at day seven after infection. However, in C57BL/6 the production of these cytokine mRNAs ceases around days 10-14 post-infection. In contrast, RIP-GP mice still produce IFNγ in islets even after 21 days.[45] These findings suggest that after initial β-cell damage caused by LCMV-(GP) specific cytotoxic CD8 lymphocytes it is IFNγ that is responsible for complete β-cell death in the final stages of diabetes (terminal stage of IDDM). The source of this pancreatic IFNγ is probably not only restricted to LCMV-(GP) specific T lymphocytes but could also involve T-cells reactive to other islet antigens at that final stage.

TNFα, like IFNγ, is classically referred to as a 'pro-inflammatory' cytokine and a vast literature is available reviewing the role of TNFα as a mediator and/or promoter of inflammation (see for example[89] or[90] for review). Anti-TNFα antibodies and TNFR-IgG fusion-proteins that neutralize biologically active TNFα are currently being successfully used for the treatment of rheumatoid arthritis patients.[91,92] However, the role of TNFα in the pathogenesis of IDDM is still controversially discussed and results obtained from several different animal models are often contradictory and range from abrogation to acceleration of disease. A summary of recent publications in this field is given in Table 12.2 and some critical observations will be discussed below. For example, constitutive TNFα expression in islets of transgenic RIP-TNFα mice leads to profound insulitis but not diabetes.[93] Only if the costimulatory molecule B7.1 is coexpressed with TNFα in the islets does clinically overt diabetes develop.[30] These experiments clearly show that expression of TNFα by itself is not sufficient to induce disease in this model. However, in connection with enhanced presentation of islet antigens, self-tolerance can be broken. It was demonstrated very recently that the duration of initial TNFα expression is essential for the progress to diabetes in the TNFα/B7.1 model. Using an inducible repression/derepression system for TNFα expression (Tet-TNFα transgenic mice) it was possible to determine the crucial time window important for the fate of diabetes pathogenesis in TNFα/B7.1 mice.[94] In the NOD mouse model, TNFα had a dual role depending on its time of expression. In neonatal transgenic RIP-TNFα NOD mice, expression of TNFα resulted in acceleration of spontaneous diabetes due to enhanced presentation of β-cell antigens to islet infiltrating CD4 as well as CD8 T lymphocytes.[31,32,95] In contrast, transgenic RIP-TNFα lines that express TNFα only later in life had a reduced activity of autoreactive T-cells (Th1 and Th2 type) and were protected from spontaneous diabetes.[39] Experiments with systemic administration of TNFα revealed similar findings. Whereas early administration of TNFα enhanced diabetes in NOD mice,[96,97] late administration during diabetes development could abrogate the disease process probably by affecting expansion, migration, and function of autoreactive lymphocytes.[98,99] Taken together these results obtained from several laboratories suggest that TNFα plays clearly a dual role in the initiation, propagation, and/or regulation of the ongoing autoimmune process that ultimately leads to IDDM and its precise function appears to critically depend on the timing of expression.

How TNFα Can Enhance or Abrogate an Ongoing Autoimmune Process in RIP-LCMV Mice

In order to dissect the function of TNFα during the pathogenesis in relation to the exact time of its expression, one is in need of a system where both the ongoing autoimmune process as well as the expression of TNFα can be precisely manipulated. By crossing RIP-LCMV-GP mice with Tet-TNFα mice[94] we were able to control (i) the onset of the autoimmune process by infection with LCMV and (ii) the expression of TNFα in the β-cells of the pancreatic islet of Langerhans by removal of doxycycline from the diet. Tet-TNFα mice express TNFα via the tTA-system under the control of a tetracycline sensitive promoter system. The resulting RIP-GP-TNFα mice[73] were therefore bred in the presence of the tetracycline derivative doxycycline (Dox) to block transgenic TNFα expression. Dox was removed at several times after the onset of the autoimmune process (infection with LCMV) to induce β-cell specific TNFα expression. Because the chronology of immunopathological events in the islets of RIP-GP mice after initiation of IDDM by infection with LCMV is know in great detail,[72] TNFα could be expressed at the times of:

1. initiation of autoimmunity,
2. expansion and propagation of autoreactive lymphocytes, and
3. clinically overt disease.[73]

As expected the exact time of TNFα expression was very important for the resulting influence of TNFα on diabetes pathogenesis in LCMV-infected RIP-GP-TNFα mice and revealed a dual role of TNFα. Early expression (at the time of LCMV infection) enhanced the frequency of

Table 12.2 Controversial role of TNFα on IDDM pathogenesis

Model	Delivery	Effect	Reference
NOD (adult)	Chronic systemic recTNFα treatment	Reduced islet infiltration and blocking of IDDM	Jacob, 1990[98]
NOD (adult)	Neutralizing anti-TNFα ab treatment	Enhanced insulitis and accelerated IDDM	Jacob, 1992[99]
NOD (adult)	Chronic systemic recTNFα treatment	Blocking of IDDM by attenuation of TcR signalling	Cope, 1997[96]
NOD (adult, diabetic)	Systemic administration of recTNFα	Prolongs survival of islet grafts by downregulation of Th1 type cytokines	Rabinovitch, 1997[20]
NOD (neonatal)	Systemic administration of recTNFα	Earlier onset and higher incidence of IDDM via potentiated development od autoreactive T cells	Yang, 1994[97]
NOD (neonatal)	Neutralizing anti-TNFα ab treatment	Blocking of IDDM due to unresponsiveness to autoantigens	Yang, 1994[97]
RIP-TNFα NOD (adult)	b-cell specific expression of tgTNFα	Blocking of IDDM by prevention of development of autoreactive T cells	Grewal, 1996[39]
RIP-TNFα NOD (neonatal)	b-cell specific expression of tgTNFα	Accelerated IDDM by enhanced islet antigen presentation	Green, 1998[32]
RIP-TNFα	b-cell specific expression of tgTNFα	Massive insulitis but no IDDM	Picarella, 1993[93]
RIP-TNFα x RIP-B7.1	b-cell specific expression of tgTNFα and B7.1	Massive insulitis and spontaneous IDDM	Guerder, 1994[30]
Tet-TNFα x RIP-B7.1 (neonatal)	inducible b-cell specific expression of tgTNFα (RIP-B7.1 background)	IDDM only when TNFα is expressed for at least 21 days after birth	Green, 2000[94]
RIP-GP-TNFα (adult)	b-cell specific expression of tgTNFα <u>*early*</u> during pathogenesis	Higher IDDM incidence because of enhanced inflammation	Christen, 2001[73]
RIP-GP-TNFα (adult)	b-cell specific expression of tgTNFα <u>*late*</u> during pathogenesis	Reversion from IDDM to normal most likely due to apoptosis of auto-reactive CD8 lymphocytes	Christen, 2001[73]

diabetic mice. Since transgenic TNFα is expressed in 85-90% of mice and appears as early as four days after removal of Dox from the diet, early expression of transgenic TNFα is superimposed onto the endogenous TNFα produced in response to the viral challenge itself, thus leading to a higher magnitude of inflammation and further enhanced infiltration of islets and subsequently to a higher IDDM incidence (Fig.12.3). In contrast, late expression (at day 10-14 post-infection—a time where most animals are already diabetic) resulted in a significant decrease in IDDM incidence. At the same time the frequency of 'revertant' mice was increased. Such 'revertant' mice were initially diabetic (blood glucose > 300 mg/dl) at weeks 2 and 3 post-infection, but reverted to nondiabetic blood glucose values (<200 mg/dl) after week 4-5.[73] The occurrence of 'revertant' mice was an exciting phenomenon since TNFα could not only reduce the frequency of autoimmune disease but also abrogate the actual ongoing autoimmune process at a very late time when its clinical features were already obvious. Such 'revertant' RIP-GP-TNFα mice were further demonstrated to have anti-self (GP) specific CD8 T lymphocytes with a significantly lower cytolytic activity towards LCMV-infected or GP-peptide coated target cells and the frequency of experienced CD8 T lymphocytes in the pancreatic draining lymph node was reduced.[73] Furthermore, the presence of cells undergoing apoptosis in the pancreatic islets of 'revertant' RIP-GP-TNFα was increased. These results suggested a possibility for TNFα to induce apoptosis in auto-aggressive, experienced CD8 T cells when expressed late during pathogenesis of autoimmune diabetes resulting in restoration of normal function in b-cells (Fig.12.3). Apoptosis is a very plausible mechanism for TNFα-induced diabetes abrogation since TNFα can directly interact with TNFR55 (TNFR1, CD120a), a potent cell surface receptor capable of mediating signaling towards apoptosis.[100-103] It was previously reported by other groups that TNFα could induce apoptosis of CD8 as well as CD4 T cells via TNFR55.[104,105]

Late expression of TNFα in the RIP-GP-TNFα mouse model reduced the incidence of diabetes to ~35% and, additionally, reversion from clinically overt diabetes to normoglycemia occurred in up to 50% of the mice. An interesting question is why not all mice 'reverted' to normal blood sugar levels. The answer is rather speculative at the moment. It is clear from some of our experiments assessing the level of TNFα expression after removal of doxycycline from the diet that not all mice expressed elevated levels of transgenic TNFα even though they expressed the TNFα transgene and the tetO regulatory element of the tTA-system.[73] In addition, the actual time-point of availability of bioactive TNFα after removal of Dox might differ among individual mice. Considering the close succession of immunopathological events within the islet and their temporal proximity to the actual onset of clinically overt diabetes, differences as little as 1-3 days in TNFα availability and/or pathogenesis might be crucial. One scenario might be that a combination of early diabetes onset and late TNFα availability would lead to enhanced physiological stress to b-cells preventing them from producing not only insulin but also transgenic TNFα and therefore reducing the possibility of TNFα induced apoptosis of autoreactive T cells.

The Role of Cytokines and Chemokines in Viral Infections and Their Potential Interference with Autoimmunity

Viral infections were implicated as the possible cause of autoimmunity[52,53,106,107] but were also reported to prevent or abrogate autoimmune disease.[65,66] One possible mechanism by which viral infections could maybe predictably influence autoimmunity is the distinct pattern of cytokines and/or chemokines that results as a direct consequence of the infection per se. For example infection with LCMV has a protective effect in regular NOD mice when given during ongoing autoimmunity. Development of diabetes was aborted possibly due to a direct effect of LCMV on a subset of CD4 T lymphocytes.[65,66] In contrast, infection with Coxsackie virus could enhance diabetes development when given to BCD2.5 transgenic NOD mice[108] or to regular NOD mice during a susceptible phase of disease progress where their auto-reactive T cells reached a critical mass.[60] This was likely due to the fact that Coxsackie infection leads to

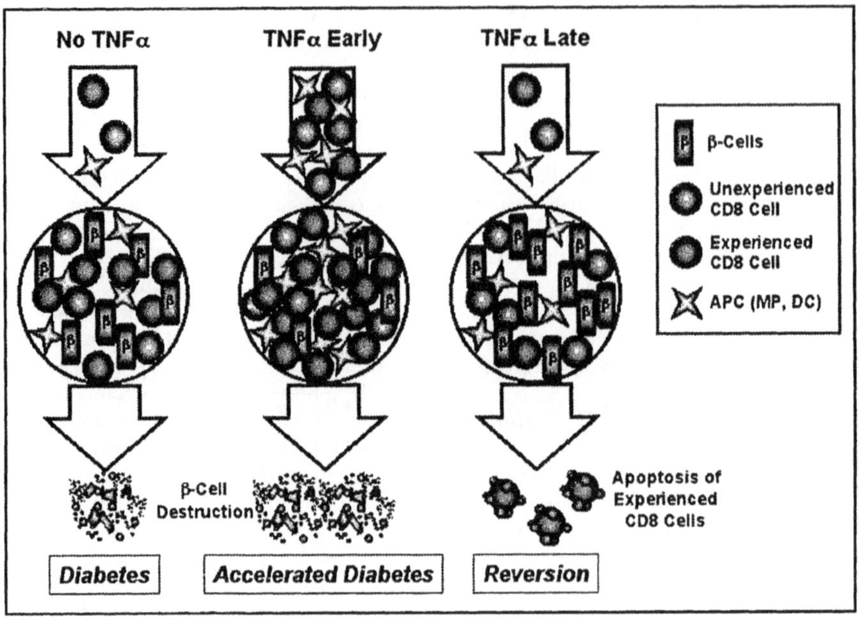

Fig. 12.3. Dual role of TNFα on the pathogenesis of type 1 diabetes. In the absence of transgenically expressed TNFα (RIP-GP-TNFα mice with blocked TNFα expression) diabetes occurs after 10-14 days in ~75-80% of mice that were infected with LCMV. Destruction of β-cells is mediated by autoreactive, inexperienced (predominantly IFNγ expressing) CD8 cells which are continuously activated to an experienced (IFNγ and TNFα expressing) state. When transgenic TNFα is expressed *early* (at the time of LCMV infection), it is superimposed onto endogenous TNFα that is expressed as a direct consequence of infection. This intensified inflammation at a critical time at which tolerance against self (viral) proteins (GP) is broken leads to enhanced islet infiltration and increased frequency of diabetic animals (90-95%). Even in the absence of transgenic TNFα, expression of endogenous TNFα at that stage is a prerequisite for autoimmunity since its neutralization by TNFR55-IgG1 completely abrogates disease. When TNFα is expressed *late* (at a time where the autoimmune process is already ongoing), diabetic RIP-GP mice can revert to a nondiabetic state by a mechanism that most likely involves direct induction of apoptosis of experienced (IFNγ and TNFα expressing) CD8 cells. The activation cycle between autoreactive CD8 cells and APC in the islets or in the pancreatic draining lymph node is broken and autoimmunity is abrogated.

the production of much higher levels of inflammatory cytokines than LCMV and thus enhances disease.[109] In addition, Coxsackie infection at an even lower dose (10^3 pfu) causes a more rapid pancreatic expression of mRNA species of several inflammatory mediators when compared to infection with LCMV (at a dose of 10^5 pfu). This acceleration (and augmentation) of inflammation includes:

1. Chemokines, such as IP-10 (Crg-2), RANTES, MIP-1α and MIP1β, MIP-2, and eotaxin,
2. Inflammatory cytokines including TNFα, IFNβ, IFNγ, and IL-1, and
3. Th1 cytokines like IL-12 and IL-15, but also
4. Regulatory (Th2 / Th3-type) cytokines such as IL-10 and TGFβ (Christen and von Herrath, unpublished observations).

On the flipside, an additional inflammatory focus caused by a viral infection may have a similar effect on an ongoing autoimmune process as the direct expression of a transgenic cytokine, such as TNFα, and therefore, apoptosis of auto-aggressive lymphocytes might be a mechanism involved in virally induced abrogation of autoimmunity as well. A major problem in the investigation of viral-induced damage or modulation in humans is usually that the causative agent (virus) might have been cleared from the organism long before its effect becomes evident. This "hit and run" event might leave no "footprints" behind in the affected organ(s) that would suggest the former presence of a virus. The RIP-LCMV-NP mouse is therefore an ideal model system to examine viral modulation of an ongoing autoimmune process. After infection with LCMV it takes in average about 2-3 months until IDDM develops[71] which is enough time for the virus to be cleared by the immune system and for testing the effect of a sequential infections on the pathogenesis of IDDM. Interestingly in our model system, the ongoing destruction of b-cells triggered by the primary infection with LCMV could be prevented by sequential infection with a very high dose of the same LCMV (Armstrong) strain ($5x10^6$ (i.v.)) after 4 weeks. In contrast, secondary infection with the same titer used for the primary infection ($1x10^5$ (i.p.)) had no effect on the disease progress.[110] Further experiments demonstrated that viral clearance was unaffected, and therefore no viral persistence occurred in such repeatedly infected RIP-NP mice. However, the local viral titer in the pancreas and draining lymph node was considerably higher after the secondary infection that abrogated autoimmune diabetes and in parallel resulted in reduction of islet infiltration. In contrast, islets of mice secondarily infected with low dose LCMV-Armstrong remained heavily infiltrated[110] along with the development of clinical diabetes. These results indicated that the higher antigenic/viral load and/or inflammation may recruit autoreactive CD8 and CD4 T lymphocytes away from the target location of the autoimmune response (islets). Additionally and as mentioned previously in this chapter, viral infections lead to the generation of inflammatory cytokines, such as TNFα and IFNγ. Therefore, as with the RIP-GP-TNFα model where TNFα expression late during pathogenesis is likely to abrogate IDDM through induction of apoptosis in autoreactive experienced CD8 T lymphocytes, the occurrence of apoptosis was examined in diabetes-prone RIP-LCMV-NP mice that were 'rescued' from autoimmune diabetes by a late secondary high-dose infection. Indeed, an increased frequency of apoptotic cells in the pancreatic draining lymph nodes, as evidenced by the well-established TUNEL staining method, was observed in both high-dose Armstrong as well as low-dose Pasteur treated RIP-NP mice. In contrast, mice infected with low-dose Armstrong had a decreased frequency of apoptotic cells similar to RIP-NP mice that did not receive a secondary infection.[110] Regulatory lymphocytes seem to have only a minor (if any) influence on the rescue from autoimmunity because protection could not be transferred to similarly infected RIP-NP recipients and no regulatory cytokines (IL-4, IL-10) were found to be increased in rescue double-infected mice. However, it will be of great interest to find out whether cytokines, such as TNFα and/or IFNβ, are involved in the process of viral-induced abrogation of autoimmunity via induction of apoptosis.

Conclusions

Cytokines can enhance as well as abrogate an ongoing autoimmune process by a variety of effector mechanisms such as their influence on islet infiltration, activation and proliferation of autoreactive lymphocytes, cellular interactions (antigen presentation, TcR signaling), or their ability to directly inactivate cells by inducing apoptosis or anergy. In this chapter we reviewed some of the possible mechanisms by focusing on new findings with the transgenic RIP-LCMV mouse model for type 1 diabetes. In the terminal preclinical stage of diabetes development, when pronounced β-cell destruction is already ongoing, cytokines may play the most important role and their effects range from enhancement to blocking of disease pathogenesis. For example TNFα, traditionally described as a 'pro-inflammatory' cytokine, expressed in the β-cells at that time can even reverse the ongoing immune process that is already clinically apparent. Direct induction of apoptosis and therefore direct physical elimination rather than indirect regulation/suppression of the auto-aggressive lymphocytes that precipitate the disease was identified as a possible mechanism for disease prevention. Similarly, enhanced apoptosis was also involved in virus-induced abrogation of IDDM in the RIP-LCMV system. A secondary LCMV infection at a time where the autoimmune destruction is already ongoing provided a high load of viral (self) antigen and an additional focus of inflammation locally in the pancreas but not in the islets and thus recruited auto-aggressive CD8 and CD4 cells away from the b-cells subsequently resulting in activation-induced cell death.

In a complex network of cellular attractions involving lymphocyte activation and suppression, it seems obvious that various factors influence the effect of one individual cytokine on the fate of an ongoing autoimmune reaction. Time and location appear to be the most important factors. Further, systemic administration of cytokines can lead to an outcome other than expression in a locally restricted area. Similarly, the change in local cytokine concentration over a specific time period and the duration of cytokine exposure may be more effective than the absolute magnitude of expression. Thus, the dissection of mechanisms influencing autoimmunity should use experimental tools, such as animal models or in vitro studies that are very well defined and carefully designed. Based on such experiments we will hopefully soon be able to better understand human diabetes and other autoimmune disorders and plan sufficiently detailed epidemiological studies that could pinpoint deleterious as well as beneficial effects of one viral strain or group of viruses characterized by common cytokine-induction profiles.

References

1. Ohashi P, Oehen S, Buerki K, Pircher H, Ohashi C et al. Ablation of tolerance and induction of diabetes by virus infection in viral antigen transgenic mice. Cell 1991; 65:305-317.
2. Oldstone MBA, Nerenberg M, Southern P, Price J, Lewicki H. Virus infection triggers insulin-dependent diabetes mellitus in a transgenic model: Role of anti-self (virus) immune response. Cell 1991; 65:319-331.
3. Zhao ZS, Granucci F, Yeh L, Schaffer PA, Cantor H. Molecular mimicry by herpes simplex virus-type 1: Autoimmune disease after viral infection. Science 1998; 279:1344-1347.
4. Fujinami RS, Oldstone MB. Amino acid homology between the encephalitogenic site of myelin basic protein and virus: Mechanism for autoimmunity. Science 1985; 230:1043-1045.
5. Baggiolini M, Dewald B, Moser B. Human chemokines: An update. Annu Rev Immunol 1997; 15:675-705.
6. Luster AD. Chemokines—Chemotactic cytokines that mediate inflammation. N Engl J Med 1998; 338:436-445.
7. Rossi D, Zlotnik A. The biology of chemokines and their receptors. Annu Rev Immunol 2000; 18:217-242.
8. Sallusto F, Mackay CR, Lanzavecchia A. The role of chemokine receptors in primary, effector, and memory immune responses. Annu Rev Immunol 2000; 18:593-620.
9. Mosmann TR, Coffman RL. Heterogeneity of cytokine secretion patterns and functions of helper T cells. Adv Immunol 1989; 46:111-147.
10. Coffman RL, Mocci S, O'Garra A. The stability and reversibility of Th1 and Th2 populations. Curr Top Microbiol Immunol 1999; 238:1-12.
11. von Herrath MG, Oldstone MBA. Interferon-γ is essential for destruction of β-cells and development of insulin-dependent diabetes mellitus. J Exp Med 1997; 185:531-539.
12. Liblau RS, Singer SM, McDevitt H. Th1 and Th2 CD4+ T cells in the pathogenesis of organ-specific autoimmune diseases. Immunol Today 1995; 16:34-38.
13. Charlton B, Lafferty KJ. The Th1/Th2 balance in autoimmunity. Curr Opin Immunol 1995; 7:793-798.
14. Han HS, Jun HS, Utsugi T, Yoon JW. Molecular role of TGF-beta, secreted from a new type of CD4+ suppressor T-cell, NY4.2, in the prevention of autoimmune IDDM in NOD mice. J Autoimmun 1997; 10:299-307.
15. Haneda K, Sano K, Tamura G, Sato T, Habu S et al. TGF-β induced by oral tolerance ameliorates experimental tracheal eosinophilia. J Immunol 1997; 159:4686-4690.
16. von Herrath MG, Dyrberg T, Oldstone MBA. Oral insulin treatment suppresses virus-induced antigen-specific destruction of b-cells and prevents autoimmune diabetes in transgenic mice. J Clin Invest 1996:1324-1331.
17. Gianani R, Sarvetnick N. Viruses, cytokines, antigens, and autoimmunity. Proc Natl Acad Sci USA 1996; 93:2257-2259.
18. Lafaille JJ, Van de Kerre F, Hsu AL, Baron JL, Haas W et al. Myelin basic protein-specific T helper 2 (Th2) cells cause experimental autoimmune encephalomyelitis in immunodeficient hosts rather than protect them from the disease. J Exp Med 1997; 186:307-312.

19. McSorley SJ, Soldera S, Malherbe L, Carnaud C, Locksley RM et al. Immunological tolerance to a pancreatic antigen as a result of local expression of TNFα by islet β-cells. Immunity 1997; 7:401-409.
20. Rabinovitch A, Suarez-Pinzon WL, Sorensen O, Rajotte RV, Power RF. TNF-α down-regulates type 1 cytokines and prolongs survival of syngeneic islet grafts in nonobese diabetic mice. J Immunol 1997; 159:6298.
21. Garside P, Steel M, Worthey A, Satoskar A, Alexander J et al. T helper 2 cells are subject to high dose oral tolerance and are not essential for its induction. J Immunol 1995; 154:5649-5655.
22. Makino S, Kunimoto K, Muraoka Y, Mizushima Y, Katagiri K et al. Breeding of a nonobese, diabetic strain of mice. Jikken Dobutsu 1980; 29:1-13.
23. Delovitch TL, Singh B. The nonobese diabetic mouse as a model of autoimmune diabetes: Immune dysregulation gets the NOD. Immunity 1997; 7:727-738.
24. Wong FS, Janeway CA, Jr. Insulin-dependent diabetes mellitus and its animal models. Curr Opin Immunol 1999; 11:643-647.
25. Rabinovitch A. An update on cytokines in the pathogenesis of insulin-dependent diabetes mellitus. Diabetes Metab Rev 1998; 14:129-151.
26. Sarvetnick N, Shizuru J, Liggitt D, Martin L, McIntyre B et al. Loss of pancreatic islet tolerance induced by beta-cell expression of interferon-gamma. Nature 1990; 346:844-847.
27. Sarvetnick N, Liggitt D, Pitts SL, Hansen SE, Stewart TA. Insulin-dependent diabetes mellitus induced in transgenic mice by ectopic expression of class II MHC and interferon-gamma. Cell 1988; 52:773-782.
28. Allison J, Oxbrow L, Miller JF. Consequences of in situ production of IL-2 for islet cell death. Int Immunol 1994; 6:541-549.
29. von Herrath MG, Allison J, Miller JF, Oldstone MBA. Focal expression of interleukin-2 does not break unresponsiveness to "self" (viral) antigen expressed in beta-cells but enhances development of autoimmune disease (diabetes) after initiation of an anti-self immune response. J Clin Invest 1995; 95:477-485.
30. Guerder S, Picarella DE, Linsley PS, Flavell RA. Costimulator B7-1 confers antigen-presenting-cell function to parenchymal tissue and in conjunction with tumor necrosis factor alpha leads to autoimmunity in transgenic mice. Proc Natl Acad Sci USA 1994; 91:5138-5142.
31. Green EA, Wong FS, Eshima K, Mora C, Flavell RA. Neonatal tumor necrosis factor alpha promotes diabetes in nonobese diabetic mice by CD154-independent antigen presentation to CD8(+) T cells. J Exp Med 2000; 191:225-238.
32. Green EA, Eynon EE, Flavell RA. Local expression of TNFα in neonatal NOD mice promotes diabetes by enhancing presentation of islet antigens. Immunity 1998; 9:733-743.
33. Allison J, McClive P, Oxbrow L, Baxter A, Morahan G et al. Genetic requirements for acceleration of diabetes in nonobese diabetic mice expressing interleukin-2 in islet beta-cells. Eur J Immunol 1994; 24:2535-2541.
34. Sanvito F, Nichols A, Herrera PL, Huarte J, Wohlwend A et al. TGF-beta 1 overexpression in murine pancreas induces chronic pancreatitis and, together with TNF-alpha, triggers insulin-dependent diabetes. Biochem Biophys Res Commun 1995; 217:1279-1286.
35. Mueller RT, Sarvetnick N. Pancreatic expression of IL-4 abrogates insulitis and diabetes in NOD mice. J Exp Med 1996; 184:1093-1099.
36. Mueller R, Bradley LM, Krahl T, Sarvetnick N. Mechanism underlying counterregulation of autoimmune diabetes by IL-4. Immunity 1997; 7:411-418.
37. Gallichan WS, Balasa B, Davies JD, Sarvetnick N. Pancreatic IL-4 expression results in islet-reactive Th2 cells that inhibit diabetogenic lymphocytes in the nonobese diabetic mouse. J Immunol 1999; 163:1696-1703.
38. Moritani M, Yoshimoto K, Wong SF, Tanaka C, Yamaoka T et al. Abrogation of autoimmune diabetes in nonobese diabetic mice and protection against effector lymphocytes by transgenic paracrine TGF-beta1. J Clin Invest 1998; 102:499-506.
39. Grewal IS, Grewal KD, Wong S, Picarella DE, Janeway CA et al. Local expression of transgene encoded TNF alpha in islets prevents autoimmune diabetes in nonobese diabetic (NOD) mice by preventing the development of auto-reactive islet-specific T cells. J Exp Med 1996; 184:1963-1974.
40. Wogensen L, Lee MS, Sarvetnick N. Production of interleukin 10 by islet cells accelerates immune-mediated destruction of beta cells in nonobese diabetic mice. J Exp Med 1994; 179:1379-1384.
41. Balasa B, Lee J, Good A, Yeung BT, Sarvetnick N et al. IL-10 impacts autoimmune diabetes via a CD8+ T cell pathway circumventing the recuirement for CD4+ T and B lymphocytes. J Immunol 1998; 161:4420-4427.

42. Balasa B, Van Gunst K, Jung N, Balakrishna D, Santamaria P et al. Islet-specific expression of IL-10 promotes diabetes in nonobese diabetic mice independent of Fas, perforin, TNF receptor-1, and TNF receptor-2 molecules. J Immunol 2000; 165:2841-2849.
43. Lo D, Freedman J, Hesse S, Palmiter RD, Brinster RL et al. A. Peripheral tolerance to an islet cell-specific hemagglutinin transgene affects both CD4+ and CD8+ T cells. Eur J Immunol 1992; 22:1013-1022.
44. Lo D, Reilly CR, Scott B, Liblau R, McDevitt HO et al. Antigen-presenting cells in adoptively transferred and spontaneous autoimmune diabetes. Eur J Immunol 1993; 23:1693-1698.
45. Seewaldt S, Thomas H, Ejrnaes M, Christen U, Wolfe T et al. Virus-induced autoimune diabetes: Most β-cells die through inflammatory cytokines and not perforin from autoreactive (anti-viral) CTL. Diabetes 2000; 49:1801-1809.
46. Degermann S, Reilly C, Scott B, Ogata L, von Boehmer H et al. On the various manifestations of spontaneous autoimmune diabetes in rodent models. Eur J Immunol 1994; 24:3155-3160.
47. Scott B, Liblau R, Degermann S, Marconi LA, Ogata L et al. A role for nonMHC genetic polymorphism in susceptibility to spontaneous autoimmunity. Immunity 1994; 1:73-83.
48. Sarukhan A, Lanoue A, Franzke A, Brousse N, Buer J et al. Changes in function of antigen-specific lymphocytes correlating with progression towards diabetes in a transgenic model. Embo J 1998; 17:71-80.
49. Bach JF. Insulin dependent diabetes mellitus as an autoimmune disease. Endocrine Reviews 1994; 15:516-542.
50. Atkinson M, Honeyan M, Peakman M, Roep B. T-cell markers in Type 1 diabetes: Progress, prospects and realistic expectations. Diabetologia 2000; 43:819-820.
51. Katz-Levy Y, Neville KL, Girvin AM, Vanderlugt CL, Pope JG et al. Endogenous presentation of self myelin epitopes by CNS-resident APCs in Theiler's virus-infected mice. J Clin Invest 1999; 104:599-610.
52. Yoon JW. The role of viruses and environmental factors in the induction of diabetes. Curr Top Microbiol Immunol 1990; 164:95-123.
53. Notkins AL, Yoon JW. Virus-induced diabetes mellitus. In: Notkins AL, Oldstone MBA, eds. Concepts in Viral Pathogenesis. New York: Springer-Verlag, 1984:241.
54. Menser MA, Forrest JM, Bransby RD. Rubella infection and diabetes mellitus. Lancet 1978; 1:57-60.
55. Yoon JW, Austin M, Onodera T, Notkin, AL. Virus-induced diabetes mellitus: Isolation of a virus from the pancreas of a child with diabetic ketoacidosis. N Engl J Med 1979; 300:1173-1179.
56. Yoon JW. A new look at viruses in type 1 diabetes. Diabetes-Metabolism Reviews 1995; 11:83-107.
57. Prince GA, Jenson AB, Billups LC, Notkins AL. Infection of human pancreatic beta cell cultures with mumps virus. Nature 1978; 271:158-161.
58. Jenson AB, Rosenberg HS, Notkins AL. Pancreatic islet-cell damage in children with fatal viral infections. Lancet 1980; 2:354-358.
59. Oldstone MBA. Molecular mimicry and autoimmune disease. Cell 1987; 50:819-820.
60. Serreze DV, Ottendorfer EW, Ellis TM, Gauntt CJ, Atkinson MA. Acceleration of type 1 diabetes by Coxsackie virus infection requires a pre-existing critical mass of autoreactive T-cells in pancreatic islets. Diabetes 2000; 49:708-711.
61. Tian J, Lehmann PV, Kaufman DL. T cell cross-reactivity between coxsackievirus and glutamate decarboxylase is associated with a murine diabetes susceptibility allele. J Exp Med 1994; 180:1979-1984.
62. Atkinson MA, Bowman MA, Campbell L, Darrow BL, Kaufman DL et al. Cellular immunity to a determinant common to glutamate decarboxylase and Coxsackie virus in insulin-dependent diabetes. J Clin Invest 1994; 94:2125-2129.
63. Patterson K, Chandra RS, Jenson AB. Congenital rubella, insulitis, and diabetes mellitus in an infant [letter]. Lancet 1981; 1:1048-1049.
64. von Herrath MG, Oldstone MBA. Virus-induced autoimmune disease. Curr Opin Immunol 1996; 8:878-885.
65. Oldstone MBA. Prevention of type I diabetes in nonobese diabetic mice by virus infection. Science 1988;239: 500-502.
66. Oldstone MBA. Viruses as therapeutic agents. I. Treatment of nonobese insulin-dependent diabetes mice with virus prevents insulin-dependent diabetes mellitus while maintaining general immune competence. J Exp Med 1990; 171:2077-2089.
67. Kappler JW, Roehm N, Marrack P. T cell tolerance by clonal elimination in the thymus. Cell 1987; 49:273-280.
68. Kisielow P, Bluthmann H, Staerz UD, Steinmetz M, von Boehmer H. Tolerance in T-cell-receptor transgenic mice involves deletion of nonmature CD4+8+ thymocytes. Nature 1988; 333:742-746.
69. Schwartz RH. A cell culture model for T lymphocyte clonal anergy. Science 1990; 248:1349-1356.

70. Sprent J, Gao EK, Webb SR. T cell reactivity to MHC molecules: Immunity versus tolerance. Science 1990; 248:1357-1363.
71. von Herrath MG, Dockter J, Oldstone MBA. How virus induces a rapid or slow onset insulin-dependent diabetes mellitus in a transgenic model. Immunity 1994; 1:231-242.
72. von Herrath MG, Holz A. Pathological changes in the islet milieu precede infiltration of islets and destruction of b-cells by autoreactive lymphocytes in a transgenic model of virus-induced IDDM. J Autoimmun. 1997; 10:231-238.
73. Christen U, Wolfe T, Moehrle U, Hughes AC, Green A et al. A dual role for TNFα in type I diabetes: Islet specific expression abrogates the ongoing autoimmune process when induced late but not early during pathogenesis. J Immunol 2001; 166:(in press).
74. Taub DD, Lloyd AR, Conlon K, Wang JM, Ortaldo JR et al. Recombinant human Interferone-inducible protein 10 is a chemoattractant for human monocytes and T lymphocytes and promotes T cell adhesion to endothelial cells. J Exp Med 1993; 177:1809-1814.
75. Loetscher M, Gerber B, Loetscher P, Jones SA, Piali L et al. Chemokine receptor specific for IP10 and Mig: Structure, function, and expression in activated lymphocytes. J Exp Med 1996; 184:963-969.
76. Lalani AS, Barrett JW, McFadden G. Modulating chemokines: More lessons from viruses. Immunol Today 2000; 21:100-106.
77. Bradley LM, Asensio VC, Schioetz LK, Harbertson J, Krahl T et al. Islet-specific Th1, but not Th2, cells secrete multiple chemokines and promote rapid induction of autoimmune diabetes. J Immunol 1999; 162:2511-2520.
78. Cameron MJ, Arreaza GA, Zucker P, Chensue SW, Streiter RM et al. IL-4 prevents insulitis and insulin-dependent diabetes mellitus in nonobese diabetic mice by potentiation of regulatory T helper-2 cell function. J Immunol 1997; 159:4686-4692.
79. Berman MA., Sandborg CI, Wang ., Imfeld KL, Zaldivar F Jr. et al. Decreased IL-4 production in new onset type 1 insulin-dependent diabetes mellitus. J Immunol 1996; 157:4690-4696.
80. Pearson CI, McDevitt HO. Redirecting Th1 and Th2 responses in autoimmune disease. Curr Top Microbiol Immunol 1999; 238:79-122.
81. Pennline KJ, Roque-Gaffney E, Monahan M. Recombinant human IL-10 prevents the onset of diabetes in the nonobese diabetic mouse. Clin Immunol Immunopathol 1994; 71:169-175.
82. Zheng XX, Steele AW, Hancock WW, Stevens AC, Nickerson PW et al. A noncytolytic IL-10/Fc fusion protein prevents diabetes, blocks autoimmunity, and promotes suppressor phenomena in NOD mice. J Immunol 1997; 158:4507-4513.
83. Moritani M, Yoshimoto K, Ii S, Kondo M, Iwahana H et al. Prevention of adoptively transferred diabetes in nonobese diabetic mice with IL-10-transduced islet-specific Th1 lymphocytes. A gene therapy model for autoimmune diabetes. J Clin Invest 1996; 98:1851-1859.
84. Lee MS, Mueller R, Wicker LS, Peterson LB, Sarvetnick N. IL-10 is necessary and sufficient for autoimmune diabetes in conjugation with NOD MHC homozygosity. J Exp Med 1996; 183:2663-2668.
85. Moritani M, Yoshimoto K, Tashiro F, Hashimoto C, Ii S et al. Transgenic expression of IL-10 in pancreatic islet A cells accelerates autoimmune insulitis and diabetes in nonobese diabetic mice. Int Immunol 1994; 6:1927-1936.
86. Lee MS, Wogensen L, Shizuru J, Oldstone MBA, Sarvetnick N. Pancreatic islet production of murine interleukin-10 does not inhibit immune-mediated tissue destruction. J Clin Invest 1994; 93:1332-1338.
87. Balasa B, La Cava A, Van Gunst K, Mocnik L, Balakrishna D et al. A mechanism for IL-10-mediated diabetes in the nonobese diabetic (NOD) mouse: ICAM-1 deficiency blocks accelerated diabetes. J Immunol 2000; 165:7330-7337.
88. Homann D, Holz A, Bot A, Coon B, Wolfe T et al. Autoreactive CD4+ T cells protect from autoimmune diabetes via bystander suppression using the IL-4/Stat6 pathway. Immunity 1999; 11:463-472.
89. Vassalli P. The pathophysiology of tumor necrosis factors. Annu Rev Immunol 1992; 10:411-452.
90. Aggarwal BB, Natarajan K. Tumor necrosis factors: Developments during the last decade. Eur Cytokine Netw 1996; 7:93-124.
91. Epstein WV, Mannik M, Wener M. Treatment of rheumatoid arthritis with a tumor necrosis factor receptor-Fc fusion protein. N Engl J Med 1997; 337:1559-1561.
92. Garrison L, McDonnell ND. Etanercept: Therapeutic use in patients with rheumatoid arthritis. Ann Rheum Dis 1999; 58 Suppl 1:I65-9.
93. Picarella DE, Kratz A, Li C, Ruddle N, Flavell RA. Transgenic tumour necrosis factor (TNF)-α production in pancreatic islets leads to insulitis, not diabetes. J Immunol 1993; 150:4136-4150.

94. Green EA, Flavell RA. Diabetes development in transgenic mice coexpressing B7-1 and TNF-alpha on pancreatic islet cells is dependent on the duration of the TNF-alpha signal. Immunity 2000; 12:459-469.
95. Green EA, Flavell RA. Tumor necrosis factor-alpha and the progression of diabetes in nonobese diabetic mice. Immunol Rev 1999; 169:11-22.
96. Cope AP, Liblau RS, Yang XD, Congia M, Laudanna C et al. Chronic tumor necrosis factor alters T cell responses by attenuating T cell receptor signaling. J Exp Med 1997; 185:1573-1584
97. Yang XD, Tisch R, Singer SM, Cao ZA, Liblau RS et al. Effect of tumour necrosis factor α on insulin-dependent diabetes mellitus in NOD mice: I. the early development of autoimmunity and the diabetogenic process. J Exp Med 1994; 180:995-1004.
98. Jacob CO, Aiso S, Michie SA, McDevitt HO, Acha-Orbea H. Prevention of diabetes in nonobese diabetic mice by tumor necrosis factor (TNF): Similarities between TNF-alpha and interleukin 1. Proc Natl Acad Sci USA 1990; 87:968-972.
99. Jacob CO, Aiso S, Schreiber RD, McDevitt HO. Monoclonal anti-tumor necrosis factor antibody renders nonobese diabetic mice hypersensitive to irradiation and enhances insulitis development. Int Immunol 1992; 4:611-614.
100. Wallach D, Varfolomeev EE, Malinin NL, Goltsev YV, Kovalenko AV et al. Tumor necrosis factor receptor and Fas signaling mechanisms. Annu Rev Immunol 1999; 17:331-367.
101. Reed JC. Mechanisms of apoptosis. Am J Pathol 2000; 157:1415-1430.
102. Banner DW, D'Arcy A, Janes W, Gentz R, Schoenfeld HJ et al. Crystal structure of the soluble human 55 kd TNF receptor-human TNFb complex: Implications for TNF receptor activation. Cell 1993; 73:431-445.
103. Brockhaus M, Schoenfeld HJ, Schlaeger EJ, Hunziker W, Lesslauer W et al. Identification of two types of tumor necrosis factor receptors on human cell lines by monoclonal antibodies. Proc Natl Acad Sci USA 1990; 87:3127-3131.
104. Speiser DE, Sebzda E, Ohteki T, Bachmann MF, Pfeffer K et al. Tumor necrosis factor receptor p55 mediates deletion of peripheral cytotoxic T lymphocytes in vivo. Eur J Immunol 1996; 26:3055-3060.
105. Zheng L, Fisher G, Miller RE, Peschon J, Lynch DH et al. Induction of apoptosis in mature T cells by tumour necrosis factor. Nature 1995; 377:348-351.
106. Yoon JW, Ihm SH, Kim KW. Viruses as a triggering factor of type I diabetes and genetic markers related to the susceptibility to the virus-associated diabetes. Diabetes Res Clin Pract 1989; 1:S47-
107. Honeyman MC, Coulson BS, Stone NL, Gellert SA, Goldwater PN et al. Association between rotavirus infection and pancreatic islet autoimmunity in children at risk of developing type 1 diabetes. Diabetes 2000; 49:1319-1324.
108. Horwitz MS, Bradley L., Harbertson J, Krahl T, Lee J et al. Diabetes induced by Coxsackie virus: Initiation by bystander damage and not molecular mimicry. Nat Med 1998; 4:781-785.
109. Mena I, Fischer C, Gebhard JR, Perry CM, Harkins S et al. Coxsackievirus infection of the pancreas: Evaluation of receptor expression, pathogenesis, and immunopathology. Virology 2000; 271:276-288.
110. von Herrath MG, Christen U, Wolfe T, Hughes AC, Oldstone MBA. How viral infections can abrogate an ongoing autoimmune process. [manuscript submitted] 2001.

Chapter 13

Cytokines and Chemokines in Human Autoimmune Skin Disorders

Dorothée Nashan and Thomas Schwarz

Autoimmune skin diseases comprise two major groups of disorders, connective tissue diseases and bullous diseases. Connective tissue diseases represent a group of autoimmune disorders with overlapping clinical features including lupus erythematodes and its subtypes, systemic sclerosis, dermatomyositis, Sjögren's syndrome, and mixed connective disease. Acquired immunobullous diseases include pemphigus vulgaris and its associated entities, bullous pemphigoid, and dermatitis herpetiformis. In the following clinical features, immunoserological profile, and immunogenetic background for each disease will be briefly discussed to focus on the pathomechanisms with special emphasis on cytokines and chemokines. Improving knowledge concerning the precise immunopathogenesis will provide the ability to target specific mediators and effector cells involved in the disease. Hence therapeutic perspectives will be discussed at the end of the Chapter.

Introduction

In the recent past, advances in both basic and clinical research have considerably contributed to the understanding of cellular and molecular events that lead to autoimmune diseases. Pathogenetic basics are the damaging cytotoxic immune reactions caused by lymphocytes and autoantibodies synthesized by "forbidden" clones of immunologically competent cells which were not eradicated by the respective control mechanisms.[1]

It is not yet known whether a specific autoantigen initiates the disease and at which time point an antigenic challenge occurs. Different antigenic stimuli are under discussion in the course of some diseases. Tumors certainly participate in the initiation of paraneoplastic autoimmune diseases. Tumors compete with defense mechanisms against the development of "forbidden" cell clones or they might deliver a cross-reacting antigen. Inclusion bodies resembling paramyxovirus and persistent parvovirus infection in lupus erythematosus and dermatomyositis and black fly antigen in pemphigus foliaceus argue for pathogenic infectious triggers.[2-4] In addition, drugs play a role in the initiation of autoimmune diseases. The list of drugs especially those inducing lupus erythematosus is steadily increasing.[4,5] Environmental factors might be of importance both in the initiation and the aggravation of autoimmune diseases. Exogenous triggers are able to cause translocation of intracellular proteins via induction of apoptosis, e.g., exposure to ultraviolet radiation induces apoptosis of epidermal cells and thereby induces the release of nuclear proteins.[6,7] This may be of importance for the pathogenesis of lupus erythematosus.

Susceptibility and manifestation of autoimmune diseases depend on antigen presentation, HLA genes and/or T cell imbalance and/or shift of cytokine patterns. Initially dendritic cells capture antigens, followed by processing and presentation to T cells. Within this scenario the expression of costimulatory molecules is essential since it influences the fine balance of tolerance and activation of T lymphocytes. Cytokines participate in the fine tuning of T lymphocyte

Cytokines and Chemokines in Autoimmune Disease, edited by Pere Santamaria.

subsets, activation of self-reactive T cells, and B cell stimulation. Human CD4+T-helper cells can be divided in four subtypes according to their cytokine secretion profile: T-helper 1 (Th1) cells release IL-2, IL-12, IL-18, IFN-γ and TNF-α which induce cell-mediated immunity, cytotoxicity, and delayed-type hypersensitivity reactions. T helper 2 (Th2) cells secrete IL-4, IL-5, IL-6, IL-10 and IL-13 and are responsible for the development of B-cell mediated humoral immunity, IgE production, and activation of eosinophils. T helper 3 (Th3) cells produce transforming growth factor-β (TGF-β) and various amounts of IL-4 and IL-10. Th3 lymphocytes, also called regulatory T cells, exert their action primarily in maintaining tolerance and terminating immune reactions.[8-10] T helper 0 (Th0) cells can release cytokines of all three categories.

Lupus Erythematosus

Lupus erythematosus is the most diverse of the autoimmune connective tissue diseases as it may affect any organ and display a broad spectrum of clinical and immunological manifestations. The three major distinct clinical entities include discoid lupus erythematosus (DLE), subacute cutaneous lupus erythematosus (SCLE), and systemic lupus erythematosus (SLE). Rarer occurring variations on the theme are lupus erythematosus profundus, lupus erythematosus tumidus, bullous lupus erythematosus, and chilblain lupus.[11] DLE presents with red scaling plaques, which heal with atrophic scars and pigment deviations, mostly localized on the face or as a disseminated form with involvement of the trunk but always without systemic manifestations. SCLE is characterized by UV-sensitivity, non-scarring, papulosquamous or annular polycyclic skin eruptions on head and trunk; systemic manifestations may occur. SLE as a systemic disease which affects skin, joints and vessels. Organs primarily involved are kidneys, lung, heart, and the nerve system. A shift from DLE, SCLE, to SLE and mixed forms is possible, but extremely rare.

Disease activity can be assessed using the Systemic Lupus Activity Measure (SLAM) index, SLE Disease Activity Index (SLEDAI), European Consensus Lupus Activity Measures (ECLAM), and British Isles Lupus Assessment Group (BILAG) (12-16), see Appendix 1 on the American College of Rheumatology world wide web site: http://www.rheumatology.org/ar/ar.html. The histopathological changes of LE are quite characteristic and include:

1. degeneration of the basement membrane,
2. degeneration of collagen tissue,
3. patchy lymphocytic infiltrates with some plasma cells and histiocytes.

Appendages can be atrophic. Hyperkeratotic plugs of the hair follicle can appear; 80% of the patients show depositions of IgG-, IgM-, rarely IgA- and complement-depositions at the dermo-epidermal junctional zone. Immunohistology is preferentially positive in lesions of the face and the upper trunk or in longer persisting lesions. Positive immunohistology of normal, non UV-exposed skin is suggestive for SLE.

Pathogenesis of SLE is incompletely understood. Multiple genetic, environmental, hormonal and immunoregulatory factors are believed to contribute to the development of the disease.[5] SLE is a genetically heterogeneous disease and multiple genes confer susceptibility to SLE expression in a cumulative manner. An increased frequency of HLA-B8, -DR3, -DR2 is found in SCLE, HLA-B8, -DR3, -A1, -DR2 in SLE. Genomic variations, e.g., of the receptor for the Fc fragment of IgG type IIA (FcγRIIA) results in a defective FcγRIIA function and decreased IgG2 binding. Result is an impaired immune-complex handling, which is clinically associated with immune-complex deposition in the kidneys and lupus nephritis.[5]

A spectrum of non-organ-specific humoral antibodies is the hallmark for SLE. The antibodies do not seem to be the primary pathogen as they are not present in each case and transfusions of antibodies through the placenta usually are not harmful for the fetus with the exception of anti Ro-antibodies causing neonatal LE. Main antinuclear antibodies in SLE are anti-DNA-, anti-dsDNA-, anti-RNP- and anti-histone antibodies. Antibodies binding with DNA or DNA-histone conjugates form circulating immune-complexes. Anti-DNA antibodies bind to the DNA receptors on white blood cells. They block the binding and sequestration of free DNA by mononuclear cells and activate the secretion of IFN-γ. Histones, basic proteins, bind the DNA helical structure and contribute to the supercoil formation. Anti-histone antibodies are characteristic in 90% of drug-induced SLE, but may also appear in up to 30% of idiopathic SLE. Anti-dsDNA antibodies, positive in 50-83% of SLE patients, are highly characteristic of SLE. Their presence is usually associated with positive direct immunofluorescence in the patient's normal skin, low circulating complement levels, renal disease, and generally poor prognosis. Anti-dsDNA antibodies are clinically correlated with disease activity and prognosis of SLE. Binding of anti-dsDNA antibodies induces production and release of IL-1β, IL-8, TNF-α and IL-10 from resting mononuclear cells and IL-1 and IL-6 release from endothelial cells.[17,18] From these in vitro studies it is suggested that anti-dsDNA antibodies exert a dual effect, first to enhance the release of pro-inflammatory cytokines and to augment inflammatory reactions and secondly to polarize the immune reaction towards the Th2 pathway, thereby supporting further (auto)antibody production.

Acute phase response components, C reactive protein (CPR) and serum amyloid protein (SAP), are thought to bind to circulating DNA and RNA, respectively, rendering them non-immunogenic. The proinflammatory cytokines IL-6, IL-1, and TNF-α induce the release of acute phase proteins from the liver and thereby promote removal of circulating autoantigens. Recent work indicates that patients with active disease have significantly higher levels of serum TNF-α and IL-6 than patients with inactive disease.[19-21] Monocytes isolated from LE patients kept in culture secrete significantly more TNF-α than monocytes from healthy donors.[22] Enhanced TNF-α production in peripheral blood mononuclear cells (PBMC) obtained from SLE patients is considerably reduced after application of methylprednisolone.[23] However, despite the increase of TNF-α CPR levels in SLE patients are normal, implying an inadequate acute phase response in LE. This phenomenon may be explained by enhanced levels of its soluble receptor (TNF-sR) which results in a relative deficiency of active TNF-α in SLE patients. The potential protective role of TNF-α in SLE is supported by the finding that anti-TNF-α therapy causes an increase in the production of dsDNA antibodies.[21] This is also supported by studies addressing the TNF-α polymorphism. "Low TNF-a producers" appear to have a higher risk for LE development.[24]

IL-6 is a pleiotropic pro-inflammatory cytokine produced in the skin by epidermal cells, fibroblasts, and dermal endothelial cells. Grondal et al. described enhanced levels of IL-6 in sera of SLE patients.[14] In cerebrospinal fluid of patients with cerebral lupus and in the urine of patients with lupus nephritis increased IL-6 levels were found.[25,26] IL-6 has been detected in 52% of kidney biopsy specimens taken from a total of 19 patients with lupus nephritis. Local production of IL-6 in the glomeruli and tubuli was confirmed by in situ hybridisation and RT-PCR.[27] IL-6 supports the survival and differentiation of B cells and plasma cells. Thus it has been suggested that IL-6 may be involved in an autocrine loop triggering B cell hyperactivity and antibody production, thereby maintaining SLE disease activity. A correlation with disease activity, however, was not always confirmed.[24,28,29] LE is a photosensitive disease, UV-exposure can lead both to the exacerbation of SLE and to the induction of DLE lesions. In addition, SCLE is highly photosensitive. In this context it is important to mention that UV activates keratinocytes to release IL-6 both in vitro and in vivo.[30] Total body UV-exposure even causes systemic release of IL-6.[31] In addition, exposure to UV-radiation can induce IL-6 production from the monocyte/macrophage fraction of PBMC taken from patients with SLE.[32] Immunolabeling of IL-6 in lesional skin in various types of lupus erythematosus revealed an intense staining in the basal layer of the epidermis.[33] In addition, UV induces apoptosis of keratinocytes,[34] leading to the production of surface blebs containing nucleosomes and other potential autoantibody targets. Released nucleosomes may also induce IL-6 production.[35] Thus, one may speculate that one reason why UV exacerbates LE is by inducing IL-6 release.

Serum IL-10 levels are higher in patients with SLE when compared with controls.[36] This increase is mainly due to an increase in IL-10 production by monocytes, B cells and possibly memory T cells. Accordingly, increased IL-10 transcription was found in PBMCs of patients with SLE.[37] Serum titers of IL-10 are positively correlated with anti-dsDNA antibody titers and the SLEDAI score.[38,39] One of the primary features of IL-10 is inhibition of Th1 responses and driving an immune response into the Th2 direction which is associated with increased antibody production. Therefore IL-10 may exacerbate LE. On the other hand, IL-10 is known to inhibit the release of IL-6 which as mentioned above triggers LE. Interestingly the inhibition of IL-6 release by IL-10 is not found in PBMC obtained from LE patients. It is still a matter of debate whether an intrinsic defect in IL-10 signaling is responsible for this alteration in LE patients.[40,41]

TGF-β is a powerful inhibitor of the production of IL-6, IL-1 and TNF-α.[42] In addition, TGF-β suppresses IgG secretion by B lymphocytes.[43] Just based on these activities TGF-β should have a beneficial effect on LE. Interestingly, both constitutive and stimulated levels of TGF-β are significantly lower in patients with SLE, suggesting that high IgG levels in SLE may be due to inadequate suppression of IgG production by TGF-β.[44]

IFN-γ production of PBMC from patients with SLE is significantly correlated with the global disease acitivity score SLAM.[45,46] IFN-γ is able to induce class switching of immunoglobulins, inducing the production of IgG2 and IgG3. In this context, it is important to mention that IgG2 and IgG3 are primarily found in the systemic form of LE, while in SCLE IgG1 subclass antibodies predominate. This suggests that IFN-γ might play a pathogenic role. This is also supported by animal studies. IFN-γ receptor-knock-out mice with a lupus prone background (B/W) develop nephritis less frequently than lupus prone control mice.[47] In addition, overexpression of IFN-γ in the suprabasal layer of the epidermis results in a systemic autoimmune phenotype reminiscent of SLE. The animals develop antinuclear antibodies, IgG deposits in the glomeruli and proliferative glomerulonephritis.[48]

In contrast to the above mentioned cytokines, much less is known about the role of IL-4 which is a classical Th2 cytokine. This may be due to the fact that measuring of IL-4 can be technically problematic. Excessive production of IL-4 was found in peripheral blood lymphocytes of a single case with a severe disease course. The marked IL-4 production decreased under high-dose corticosteroid treatment.[49] By immunohistochemistry and RT-PCR an increased expression TNF-α, IL-2 and IFN-γ together with up-regulation of the IL-2 receptor and IFN-γ receptor were detected in DLE lesions, a pattern which favours a Th1 response.[50] No significant increase and sometimes even decreased production of IL-4 was found in stimulated and unstimulated T lymphocytes from blood samples of LE patients.[49,51]

Numerous other studies report alterations of the levels of a variety of cytokines in LE.[45,52-57] The pathogenetic relevance of these findings is not yet clear and the results are not always concordant. These discrepancies may be due to differences in sample collection, heterogenous patients groups, and variations in disease course and activity.

Systemic Sclerosis (SSc)

Systemic sclerosis (SSc) is a rare disease with an incidence between 2.3 and 10 per 10^6 population, ratios of female to male are between 3 and 6 : 1. Diagnostic criteria were determined by the Subcommitee for Scleroderma Criteria of the ARA. The classification divides in to diffuse cutaneous systemic sclerosis (SSc) and limited cutaneous systemic sclerosis. SSc is characterized by fibrosis of skin, lung, and gastrointestinal tract and by microvascular abnormalities of the skin and visceral organs.

Endothelial cells appear to be an initial target. Serum factors are cytotoxic to endothelial cells in a direct or indirect way, the latter mediated in an antibody-dependent way. Increased plasma viscosity causes thrombus formation and reduced microvascular blood flow. Fibrotic processes accompanied by an accumulation of collagen are initiated either through the effects of ischemia or via fibrogenic mediators released from platelets and inflammatory cells.

The triggering events are still unclear. There is mounting evidence that extracellular antigens, e.g., anti-endothelial cell antibodies, are involved in endothelial activation, up-regulation of cell adhesion molecules, and monocyte recruitment, possibly induced via autocrine IL-1 release.[58] Tissue damage in general might induce the presentation of cryptic epitopes. Upon effective presentation by antigen presenting cells according to the HLA molecules, the subsequent

HLA-restricted antibody production gets inititated. HLA alleles with an increased frequency for SSc belong both to the MHC class I alleles (A1, A3, B8) and MHC class II alleles (DR2, DR3, DR5, DR28, DR52). Accordingly, a close association of several HLA alleles with certain SSc autoantibodies exists.[59] The majority of SSc sera contains autoantibodies against intracellular antigens. The particular autoantibody present is often indicative for clinical expression, disease course and overall severity. Anti-centromere antibodies (ACA) are typical in the limited cutaneous disease. Anti-Scl-70 antibodies and also anti-RNA polymerase III antibodies are characteristic of diffuse cutaneous involvement.[60]

IL-17, IL-1, IL-6, TGF-β and IL-4 appear to be involved in the pathophysiology of sclerosis.[61-66] IL-17 is a T-cell-derived cytokine, which stimulates fibroblasts to proliferate, to produce IL-6, IL-8 and to express ICAM-1. Increased mRNA expression of IL-17 was shown in unstimulated peripheral blood lymphocytes and lymphocytes from skin and lung samples of SSc patients. IL-17 and IFN-γ or TNF-α reveal a marked synergism in the stimulation of IL-6 and IL-8 protein secretion from keratinocytes.[67-69] IL-17 also induces the release of IL-1 in endothelial cells.[70] IL-1 again activates fibroblasts to secrete IL-6 which stimulates proliferation and collagen synthesis of fibroblasts.[71] Accordingly, elevated IL-6 levels were found in sera of SSc patients suggesting that IL-6 might be responsible for increased collagen synthesis in SSc.[72] On the other hand, IL-6 also induces the expression of matrix metalloproteinases, including collagenase. There is recent evidence that UV exposure, especially UVA1, has a beneficial effect on SSc, which may be due to up-regulation of MMPs.[73] However, UV might not directly induce MMPs, but rather indirectly via inducing IL-1 and IL-6 which ultimately mediate MMP up-regulation.[74]

A crucial cytokine in tissue remodelling is TGF-β which up-regulates the mRNA expression of collagen, fibronectin and monocyte chemoattractant protein-1 (MCP-1) which on its own up-regulates type I collagen mRNA expression in cultured fibroblasts.[75] TGF-β appears as an ideal mediator of fibrosis in SSc. It is produced by fibroblasts and endothelial cells.[76] TGF-β receptor levels are significantly elevated and correlate with elevated collagen mRNA levels in scleroderma fibroblasts.[77] Genetic susceptibility to SSc and degree of cutaneous fibrosis appear to be dictated by genetic microsatellite markers for TGF-β.[78] TGF-β1 prevents fibroblasts from undergoing apoptosis, induces myofibroblasts to proliferate, and enhances extracellular matrix synthesis.[79] In addition, TGF-β induces connective tissue growth factor (CTGF) which is found overexpressed in fibrotic disorders. Both constitutive and TGF-β-induced expression of CTGF is found in scleroderma fibroblasts.[80] Furthermore, fibroblasts from scleroderma skin are more resistant to TNF-α-mediated repression of TGF-β-induced CTGF synthesis. Therefore, pro-collagen stimuli mediated by IL-6 and TGF-β and the inability of negative regulatory cytokines like TNF-α participate in the development of the disease.[81]

Tenascin is a large extracellular matrix glycoprotein which is present in embryonic but absent from most normal adult tissues. IL-4 but not other cytokines strongly induce tenascin both at the protein and at the mRNA level. High levels of both IL-4 and tenascin are present in the affected skin of patients with SSc. These results suggest that high levels of tenascin result from increased IL-4 expression in SSc tissue.[82]

Dermatomyositis

Inflammatory myopathies, dermatomyositis, polymyositis and the sporadic idiopathic inclusion body myositis are acquired immune-mediated myopathies. Characteristics of dermatomyositis (DM) are skin changes with red swelling of the face, Gottron papules and muscle weakness. The distribution female to male is 2:1. The disease can be idiopathic, drug-induced or of paraneoplastic origin. Clinical diagnosis is accompanied by laboratory investigations (especially serum creatine phosphokinase) which may also reflect disease activity and thus allows one to monitor therapy efficiency. Diagnosis is confirmed by muscle biopsy and electromyography.

Lymphocytes show increased responses to muscle antigens. In paraneoplastic forms skin reactions to autologous tumor extracts can be positive. Genetic background is based on HLA-B8 in children and HLA-DR3, -B14 in adults. ANA are detected in 40 to 60% of patients with

DM. More specific antibodies are anti-Mi-2 and anti-KU, anti-Jo-1, anti-PL12 and antibodies directed against tRNA-synthetases.[83] Histopathology is characterized by inflammatory infiltrates, fibrosis, and muscle fiber loss. In inflammatory lesions, muscle fibers express various cytoplasmic and surface molecules that are not detectable in normal fibers. These molecules, which include HLA antigens, heat-shock proteins, adhesion molecules and the death receptor Fas (CD95), are probably induced by locally secreted cytokines. Conversely expression and interaction of antigenic determinants may stimulate cytokine production.[84] For instance cultured human myoblasts do not constitutively express the accessory molecule CD40. IFN-γ induces CD40 expression on muscle cells. Activation of CD40 by its cognate ligand CD40L increases the production of IL-6, IL-8, IL-15 and monocyte chemoattractant protein-1 (MCP-1) in cultured myoblasts.

Key inflammatory molecules in DM are IL-1α, IL-1β, TGF-α and adhesion molecules ICAM-1, LFA-1, VLA-4. Decrease in production and expression of these molecules correlates with clinical improvement of the disease. Persistent expression of IL-1α, ICAM-1 and VCAM-1 is observed in patients who after steroid therapy still had muscle weakness.[85] Muscle function showed a consistent but not complete concordance of those molecules, while inflammatory cells had already disappeared. These parameters seem more sensitive in the evaluation of the efficacy of the therapy compared with routine histology.

Despite clear cell damage and loss of muscle cells, apoptosis is rather rarely observed in myopathies including DM. This is an intriguing observation since muscle cells express rather high levels of the death receptor Fas and thus should be prone to apoptosis. In this case muscle cells appear to be protected from Fas-induced apoptosis by enhanced expression of the antiapoptotic protein FLIP (Fas-associated death domain-like IL-1-converting enzyme inhibitory protein), which inhibits Fas-triggered death signaling.[86] Up-regulation of FLIP is mediated by the expression of TNF-α and IFN-γ, a process which is induced in DM. In addition, muscle cells might be protected from apoptosis by expression of Bcl-2, another antiapoptotic protein.[87] Although it is not absolutely clear how muscle cells die, degradation of collagen IV, an essential component of the muscle membrane, by matrix metalloproteinases (MMP-2, MMP-9) may be of relevance.[88] Cytokines including TNF-α, IL-1, TGF-β which are variably increased in myositis are well-known inducers of MMP production and may serve as an additional source for up-regulating MMP and facilitating T cell mediated cytotoxicity.[89]

Autoimmune Bullous Diseases

Autoimmune bullous skin diseases comprise a group of severe dermatosis characterized clinically by blisters or erosions of the skin and/or mucous membranes. The major pathophysiological event in bullous diseases is the occurrence of autoantibodies that target specific antigens which represent important structural components of normal human skin.

Pemphigus

The term pemphigus refers to a group of autoimmune diseases which are associated with blisters on the skin and mucous membranes caused by autoantibodies against adhesion components of keratinocytes. Pemphigus can be divided in two major types: pemphigus vulgaris and pemphigus foliaceus. Recently a new type of pemphigus has been described, paraneoplastic pemphigus which exhibits unique clinical and immunopathological features.

Pemphigus Vulgaris

Pemphigus vulgaris (PV) is a cutaneous autoimmune disease characterized by blister formation in the suprabasilar layers of skin and mucosa. Autoantibodies are directed against an extracellular domain of desmoglein3 (Dsg3), a desmosomal cadherin expressed in the basal epidermal layers.[90] Autoantibodies bind to the surface of keratinocytes and circulate in the serum of patients. Additionally sera from patients with pemphigus vulgaris might also contain antibodies against desmoglein1 and desmocollins, another family of desmosomal cadherins.[91,92]

Passive transfer of patients' IgG into neonatal mice reproduces the disease, whereby complement is not required but enhances blistering.[93] The diagnosis of PV is confirmed by direct immunofluorescence, which shows IgG deposited on the surface of keratinocytes in both lesional and unaffected skin. The most common immunoglobulin subclass is IgG. Positive staining with IgM, IgA and C3 is less frequent. Indirect immunofluorescence tests reveal antibodies in 80-90% of patients with PV. Predisposition is linked to genetic factors, an association with HLA-DR4 is common in patients with PV. Concerning susceptibility it is hypothesized that subregions in the DR peptide-binding groove influence the binding of an antigenic peptide and its recognition by T cells. In the acute phase clinical signs can be used for the monitoring of the disease. Changes in the titer of circulating pemphigus antibodies may be helpful and direct immunofluorescence of normal skin has also been recommended to predict remission or relapse. A Th2-dependent IgG4 subtype is preferentially seen in active stages of PV, while autoantibodies of the Th1-dependent IgG1 subclass are predominant upon remission of PV.[94,95]

Biological modifications leading to production of pathogenic IgG autoantibodies are based on autoreactive CD4+ T lymphocytes which respond to Dsg3 peptides.[94,96-98] Healthy subjects might also show T cell responses to Dsg3. Especially in relatives of patients with PV but also in control groups PV-IgG was detectable by indirect immunofluorescence staining or western blot analysis.[99] Those antibody-positive healthy subjects showed an MHC class II allele pattern identical or similar to those highly prevalent in PV.[100] Thus further immunological abnormalities seem to be required for the final manifestation of PV. Especially animal models with the injection of PV autoantibodies into knock-out mice confirm the major importance of cytokines in the development of PV. Decreased disease susceptibility was seen in IL-1-deficient mice and in TNF-α receptor-deficient mice.[101] Similarly acantholysis induced by PV-IgG could be inhibited by antibodies blocking IL-1α and TNF-α. Accordingly Feliciani and colleagues could demonstrate increased TNF-α together with IL-1α activity in lesional and perilesional skin.[101] TNF-α and IL-6 are significantly increased in sera of PV patients. Both TNF-α and IL-6 correlate with disease severity and decrease with corticosteroid therapy.[102]

Diverging oberservations concern the Th2 cytokine IL-10. Application of rIL-10 together with PV-IgG significantly suppressed disease activity.[101,103] Accordingly IL-10-deficient mice displayed a stronger susceptibility to the development of pemphigus vulgaris. This is also supported by studies using CD28 -/- mice. Since CD28 signaling plays an important role in the induction of Th2 cytokines, these animals are more prone to a Th1 response with decreased levels of IL-10. Passive transfer of PV IgG caused more severe blisters in CD28 -/- mice than in wild type mice. Not concurring with these findings is the detection of increased IL-10 levels in sera of PV patients.[104]

Pemphigus Foliaceus

Ten to 20% of cases with pemphigus belong to this group. Blistering is high in the epidermis, either in the granular layer or just beneath the stratum corneum. Endemic pemphigus foliaceus, also known as fogo sevalgem (wild fire), is common in South America. Epidemiological evidence suggests black fly bites as risk factor for the disease, which implies black fly antigens as a trigger for the formation of antibodies that cross-react with epidermal antigens. Genetic susceptibility is linked to the HLA-DR1 locus. The antigenic determinant is the extracellular amino-terminal domain of desmoglein 1 (Dsg1), a 100 kDa desmosomal cadherin. Dsg1 is expressed in the upper epidermal layers. Sera of patients with fogo sevalgem sometimes also contain autoantibodies against desmocollins 1 and 2. The pathogenic autoantibodies in all forms of pemphigus foliaceus are predominantly in the IgG4 subclass. Pemphigus foliaceus is transferable in mice with autoantibodies against Dsg1.[105] Although pemphigus foliaceus and pemphigus vulgaris differ in their antigens and thus represent distinct clinical entities, concerning the pathogenic role of cytokines no major differences appear to exist between these two diseases.

Bullous Pemphigoid

Bullous pemphigoid (BP), linear IgA bullous dermatosis (LAD) and cicatricial pemphigoid (CP) are clinically distinct autoimmune bullous skin diseases characterized by autoantibodies against components of the epidermal basement membrane. With an incidence of 6 per 10^6 , bullous pemphigoid is the most frequent bullous autoimmune disease, males and females are equally affected. Genetic predispostion is associated with HLA-DQ7.

Target antigens of BP are a cytoplasmic 230 kDa plaque protein which belongs to the family of desmoplakins (BP230 or BP antigen 1) and a 180 kDa transmembrane protein (BP180 or BP antigen 2 or type XVII collagen) which localizes at the hemidesmosomes. The immunodominant extracellular ectodermal region of BP180 is the NC16A.[106,107] Recently different groups could show comparable autoantibody responses to BP180 in BP, lichen planus pemphigoides, pemphigoid gestationis, LAD and CP, clinically distinct autoimmune bullous diseases.[108-111] They speculate that variable epitopes of BP 180 are targeted which results in discernible clinical pictures

Bullous pemphigoid (BP) is characterized by subepidermal bullae and circulating autoantibodies, predominantly directed against BP180 and/or BP230.[95] Complement activation with linear deposition of C3 is a further characteristic. In a murine model BP-like lesions can be induced by passive transfer of autoantibodies against the immunodominant antigenic determinant NC16A of BP180. The same circulating antibody type is useful in monitoring disease activity.[93,112,113] But there is unanimous agreement that autoantibodies alone are not capable of blister formation in BP. The immunopathological reaction of dermoepidermal separation is critically dependent of the inflammatory infiltrate consisting predominantly of eosinophils, proteases and cytokines. Accordingly, the binding of autoantibodies to BP180 (BP180 NC16A domain) on cultured keratinocytes from BP patients induces increased mRNA expression, protein synthesis and secretion of IL-6 and IL-8, while the expression of IL-1α, IL-1β, TNF-α, IL-10, and monocyte chemoattractant protein-1 (MCP-1) is not affected. This cytokine secretion pattern was not found in keratinocytes deficient in BP 180.[114]

Autoreactive T cells against BP180 preserved as cell lines or clones reveal both a Th1- and a Th2-cytokine pattern, releasing IL-2, IL-4, IL-5, IL-6, IL-13 and IFN-γ.[105] The Th1 cytokine IFN-γ induces the secretion of IgG1, which is primarily detected in chronic courses of BP. But IgG1 against BP180 was also found in serum of healthy subjects. The Th2 cytokines IL-4 and IL-13 regulate the secretion of IgG4 and IgE by activated B cells. The role of IgG4 autoantibodies in BP is associated with acute onset and severe disease. Interestingly CD4+ T cells reactive against BP antigen 2 from five BP patients produced both Th1 and Th2 cytokines, whereas three autoreactive T cell lines from three normal subjects produced exclusively IFN-γ, a Th1 cytokine. This strongly suggests that autoreactive Th2 responses to BP antigen 2 are restricted to BP patients and may thus be critical in the pathogenesis of BP.[115]

Detection of soluble IL-2R, IL-1β, IL-1 receptor antagonist and IFN-γ in blister fluid strongly indicates the involvement of activated T lymphocytes in active BP lesions.[116-120] The importance of TNF-α in blister fluids of different origin was already shown in the early 1990s by Zillikens and coworkers.[116] TNF-α may represent an important inflammatory stimulus since it promotes chemotaxis, activation of macrophages, eosinophils and neutrophils. Both Rhodes et al and Zillikens et al found increased blister levels of TNF-?, but also of IL-6. This pattern in BP (very high IL-6, high TNF-α) differs from blister levels in toxic epidermal necrolysis (very high IL-6, low TNF-α) and from allergic contact dermatitis and burns in which both cytokines were much less elevated.[121] In addition, increased levels of IL-8 and IL-10 were found in blister fluids, but not in sera, suggesting local production of these mediators. An elevation of IL-4 and IL-6 in serum of BP patients is possibly but not necessarily associated with disease activity.[120-125]

IL-5 which plays a major role in recruitment, differentiation and proliferation of eosinophils appears to be particularly increased in blister fluid and in serum of BP patients.[126,127] In contrast, Wakugawa and colleagues did not detect a positive correlation of IL-5 levels in blister fluid and dermal eosinophilic infiltrate, but observed that eotaxin, α-chemokine expressed in epidermal keratinocytes, correlates with the number of dermal infiltrating eosinophils.[128,129] But again production of IL-5 is increased in peripheral T cells of BP patients. In remission or under immunosuppressive therapy, IL-5 production of peripheral T cells is reduced.[130] Taken together, the overall cytokine patterns found in sera of BP patients resembles the Th2 type. This is additionally demonstrated by increased levels of sCD30, a member of TNF/NGF receptor family, which is regarded as activation marker of Th2 cells.[125]

Dermatitis Herpetiformis

Dermatitis herpetiformis (Duhring's disease, DH) is a chronic subepidermal vesicular blistering disease. Characteristic distribution of skin lesions include the elbows and the sacral region. Due to intensive itching, the primary lesions are not blisters but erosions yielding a polymorphous appearance. A strong association with the HLA-A1, -B8, -DR3 and -DQ2 haplotype is described.[131] IgA is deposited at the dermo-epidermal junction. The dermal and perivascular infiltrate which is mainly composed of CD4+ lymphocytes, neutrophils and eosinophils, is believed to play an important part in the pathogenesis of the disease. First studies depicted IL-8 and granulocyte-macrophages colony-stimulating factor (GM-CSF) in lesional skin which promote infiltration and activation of mononuclear cells.[132] Further tissue inflammation is caused by IL-4 and IL-5. IL-4 mRNA expression was found in cultured T cell obtained from intestinal mucosa and skin lesions of DH patients.[133-135] IL-3, IL-5 and GM-CSF, all of which are involved in activation of eosinophils were detected in situ in the jejunal mucosa and in skin infiltrates.[135,136] Furthermore IL-13 and TNF-α were detected in the inflammatory infiltrate by immunohistochemistry.[137] In addition, the chemotactic protein eotaxin which is induced by IL-4, IL-13, TNF-α and IFN-γ, was detected at the tips of the dermal papillae and within the dermal lymphomonocytic infiltrate. Taken together, these studies suggest that the recruitment of eosinophils and neutrophils in DH may be induced not only by GM-CSF but also by Th2 type cytokines.

Therapeutic Perspectives

The major basis for the treatment of autoimmune skin disorders is immunosuppression. The most frequently applied drugs include systemic corticosteroids, azathioprine, cyclophosphamide, methotrexate, cyclosporin A and recently mycophenolate mofetil. Drugs administered which do not act primarily by immunosuppression comprise thalidomide, sulfones, chloroquine, plasmapheresis, photopheresis and intravenous immunoglobulins.[138-142] Although the mode of action of the majority of these medications is not clear at all, they can be quite beneficial for certain indications. The classical immunosuppressive drugs target cellular components, primarily leukocytes rather than distinct cytokines. An exception is cyclosporin A which acts via inhibiting the release of IL-2, but ultimately through this activity primarily hits the lymphocytes.[143,144] Through inhibiting leukocyte function in general, the classical immunosuppressive drugs exhibit a broad and strong activity which is associated with the occurrence of side effects. Toxicity of long-term immunosuppression demands the search for more selective strategies of immunointervention causing fewer side effects. In this respect, targeting of specific, pathologically relevant cytokines has been regarded as an attractive alternative approach.

Since cytokines amplify inflammatory reactions, mediate pleiotropic tissue damage, modulate the Th1/Th2 balance and contribute to the induction of humoral immunity, they should be ideal candidates for specific immunointervention. This concept has been impressively confirmed in a variety of animal studies. The administration of anti-IL-6 antibodies significantly prevented production of anti-dsDNA, reduced proteinuria, and prolonged life of autoimmune prone mice.[145] Similarly application of anti IL-4 antibodies was effective in preventing the onset of lupus nephritis in lupus-prone mice.[146] IFN-γ, a key effector molecule in pathogenesis of several autoimmune diseases, is another potential target, e.g., by neutralizing its activity through soluble receptors. Accordingly, intramuscular injection of cDNA-plasmids encoding an IFN-γ receptor Fc-fusion protein retarded development and progression of lupus in MRL-Fas/lpr mice.[147] Other strategies besides the application of antibodies and soluble receptors include the application of natural cytokine antagonists (e.g., IL-1 receptor antagonist), of antagonizing cytokines (IL-10 versus IL-12 and vice versa) and inhibitors of signal transduction. Despite these promising experimental data the usage of these applications in daily practice is still limited. An exception may be the recent introduction of humanized anti-TNF, anti-TNF-receptor and anti-IL-2 antibodies.[148,149] The major indications include rheumatoid arthritis, chronic inflammatory bowel diseases and psoriasis. Future will certainly add more indications.

Conclusions

The reasons why anti-cytokine strategies have not yet been the major breakthrough in the treatment of autoimmune diseases may be multiple. Although cytokines play an important role in the pathogenesis of a variety of diseases, there is really no good example of a disease whose pathogenesis can be exclusively explained by the dysregulation of a single cytokine. This may be due to the fact that the cytokine system is an extremely complex one. Cytokines can induce or suppress their release and synergize or antagonize each other. In addition, cytokines exhibit multiple effects and it is yet not possible to block only certain activities. In addition, due to the multitude of cytokines there is a redundancy and neutralization of a particular cytokine may be compensated by other "fellow" mediators. Although the direct therapeutic implications in the sense of anti-cytokine therapy may be limited, studying the role of cytokines in diseases still remains important since it will tremendously increase our understanding of the pathogenetic mechanisms and thereby indirectly guide us to new therapeutic approaches. This certainly applies also for autoimmune diseases of the skin.

References

1. Mutasim DF, Adams BB. A practical guide for serologic evaluation of autoimmune connective tissue diseases. J Am Acad Dermatol 2000; 42:159-174.
2. Crowson AN, Magro CM, Dawood MR. A causal role for parvovirus B19 infection in adult dermatomyositis and other autoimmune syndromes. J Cutan Pathol 2000; 27:505-515.
3. Chou TN, Hsu TC, Chen RM et al. Parvovirus B19 infection associated with the production of anti-neutrophil cytoplasmic antibody (ANCA) and anticardiolipin antibody (aCL). Lupus 2000; 9:551-554.
4. Rowell NR, Goodfield MJD. The connective tissue diseases. In: Champion RH, Burton JL, Burns DA, Breathnach SM, eds. Rook Wilkinson Ebling. Textbook of Dermatology. 6th ed. Oxford: Blackwell Scientific Publications, 1998:2437-2575.
5. Liossis SN, Tsokos GC. Molecular aspects in the pathogenesis of human systemic lupus erythematosus. Arch Immunol Ther Exp 2000; 48:11-19.
6. Casciola-Rosen LA, Anhalt G, Rosen A. Autoantigens targeted in systemic lupus erythematosus are clustered in two populations of surface structures on apoptotic keratinocytes. J Exp Med 1994; 179:1317-130.
7. Casciola-Rosen LA, Anhalt GJ, Rosen A. DNA-dependent protein kinase is one of a subset of autoantigens specifically cleaved early during apoptosis. J Exp Med 1995; 182:1625-1634.
8. Inobe J, Slavin AJ, Komagata Y et al. IL-4 is a differentiation factor for transforming growth factor-beta secreting Th3 cells and oral administration of IL-4 enhances oral tolerance in experimental allergic encephalomyelitis. Eur J Immunol 1998; 28:2780-2790.
9. Kitani A, Chua K, Nakamura K et al. Activated self-MHC-reactive T cells have the cytokine phenotype of Th3/T regulatory cell 1 T cells. J Immunol 2000; 165:691-702.
10. Prud'homme GJ, Piccirillo CA. The inhibitory effects of transforming growth factor-beta-1 (TGF-beta1) in autoimmune diseases. J Autoimmun 2000; 14:23-42.
11. Crowson AN, Magro C. The cutaneous pathology of lupus erythematosus: A review. J Cutan Pathol 2001; 28:1-23.
12. Corzillius M, Fortin P, Stucki G. Responsiveness and sensitivity to change of SLE disease activity measures. Lupus 1999; 8:655-659.
13. Brunner HI, Feldman BM, Bombardier C et al. Sensitivity of the Systemic lupus erythematosus Disease Activity Index, British Isles Lupus Assessment Group Index, and Systemic Lupus Activity Measure in the evaluation of clinical change in childhood-onset systemic lupus erythematosus. Arthritis Rheum 1999; 42:1354-1360.

14. Grondal G, Guanarsson I, Ronnelid J et al. Cytokine production, serum levels and disease activity in systemic lupus erythematosus. Clin Exp Rheumatol 2000; 18:565-570.
15. Fortin PR, Abrahamowicz M, Clarke AE et al. Do lupus disease activity measures detect clinically important change? J Rheumatol 2000; 27:1421-1428.
16. Gladman DD, Urowitz MB, Kagal A et al. Accurately describing changes in disease activity in systemic lupus erythematosus. J Rheumatol 2000; 27:377-379.
17. Sun KH, Yu CL, Tang SJ et al. Monoclonal anti-double-stranded DNA autoantibody stimulates the expression and release of IL-1β, IL-5, IL-8, IL-10 and TNF-α from normal human mononuclear cells involving in the lupus pathogenesis. Immunology 2000; 99:352-360.
18. Neng-Lai KN, Leung JCK, Lai KB et al. Anti-DNA autoantibodies stimulate the release of interleukin-1 and interleukin-6 from endothelial cells. J Pathol 1996; 178:451-475.
19. Davas EM, Tsirogianni A, Kappou I et al. Serum IL-6, TNFalpha, p55, srTNFalpha, p75srTNFalpha, srIL-2alpha levels and disease activity in systemic lupus erythematosus. Clin Rheumatol 1999; 18:17-22.
20. Emilie D, Llorente L, Galanaud P. Cytokines and lupus. Ann Med Interne (Paris) 1996; 147:480-484.
21. Dean GS, Tyrell-Price J, Crawley E et al. Cytokines and systemic lupus erythematosus. Ann Rheum Dis 2000; 59:243-251.
22. Steinbach F, Henke F, Krause B et al. Low levels of interleukin-1 receptor antagonist coincide with kidney involvement in systemic lupus erythematosus. Br J Rheumatol 2000; 26:1283-1289.
23. Brink I, Thiele B, Burmester GR et al. Effects of anti-CD4 antibodies on the release of IL-6 and TNF-alpha in whole blood samples from patients with systemic lupus erythematosus. Lupus 1999; 8:723-730.
24. Jacob CO, Fronik Z, Lewis GD et al. Heritable major histocompatibility complex class II-associated differences in production of tumor necrosis factor alpha: Relevance to genetic predisposition of systemic lupus erythematosus. Proc Natl Acad Sci USA 1990; 87:1233-1237.
25. Trysberg E, Carlsten H, Tarkowski A. Intrathecal cyokines in systemic lupus erythematosus with central nervous system involvement. Lupus 2000; 9:498-503.
26. Horii Y, Iwano M, Hirata E et al. Role of interleukin-6 in the progression of mesangial proliferative glomerulonephritis. Kidney Int Suppl 1993; 39:S71-75.
27. Herrera-Esparza R, Barbosa-Cisneros O, Villalobos-Hurtando R et al. Renal expression of IL-6 and TNF alpha genes in lupus nephritis. Lupus 1998; 7:154-158.
28. Waszczykowska E, Robak E, Wozniacka A et al. Estimation of SLE activity based on the serum level of chosen cytokines and superoxide radical generation. Med Inflamm 1999; 8:93-100.
29. Robak E, Sysa-Jedrzejewska A, Dziankowska B et al. Association of interferon gamma, tumour necrosis factor alpha and interleukin 6 serum levels with systemic lupus erythematosus activity. Arch Immunol Ther Exp 1998; 46:375-380.
30. Kirnbauer R, Kock A, Neuner P et al. Regulation of epidermal cell interleukin-6 production by UV light and corti. J Invest Dermatol 1991; 96:484-489.
31. Urbanski A, Schwarz T, Neuner P et al. Ultraviolet light induces increased circulating interleukin-6 in humans. J Invest Dermatol 1990; 94:808-811.
32. Pelton BK, Hylton W, Denman AM. Activation of IL-6 production by UV irradiation of blood mononuclear cells from patients with systemic lupus erythematosus. Clin Exp Immunol 1992; 89:251-254.
33. Nürnberg W, Haas N, Schadendorf D et al. Interleukin-6 expression in the skin of patients with lupus erythematosus. Exp Dermatol 1995; 4:52-57.
34. Kulms D, Schwarz T. Molecular mechanisms of UV-induced apoptosis. Photodermatol Photoimmunol Photomed 2000; 16:195-201.
35. Hefeneider SH, Cornell KA, Brown Le et al. Nucleosomes and DNA bind to specific cell surface molecules on murine cells and induce cytokine production. Clin Immunol Immunopathol 1992; 63:245-251.
36. Lacki JK, Leszczynski P, Kelemen J et al. Cytokine concentration in serum of lupus erythematosus patients: the effect on acute phase response. J Med 1997; 28:99-107.
37. Csiszar A, Nagy G, Gergely P et al. Increased interferon-gamma (IFN-gamma), IL-10 and decreased IL-4 mRNA expression in peripheral blood mononuclear cells (PBMC) from patients with systemic lupus erythematosus. Clin Exp Immunol 2000; 122:464-470.
38. Houssiau FA, Lefebvre C, Vanden Berghe M et al. Serum interleukin 10 titres in systemic lupus erythematosus reflect disease activity. Lupus 1995; 4:393-395.
39. Park YB, Kim DS, Lee WK et al. Elevated serum interleukin-15 levels in systemic lupus erythematosus. Yonsei Med J 1999; 40:343-348.

40. Linker-Israeli M, Honda M, Nand R et al. Exogenous IL-10 and IL-4 down-regulate IL-6 production by SLE-derived PBMC. Clin Immunol 1999; 91:6-16.
41. Mongan EA, Ramdahin S, Warrington RJ. Interleukin-10 response abnormalities in systemic lupus erythematosus. Scand J Immunol 1997; 46:406-412.
42. Kitamura M, Suto T, Yokoo T et al. Transforming growth factor-beta 1 is the predominant paracrine inhibitor of macrophage cytokine synthesis produced by glomerular mesangial cells. J Immmunol 1996; 156:2964-2971.
43. Horwitz DA, Gray JD, Ohtsuka K et al. The immunoregulatory effects of NK cells: The role of TGF-beta and implications for autoimmunity. Immunol today 1979; 18:538-542.
44. Ohtsuka K, Gray JD, Stimmler MM et al. Decreased production of TGF-beta by lymphocytes from patients with systemic lupus erythematosus. J Immunol 1998; 160:2539-2545.
45. Viallard JF, Pellegrin JL, Ranchin V et al. Th1 (IL-2, interferon-gamma (IFN-gamma)) and Th2 (IL-10, IL-4) cytokine production by peripheral blood mononuclear cells (PBMC) from patients with systemic lupus-erythematosus (SLE). Clin Exp Immunol 1999; 115:189-195.
46. Parodi A, Massone C, Cacciapuoti M et al. Measuring the activity of the disease in patients with cutaneous lupus erythematosus. Br J Dermatol 2000; 142:457-460.
47. Haas N, Ryffel B, Le Hir M. IFN-gamma receptor deletion prevents autoantibody production and glomerulonephritis in lupus-prone (NZB x NZW) F1 mice. J Immunol 1998; 160:3713-3718.
48. Seery JP, Carroll JM, Cattel V et al. Antinuclear autoantibodies and lupus nephritis in transgenic mice expressing epidermis. J Exp Med 1997; 186:1451-1459.
49. Nagy G, Pallinger E, Antal-Syalmas P et al. Measurement of intracellular interferon-gamma and interleukin-4 in whole blood T lymphocytes from patients with systemic lupus erythematosus. Immunol Lett 2000; 74:207-210.
50. Toro JR, Finlay D, Dou X et al. Detection of type 1 cytokines in discoid lupus erythematosus. Arch Dermatol 2000; 136:1497-1501.
51. Jones BM, Liu T, Wong RW. Reduced in vitro production of interferon-gamma, interleukin-4 and interleukin-12 and increased production of interleukin-6, interleukin-10 and tumour necrosis factor-alpha in systemic lupus erythematosus. Weak correlations of cytokine production with disease activity. Autoimmunity 1999; 31:117-124.
52. Liu TF, Jones BM. Impaired production of IL-12 in systemic lupus erythematosus. II IL-12 production in vitro is correlated negatively with serum IL-12, positively with serum IFN-gamma and negatively with disease activity in SLE. Cytokines 1998; 10:148-153.
53. Weber JC, Korganow AS, Ginsbourger M et al. Soluble tumour necrosis factor receptors in patients with systemic lupus erythematosus. Presse Med 1998; 27:1941-1945.
54. Wolf RE, Brelsford WG. Soluble interleukin-2 receptors in systemic lupus erythematosus. Arthritis Rheum 1988; 31:729-735.
55. Wong CK, Ho CY, Li EK et al. Elevation of proinflammatory cytokine (IL-18, IL-17, IL-12) and Th2 cytokine (IL-4) concentrations in patients with systemic lupus erythematosus. Lupus 2000; 9:589-593.
56. Wong CK, Li EK, Ho CY et al. Elevation of plasma interleukin-18 concentration is correlated with disease activity in systemic lupus erythematosus. Rheumatology 2000; 39:1078-1081.
57. Funauchi M, Ikoma S, Enomoto H et al. Decreased Th1-like and increased Th2-like cells in systemic lupus erythematosus. Scand J Rheumatol 1998; 27:219-224.
58. Carvalho D, Savage Cos, Black CM et al. IgG antiendothelial cell autoantibodies from scleroderma patients induce leukocyte adhesion to human vascular endothelial cells in vitro. Induction of adhesion molecules expression and involvement of endothelium-derived cytokines. J Clin Invest 1996: 101:2029-2035.
59. Harvey GR, McHugh NJ. Serologic abnormalities in systemic sclerosis. Curr Opin Rheumatol 1999; 11:495-502.
60. Falkner D, Wilson J, Fertig N et al. Studies of HLA-DR and DQ alleles in systemic sclerosis patients with autoantibodies to RNA polymerases and U3-RNP (fibrillarin). J Rheumatol 2000; 27:1196-1202.
61. Ignotz R, Massague J. Transforming growth factor beta stimulates the expression of fibronectin and collagen and the incorporation into the extracellular matrix. J Biol Chem 1986; 261:4337-4345.
62. Roberts AB, Sporn MB, Assoian RK et al. Transforming growth factor type beta: Rapid induction of fibrosis and angiogenesis in vivo and stimulation of collagen formation in vitro. Proc Natl Acad Sci USA 1986; 83:4167-4171.
63. Fertin C, Nicolas JF, Gillery P et al. Interleukin-4 stimulates collagen synthesis by normal and scleroderma fibroblasts in dermal equivalents. Cell Mol Biol 1991; 37:823-329.
64. Postlethwaite AE, Holness MA, Katai H et al. Human fibroblasts synthesize elevated levels of extracellular matrix proteins in response to interleukin 4. J Clin Invest 1992; 90:1479-1485.

65. Le Roy EC. The pathogenesis of systemic sclerosis. Clin Exp Rheumatol 1989; 7:S135-137.
66. Kulozik M, Hogg A, Lankat-Buttgereit B et al. Colocalization of transforming growth factor beta2 with alpha 1(I) procollagen mRNA in tissue sections of patients with systemic sclerosis. J Clin Invest 1990; 86:917-922.
67. Teunisen MB, Koomen CW, de Waal Malefyt R et al. Interleukin-17 and interferon-gamma synergize in the enhancement of proinflammatory cytokine production by human keratinocytes. J Invest Dermatol 1998; 111:645-649.
68. Albanesi C, Cavani A, Girolomoni G. IL-17 is produced by nickel-specific T lymphocytes and regulates ICAM-1 expression and chemokine production in human keratinocytes: Synergistic or antagonistic effects with IFN-gamma and TNF-alpha. J Immunol 1999; 162:494-502.
69. Albanesi C, Scarponi C, Cavani A et al. Interleukin-17 is produced by both Th1 and Th2 lymphocytes, and modulates interferon-gamma- and interleukin-4 induced activation of human keratinocytes. J Invest Dermatol 2000; 115:81-87.
70. Kurasawa K, Hirose K, Sano H et al. Increased interleukin-17 production in patients with systemic sclerosis. Arthritis Rheum 2000; 43:2455-2460.
71. Kawaguchi Y, Hara M, Wright TM. Endogenous IL-1alpha from systemic sclerosis fibroblasts induces IL-6 and PDGF-A. J Clin Invest 1999; 103:1253-1260.
72. Needleman BW, Wigley FM, Stair RW. Interleukin-1, interleukin-2, interleukin-4, interleukin 6, tumor necrosis factor alpha and interferon-gamma levels in sera from patients with scleroderma. Arthritis Rheum 1992; 35:67-72.
73. Gruss C, Reed JA, Altmeyer P et al. Induction of interstitial collagenase (MMP-1) by UVA-1 phototherapy in morphea fibroblasts. Lancet 1997; 350:1295-1296.
74. Wlaschek M, Heinen G, Poswig A et al. UVA-induced autocrine stimulation of fibroblast-derived collagenase/MMP-1 by interrelated loops of interleukin-1 and interleukin-6. Photochem Photobiol 1994; 59:550-556.
75. Yamamoto T, Eckes B, Krieg T. Effect of interleukin-10 on the gene expression of type I collagen, fibronectin, and decorin in human skin fibroblasts: differential regulation by transforming growth factor-beta and monocyte chemoattractant protein-1. Biochem Biophys Res Commun 2001; 281:200-205.
76. Cotton SA, Herrick AL, Jayson MI et al. TGF beta—A role in systemic sclerosis? J Pathol 1998; 184:4-6.
77. Kawakami T, Ihn H, Xu W et al. Increased expression of TGF-beta receptors by scleroderma fibroblasts: Evidence for contribution of autocrine TGF-beta signaling to scleroderma phenotype. J Invest Dermatol 1998; 110:47-51.
78. Susol E, Rands AL, Herrick A et al. Association of markers for TGFbeta3, TGFbeta2 and TIMP1 with systemic sclerosis. Rheumatology 2000; 39:1332-1336.
79. Jelaska A, Korn JH. Role of apoptosis and transforming growth factor beta 1 in fibroblast selection and activation in systemic sclerosis. Arthritis Rheum 2000; 43:2230-2239.
80. Abraham DJ, Shiwen X, Black CM et al. Tumour necrosis factor alpha suppresses the induction of connective tissue growth factor by transforming growth factor-beta in normal and scleroderma fibroblasts. J Biol Chem 2000; 275:15220-15225.
81. Haustein UF, Anderegg U. Pathophysiology of scleroderma: An update. J Eur Acad Dermatol Venereol 1998; 11:1-8.
82. Makhluf HA, Stepniakowska J, Hoffman S et al. IL-4 upregulates tenascin synthesis in scleroderma and healthy skin fibroblasts. J Invest Dermatol 1996; 107:856-859.
83. Rozman B, Bozic B, Kos-Golja M et al. Immunoserological aspects of idiopathic inflammatory muscle disease. Wien Klin Wochenschr 2000; 112:722-727.
84. Sugiura T, Kawaguchi Y, Harigai M et al. Increased CD40 expression on muscle cells of polymyositis and dermatomyositis: Role of CD40-CD40 ligand interaction in IL-6, IL-8, IL-15, and monocyte chemoattractant protein-1 production. J Immunol 2000; 164:6503-6600.
85. Lundberg I, Kratz AK, Alexanderson H et al. Decreased expression of interleukin-1alpha, interleukin-1beta, and cell adhesion molecules in muscle tissue following corticosteroid treatment in patients with polymyositis and dermatomyositis. Arthritis Rheum 2000; 43:336-348.
86. Nagaraju K, Casciola-Rosen L, Rosen A et al. The inhibition of apoptosis in myositis and in normal muscle cells. J Immunol 2000; 164:5459-5465.
87. Behrens L, Bender A, Johnson MA et al. Cytotoxic mechanisms in inflammatory myopathies: Co-expression of Fas and protective Bcl-2 in muscle fibres and inflammatory cells. Brain 1997; 120:929-938.
88. Choi YC, Dalakas MC. Expression of matrix metalloproteinases in the muscle of patients with inflammatory myopathies. Neurology 2000; 54:65-71.

89. Dalakas MC. Molecular immunology and genetics of inflammatory muscle diseases. Arch Neurol 1998; 55:1509-1512.
90. Fan JL, Memar O, McCormick DJ et al. BALB/c mice produce blister-causing antibodies upon immunization with a recombinant human desmoglein3. J Immunol 1999; 163:6228-6235.
91. Hashimoto T, Amagai M, Watanabe K et al. A case of pemphigus vulgaris showing reactivity with pemphigus antigens (Dsgl and Dsg3) and desmocollins. J Invest Dermatol 1995; 104:541-544.
92. Harman KE, Gratian MJ, Bhogal BS et al. A study of desmoglein 1 antoantibodies in pemphigus vulgaris: Racial differences in frequency and the association with a more severe phenotype. Br J Dermatol 2000; 143:343-348.
93. Yancey KB. From bedside to bench and back. Arch Dermatol 1994; 130:983-987.
94. Hertl M, Riechers R. Analysis of the T cells that are potentially involved in autoantibody production in pemphigus vulgaris. J Dermatol 1999; 26:748-752.
95. Hertl M. Humoral and cellular autoimmunity in autoimmune bullous skin disorders. Int Arch Allergy Immunol 2000; 122:91-100.
96. Wucherpfennig KW, Yu B, Bhol K et al. Structural basis for major histocompatibility complex (MHC)-linked susceptibility to autoimmunity: charged residues of a single MHC binding pocket confer selective presentation of self-peptides in pemphigus vulgaris. Immunol 1995; 92:11935-11939.
97. Lin MS, Swartz SL, Lopez A et al. Development and characterization of desmoglein-3 specific T cells from patients with pemphigus vulgaris. J Clin Invest 1997; 99:31-40.
98. Nishifuji K, Amagai M, Kuwana M et al. Detection of antigen-specific B cells in patients with pemphigus vulgaris by enzyme-linked immunospot assay: Requirement of T cell collaboration for autoantibody production. J Invest Dermatol 2000; 112:88-94.
99. Kricheli D, David M, Frusic-Zlotkin M et al. The distribution of pemphigus vulgaris-IgG subclasses and their reactivity with desmoglein 3 and 1 in pemphigus patients and their first-degree relatives. Br J Dermatol 2000; 143:337-342.
100. Hertl M, Amagai M, Sundaram H et al. Recognition of desmoglein 3 by autoreactive T cells in pemphigus vulgaris patients and normals. J Invest Dermatol 1998; 110:62-66.
101. Feliciani C, Toto P, America P et al. In vitro and in vivo expression of interleukin-1alpha and tumor necrosis factor-alpha mRNA in pemphigus vulgaris: Interleukin-1alpha and tumor necrosis factor-alpha are involved in acantholysis. J Invest Dermatol 2000; 114:71-77.
102. D'Auria L, Bonifati C, Mussi A et al. Cytokines in the sera of patients with pemphigus vulgaris: interleukin-6 and tumour necrosis factor-alpha levels are significantly increased as compared to healthy subjects and correlate with disease activity. Eur Cytokine Network 1997; 8:383-387.
103. Toto P, Feliciani C, Amerio P et al. Immune modulation in Pemphigus vulgaris: Role of CD28 and IL-10. J Immunol 2000; 164:522-529.
104. Bhol KC, Rojas AI, Khan IU et al. Presence of interleukin 10 in the serum and blister fluid of patients with pemphigus vulgaris and pemphigoid. Cytokine 2000; 12:1076-1083.
105. Lin MS, Fu CL, Aoki Vet al. Desmoglein-1-specific T lymphocytes from patients with endemic pemphigus foliaceus (fogo selvagem). J Clin Invest 2000; 105:207-213.
106. Lin MS, Fu CL, Giudice GJ et al. Epitopes targeted by bullous pemphigoid T lymphocytes and autoantibodies map to the same sites on the bullous pemphigoid 180 ectodomain. J Invest Dermatol 2000; 115:955-961.
107. Hata Y, Fujii Y, Tsunoda K et al. Production of the entire extracellular domain of BP180 (type XVII collagen) by baculovirus expression. J Dermatol Sci 2000; 23:183-190.
108. Christophoridis S, Budinger L, Borradori L et al. IgG, IgA and IgE autoantibodies against the ectodomain of BP180 in patients with bullous and cicatricial pemphigoid and linear IgA bullous dermatosis. Br J Dermatol 2000; 143:349-355.
109. Roh JY, Yee C, Lazarova Z et al. The 120kDa soluble ectodomain of type XVII collagen is recognized by autoantibodies in patients with pemphigoid and linear IgA dermatosis. Br J Dermatol 2000; 143:104-111.
110. Kromminga A, Scheckenbach C, Georgi M et al. Patients with bullous pemphigoid and linear IgA disease show a dual IgA and IgG autoimmune response to BP180. J Autoimmun 2000; 15:293-300.
111. Zillikens D, Herzele K, Georgi M et al. Autoantibodies in a subgroup of patients with linear IgA disease react with the NC16A domain of BP180. J Invest Dermatol 1999; 113:947-953.
112. Zillikens D, Mascaro JM, Rose PA et al. A highly sensitive enzyme-linked immunosorbent assay for the detection of circulating anti-BP180 autoantibodies in patients with bullous pemphigoid. J Invest Dermatol 1997; 109:679-683.
113. Liu Z, Diaz LA, Troy JL et al. A passive transfer model of the organ-specific autoimmune disease, bullous pemphigoid, using antibodies generated against the hemidesmosomal antigen, BP180. J Clin Invest 1993; 92:2480-2488.

114. Schmidt E, Reimer S, Kruse N et al. Autoantibodies to BP180 associated with bullous pemphigoid release interleukin-6 and interleukin-8 from cultured human keratinocytes. J Invest Dermatol 2000; 115:842-848.
115. Büdinger L, Borradori L, Yee C et al. Identification and characterization of autoreactive T cell responses to bullous pemphigoid antigen 2 in patients and healthy controls. J Clin Invest 1998; 102:2082-2089.
116. Zillikens D, Schuessler M, Dummer R et al. Tumour necrosis factor in blister fluids of bullous pemphigoid. Eur J Dermatol 1992; 2:429-431.
117. Zillikens D, Ambach A, Schuesller M et al. The interleukin-2 receptor in lesions and serum of bullous pemphigoid. Arch Dermatol Res 1992; 184:141-145.
118. Grando SA, Glukhenky BT, Drannik GN et al. Mediators of inflammation in blister fluids from patients with pemphigus vulgaris and bullous pemphigoid. Arch Dermatol 1989; 125:925-930.
119. Kaneko F, Minagawa T, Takiguchi Y et al. Role of cell-mediated immune reaction in blister formation of bullous pemphigoid. Dermatology 1992; 184:34-39.
120. Schmidt E, Mittnacht A, Schömig H et al. Detection of IL-1 alpha, IL-1 beta and IL-1 receptor antagonist in blister fluid of bullous pemphigoid. J Dermatol Science 1996; 11:142-147.
121. Rhodes LE, Hashim IA, McLaughlin PJ et al. Blister fluid cytokines in cutaneous inflammatory bullous disorders. Acta Derm Venerol 1999; 79:288-290.
122. Schmidt E, Bastian B, Dummer R et al. Detection of elevated levels of IL-4, IL-6, and IL-10 in blister fluid of bullous pemphigoid. Arch Dermatol Res 1996; 288:353-357.
123. Schmidt E, Ambach A, Bastian B et al. Elevated levels of interleukin-8 in blister fluid of bullous pemphigoid compared with suction blisters of healthy control subjects. J Am Acad Dermatol 1996; 34:310-312.
124. Sun CC, Wu J, Wong TT et al. High levels of interleukin-8, soluble CD4 and soluble CD8 in bullous pemphigoid blister fluid. The relationship between local cytokine production and lesional T-cell activities. Br J Dermatol 2000; 143:1235-1240.
125. De Pita O, Frezzolini A, Cianchini G et al. T-helper 2 involvement in the pathogenesis of bullous pemphigoid: role of soluble CD30 (sCD30). Arch Dermatol Res 1997; 289:667-670.
126. Endo H, Iwamoto I, Fujita M et al. Increased immunoreactive interleukin-5 levels in blister fluids of bullous pemphigoid. Arch Dermatol Res 1992; 284:312-314.
127. Engineer L, Bhol K, Kumari S et al. Bullous pemphigoid: Interaction of interleukin5, anti-basement membrane zone antibodies and eosinophils. A preliminary observation. Cytokine 2001; 13:32-38.
128. Wakugawa M, Nakamura K, Hino H et al. Elevated levels of eotaxin and interleukin-5 in blister fluid of bullous pemphigoid: Correlation with tissue eosinophilia. Br J Dermatol 2000; 143:112-116.
129. Rico MJ, Benning C, Weingart ES et al. Characterization of skin cytokines in bullous pemphigiod and pemphigus vulgaris. Br J Dermatol 1999; 140:1079-1086.
130. Eming R, Budinger L, Riechers R et al. Frequency analysis of autoreactive T-helper 1 and 2 cells in bullous pemphigoid and pemphigus vulgaris by enzyme-linked immunospot assay. Br J Dermatol 2000; 143:1279-1292.
131. Wilson AG, Clay FE, Crane AM et al. Comparative genetic association of human leukocyte antigen class II and tumour necrosis factor-alpha with dermatitis herpetiformis. J Invest Dermatol 1995; 104:856-858.
132. Graeber M, Baker BS, Garioch JJ et al. The role of cytokines in the generation of skin lesions in dermatitis herpetiformis. Br J Dermatol 1993; 129:530-532.
133. Smith AD, Baheri B, Steilein RD et al. Expression of interleukin-4 and interferon-gamma in the small bowel of patients with dermatitis herpetiformis and isolated gluten-sensitive enteropathy. Dig Dis Sci 1999; 44:2124-2132.
134. Hall RP, Smith AD, Streilein RD. Increased production of IL-4 by gut T-cell lines from patients with dermatitis herpetiformis compared to patients with isolated gluten-sensitive enteropathy. Dig Dis Sci 2000; 45:2036-2043.
135. Caproni M, Feliciani C, Fuligni A et al. Th2-like cytokine activity in dermatitis herpetiformis. Br J Dermatol 1998; 138:242-247.
136. Desreumaux P, Delaporte E, Colombel JF et al. Similar IL-5, IL-3,and GM-CSF syntheses by eosinophils in the jejunal mucosa of patients with celiac disease and dermatitis herpetiformis. Clin Immunol Immunopathol 1998; 88:14-21.
137. Amerio P, Verdolini R, Giangiacomi M et al. Expression of eotaxin, interleukin 13 and tumour necrosis factor-alpha in dermatitis herpetiformis. Br J Dermatol 2000; 143:974-978.
138. Jonsson CA, Erlandsson M, Svensson L et al. Mycophenolate mofitil ameliorates perivasular T lymphocyte inflammation and reduces the double-negative T cell population in SLE-prone MRLlpr/lpr mice. Cell Immunol 1999; 197:136-144.

139. Turner MS, Sutton D, Sauder DN. The use of plasmapheresis and immunosuppression in the treatment of pemphigus vulgaris. J Am Acad Dermatol 2000; 43:1058-1064.
140. Wolf R, Orni-Wasserlauf R. A century of the synthesis of dapsone: its anti-infective capacity now and then. Pharmacology and therapeutics 2000; 39:779-783.
141. Walchner M, Meurer M, Plewig G et al. Clinical and immunologic parameters during thalidomide treatment of lupus erythematosus. Pharmacology and therapeutics 2000; 39:383-388.
142. Engineer L, Bhol K, Razzaque Ahmed A. Analysis of current data on the use of intravenous immunoglobulins in management of pemphigus vulgaris. J Am Acad Dermatol 2000; 43:1049-1057.
143. Faulds D, Goa KL, Benfield P. Cyclosporin. A Rewiew of its pharmacodynamic and pharmacokinetic properties, and therapeutic use in immunoregulatory disorders. Drugs 1993; 45:953-1040.
144. Follath F, Fontana A, Leumann E et al. Ciclosporin bei Autoimmmunkrankheiten. Schweiz Med Wochenschr 1994; 124:1232-1239.
145. Finck BK, Chan B, Wofsy D. Interleukin 6 promotes murine lupus in NZB/NZW F1 mice. J Clin Invest 1994; 94:585-591.
146. Peng SL, Moslehi J, Craft J. Role of interferon-gamma and interleukin-4 in murine lupus. J Clin Invest 1997; 99:1936-1940.
147. Lawson BR, Prud'homme GJ, Chang Y et al. Treatment of murine lupus with cDNA encoding IFN-gammaR/Fc. J Clin Invest 2000; 106:207-215.
148. Leeb BF, Sautner J. Anti-TNF-alpha therapy as a new option in treatment of rheumatoid arthritis? Wien Med Wochenschr 1999; 149:554-557.
149. Prud'homme GJ. Gene therapy of autoimmune diseases with vectors encoding regulatory cytokines or inflammatory cytokine inhibitors. J Gene Med 2000; 2:222-232.

Chapter 14

Involvement of Cytokines in the Pathogenesis of Systemic Lupus Erythematosus

B.R. Lauwerys and F.A. Houssiau

Introduction

Systemic lupus erythematosus (SLE) is characterized by overt polyclonal B-cell activation and autoantibody (Ab) production. By contrast, cellular immune responses against allo- or recall antigens are significantly impaired. Many evidences indicate that IL-10 overproduction plays a pivotal role in the disease and the contribution of the IL-10/IL-12 imbalance to the pathophysiology of SLE will be extensively discussed. The authors will further summarize the available data about the involvement of IFN-γ, TNF-α, TGF-β and TALL-1. Other cytokines (IL-1, IL-2, IL-4, IL-6, IL-16, IL-17 and IL-18) will be briefly discussed.

Numerous abnormalities of the cytokine network have been described in patients suffering from systemic lupus erythematosus as well as in murine lupus models. Some of them were shown to play a pivotal physiopathological role in certain T-cell, B-cell or antigen-presenting cell (APC) dysfunctions characteristic of the disease, while others are more likely to be innocent bystanders. It should be stressed, moreover, that not all the data discussed herein fit into a single picture. The heterogeneity of human SLE, the influence of disease activity and therapy on cytokine production and function, the relative shortage of human lupus peripheral blood-derived mononuclear cells (PBMC), together with the lack of a perfectly-matched animal model, further sophisticate the issue. For these very reasons, this review will focus on the cytokines that the authors—wrongly or wrightly—consider as the major players in the game, namely IL-10, IL-12, TNF-α, IFN-γ and the recently-described TALL-1/BLys/BAFF. Before doing so, we felt worthwhile to briefly glance through the major immune abnormalities observed in SLE.

Dysregulation of B-, T- and APC Function in SLE

SLE is a prototypical systemic autoimmune disorder occurring mostly in young females and characterized by pleiotropic symptoms grading from benign (e.g., cutaneous rash, fever, arthritis, serositis) to severe (e.g., glomerulonephritis, pancytopenia, seizures). The serological hallmarks are a striking hypergammaglobulinaemia due to polyclonal B cell activation and the presence of specific autoantibodies targetting chromatin constituents (DNA, histone proteins) and cytoplasmic ribonucleoproteins (Ro, La, Sm, etc.).[1]

Anti-double stranded (ds)DNA Ab are the most specific markers of the disease. Their pathogenic role has been extensively demonstrated, in particular in lupus glomerulonephritis.[2] In mice as in humans, the onset of glomerulonephritis is accompanied by a striking rise in high affinity anti-dsDNA Ab serum titers.[3,4,5,6] Moreover, these Ab can be eluted from post-mortem kidney specimens[7]. Finally, the injection of anti-dsDNA Ab producing hybridomas into SCID (Severe Combined Immunodeficiency) mice induces a lupus-like glomerulonephritis.[8]

Studies performed in murine models of SLE have dramatically improved our knowledge on the physiopathology of the human disease. Female (NZB x NZW)F_1 (BWF1) and (NZB x

Cytokines and Chemokines in Autoimmune Disease, edited by Pere Santamaria.

SWR)F_1 (SWF1) hybrid mice spontaneously develop an autoimmune glomerulonephritis associated with anti-DNA Ab deposits. Both models share many features with the human disease and are extensively used for pathogenic or therapeutical studies. MRL/*lpr* mice also produce high amounts of autoAb, including nephritogenic anti-DNA Ab and are therefore commonly used as a murine model of SLE. The *lpr* trait result from an autosomal recessive mutation in Fas leading to accumulation of nondeleted autoimmune T and B lymphocytes (*lpr*: lymphoproliferation). Although a similar defect has been described in a rare pediatric lymphoproliferative disorder, named Canale-Smith syndrome,[9] it has not been found in human SLE. Finally, chronic Graft-*Versus*-Host Disease (cGVHD), induced by the injection of parental DBA/2 splenocytes into (C57Bl/6 x DBA/2)F_1 hybrids, is also considered as a lupus murine model. This allogeneic reaction is characterized by preferential activation of donor Th2 lymphocytes leading to increased IL-4 and IL-10 production, polyclonal B cell activation (mainly IgG1 and IgE) and production of nephritogenic autoAb.[10,11,12] It should be stressed, however, that such a caricatural type-2 bias is not observed in the human disease.

B Cell Dysfunction in SLE

Overt polyclonal B cell activation resulting in serum hypergammaglobulinaemia is a hallmark of SLE.[13] This intense stimulation of the B cell population is accompanied by a marked skewing in V_L and V_H gene usage of peripheral CD19+ B cells.[14] Numerous defects in B cell phenotype and function have further been described both in affected humans and mice: increased cell surface expression of CD40L,[15,16] increased expression of CD86/B7.2,[17,18] decreased expression of complement receptors CD35 and CD21,[19] higher intracellular calcium concentrations and phosphorylated tyrosine residues after stimulation of SLE B cells with an anti-IgM Ab.[20,21] Whether these findings are primary B cell defects or secundary to aberrant stimulatory signals is still under investigation.[18,22]

SLE B cells can function as APC for autoreactive T cells, probably via over-expression of costimulatory molecules.[23,24,25] Interestingly, a subset of CD1c$^+$ circulating B cells were found to interact with CD1c-restricted double-negative T cells, that are present at a higher frequency in SLE. While coculture of CD1c-reactive T cells from healthy donors with CD1c$^+$ B cells results in the production of IgM Ab and little or no IgG, CD1c-restricted double negative T cells from SLE patients induce isotype switching and a striking increase in IgG production.[26]

T Cell Dysfunction in SLE

T cells play a pivotal role in the pathogenesis of the disease: the production of anti-DNA autoAb has been proven to be T-cell dependent. Thus, the pattern of somatic mutations in the V_H regions of both murine or human anti-DNA mAb has indicated that their production is antigen-driven.[27,28,29] Moreover, the injection of depleting anti-CD4 Ab to BWF1 mice decreases serum anti-DNA Ab titers and delays disease onset.[30] Recently, Mohan et al isolated CD4$^+$ T cell clones from SWF1 mice on the basis of their capacity to drive anti-DNA Ab production.[31] They found that some of these clones were specifically stimulated by nucleosomes or by peptides derived from their histone proteins, thereby suggesting that nucleosomes, complexes made of a 146-bp-DNA loop wound round a histone octamer core, could be one of the autoantigens involved in SLE.[32,33,34,35,36] The following picture can be sketched: anti-DNA Ab-bearing B cells trap circulating DNA-binding proteins, such as nucleosomal DNA. The complex is endocytosed, processed and peptides derived from the binding proteins are presented by anti-DNA B cells to specific pathogenic T helper cells which in turn drive B cells to make pathogenic autoAb, by secreting B cell stimulating cytokines and by providing costimulation through accessory molecules.

Paradoxically, cellular immune responses are significantly impaired in SLE. PBMC from lupus patients proliferate less in vitro than control PBMC in response to allogeneic targets. Similarly, their IFN-γ production in the presence of recall antigens (such as tetanus toxoid) is decreased as compared to controls.[37] The clinical significance of these observations is unknown.

However, impaired cellular imunity might contribute to increase the susceptibility to infection in SLE patients.

APC Dysfunction in SLE

Tsokos et al found that the expression of CD80/B7.1 at the surface of IFN-γ stimulated monocytes was decreased in lupus patients as compared to controls.[38] Interestingly, they demonstrated that this defect was involved in the impaired cellular immune responses typical of SLE since they evidenced that IFN-γ production by SLE T cells stimulated with recall antigens could be restored to the levels of controls by the addition of CD80/B7.1 transfected cells. Similarly, Horwitz et al showed that addition of an anti-CD28 mAb (mimicking the costimulatory signals constituted by CD80/B7.1 and CD86/B7.2) restored the proliferation of lupus PBMC in response to an anti-CD2 mAb.[39]

Many evidences have been accumulated that phagocyte function is impaired in SLE.[40,41,42,43] Of note, Herrmann et al recently found that phagocytosis of apoptotic cells by macrophages derived in vitro from SLE PBMC was reduced when compared to controls, thereby suggesting that impaired APC function and defective clearance of apoptotic waste could play a role in the initiation of the autoimmune responses characteristic of SLE.[44]

Role of IL-10 in the Pathogenesis of SLE

Physiological Functions of IL-10

hIL-10 is a 18 kDa protein merely produced by monocytic cells (macrophages, dendritic cells), B lymphocytes, activated Th cells, mast cells and keratinocytes.[45] It displays strong regulatory activities on B and T lymphocytes, as well as on APC.

hIL-10 is a potent growth and activation factor for B lymphocytes. It stimulates the expression of HLA class II molecules on resting B cells. Moreover, IL-10 induces the proliferation of activated B lymphocytes (stimulated with an anti-CD40 mAb or by cross-linking of the Ig receptor) and their differentiation in IgA-, M- or G-secreting plasmocytes.[46] Noteworthy, IL-10 displays differential effects on the apoptosis of *Staphylococcus aureus* Cowan I (SAC) activated lymphocytes according to their stage of activation. In the initial phase, IL-10 induces the apoptosis of B lymphocytes, a phenomenon that is reverted by addition of IL-2. In a later stage, IL-10 inhibits apoptosis of SAC-activated B lymphocytes and increases their expression of *bcl-2*.[47,48]

By contrast with its positive effects on B cell proliferation and activation, IL-10 displays strong inhibitory activities on the function of T lymphocytes. Thus, IL-10 inhibits IFN-γ production by $CD4^+$ Th1 and $CD8^+$ cytolytic clones. Interestingly, IL-10 does not act directly on T cells but via down-regulation of monocytes/macrophages functions.[49] In mice, inhibition of IFN-γ production by Th1 clones could only been evidenced in case of antigen presentation by macrophages or total splenocytes but not in case of presentation by purified B cells or aspecific T cell stimulation with anti-CD3 mAb.[50]

Finally, IL-10 inhibits HLA class II molecule expression, IL-1α, TNF-α, GM-CSF and NO production by activated monocytes[51,52,53,54] and has similar inhibitory effects on the maturation and function of dendritic cells (DC)[55,56,57,58,59]. In particular, it inhibits IL-12 production by LPS or anti-CD40 stimulated DC[60] and down-regulates HLA class II and CD86/B7.2 expression by LPS stimulated DC. Surprisingly, only immature or maturing DC are sensitive to these inhibitory effects of IL-10 but not mature DC.[61]

Excessive Production of IL-10 in SLE

In 1993, Llorente et al discovered that PBMC from SLE patients spontaneously produced much more IL-10 than controls.[62,63] This observation was rapidly confirmed by others who evidenced that serum levels of IL-10 were elevated in SLE patients, commensurate with clinical and/or biological indices of disease activity (SLEDAI, anti-DNA Ab).[64,65] Interestingly, monocytes and B cells, rather than T cells, are responsible for this increased production of IL-10.

The reasons for this abnormal IL-10 production is currently unknown. A genetic predisposition is suggested by the observation that healthy members of SLE *multiplex* families (families in which more than one member is affected) have higher serum IL-10 titers than controls.[66,67] Extensive studies of the hIL-10 promoter gene polymorphism failed to reveal convincing associations of any haplotype with disease expression. Single base substitutions in positions –1082, -819 and -592 from the transcriptional start site characterize 3 different haplotypes (due to linkage desequilibrium): GCC, ACC or ATA.[68] Noteworthy, A in position -1082 is associated with enhanced in vitro IL-10 production by PBMC from healthy subjects. None of these haplotypes could be associated with SLE.[69] However, some of them are found more frequently in subgroups of patients, e.g., GCC in patients with anti-Ro Ab,[66] ATA in patients with renal disease,[69] etc. Other polymorphic variants—IL-10G and IL-10R—characterized by the distribution of microsatellite markers, have been studied by Eksdale et al who found a skewing in their allelic distribution amongst SLE patients (increased representation of IL-10G13 and decreased representation of IL-10G9).[70] However, their study only evaluated a small number of individuals and needs to be confirmed on a larger scale.

Involvement of IL-10 in SLE Pathogenesis

Murine models of the disease have brought considerable evidence confirming the central role of IL-10 in autoAb production. Thus, Ishida et al demonstrated that administration of blocking anti-mIL-10 Ab to BWF1 mice significantly inhibits their anti-DNA Ab serum levels.[71] Of note, the onset of proteinuria and glomerulonephritis was delayed in the treated group. Interestingly, the protective effect of the anti-IL10 Ab regimen was suppressed by concurrent administration of blocking anti-TNF-α Ab. Conversely, injection of IL-10 in BWF1 mice accelerates disease onset and proteinuria.

Unexpectedly, Yin et al recently found that IL-10-deficient MRL/*lpr* mice develop a more severe disease than their control littermates.[72] This finding paralleled increased numbers of IFN-γ producing $CD4^+$ and $CD8^+$ T cells and increased IgG2a anti-DNA autoAb in IL-10-deficient animals. These contrasting data probably result from the distinct pathophysiological mechanisms involved in both strains of mice.

In human SLE, addition of neutralizing anti-hIL-10 Ab to SLE PBMC cultures inhibits in vitro total Ig and IgG production, whereas addition of blocking anti-hIL-6 Ab has no similar effect. Both anti-hIL-10 and anti-hIL-6 Ab do not inhibit Ig production by control PBMC.[73] More strikingly, administration of anti-hIL-10 Ab to SCID mice reconstituted with human SLE PBMC, inhibits in vivo IgG and anti-DNA Ab production. Finally, in an open trial, Llorente et al injected murine anti-hIL-10 mAb to 6 SLE patients with active disease. Although serum levels of anti-DNA Ab were unaffected, 5 of the 6 patients achieved complete clinical remission and their corticosteroid treatment could be significantly tapered.[74]

We recently studied the role of IL-10 in the impaired cellular immune responses of SLE PBMC against allogeneic targets. We confirmed that a subgroup of patients displayed strongly impaired in vitro proliferative responses against allogeneic DC. These patients had higher serum IL-10 titers and their deficient response against allogeneic DC could be restored in vitro by anti-IL-10 blocking Ab. IL-12 supplementation displayed similar effects, thereby suggesting that dysregulation of the IL-10/IL-12 balance plays a critical role in the impaired cellular immune responses observed in SLE patients.[75]

Role of IL-12 in the Pathogenesis of SLE

Physiological Functions of IL-12

IL-12 is a 70 kDa heterodimeric cytokine made of disulfide-linked α (p35) and β (p40) chains. The α chain is constitutively produced by almost all cell types while only monocytes, dendritic cells, neutrophils, keratinocytes and B-EBV transformed B cells also secrete the β chain and the bioactive heterodimer (IL-12 p70).[76] Both in mice and in humans, the β chain is produced in 10- to 100-fold excess over the α chain. These β chain monomers do not display any biological effect but some weak in vitro antagonism of IL-12 p70, when added at supraphysiological concentrations. Interestingly, recent work by Oppmann et al demonstrated that the α chain could also bind a p19 protein. The biological activities of the resulting cytokine, called IL-23, are very similar to those of IL-12 p70.[77] Incidentally, murine cells, but not human cells, also produce α chain homodimers $(p40)_2$ that are strong antagonists of IL-12 p70.[78-80]

IL-12 p70 stimulates the proliferation of activated T lymphocytes and their IFN-γ secretion.[81] In case of antigen-specific stimulation, IL-12 p70 plays a major role in driving the T cell response towards a Th1 pattern of cytokine secretion.[82,83,84] Further, IL-12 p70 enhances NK cell toxicity against tumoral or infected targets and their secretion of IFN-γ or TNF-α.[85-91] IL-12 p70 inhibits IgG1 and IgE production by B cells and stimulates their secretion of IgG2a, mainly via stimulation of IFN-γ and inhibition of IL-4 production by T cells[92,93]. Finally, IL-12 p70 displays striking synergies with IL-18, a recently discovered IFN-γ inducing cytokine, for the proliferation and activation of murine NK cells and for the production of IFN-γ by T, NK and B cells.[94-101]

Impaired Production of IL-12 p70 in SLE

Due to the distinct regulation of the α, β monomers and IL-12 p70 heterodimer, there is some confusion in the literature about the production of IL-12 in systemic lupus. Many evidences indicate that IL-12 p70 production is impaired in SLE. Liu et al have shown that SAC-stimulated monocytes from SLE patients produce in vitro significantly less IL-12 p70 than control monocytes. Interestingly, the IL-12 p70 concentrations in the culture supernatants are negatively correlated to IL-10 concentrations. Moreover, addition of neutralizing anti-IL-10 Ab increases the in vitro production of IL-12 p70 but has no effect on control monocytes.[102-104] Impaired production of IL-12 p70 has been confirmed by other groups[105] as well as in murine models of the disease. Thus, LPS stimulated peritoneal macrophages from BWF1 or MRL/*lpr* mice produce less IL-12 p70, TNF-α, IL-1 and IL-6 than macrophages obtained from control mice. Addition of TNF-α restores in vitro the production of IL-1 and IL-6 but has no effect on IL-12 p70 expression.[106]

In contrast to IL-12 p70, serum levels of IL-12 β chain monomers are significantly more elevated in SLE patients than in controls or in patients with rheumatoid arthritis (Lauwerys BR et al, unpublished data and ref. 107). The significance of this increased IL-12 β chain production is currently unknown.

Involvement of IL-12 in SLE Pathogenesis

IL-12 p70 displays prominent regulatory effects on immunoglobulin (Ig) and autoAb production in human and murine lupus. Thus, we found that IL-12 p70 inhibits in vitro Ig production by unstimulated SLE PBMC while it has no effect on control PBMC. In parallel, IL-12 inhibits the number of anti-DNA Ab-producing cells. The mechanism of the inhibition is still under investigation. Although IL-12 induces IFN-γ production and inhibits IL-10 synthesis by SLE PBMC, its inhibitory effect on Ig production is not dependent upon IFN-γ or IL-10. Thus, addition of a neutralizing anti-IFN-γ Ab reverts the inhibitory effect of IL-12 p70 on IL-10 synthesis but has no effect on Ig production.[108]

The regulatory role of IL-12 p70 on Ig production has been studied in murine models of the disease. Nakajima et al injected BWF1 mice with autologous splenocytes that had been incubated in vitro with IL-12 or IL-4. In both cases, the injection of autologous splenocytes stimulated the production of anti-DNA autoAb. However, the isotypes of the induced Ab were different according to the used cytokine: nephritogenic IgG1 and IgG3 autoAb in case of IL-4 whereas splenocytes incubated with IL-12 stimulated the production of IgG2a and IgG2b anti-DNA autoAb but had no effect on the glomerular lesions. Conversely, injection of anti-IL-4 Ab to BWF1 mice inhibited serum levels of IgG1 and IgG3 anti-DNA Ab and delayed the onset of glomerulonephritis while injection of anti-IL-12 Ab had no effect on clinical symptoms. Noteworthy, concurrent administration of both Ab suppressed the inhibitory effect of the anti-IL4 Ab.[109]

Administration of mercuric chloride ($HgCl_2$) in A.SW mice induces an auto-immune disease characterized by an elevation of serum levels of IgG1 and IgE, production of anti-DNA autoAb and the development of an autoimmune glomerulonephritis with Ig deposits. Concurrent injection of IL-12 inhibits serum IgG1 titers but has no effect on glomerular lesions, IgE titers or *ex vivo* production of IL-4 by PMA/ionomycine stimulated splenocytes.[110]

In MRL/*lpr* mice, injection of an IL-12 coding plasmid has been found to inhibit anti-DNA Ab production and accumulation of splenic and nodal double negative $CD4^-$ $CD8^-$ T cells. Serum levels of IFN-γ were enhanced. Clinical improvement was only modest with a slight reduction of glomerular lesions and proteinuria.[111] In another study, administration of IL-12 to MRL/*lpr* mice led to increased serum levels of IFN-γ and NO metabolites, and accelerated glomerulonephritis.[112] Accordingly, implantation of IL-12 secreting cells under the kidney capsule of MRL/*lpr* mice induced local accumulation of $CD4^+$, $CD8^+$ and $CD4^-CD8^-$ T cells and acceleration of the renal pathology that was IFN-γ dependent.[113] Again, these results should be interpreted in view of the specific immunopathological context of these mice and may not be relevant for the pathogenesis of human SLE.

Finally, Via et al found that administration of IL-12 in cGVHD mice leads to a profound drop in Ig and autoAb levels due to the stimulation of donor-derived anti-host $CD8^+$ cytolytic T cells. This effect is mediated via the induction of IFN-γ and is partially reverted by the concurrent administration of neutralizing anti-IFN-γ Ab.[114] Interestingly, we found that IL-12 also inhibits ex vivo Ig production by $CD8^+$ depleted splenocytes from cGVHD mice via the induction of IFN-γ. Moreover, we evidenced that IL-12 and IL-18 display striking synergistic and direct effects on highly purified B cells from cGVHD mice, inhibiting their Ig production and stimulating their IFN-γ synthesis.[99]

Other Cytokines

Involvement of TNF-α in SLE Pathogenesis

Data collected from animal models of SLE have brought considerable evidences that defective production of TNF-α by monocytes play a facilitating role in the disease. In 1988, Jacob et al found that the TNF-α gene of NZW mice was characterized by an unique restriction length polymorphism associated with decreased TNF-α production by peritoneal exudate cells from both NZW mice and BWF1 hybrids.[115] Mutations in the 3'-untranslated region of the gene were subsequently shown to participate in the low TNF-α secretion profile of these mice.[116] Later work by Alleva et al confirmed that monocytes from both BWF1 and MRL/lpr lupus strains produce less TNF-α than control mice.[106] By contrast, Brennan et al found enhanced levels of TNF-α mRNA in the renal cortices of BWF1 mice with glomerulonephritis, a finding that probably reflects local inflammation rather than systemic dysregulation of the cytokine.[117] In vivo experiments showed that early administration of high dose TNF-α to BWF1 mice delays the onset of glomerulonephritis and improves survival.[115,118] When injected later in the disease course, TNF-α has no beneficial effect and may even have a negative impact[117,119]. Conversely, F_1 hybrids from NZB and TNF deficient mice—(NZB x B6, 129 TNF-/-)F_1 hybrids—develop enhanced anti-DNA antibodies and severe renal disease similar to BWF1 mice, unlike (NZB x B6, 129 TNF+/+)F_1 control mice.[120] Moreover, as indicated earlier in this section, injection of anti-TNF-α Ab could suppress the positive effect induced by anti-IL-10 Ab on the disease course of BWF1 mice.[71]

In humans, several studies have indicated that serum levels of TNF-α are more elevated in SLE patients than in controls.[121-124] However, circulating titers of both p55 and p75 soluble TNF receptors (sTNFR) are similarly increased,[125-124] thereby probably inhibiting the biological activity of TNF-α in the serum. Accordingly, sera from SLE patients were found to inhibit the in vitro cytotoxic activity of TNF, due to increased sTNFR concentrations.[125] Genetic studies have found a close association between the TNF2 variant (A in position –308) of the TNF-α promoter gene, that is associated with increased TNF-α production, and SLE in

Caucasian or Afro-American populations.[127-130] In two studies, the association was due to linkage desequilibrium with HLA-DR3 alleles,[129,130] while two other studies described HLA-DR3 and TNF-308A as two independent susceptibility factors for SLE.[127,128] In South-African patients, in whom the HLA-DR2 allele is increased but not HLA-DR3, the TNF-308A allele is reduced rather than increased.[130] Komata et al focused on the TNF receptor 2 gene (TNFR2) polymorphism in SLE and found an association between the 196R allele (coding for an arginine in position 196) and disease in Japanese patients.[131] However, their findings could not be confirmed by other groups, nor in Asian, nor in European SLE populations.[132,133]

There is no data available about the (dys)regulatory role of TNF-α in human SLE. Anecdotally, Charles et al reported that, amongst 156 RA patients treated with an anti-TNF-α mAb, 21 developed low-affinity IgM anti-DNA Ab while one patient developed high titers IgG anti-DNA Ab and a self-limited clinical lupus syndrome.[134] Finally, it should be noted that TNF-α expression was detected by immunochemistry in 50% of the kidney biopsy specimens obtained from SLE patients, thereby raising intriguing therapeutical issues.[135]

Involvement of IFN-γ in SLE Pathogenesis

PBMC from lupus patients secrete less IFN-γ than control PBMC, either in basal conditions or after stimulation with IL-2.[105,136] Moreover, results of ELISPOT assays performed on PBMC from SLE patients indicate that the IL-10/IFN-γ ratio significantly correlates with disease activity.[137]

In BWF1 mice, IFN-γ administration increases the incidence and severity of disease manifestations.[138] Conversely, IFN-γR-deficient BWF1 mice or mice treated with anti-IFN-γ mAb or soluble IFN-γR have decreased rates of glomerulonephritis and mortality.[138-140] In IFN-γR-deficient BWF1 mice, serum titers of anti-dsDNA and anti-histone Ab were dramatically reduced, a finding that was not accounted for defective class switching to IgG2a, since levels of IgG1 and IgM were also reduced.[140] Incidentally, the frequency of B cell lymphoma in these mice was abnormally high.

In the MRL/*lpr* model, administration of IFN-γ in young animals displays protective effects on the disease course, lymph node enlargement and serum levels of anti-dsDNA Ab.[141] By contrast, treatment of 12-18-week-old animals results in higher IgG2a, IgG3 and autoAb levels, more aggressive glomerulonephritis and earlier mortality.[141] Similarly, implantation of IFN-γ-secreting cells under the renal capsule of MRL/*lpr* mice induces more severe kidney disease.[113] Conversely, IFN-γ or IFN-γR deficient MRL/*lpr* mice have lower levels of IgG2a and IgG3 anti-dsDNA levels and less severe glomerulonephritis.[142,143] Results of recent work by Lawson et al are well in line with these observations, by demonstrating that intramuscular injections of plasmids encoding an IFN-γR/Fc fusion protein reduced disease manifestations and mortality in MRL/*lpr* mice.[144]

Involvement of TALL-1 in SLE Pathogenesis

TALL-1 (TNF and Apoptosis Ligand-related Leucocyte-expressed Ligand 1), also called BAFF (B cell Activation Factor From the TNF family) or zTNF4, is a new cytokine from the TNF family that induces the activation and proliferation of B cells. Mice transgenic for zTNF4 display strong expansion of their peripheral B220$^+$ mature B lymphocyte population, increased number of B-1a lymphocytes (that are involved in autoantibody production), hypergammaglobulinaemia and production of nephritogenic anti-DNA Ab.[145,146]

Serum titers of zTNF4 are higher in autoimmune BWF1 and MRL/*lpr* murine strains as compared to parental or control mice and correlate positively with disease activity.[147] Similarly, Zhang et al recently showed that serum levels of the protein were increased in SLE patients and correlated with autoAb titers.[148] Involvement of TALL-1 in murine SLE pathogenesis was recently demonstrated by blocking experiments, using a fusion protein made of a TALL-1 receptor and a Ig Fc fragment (TACI-Ig). Thus, administration of TACI-Ig to BWF1 mice significantly delayed the onset of proteinuria and improved survival, thereby opening potential new therapeutical perspectives.[147]

Varia

IL-2 deficient Balb/c mice exhibit lymphoid hyperplasia and autoimmune features such as haemolytic anaemia or ulcerative bowel disease.[149,150] The contribution of IL-2 deficiency to autoimmune manifestations might be related to regulatory activities of IL-2 on apoptosis of T lymphocytes.[151] Accordingly, MRL/*lpr* and BWF1 mice have been shown to display significant defects in IL-2 production.[152,153] However, in vivo substitution experiments gave very contrasting results according to the timing and way of administration.[154-157] Phytohemagglutinin (PHA)-stimulated T cells from SLE patients with active disease also produce less IL-2 than control PBMC.[158] This defect is not primary but results from environmental factors. Thus, resting of T cells for 3 days[159] or adequate stimulation with PHA and PMA[160] or PHA and anti-CD28 mAb[161] restores their ability to produce normal amounts of IL-2. Solomou et al recently evidenced that defective production of IL-2 by SLE T cells was due to transcriptional repression mediated by an increase in phosphorylated cAMP-responsive element modulator (p-CREM) that binds the IL-2 promoter.[162]

Serum levels of IL-4 are slightly increased in SLE patients as compared to controls.[163,164] Funauchi et al found that SLE patients had increased numbers of IL-4-producing T/NK cells.[165] IL-4 deficient MRL/*lpr* mice have decreased levels of IgG1 and IgE but comparable levels of IgG2a, IgG2b and autoantibodies as wild-type mice. Clinically, they display reduced lymphadenopathy and glomerulonephritis.[142] Administration of IL-4-treated BWF1 splenocytes to autologous mice induces the production of nephritogenic IgG1 and IgG3 autoAb. Conversely, injection of anti-IL-4 Ab prevents the onset of glomerulonephritis.[109]

PBMC from lupus patients produce in vitro more IL-6 than controls.[166] Administration of IL-6 to BWF1 mice aggravates the clinical manifestations,[167] while IL-6 blocking Ab have a beneficial effect.[168] Noteworthy, IL-6 blocking Ab have no effects on the in vitro production of IgG by SLE PBMC as it did not affect serum levels of human IgG in SCID mice reconstituted with human SLE PBMC.[73]

Ohtsuka et al found that TGF-β production by SLE PBMC or NK cells is reduced as compared to controls[169,170]. This observation might be of pathophysiological interest as addition of IL-2 and TGF-β to SLE PBMC cultures inhibits spontaneous Ig production[171]. Previous work by the same group had identified a regulatory pathway of Ig production involving suppressor $CD8^+$ T cells that are activated by TGF-β-secreting NK cells[172].

Finally, serum levels of IL-16,[173] IL-17,[164] IL-18[164,174] and sIL-2R[163] are all increased in SLE patients and correlate with disease activity. Murine lupus strains also display increased production of IL-1,[175,176] whose administration to BWF1 mice resulted in increased disease severity.[117]

Conclusions

Over the last decade, studies on the involvement of cytokines in SLE have clearly opened new avenues in our understanding of some physiopathological aspects of the disease. Thus, we believe that the imbalance between the production of IL-10 and IL-12 participates to both the overt B-cell activation and the impaired T-cell responses observed in SLE. One should, however, beware of oversimplification: other cytokines, in particular IFN-γ and TGF-β play a critical role; the recently described TALL-1 molecule raises intriguing issues; and the purported "protective" role of TNF-α may need to be revisited.

Biotechnology provides us today with new therapeutical strategies aimed at correcting exagerated or defective cytokine production. It will be one of the challenges of the forthcoming years to evaluate the feasibility, activity and toxicity of these new therapeutical approaches, keeping in mind that any interference within the cytokine network may have unexpected side-effects.

References

1. Kotzin BL. Systemic lupus erythematosus. Cell 1996; 85:303-306.
2. Isenberg DA, Ravirajan CT, Rahman A et al. The role of antibodies to DNA in systemic lupus erythematosus. A review and introduction to an international workshop on DNA antibodies held in London, May 1996. Lupus 1997; 6: 290-304.

3. Okamura M, Kanayama Y, Amatsu K et al. Significance of enzyme linked immunosorbent assay (ELISA) for antibodies to double standed and single stranded DNA in patients with lupus nephritis: Correlation with severity of renal histology. Ann Rheum Dis 1993; 52:14-20.
4. Bootsma H, Spronk P, Derksen R et al. Prevention of relapses in systemic lupus erythematosus. Lancet 1995; 345:1595-1599.
5. Papoian R, Pillarisetty R, Talal N. Immunological regulation of spontaneous antibodies to DNA and RNA. II. Sequential switch from IgM to IgG in NZB/NZW F1 mice. Immunology 1997; 32:75-79.
6. Datta SK, Patel H, Berry D. Induction of a cationic shift in IgG anti-DNA autoantibodies. Role of T helper cells with classical and novel phenotypes in three murine models of lupus nephritis. J Exp Med 1987; 165:1252-1268.
7. Koffer D, Schur PH, Kunkel HG. Immunological studies concerning the nephritis of systemic lupus erythematosus. J Exp Med 1967; 126:607-624.
8. Ehrenstein MR, Katz DR, Griffiths MH et al. Human IgG anti-DNA antibodies deposit in kidneys and induce proteinuria in SCID mice. Kidney Int 1995; 48:705-711.
9. Drappa J, Vaishnaw AK, Sullivan KE et al. Fas gene mutations in the Canale-Smith syndrome, an inherited lymphoproliferative disorder associated with autoimmunity. N Engl J Med 1996; 335:1643-1649.
10. Gleichmann E, Pals ST, Rolink AG et al. Graft-versus-host reactions: Clues to the etiopathology of a spectrum of immunological diseases. Immunology Today 1984; 5:324-332.
11. De Wit D, Van Mechelen M, Zanin C et al. Preferential activation of Th2 cells in chronic graft-versus-host reaction. J Immunol 1993; 150:361-366.
12. Doutrelepont JM, Moser M, Leo O et al. Hyper IgE in stimulatory graft-*Versus*-Host Disease: Role of interleukin-4. Clin Exp Immunol 1991; 83:133-136.
13. Merino R, Iwamoto M, Fossati L et al. Polyclonal B cell activation arises from different mechanisms in lupus-prone (NZB x NZW)F1 and MRL/MpJ-*lpr/lpr* mice. J Immunol 1993; 151:6509-6516.
14. Dorner T, Farner NL, Lipsky PE. Ig lambda and heavy chain gene usage in early untreated systemic lupus erythematosus suggests intensive B cell stimulation. J Immunol 1999; 163:1027-1036.
15. Desai-Mehta A, Lu L, Ramsey-Goldman R et al. Hyperexpression of CD40 ligand by B and T cells in human lupus and its role in pathogenic autoantibody production. J Clin Invest 1996; 97:2063-2073.
16. Blossom S, Chu EB, Weigle WO et al. CD40 ligand expressed on B cells in the BXSB mouse model of systemic lupus erythematosus. J Immunol 1997; 159:4580-4586.
17. Folzenlogen D, Hofer MF, Leung DY et al. Analysis of CD80 and CD86 expression on peripheral blood B lymphocytes reveals increased expression of CD86 in lupus patients. Clin Immunol Immunopathol 1997; 83:199-204.
18. Wither JE, Paterson AD, Vukusic B. Genetic dissection of B cell traits in New Zealand black mice. The expanded population of B cells expressing up-regulated costimulatory molecules shows linkage to Nba2. Eur J Immunol 2000; 30:356-365.
19. Takahashi K, Kozono Y, Waldschmidt TJ et al. Mouse complement receptors type 1 (CR1; CD35) and type 2 (CR2; CD21): Expression on normal B cell subpopulations and decreased levels during the development of autoimmunity in MRL/lpr mice. J Immunol 1997; 159:1557-1569.
20. Liossis SN, Kovacs B, Dennis G et al. B cells from patients with systemic lupus erythematosus display abnormal antigen receptor-mediated early signal transduction events. J Clin Invest 1996; 98:2549-2557.
21. Jongstra-Bilen J, Vukusic B, Boras K et al. Resting B cells from autoimmune lupus-prone New Zealand Black and (New Zealand Black x New Zealand White)F1 mice are hyper-responsive to T cell-derived stimuli. J Immunol 1997; 159:5810-5820.
22. Reininger L, Winkler TH, Kalberer CP et al. Intrinsic B cell defects in NZB and NZW mice contribute to systemic lupus erythematosus in (NZB x NZW)F1 mice. J Exp Med 1996; 184:853-861.
23. Roth R, Nakamura T, Mamula MJ. B7 costimulation and autoantigen specificity enable B cells to activate autoreactive T cells. J Immunol 1996; 157:2924-2931.
24. Mamula MJ, Fatenejad S, Craft J. B cells process and present lupus autoantigens that initiate autoimmune T cell responses. J Immunol 1994; 152:1453-1461.
25. Mohan C, Morel L, Yang P et al. Accumulation of splenic B1a cells with potent antigen-presenting capability in NZM2410 lupus-prone mice. Arthritis Rheum 1998; 41:1652-1662.
26. Sieling PA, Porcelli SA, Duong BT et al. Human double-negative T cells in systemic lupus erythematosus provide help for IgG and are restricted by CD1c. J Immunol 2000; 165:5338-5344.
27. Diamond B, Katz JB, Paul E et al. The role of somatic mutation in the pathogenic anti-DNA response. Annu Rev Immunol 1992; 10:731-757.

28. Hirose S, Wakiya M, Kawano-Nishi Y et al. Somatic diversification and affinity maturation of IgM and IgG anti-DNA antibodies in murine lupus. Eur J Immunol 1993; 23:2813-2820.
29. van Es JH, Gmelig Meyling FH, van de Akker WR et al. Somatic mutations in the variable regions of a human IgG anti-double-stranded DNA autoantibody suggest a role for antigen in the induction of systemic lupus erythematosus. J Exp Med 1991; 173:461-470.
30. Connoly K, Roubinian JR, Wofsy D. Development of murine lupus in CD4-depleted NZB/NZW mice. Sustained inhibition of residual CD4+ T cells is required to suppress autoimmunity. J Immunol 1992; 149:3083-3088.
31. Sainis K, Datta SK. CD4+ T cell lines with selective patterns of autoreactivity as well as CD4-CD8-T helper cell lines augment the production of idiotypes shared by pathogenic anti-DNA autoantibodies in the NZB x SWR model of lupus nephritis. J Immunol 1988; 140:2215-2224.
32. Kaliyaperumal A, Mohan C, Wu W et al. Nucleosomal peptide epitopes for nephritis-inducing T helper cells of murine lupus. J Exp Med 1996; 183:2459-2469.
33. Mohan C, Adams S, Stanik V et al. Nucleosome: A major immunogen for pathogenic autoantibody-inducing T cells of lupus. J Exp Med 1993; 177:1367-1381.
34. Kaliyaperumal A, Michaels MA, Datta SK. Antigen-specific therapy of murine lupus nephritis using nucleosomal peptides: Tolerance spreading impairs pathogenic function of autoimmune T and B cells. J Immunol 1999; 162:5775-5783.
35. Lu L, Kaliyaperumal A, Boumpas DT et al. Major peptide autoepitopes for nucleosome-specific T cells of human lupus. J Clin Invest 1999; 104:345-355.
36. Shi Y, Kaliyaperumal A, Lu L et al. Promiscuous presentation and recognition of nucleosomal autoepitopes in lupus: role of autoimmune T cell receptor α chain. J Exp Med 1998; 187:367-378.
37. Bermas BL, Petri M, Goldman D et al. T helper cell dysfunction in systemic lupus erythematosus (SLE): relation to disease activity. J Clin Imumnol 1994; 14:169-177.
38. Tsokos GC, Kovacs B, Sfikakis PP et al. Defective antigen-presenting cell function in patients with systemic lupus erythematosus. Arthritis Rheum 1996; 39:600-609.
39. Horwitz DA, Tang FL, Stimmler MM et al. Decreased T cell response to anti-CD2 in systemic lupus erythematosus and reversal by anti-CD28: Evidence for impaired T cell-accessory cell interaction. Arthritis Rheum 1997; 40:822-833.
40. Kimberly RP, Meryhew NL, Runquist OA. Mononuclear phagocyte function in SLE. I. Bipartite Fc- and complement-dependent dysfunction. J Immunol 1986; 137:91-96.
41. Shirakawa F, Yamashita U, Suzuki H. Reduced function of HLA-DR-positive monocytes in patients with systemic lupus erythematosus (SLE). J Clin Immunol 1985; 5:396-403.
42. Phillips R, Lomnitzer R, Wadee Aa et al. Defective monocyte function in patients with systemic lupus erythematosus. Clin Immunol Immunopathol 1985; 34:69-76.
43. Salmon JE, Kimberly RP, Gibofsky A et al. Defective mononuclear phagocyte function in systemic lupus erythematosus: Dissociation of Fc receptor-ligand binding and internalization. J Immunol 1984; 133:2525-2531.
44. Herrmann M, Voll RE, Zoller OM et al. Impaired phagocytosis of apoptotic cell material by monocyte-derived macrophages from patients with systemic lupus erythematosus. Arthritis Rheum 1998; 41:1241-1250.
45. Mosmann TR. Properties and functions of interleukin-10. Adv Immunol 1994; 56:1-26.
46. Rousset F, Garcia E, Defrance T et al. Interleukin 10 is a potent growth and differentiation factor for activated human B lymphocytes. Proc Natl Acad Sci USA 1992; 89:1890-1893.
47. Itoh K, Hirohata S. The role of IL-10 in human B cell activation, proliferation, and differentiation. J Immunol 1995; 154:4341-4350.
48. Levy Y, Brouet JC. Interleukin-10 prevents spontaneous death of germinal center B cells by induction of the Bcl-2 protein. J Clin Invest 1994; 93:424-428.
49. de Waal Malefyt R, Haanen J, Spitz H et al. Interleukin 10 (IL-10) and viral IL-10 strongly reduce antigen-specific human T cell proliferation by diminishing the antigen-presenting capacity of monocytes via downregulation of class II major histocompatibility complex expression. J Exp Med 1991; 174:915-924.
50. Fiorentino DF, Zlotnik A, Vieira P et al. IL-10 acts on the antigen-presenting cell to inhibit cytokine production by Th1 cells. J Immunol 1991; 146:3444-3451.
51. de Waal Malefyt R, Abrams J, Bennett B et al. Interleukin 10(IL-10) inhibits cytokine synthesis by human monocytes: An autoregulatory role of IL-10 produced by monocytes. J Exp Med 1991; 174:1209-1220.
52. Fiorentino DF, Zlotnik A, Mosmann TR et al. IL-10 inhibits cytokine production by activated macrophages. J Immunol 1991; 147:3815-3822.
53. Oswald IP, Wynn TA, Sher A et al. Interleukin 10 inhibits macrophage microbicidal activity by blocking the endogenous production of tumor necrosis factor alpha required as a costimulatory factor for interferon gamma-induced activation. Proc Natl Acad Sci USA 1992; 89:8676-8680.

54. Vouldoukis I, Becherel PA, Riveros-Moreno V et al. Interleukin-10 and interleukin-4 inhibit intracellular killing of Leishmania infantum and Leishmania major by human macrophages by decreasing nitric oxide generation. Eur J Immunol 1997; 27:860-865.
55. Allavena P, Piemonti L, Longoni D et al. IL-10 prevents the differentiation of monocytes to dendritic cells but promotes their maturation to macrophages. Eur J Immunol 1998; 28:359-369.
56. Buelens C, Verhasselt V, De Groote D et al. Interleukin-10 prevents the generation of dendritic cells from human peripheral blood mononuclear cells cultured with interleukin-4 and granulocyte/macrophage-colony-stimulating factor. Eur J Immunol 1997; 27:756-762.
57. De Smedt T, Van Mechelen M, De Becker G et al. Effect of interleukin-10 on dendritic cell maturation and function. Eur J Immunol 1997; 27:1229-1235.
58. Steinbrink K, Wölfl M, Jonuleit H et al. Induction of tolerance by IL-10 treated dendritic cells. J Immunol 1997; 159:4772-4780.
59. Caux C, Massacrier C, Vanbervliet B et al. Interleukin 10 inhibits T cell alloreaction induced by human dendritic cells. Int Immunol 1994; 6:1177-1185.
60. Buelens C, Verhasselt V, De Groote D et al. Human dendritic cell responses to lipopolysaccharide and CD40 ligation are differentially regulated by interleukin-10. Eur J Immunol 1997; 27:1848-1852.
61. Kalinski P, Schuitemaker JHN, Hilkens CMU et al. Prostaglandin E2 induces the final maturation of IL-12 deficient $CD1a^{+}CD83^{+}$ dendritic cells: The levels of IL-12 are determined during the final dendritic cell maturation and are resistant to further modulation. J Immunol 1998; 161:2804-2809.
62. Llorente L, Richaud-Patin Y, Wijdenes J et al. Spontaneous production of interleukin-10 by B lymphocytes and monocytes in systemic lupus erythematosus. Eur Cytokine Netw 1993; 4:421-430.
63. Llorente L, Richaud-Patin Y, Fior R et al. In vivo production of interleukin-10 by nonT cells in rheumatoid arthritis, Sjögren's syndrome, and systemic lupus erythematosus. Arthritis Rheum 1994; 37:1647-1655.
64. Houssiau FA, Lefebvre C, Vanden Berghe M et al. Serum interleukin 10 titers in systemic lupus erythematosus reflect disease activity. Lupus 1995; 4:393-395.
65. Park YB, Lee SK, Lim DS et al. Elevated interleukin-10 levels correlated with disease activity in systemic lupus erythematosus. Clin Exp Rheumatol 1998; 16:283-288.
66. Gröndal G, Kristjansdottir H, Gunnlaugsdottir B et al. Increased number of interleukin-10-producing cells in systemic lupus erythematosus patients and their first-degree relatives and spouses in Icelandic multicase families. Arthritis Rheum 1999; 42:1649-1654.
67. Llorente L, Richaud-Patin Y, Couderc J et al. Dysregulation of interleukin-10 production in relatives of patients with systemic lupus erythematosus. Arthritis Rheum 1997; 40:1429-1435.
68. Turner DM, Williams DM, Sankaran D et al. An investigation of polymorphism in the interleukin-10 gene promoter. Eur J Immunogenet 1997; 24:1-8.
69. Lazarus M, Hajeer AH, Turner D et al. Genetic variation in the interleukin 10 gene promoter and systemic lupus erythematosus. J Rheumatol 1997; 24:2314-2317.
70. Eskdale J, Wordsworth P, Bowman S et al. Association between polymorphisms at the human IL-10 locus and systemic lupus erythematosus. Tissue Antigens 1997; 49:635-639.
71. Ishida H, Muchamuel T, Sakaguchi S et al. Continuous administration of anti-interleukin 10 antibodies delays onset of autoimmunity in NZB/W F_1 mice. J Exp Med 1994; 179:305-310.
72. Yin Z, Bahtiyar G, Lanzhen L et al. IL-10 down-modulates murine lupus. Arthritis Rheum 2000; 43; 9:(supplement) (abstract).
73. Llorente L, Zou W, Levy Y et al. Role of interleukin 10 in the B lymphocyte hyperactivity and autoantibody production of human systemic lupus erythematosus. J Exp Med 1995; 181:839-844.
74. Llorente L, Richaud-Patin Y, Garcia-Padilla C et al. Clinical and biologic effects of anti-interleukin-10 monoclonal antibody administration in systemic lupus erythematosus. Arthritis Rheum 2000; 43:1790-1800.
75. Lauwerys BR, Garot N, Renauld J-C et al. Interleukin-10 blockade corrects impaired in vitro cellular immune responses of systemic lupus erythematosus patients. Arthritis Rheum 2000; 43:1976-1981.
76. Ma X, Riemann H, Gri G, Trinchieri G. Positive and negative regulation of interleukin-12 gene expression. Eur Cytokine Netw 1998; 9 (suppl 3):54-64.
77. Oppmann B, Lesley R, Blom B et al. Novel p19 protein engages IL-12p40 to form a cytokine, IL-23, with biological activities similar as well as distinct from IL-12. Immunity 2000; 13:715-725.
78. Heinzel FP, Hujer AM, Ahmed FN et al. In vivo production and function of IL-12 p40 homodimers. J Immunol 1997; 158:4381-4388.
79. Ling P, Gately MK, Gubler U et al. Human IL-12 p40 homodimer binds to the IL-12 receptor but does not mediate biologic activity. J Imunol 1995; 154:116-127.

80. Gillessen S, Carvajal D, Ling P et al. Mouse interleukin-12 (IL-12) p40 homodimer: A potent IL-12 antagonist. Eur J Immunol 1995; 25:200-206.
81. Perussia B, Chan SH, D'Andrea A et al. Natural killer (NK) cell stimulatory factor or IL-12 has differential effects on the proliferation of TCR-alpha beta+, TCR-gamma delta+ T lymphocytes, and NK cells. J Imunol 1992; 149:3495-3502.
82. Magram J, Connaughton SE, Warrier RR et al. IL-12 deficient mice are defective in IFN gamma production and type 1 cytokine responses. Immunity 1996; 4:471-481.
83. McKnight AJ, Zimmer GJ, Fogelman I et al. Effects of IL-12 on helper T cell-dependent immune responses in vivo. J Immunol 1994; 152:2172-2179.
84. Constantinescu CS, Hondowicz BD, Elloso MM et al. The role of IL-12 in the maintenance of an established Th1 immune response in experimental leishmaniasis. Eur J Immunol 1998; 28:2227-2233.
85. Chehimi J, Valiante NM, D'Andrea A et al. Enhancing effect of natural killer cell stimulatory factor (NKSF/interleukin-12) on cell-mediated cytotoxicity against tumor-derived and virus-infected cells. Eur J Immunol 1993; 23:1826-1830.
86. Naume B, Gately L, Espevik T. A comparative study of IL-12 (cytotoxic lymphocyte maturation factor)-, IL-2-, and IL-7-induced effects on immunomagnetically purified CD56+ NK cells. J Immunol 1992; 148:2429-2436.
87. Naume B, Johnsen AC, Espevik T et al. Gene expression and secretion of cytokines and cytokine receptors from highly purified CD56+ natural killer cells stimulated with interleukin-2, interleukin-7 and interleukin-12. Eur J Immunol 1993; 23:1831-1838.
88. DeBlaker-Hohe DF, Yamauchi A, Yu CR et al. IL-12 synergizes with IL-2 to induce lymphokine-activated cytotoxicity and perforin and granzyme gene expression in fresh human NK cells. Cell Immunol 1995; 165: 33-43.
89. Bonnema JD, Rivlin KA, Ting AT et al. Cytokine-enhanced NK cell-midated cytotoxicity. Positive modulatory effects of IL-2 and IL-12 on stimulus-dependent granule exocytosis. J Immunol 1994; 152:2098-2104.
90. Salcedo TW, Azzoni L, Wolf SF et al. Modulation of perforin and granzyme messenger RNA expression in human natural killer cells. J Imumnol 1993; 151:2511-2520.
91. Hyodo Y, Matsui K, Hayashi N et al. IL-18 up-regulates perforin-mediated NK activity without increasing perforin messenger RNA expression by binding to constitutively expressed IL-18 receptor. J Immunol 1999; 162:1662-1668.
92. Gagro A, Gordon J. The interplay between T helper subset cytokines and IL-12 in directing human B lymphocyte differentiation. Eur J Immunol 1999; 29:3369-3379.
93. Yoshimoto T, Okamura H, Tagawa YI et al. Interleukin 18 together with interleukin 12 inhibits IgE production by induction of interferon-γ production from activated B cells. Proc Natl Acad Sci USA 1997; 94:3948-3953.
94. Okamura H, Kashiwamura S, Tsutsui H et al. Regulation of interferon-gamma production by IL-12 and IL-18. Curr Opin Immunol 1998; 10:259-264.
95. Tomura M, Zhou XY, Maruo S et al. A critical role for IL-18 in the proliferation and activation of $NK1.1^+$ $CD3^-$ cells. J Immunol 1998; 160:4738-4746.
96. Akira S. The role of IL-18 in innate immunity. Curr Opinion Immunol 2000; 12:59-63.
97. Yoshimoto T, Takeda K, Tanaka T et al. IL-12 up-regulates IL-18 receptor expression on T cells, Th1 cells, and B cells: synergism with IL-18 for IFN-γ production. J Immunol 1998; 161:3400-3407.
98. Fehniger TA, Shah MH, Turner MJ et al. Differential cytokine and chemokine gene expression by human NK cells following activation with IL-18 or IL-15 in combination with IL-12: Implications for the innate immune response. J Immunol 1999; 162:4511-4520.
99. Lauwerys BR, Renauld JC, Houssiau FA. Inhibition of in vitro immunoglobulin production by IL-12 in murine chronic graft-vs.-host disease: Synergism with IL-18. Eur J Immunol 1998; 28:2017-2024.
100. Lauwerys BR, Renauld JC, Houssiau FA. Synergistic proliferation and activation of natural killer cells by interleukin 12 and interleukin 18. Cytokine 1999; 11:822-830.
101. Lauwerys BR, Garot N, Renauld JC et al. Cytokine production and killer activity of NK/T-NK cells derived with IL-2, IL-15, or the combination of IL-12 and IL-18. J Immunol 2000; 165:1847-1853.
102. Liu TF, Jones BM. Impaired production of IL-12 in systemic lupus erythematosus. I. Excessive production of IL-10 suppresses production of IL-12 by monocytes. Cytokine 1998; 10:140-147.
103. Liu TF, Jones BM. Impaired production of IL-12 in system lupus erythematosus. II: IL-12 production in vitro is correlated negatively with serum IL-10, positively with serum IFN-gamma and negatively with disease activity in SLE. Cytokine 1998; 10:148-153.

104. Liu TF, Jones BM, Wong RW et al. Impaired production of IL-12 in systemic lupus erythematosus. III: Deficient IL-12 p40 gene expression and cross-regulation of IL-12, IL-10 and IFN-gamma gene expression. Cytokine 1999; 11:805-811.
105. Horwitz DA, Gray JD, Behrendsen SC et al. Decreased production of interleukin-12 and other Thl-type cytokines in patients with recent-onset systemic lupus erythematosus. Arthritis Rheum 1998; 41:838-844.
106. Alleva DG, Kaser SB, Beller DI. Intrinsic defects in macrophage IL-12 production associated with immune dysfunction in the MRL/++ and New Zealand black/white F_1 lupus-prone mice and the *Leishmania major*-susceptible BALB/c strain. J Immunol 1998; 161:6878-6884.
107. Wong CK, Li EK, Ho CY et al. Elevation of plasma interleukin-18 concentration is correlated with disease activity in systemic lupus erythematosus. Rheumatology 2000; 39:1078-1081.
108. Houssiau FA, Mascart-Lemone F, Stevens M et al. IL-12 inhibits *in vitro* immunoglobulin production by human lupus peripheral blood mononuclear cells (PBMC). Clin Exp Immunol 1997; 108:375-380.
109. Nakajima A, Hirose S, Yagita H et al. Roles of IL-4 and IL-12 in the development of lupus in NZB/W F_1 mice. J Immunol 1997; 158:1466-1472.
110. Bagentose LM, Salgame P, Monestier M. IL-12 down regulates autoantibody production in mercury-induced autoimmunity. J Immunol 1998; 160:1612-1617.
111. Hagiwara E, Okubo T, Aoki I et al. IL-12 encoding plasmid has a beneficial effect on spontaneous autoimmune disease in MRL/MP-Ipr/Ipr mice. Cytokine 2000; 12:1035-1041.
112. Huang FP, Feng GJ, Lindop G et al. The role of interleukin 12 and nitric oxide in the development of spontaneous autoimmune disease in MRL/MP-*lpr/lpr* mice. J Exp Med 1996; 183:1447-1459.
113. Schwarting A, Tesch G, Kinoshita K et al. IL-12 drives IFN-gamma-dependent autoimmune kidney disease in MRL-Fas(lpr) mice. J Immunol 1999; 163:6884-6891.
114. Via CS, Rus V, Gately MK, Finkelman FD. IL-12 stimulates the development of acute graft-*Versus*-Host Disease in mice that normally would develop chronic, autoimmune graft-*Versus*-Host Disease. J Immunol 1994; 153:4040-4047.
115. Jacob CO, McDevitt HO. Tumour necrosis factor-α in murine autoimmune "lupus" nephritis. Nature 1988; 331:356-358.
116. Jacob CO, Lee SK, Strassmann G. Mutational analysis of TNF-α gene reveals a regulatory role for the 3'-untranslated region in the genetic predisposition to lupus-like autoimmune disease. J Immunol 1996; 156:3043-3050.
117. Brennan DC, Yui MA, Wuthrich RP et al. Tumor necrosis factor and IL-1 in New Zealand black/white mice. J Immunol 1989; 143:3470-3475.
118. Gordon C, Ranges GE, Greenspan JS et al. Chronic therapy with recombinant tumor necrosis factor-alpha in autoimmune NZB/NZW F1 mice. Clin Immunol Immunopathol 1989; 52:421-434.
119. Jacob CO, Hwang F, Lewis GD et al. Tumor necrosis factor alpha in murine systemic lupus erythematosus disease models: Implications for genetic predisposition and immune regulation. Cytokine 1991; 3:551-561.
120. Kontoyiannis D, Kollias G. Accelerated autoimmunity and lupus nephritis in NZB mice with an engineerd heterozygous deficiency in tumor necrosis factor. Eur J Immunol 2000; 30:2038-2047.
121. Maury CP, Teppo AM. Tumor necrosis factor in the serum of patients with systemic lupus erythematosus. Arthritis Rheum 1989; 32:146-150.
122. Meijer C, Huysen V, Smeenk RT et al. Profiles of cytokines (TNF alpha and IL-6) and acute phase proteins (CPR and alpha 1AG) related to the disease course in patients with systemic lupus erythematosus. Lupus 1993; 2:359-365.
123. Studnicka-Benke A, Steiner G, Petera P et al. Tumour necrosis factor alpha and its soluble receptors parallel clinical disease and autoimmune activity in systemic lupus erythematosus. Br J Rheumatol 1996; 35:1067-1074.
124. Davas EM, Tsirogianni A, Kappou I et al. Serum IL-6, TNFalpha, p55 srTNFalpha, p75 srTNFalpha, srIL-2alpha levels and disease activity in systemic lupus erythematosus. Clin Rheumatol 1999; 18:17-22.
125. Aderka D, Wysenbeek A, Engelmann H et al. Correlation between serum levels of soluble tumor necrosis factor receptor and disease activity in systemic lupus erythematosus. Arthritis Rheum 1993; 36:1111-1120.
126. Gabay C, Cakir N, Moral F et al. Circulating levels of tumor necrosis factor soluble receptors in systemic lupus erythematosus are significantly higher than in other rheumatic diseases and correlate with disease activity. J Rheumatol 1997; 24:303-308.
127. Wilson AG, Gordon C, di Giovine FS et al. A genetic association between systemic lupus erythematosus and tumor necrosis factor alpha. Eur J Immunol 1994; 24:191-195.

128. Rood MJ, van Krugten MV, Zanelli E et al. TNF-308A and HLA-DR3 alleles contribute independently to susceptibility to systemic lupus erythematosus. Arthritis Rheum 2000; 43:129-134.
129. Sullivan KE, Wooten C, Schmeckpeper BJ et al. A promoter polymorphism of tumor necrosis factor alpha associated with systemic lupus erythematosus in African-Americans. Arthritis Rheum 1997; 40:2207-2211.
130. Rudwaleit M, Tikly M, Khamashta M et al. Interethnic differences in the association of tumor necrosis factor promoter polymorphisms with systemic lupus erythematosus. J Rheumatol 1996; 23:1725-1728.
131. Komata T, Tsuchiya N, Matsushita M et al. Association of tumor necrosis factor receptor 2 (TNFR2) polymorphism with susceptibility to systemic lupus erythematosus. Tissue Antigens 1999; 53:527-533.
132. Takahashi M, Hashimoto H, Akizuki M et al. Lack of association between the Met196Arg polymorphism in the TNFR2 gene and autoimmune diseases accompanied by vasculitis including SLE in Japanese. Tissue Antigens 2001; 57:66-69.
133. Al-Ansari AS, Ollier WE, Villarreal J et al. Tumor necrosis factor receptor II (TNFRII) exon 6 polymorphism in systemic lupus erythematosus. Tissue Antigens 2000; 55:97-99.
134. Charles PJ, Smeenk RJ, De Jong J et al. Assessment of antibodies to double-stranded DNA induced in rheumatoid arthritis patients following treatment with infliximab, a monoclonal antibody to tumor necrosis factor alpha: Findings in open-label and randomized placebo-controlled trials. Arthritis Rheum 2000; 43:2383-2390.
135. Herrera-Esparza R, Barbosa-Cisneros O, Villalobos-Hurtando R et al. Renal expression of IL-6 and TNF alpha genes in lupus nephritis. Lupus 1998; 137:271-275.
136. Tsokos GC, Boumpas DT, Smith PL et al. Deficient gamma-interferon production in patients with systemic lupus erythematosus. Arthritis Rheum 1986; 29:1210-1215.
137. Hagiwara E, Gourley MF, Lee S et al. Disease severity in patients with systemic lupus erythematosus correlates with an increased ratio of interleukin-10: Interferon-gamma-secreting cells in the peripheral blood. Arthritis Rheum 1996; 39:379-385.
138. Jacob CO, van der Meide PH, McDevitt HO. In vivo treatment of (NZB X NZW)F1 lupus-like nephritis with monoclonal antibody to gamma interferon. J Exp Med 1987; 166:798-803.
139. Ozmen L, Roman D, Fountoulakis M et al. Experimental therapy of systemic lupus erythematosus: the treatment of NZB/W mice with mouse soluble interferon-gamma receptor inhibits the onset of glomerulonephritis. Eur J Immunol 1995; 25:6-12.
140. Haas C, Ryffel B, Le Hir M. IFN-gamma receptor depletion prevents autoantibody production and glomerulonephritis in lupus-prone (NZB x NZW)F1 mice. J Immunol 1998; 160:3173-3178.
141. Nicoletti F, Di Marco R, Zaccone P et al. Dichotomic effects of IFN-gamma on the development of systemic lupus erythematosus-like syndrome in MRL-lpr/lpr mice. Eur J Immunol 2000; 30:438-447.
142. Peng SL, Moslehi J, Craft J. Roles of interferon-gamma and interleukin-4 in murine lupus. J Clin Invest 1997; 99:1936-1946.
143. Haas C, Ryffel B, Le Hir M. IFN-gamma is essential for the development of autoimmune glomerulonephritis in MRL/lpr mice. J Immunol 1997; 158:5484-5491.
144. Lawson BR, Prud'homme GJ, Chang Y et al. Treatment of murine lupus with cDNA encoding IFN-gamma R/Fc. J Clin Invest 2000; 106:207-215.
145. Mackay F, Woodcock SA, Lawton P et al. Mice transgenic for BAFF develop lymphocytic disorders along with autoimmune manifestations. J Exp Med 1999; 190:1697-1710.
146. Khare SD, Sarosi I, Xia XZ et al. Severe B cell hyperplasia and autoimmune disease in TALL-1 transgenic mice. Proc Natl Acad Sci USA 2000; 97:3370-3375.
147. Gross JA, Johnston J, Mudri S et al. TACI and BCMA are receptors for a TNF homologue implicated in B-cell autoimmune disease. Nature 2000; 404:995-999.
148. Zang J, Roschke V, Baker KP et al. Cutting Edge: A role for B lymphocyte stimulator in systemic lupus erythematosus. J Immunol 2001; 166:6-10.
149. Sadlack B, Lohler J, Schorle H et al. Generalized autoimmune diseas in interleukin-2-deficient mice is triggered by an uncontrolled activation and proliferation of CD4+ T cells. Eur J Immunol 1995; 25:3053-3059.
150. Horak I, Lohler J, Ma A et al. Interleukin-2 deficient mice: a new model to study autoimmunity and self-tolerance. Immunol Rev 1995; 148:35-44.
151. Wang R, Rogers AM, Rush BJ et al. Induction of sensitivity to activation-induced death in primary CD4+ cells: a role for interleukin-2 in the negative regulation of responses by mature CD4+ T cells. Eur J Immunol 1996; 26:2263-2270.

152. Altman A, Theofilopoulos AN, Weiner R et al. Analysis of T cell function in autoimmune murine strains. Defects in production of, and responsiveness to, interleukin-2. J Exp Med 1981; 154:791-808.
153. Dauphinee MS, Kipper SB, Wofsy D et al. Interleukin 2 deficiency is a common feature of autoimmune mice. J Immunol 1981; 127:2483-2487.
154. Guttierez-Ramos JC, Andreu JL, Revilla Y et al. Recovery from autoimmunity of MRL/*lpr* mice after infection with an interleukin-2/vaccinia recombinant virus. Nature 1990; 346:271-274.
155. Huggins ML, Huang F-P, Xu D et al. Modulation of autoimmune disease in the MRL-*lpr/lpr* mouse by IL-2 and TGF-β1 gene therapy using attenuated *Salmonella typhimurium* as gene carrier. Lupus 1999; 8:29-38.
156. Raz E, Dudler J, Lotz M et al. Modulation of disease activity in murine systemic lupus erythematosus by cytokine gene delivery. Lupus 1995; 4:286-292.
157. Kelley VE, Gaulton GN, Hattori M et al. Anti-interleukin-2 receptor antibody suppresses murine diabetic insulitis and lupus nephritis. J Immunol 1988; 140:59-68.
158. Alcocer-Varela J, Alarcon-Segovia D. Decreased production of and response to interleukin-2 by cultured lymphocytes from patients with systemic lupus erythematosus. J Clin Invest 1982; 69:1388-1392.
159. Huang Y-P, Miescher PA, Zubler RH. The interleukin-2 secretion defect in vitro in systemic lupus erythematosus is reversible in rested cultured cells. J Immunol 1986; 137:3515-3520.
160. Murakawa Y, Sakane T. Deficient phytohemagglutinin-induced interleukin-2 activity in patients with nactive systemic lupus erythematosus is correctable by the addition of phorbom myristate acetate. Arthritis Rheum 1988; 31:826-833.
161. Garcia-Cozar FJ, Molina IJ, Cuadrado MJ et al. Defective B7 expression on antigen-presenting cells underlying T cell activation abnormalities in systemic lupus erythematosus (SLE) patients. Clin Exp Immunol 1996; 104:72-79.
162. Solomou EE, Juang YT, Gourley MF et al. Molecular basis of deficient IL-2 production in T cells from patients with systemic lupus erythematosus. J Immunol 2001; 166:4216-4222.
163. Cuadrado MJ, Marubayashi M, Ortega C et al. Relationship of IL-2, IL-2R (CD25+), soluble IL-2R and IL-4 with disease activity in SLE patients. Lupus 1993; 2:257-260.
164. Wong CK, Ho CY, Li EK et al. Elevation of proinflammatory cytokine (IL-18, IL-17, IL-12) and Th2 cytokine (IL-4) concentrations in patients with systemic lupus erythematosus. Lupus 2000; 9:589-593.
165. Funauchi M, Yu H, Sugiyama M et al. Increased interleukin-4 production by NK T cells in systemic lupus erythematosus. Clin Immunol 1999; 92:197-202.
166. Jones BM, Liu T, Wong RW. Reduced in vitro production of interferon-gamma, interleukin-4 and interleukin-12 and increased production of interleukin-6, interleukin-10 and tumour necrosis factor-alpha in systemic lupus erythematosus. Weak correlation of cytokine production with disease activity. Autoimmunity 1999; 31:117-124.
167. Ryffel B, Car BD, Gunn H et al. Interleukin-6 exacerbates glomerulonephritis in (NZBxNZW)F1 mice. Am J Pathol 1994; 144:927-937.
168. Finck BK, Chan B, Wofsy D. Interleukin-6 promotes murine lupus in NZB/NZW F1 mice. J Clin Invest 1994; 94:585-591.
169. Ohtsuka K, Gray JD, Stimmler MM et al. Decreased production of TGF-beta by lymphocytes from patients with systemic lupus erythematosus. J Immunol 1998; 160:2539-2545.
170. Ohtsuka K, Gray JD, Stimmler MM et al. The relationship between defects in lymphocyte production of transforming growth factor-beta1 in systemic lupus erythematosus and disease activity or severity. Lupus 1999; 8:90-94.
171. Ohtsuka K, Gray JD, Quismorio FP et al. Cytokine-mediated down-regulation of B cell activity in SLE: effects of interleukin-2 and transforming growth factor-beta. Lupus 1999; 8:95-102.
172. Gray JD, Hirokawa M, Horwitz DA. The role of transforming growth factor beta in the generation of suppression: an interaction between CD8+ T and NK cells. J Exp Med 1994; 180:1937-1942.
173. Lee S, Kaneko H, Sekigawa I et al. Circulating interleukin-16 in systemic lupus erythematosus. Br J Rheumatol 1998; 37:1334-1337.
174. Wong CK, Li EK, Ho CY et al. Elevation of plasma interleukin-18 concentration is correlated with disease activity in systemic lupus erythematosus. Rheumatology (Oxford) 2000: 39:1078-1081.
175. Mao C, Singh AK. IL-1 beta gene expression in B cells derived from the murine MRL/*lpr* model of lupus. Autoimmunity 1996; 24:71-79.
176. Lebedeva TV, Singh AK. Increased responsiveness of B cells in the murine MRL/*lpr* model of lupus nephritis to interleukin-1 beta. J Am Soc Nephrol 1995; 5:1530-1534.

Chapter 15

Cytokines, Chemokines and Growth Factors in the Pathogenesis and Treatment of Inflammatory Bowel Disease

Deborah O'Neil and Lothar Steidler

Introduction

Chronic inflammatory bowel disease is the term applied to a spectrum of gastrointestinal immunopathologies of which Crohn's disease and ulcerative colitis are the most common. Ulcerative colitis, as the name suggests, is specific to the colon while Crohn's is somewhat more insidious in that it can affect any part of the gastrointestinal tract, from the oral cavity to the rectum. Both are chronic relapsing-remitting diseases and while, in the 21st century, they are no longer significant in terms of mortality, Crohn's disease and ulcerative colitis greatly impact quality of life. Patients present with a spectrum of symptoms of which diarrhea, nausea, abdominal pain, appetite loss, fever and lethargy are common. Malnutrition, dehydration and anemia are therefore become significant issues if the condition remains untreated. Treatment depends on disease severity, but the age of the patient and the rate of disease relapse are also key considerations. Conventional therapy consists of anti-inflammatory or immunosuppressive drug regimes. Surgical intervention, either to remove sections of affected intestine and/or to divert the faecal stream, is often indicated. Roughly 1:1000 individuals in the developed world are affected, with focused pools in north-western Europe and northern America. A recent increase in the rate of disease incidence in these areas, particularly in children, now appears to have slowed, whereas the number of cases worldwide is increasing.[1-3]

The initiating factor, or factors, that cause Crohn's disease and ulcerative colitis remain undefined. Current consensus is that environmental factors and a predisposing genetic component interact to trigger excessive and inappropriate immune activity in the gut wall.[1,4,5] A susceptibility locus, termed the IBD1 region, on chromosome 16 has been linked to both Crohn's disease and ulcerative colitis and a region of chromosome 12 has been identified as a susceptibility locus for ulcerative colitis.[4,6,7] Growing evidence points towards a component of the resident intestinal microflora as being the environmental factor, although smoking, certain dietary factors and primary infections have also been implicated as causative agents in the past. That the most commonly affected areas of intestine in inflammatory bowel disease are the most microbe-rich (i.e., the distal and terminal ileum and the colon) and that diversion of the faecal stream and treatment with antibiotics have a beneficial impact on Crohn's disease and ulcerative colitis is circumstantial evidence of a role for the commensal flora. The finding that T cells isolated from Crohn's and ulcerative colitis patients proliferate in vitro in response to fractions of autologous commensal bacteria, while T cells from unaffected individuals show no such proliferative response to autologous bacterial antigens adds experimental weight to this hypothesis.[8-14] Whatever the cause of inflammatory bowel disease, there is little doubt that the pathophysiology of these diseases results from an increase in pro-inflammatory cytokines.

Cytokines and Chemokines in Autoimmune Disease, edited by Pere Santamaria.

Soluble Regulators of Immunity

Cytokines are small soluble proteins that, together with numerous growth factors and chemokines, provide the means by which cells of the immune system communicate with each other and with most other tissues in the body. As such, these molecules are able to regulate many aspects of the immune response in which numerous cells and tissues may be involved at any one time.

Cytokines in the Normal versus the Inflammatory State

Cytokines can be divided into those responsible for mediating immunogenic or inflammatory responses and those responsible for inhibiting or dampening immunogenic activity or facilitating tolerogenic responses. The pro-inflammatory cytokines include interleukin-1 (IL-1α/β),[15-18] interleukin-12 (IL-12),[19-21] interferon-gamma (IFN-γ)[22-25] and tumour necrosis factor-alpha (TNF-α).[26-30] Interleukin-4 (IL-4)[31-35] interleukin-10 (IL-10),[36-38] interleukin-11 (IL-11),[39] interleukin-13 (IL-13)[32-35,40] and transforming growth factor-beta (TGF-β)[41-43] are the major regulatory, anti-inflammatory immune mediators. Their presence will skew the immune process towards a response in which cell mediated immunogenic actions are effectively suppressed[31,36,40,43] and in which immunological tolerance to antigen can be established.[44-46]

Evidence points towards the existence of at least two distinct, functionally polarised sets of CD4 positive ($CD4^+$) T 'helper' T cells, based on which of these cytokines they secrete and as a consequence, their role in the immune response. These are termed either Type 1 T helper (Th1) or Type 2 T helper (Th2) responses. Th1 responses are characterised by the production of IL-2 and the inflammatory mediators, TNF-α, lymphotoxin (TNF-β), IL-12 and IFN-γ and are associated with delayed-type hypersensitivity reactions. Th1 cells are central in promoting cell-mediated and also phagocyte-dependent immunity and develop at sites of infection or inflammation very rapidly, their presence therein driving immunogenic, inflammatory responses. Th2 cells are those that secrete IL-4, IL-5, IL-10 and IL-13 which, along with TGF-β and IL-11, are largely responsible for driving humoral immunity by promoting antibody production and providing phagocyte-independent protective responses. Th2 cells also mediate eosinophil activation and inhibit macrophage function. Th2 responses inhibit Th1 cell development and the activation of existing Th1 cells. Taken together their actions are geared towards dampening or inhibiting immunogenic or inflammatory responses.

A Balancing Act

The regulation of the immune response relies, in part, on the cytokines to which the cells involved are exposed. In any tissue, a balance of cytokines, chemokines and growth factors is required in order to maintain immune homeostasis. During an inflammatory response, following infection or tissue damage, pro-inflammatory, mainly Th1 cytokines will predominate. Under normal circumstances, the Th2-associated and other regulatory mechanisms that are in place ensure that once an effective and appropriate immune response to any such challenge has been mounted, the normal state will be restored. Failure to regulate the immunogenic or Th1 response and restore immune homeostasis is implicated in many autoimmune disease immunopathologies. There are a number of areas in the body where the maintenance of a specific cytokine microenvironment is particularly important, one such region being the gastrointestinal mucosa. Herein, the effector cells that are responsible for the capture of foreign material (antigens), otherwise known as the antigen presenting cell (APC) population (comprising dendritic cells, macrophages and B cells) and the effector T and B lymphocytes to which the antigens are presented, must distinguish between two distinct sets of foreign antigen that enter across the intestinal epithelial barrier. First are those derived from harmless nutrient macromolecules and the resident commensal microflora, and second are those from any number of pathogens that attempt to breach the gut wall as a route of infection. The effector cell population must therefore not react against, or in other words, remain immunologically tolerant of the harmless component of this foreign matter and the maintenance of such an environment

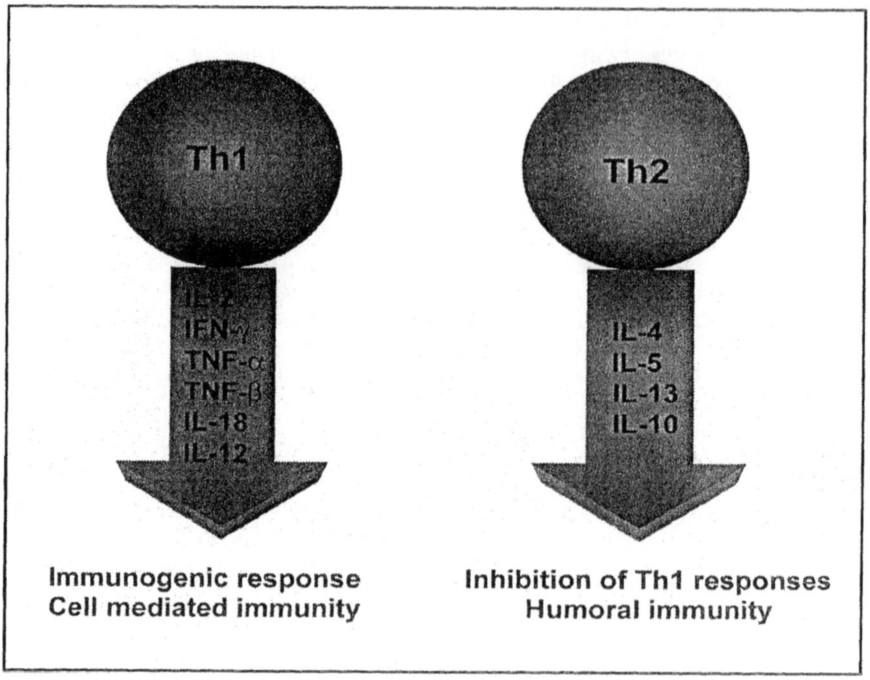

Fig. 15.1. The Th1/Th2 paradigm.

appears to be facilitated by TGF-β,[45,47] IL-10,[46,48] IL-4,[49-51] IL-11[39,52] and IL-13.[50,51] However, wherever a breach of the barrier by pathogens and/or tissue damage occurs, immunogenic Th1 cytokines such as IL-12[19-21] and IFN-γ[22,53 24,25,54] are still required to drive immunogenic responses. A complex balance of pro-inflammatory and regulatory cytokines, chemokines and growth factors is therefore required within the numerous and distinct microenvironments of immune activity throughout the gut wall, in both the organised lymphoid tissue and also the underlying lamina propria, to ensure that nonpathogenic and pathogenic material is dealt with appropriately, enabling the gut to maintain its dual roles as both an absorptive membrane for nutrient-derived macromolecules and as the body's first line of defence again luminal pathogens.

When the Balance Tips: Chronic Inflammation

Evidently, this balance is lost in the inflammatory bowel disease process, the result being the prevalence of inflammatory cytokines. In the mucosa of Crohn's disease and ulcerative colitis patients there is a marked elevation in the levels of IL-1,[55,56] IL-6,[57-59] IL-8[57,60-62] and TNF-α[63-66] produced therein. This is the limit of the similarities between the diseases in terms of cytokine profiles as the inflammation in each condition differs quite markedly. In Crohn's disease, there is predominantly a Th1-like environment, with exaggerated production of IL-12,[67-70] IL-15,[71,72] IL-16,[73] IL-18[26,74,75] and IFN-γ.[76-79] In ulcerative colitis however, these powerful Th1-skewing immune mediators are not present and the inflammation appears to be more of an antibody-mediated hypersensitivity reaction.[65,80-83] As well as recruiting and activating the T cell and APC component of the mucosal immune system, the sustained inflammatory response leads to the nonspecific recruitment and activation of granulocytes, mononulcear cells and keratinocytes, and the consequent release from these cells of the mediators responsible for the degradation (e.g., matrix metalloproteinases, cathepsin G) and modification (e.g., keratinocyte growth factor) of bowel wall structure that is characteristic of the inflammatory

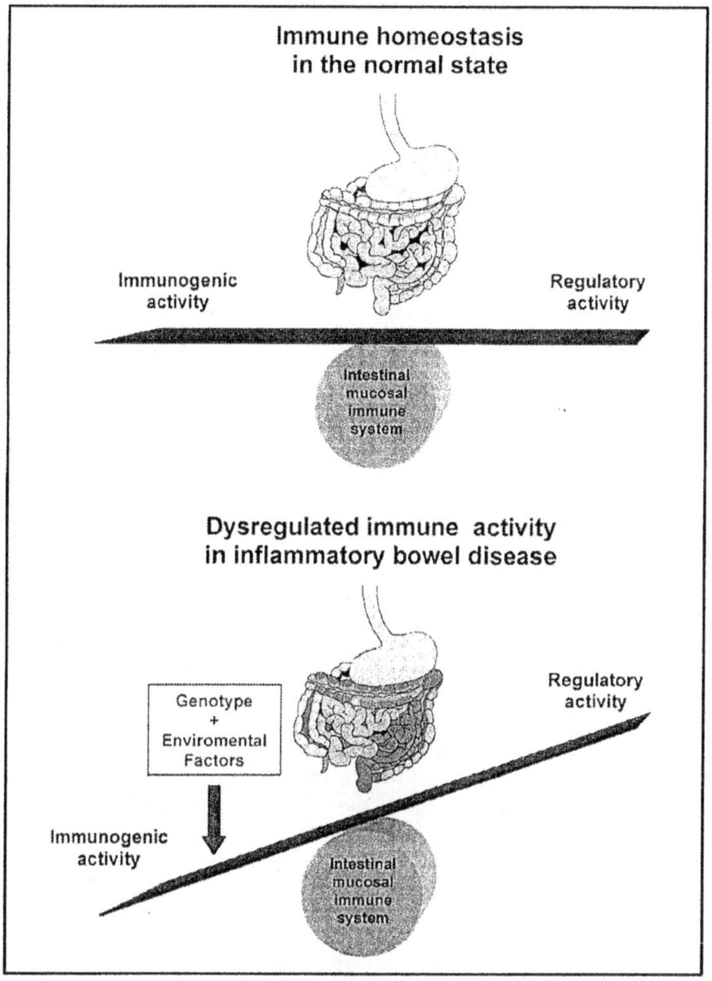

Fig. 15.2. Dysregulation of the intestinal mucosal immune system in inflammatory bowel disease.

bowel disease process. As well as mediating these deleterious local effects, the continued and elevated presence of soluble inflammatory factors contributes to the systemic effects of inflammatory bowel disease, which include, including growth suppression, joint inflammation, anorexia and anemia.[82-85]

Tumor Necrosis Factor alpha (TNF-α)

TNF-α is a potent prototypical pleiotropic inflammatory cytokine produced by activated macrophages and monocytes and Th1 T cells. TNF-α is responsible for a diverse range of biological effects and was originally identified as a systemic mediator of endotoxemic shock, cachexia, and tumour regression.[26-30] TNF-α belongs to a large family of proteins, including Fas ligand, whose actions, primarily paracrine in nature, are to regulate cell proliferation and apoptotic death. Thus, TNF-α is both a powerful inflammatory factor and a key regulator of cell death. These distinct ends are achieved by TNF-α binding, in trimeric form, to one of two functionally distinct, members of the TNF receptor superfamily, a 55kDa cell membrane

Table 15.1. Overview of cytokines targeted for therapeutic intervention in inflammatory bowel disease to date

Target Cytokine	Efficacy Demonstrated in: Human Trials	Animal Models	Therapeutic Protocol
IL-4		*	Administration of neutralising monoclonal antibodies or against target cytokine or its receptor, administration of soluble receptors.
IL-6		*	
IL-12		*	
IL-18		*	
IFN-γ		*	
TNF-α	*	*	
IL-4	*	*	Administration of exogenous cytokine protein or DNA.
IL-10	*	*	
IL-11	*	*	
TGF-β		*	

receptor termed TNF receptor 1 (TNFR-1) or a 75kDa cell membrane receptor, TNFR-2. Other members of the TNFR family, the defining trait of which is an extracellular domain comprised of two to six repeats of cysteine rich motifs, are FAS, CD40, CD27 and RANK. Ligation of TNFR-2 leads, via TNFR-associated factor (TRAF)-2, to the activation of the serine kinase complex, IKKα/β. IKKα/β which facilitates phosphorylation of the nulcear factor-kappa B (NF-κB)-binding, inhibitory proteins, or IKBs and their subsequent polyubiquitination and proteolytic degradation, freeing the transcription factor NF-κB for translocation to the nucleus, wherein it binds the relevant DNA response elements and initiates transcription of a large panel of pro-inflammatory immune mediator genes. Concomitant to IKK complex activation via the TNF-α receptors, JNK and p38 MAP kinase cascades are also initiated upon IL-1 receptor ligation. These result being the induction of the AP-1 transcription factor.

Ligation of TNFR-1 can also trigger this pathway of NF-κB activation, or alternatively, can lead to apoptosis. This is because, like FAS, the cytoplasmic region of TNFR-1 contains death domain, the presence of which allows TNFR-1 to play a dual role in determining the fate of a cell. If exposed to other apoptotic signals in conjunction with TNF-α, then signaling intermediates are such as TRADD, FADD and RIP can interact with the TNFR-1 death domain to trigger programmed cell death. If no other apoptotic signals are present then the NF-κB activation signaling cascade proceeds. A key point in determining the outcome of TNFR-1 ligation is the association of TRADD with the TNFR-1 death domain. Both TRAF-2 and FADD can associate with TRADD via TNFR-1. If TRAF-2 is recruited to TNFR-1, then the NF-κB activation pathway is initiated, but if FADD associates then the apoptotic cascade is triggered. TNFR-2 binds TRAF-2 directly and does not associate with TRADD.[86,87] A number of "decoy receptors", structurally related to TNFR-1 and TNFR-2 exist, that act to sequester TNF-α molecules, thereby rescuing cells from apoptosis and/or preventing the immune function. Soluble type I (p55) and type II (p75) TNF receptor (sTNFR) act as specific inhibitors of TNF-α in the normal state, by binding to and thus inactivating TNF-α. Levels of sTNFR are elevated in inflammation and in nonlethal infections, appear to circulate at levels that can block the quantities of TNF-α produced therein, but in chronic inflammatory disease and lethal septic shock, such concentrations of inhibitors appear inadequate to prevent the deleterious host responses to exaggerated TNF-α production.[26,29]

TNF-α is produced as a 233-amino-acid membrane-anchored precursor, which is proteolytically processed by metalloproteinases to yield the mature, 157-amino-acid cytokine.[88] The enzyme responsible is TNF-α converting enzyme or TACE, a member of the adamalysin family of metzicinin metalloproteinases.[89,90] Besides the potentially lethal effects of TNF-α overproduction in sepsis, its sustained production is a major cause of the pathophysiology of chronic inflammatory disease, including Crohn's and ulcerative colitis[65,91] wherein TNF-α is not only responsible for perpetuating local damage to the gut wall by the induction of focal vascular thrombosis and matrix degradation via matrix metalloproteinases,[85] but also contributes, more perhaps than any other cytokine to the systemic effects of these diseases, such as fever, lethargy, loss of appetite, growth suppression, and chronic anaemia.[84]

Extensive studies have confirmed an increase or overexpression of TNF-αmRNA and protein, both in the mucosa and systemically, in patients with Crohn's disease and ulcerative colitis.[63-66] The lamina propria monocytic and macrophage population are the major source of local TNF-α production in inflammatory bowel disease.[56,64,92] The TNF-α-producing lamina propria macrophages isolated from the inflamed mucosa of Crohn's disease and ulcerative colitis display high levels of DNA-binding activity towards the transcription factor nuclear factor kappa-B (NF-κB) and demonstrate a marked increase in expression of the p65 sub-unit (and to a lesser extent, also the p50 and c-rel sub-units) of NF-κB.[56] Ex vivo organ culture of colonic mucosal biopsy tissue obtained from ulcerative colitis patients reveals that the levels of TNF-α secreted by the macrophage and monocytic population therein correlates to the endoscopic grade of inflammation and that the production of TNF-α is significantly higher in patients with intractable disease receiving corticosteroids than in patients with nonintractable disease receiving corticosteroids.[57] Furthermore, increased TNF-α production by cultured lamina propria mononuclear cells (LPMC) isolated from Crohn's disease patients, even when in steroid-induced remission, appears to be a predictive measure for acute relapse within the short-term.[92] Polymorphonuclear granulocytes, the most abundant cell type within the lesions characteristic of Crohn's disease and colitis, are another source of TNF-α, both locally and in the periphery.[63] In addition, it appears that mast cells within the sub-mucosa and muscularis propria of inflamed ileal intestine are an additional source of TNF-α in Crohn's disease.[93]

Discussed previously, endogenous soluble TNFR is a natural inhibitor of TNF-α. Comparing the ratio of secretion of TNF-α to sTNFR in culture supernatants of colonic biopsy specimens and from LPMC isolated from patients with active colonic inflammatory bowel disease reveals an that the enhanced secretion of TNF-α in mucosal mononulcear cells in Crohn's disease and colitis occurs without a concomitant release of enhanced amounts of sTNFR, suggesting that an imbalance in secretion between TNF-α and one of its natural inhibitors, sTNFR, may be implicated in the pathogenesis of inflammatory bowel disease.[64]

Early work in a number of animal models in which neutralising the bioactivity of TNF-α with the administration of anti-TNF-α antibodies showed striking beneficial effects, provided the rationale for the use of such anti-TNF-α therapies as a potential treatment for Crohn's disease. Since the inception of their use in trials and as an experimental therapy in the clinic, Infliximab®, a chimeric anti-TNF-α antibody, has shown marked clinical benefit in the some, but not all, patients with Crohn's disease who had failed to respond to conventional immunosuppressive therapies.[83,94-96] These studies have shown that administration of Infliximab® anti-TNF-α antibodies to such patients can achieve a rapid reduction in the clinical signs and symptoms of the disease, benefits which are substantiated by both endoscopic and microscopic evaluation. Infleximab therapy has resulted in complete remission in some cases, an effect that appears to maintained with continued dosage of the antibody.[94,95,97,98] Infliximab® mediates rapid healing of intestinal ulceration and also fistulae closure in Crohn's inflammation.[99,100] In terms of inflammatory cascades, a rapid reduction of circulating IL-6, and the chemokines RANTES (Regulated upon Activation, Normal T cell Expressed and Secreted) and macrophage inflammatory protein (MIP)-1 has been observed in a number of patients following administration of Infleximab.[94]

Although these preliminary studies appear to be promising, the efficacy and safety of long-term anti-TNF-α therapy remains to be determined. One must remember that while its central role in the pathophysiology of the disease is undisputed, TNF-α is not the cause of Crohn's disease. Crudely blocking the bioactivity of a powerful and central component in the normal inflammatory and apoptotic process carries with it significant and very obvious risks. Of the reported side effects and complications associated with the use of Infliximab®, an enhanced susceptibility to bacterial infections, serum sickness, anaphylaxis, lymphoma and systemic lupus erythematosus have been reported to date.[101-103] Furthermore, although beneficial effects of Infliximab® therapy have been demonstrated in adult patients, its use is associated with only marginal, short-term clinical improvement in children and adolescents with severe Crohn's disease. The rapid return of disease activity in some patients following treatment suggesting that additional dosing strategies may be required in younger patients, with which comes greater risk.[104]

Interleukin-1 (IL-1)

IL-1 (IL-1α or IL-1β) is among the most pleiotropic of cytokines, affecting almost every cell type in the body, and is a key factor in both innate and acquired immune responses. IL-1 (α and β), along with TNF-α is a prototypical inflammatory cytokine, and the margin between biological benefit and unacceptable toxicity in humans is exceedingly narrow. There is growing evidence that the production and activity of IL-1, particularly IL-1β, are tightly regulated events with a number of biological roadblocks being in place as an adaptation to limit the toxicity of IL-1 during the inflammatory process. This regulation occurs at the level of gene transcription, protein synthesis and secretion, not just of IL-1 itself, but of its surface receptors, soluble receptors and its naturally occurring receptor antagonist, IL-Ra. IL-1β is produced as biologically inactive 269 amino acid polypeptide precursor and is cleaved to the mature form by caspase-1. IL-1 mediates its effects by ligating the IL-R type I complex, comprising the IL-1RI IL-1 receptor accessory protein (IL-1Rap) transmembrane glycoproteins. IL-1RI is the binding component and IL-1Rap is the signaling component of the complex. A second IL-1R, IL-1RII cannot mediate IL-1 signaling (although IL-1β preferentially binds to IL-1RII over IL-1RI), as it does not complex with IL-1Rap when ligated. Subsequently, IL-1R-activating kinases (IRAKs) are recruited to the receptor complex, followed by activation of TNF receptor-associated factor-6 (TRAF-6). From here, a number of specific protein kinases are activated, including the NF kappa B (NF-κB) inducing kinase (NIK) and three distinct mitogen-activated protein (MAP) kinase cascades. These modulate a number of transcription factors including NF-κB and AP-1, each of which rapidly regulate the transcription of a plethora of inflammatory genes, among which are other cytokines, cytokine receptors, antimicrobial peptides, acute-phase reactants, growth factors, adhesion molecules and tissue remodelling enzymes. The powerful inflammatory properties of IL-1 are a determining factor in the pathogenesis of most inflammatory diseases, Crohn's disease and ulcerative colitis being among them. (Reviewed in refs. 15-18).

IL-1β mRNA is present in biopsy tissue obtained from uninflamed colonic mucosa. However, levels are significantly higher in biopsies from individuals with active inflammatory bowel disease compared to such controls. Interestingly, significantly higher levels of IL-1β mRNA have been found in uninvolved mucosa of Crohn's disease patients who present with a relapse of disease activity, as compared to newly diagnosed cases with characteristic early histological features of the disease.[105] Similarly, colonic explants of Crohn's and ulcerative colitis mucosal tissue secrete significantly more IL-1β protein when cultured ex vivo, compared to control, unaffected tissue. Furthermore, the levels of IL-1β secreted into such culture supernatants from inflamed tissue correlate with the disease severity in the mucosa from which the biopsies were removed.[55] The mononuclear-phagocytic population appears to be the source for the increased production of IL-1 in the mucosa in inflammatory bowel disease. Lamina propria macrophages isolated from Crohn's disease and ulcerative colitis patients display higher levels of IL-1 production when cultured in vitro than cells obtained from control, healthy tissue[56] as do total peripheral blood mononuclear cells (PBMC) and lamina propria mononuclear cells (LPMC) which, when cultured for 48h in the presence of poke-weed mitogen (PMA), secrete significantly higher levels of IL-1β protein into culture supernatants than cells isolated from control subjects.[65]

Parallel to the increase in plasma IL-1β in Crohn's and ulcerative colitis patients, there is concomitant increase in plasma levels of IL-1Ra, the natural inhibitor of IL-1.[106] An increase in the secretion of IL-1RA has also been observed in supernatants from ex vivo biopsy cultures from Crohn's and ulcerative colitis explants. This increase correlates with an increase in IL-1β production and does not occur in tissue from control, uninflamed intestinal tissue.[55] IL-Ra is an endogenous modulator of IL-1 and acts to limit the potential cytotoxic effects of this potent inflammatory cytokine. Many studies have investigated the possibility of an imbalance between the expression of IL-1β and the receptor antagonist as being a pathogenic factor in inflammatory bowel disease. That IL-Ra administration is beneficial in modulating intestinal inflammation in animal models of inflammatory bowel disease[106] adds weight to this theory and highlights the importance of IL-1Ra as a modulator of inflammation in the intestine. In some studies, relative analysis of IL-1 and IL-Ra protein levels secreted from colonic biopsy explants cultured *ex vivo* reveals that the ratio of IL-1Ra: IL-β secretion appears to be significantly lower in ulcerative colitis biopsy specimens compared to healthy intestinal tissue.[55] Certain alleles within the polymorphic regions of exon 5 of the IL-1β gene and in intron 2 of the IL-1Ra gene have been associated with susceptibility to inflammatory bowel disease and this imbalance of IL-1β: IL-1Ra ratios. However, the association studies carried out to date have been obtained from different ethnic populations and have generated conflicting data.[107] Taken together however, these studies demonstrate that carriage of allele 1 of the IL-1β gene and/or allele 2 of the IL-Ra, (IL-1RN*2) gene is associated with incidence of inflammatory bowel disease in certain ethnic populations.[108-110] That the IL-1RN*2 carriage is linked to a phenotype of imbalanced IL-1β/IL-1Ra levels is suggested by the finding that PBMCs isolated from both control subjects and ulcerative colitis patients carrying the IL-1RN*2 exhibit decreased production of IL-1ra when cultured in vitro compared to PMCs that do not carry the IL-1RN*2 allele.[110]

In addition to IL-Ra, another level at the activities of IL-1β are regulated is in the cleavage of the inactive pro-IL-β precursor to the mature, bioactive polypeptide. Evidence suggests that in chronic inflammatory bowel disease, this biological check point for IL-1β regulation might be by-passed, in part contributing to the inflammatory process. Increased levels of p20 subunit of caspase-1 are present in biopsy samples from both Crohn's and ulcerative colitis patients, indicating the presence of mature, active caspase-1. In colonic mucosa from uninflamed control tissue however, only the p45 sub-unit of the precursor form of caspase-1 is present.[111] Consistent with a lack of caspase-1 activity, LPS stimulation of colon macrophages isolated from control, healthy individuals elicits no activation of caspase-1 and as expected, these cells produce only the inactive precursor form of IL-1β. In contrast, colonic macrophages isolated from Crohn's and ulcerative colitis patients express the mature form of caspase-1 and hence release mature IL-1β in a manner similar to circulating monocytes [112]. Targeted inhibition of caspase-1 may represent a novel form of therapy in inflammatory bowel disease and may act to limit or block the contribution of two powerful inflammatory mediators, IL-1β and IL-18 (see later section) in to the disease process.

Interferon-gamma (IFN-γ)

IFN-γ is a major inflammatory cytokine produced predominantly by activated Th1, CD4 T cells and natural killer (NK) cells. IFN-γ expression is induced/inhibited in these cell types by a wide variety of extracellular signals, thus implicating a number of diverse, yet convergent signal transduction pathways in the transcriptional control of IFN-γ production. Interferons are cytokines that play a complex and central role in the resistance of mammalian hosts to pathogens. Originally described by Wheelock et al in 1965 as an anti-viral agent, IFN-γ (or type II interferon) is in fact only weakly anti-viral in comparison to other members of the interferon family, and is distinguished from the antiviral type I interferons, IFN-α and IFN-β by its ability to regulate many other aspects of the immune response in addition to protection against viral pathogens. IFN-γ promotes the bactericidal activity of phagocytes, stimulates antigen presentation through class I and class II major histocompatability complex (MHC I) and class II (MHC II) molecules, orchestrates leukocyte-endothelium interactions, and effects cell proliferation and apoptosis. IFN-γ up-regulates MHC I APC function in all somatic cells, thus affording protection against intracellular pathogens, and promotes MHC II associated APC function in cells derived from the myeloid lineage, namely DC monocytes and macrophages (otherwise known as professional APC) by up-regulating the expression of MHC I and MHC II molecules and associated components of the antigen processing and presentation machinery such as accessory molecules, processing enzymes, proteosome components and peptide transporting chaperones. In its potent induction of MCH II molecules, IFN-γ is also responsible for the 'switching on' of MHC II-restricted antigen presentation in nonprofessional APC such as epithelial and endothelial cells, astrocytes, microglia and thymocytes. The effects of IFN-γ described to date to date concerning B cells are quite distinct from those on cells of the myeloid lineage, in that IFN-γ inhibits antigen and cytokine driven proliferation of B cells and to downregulate MCH II-restricted antigen processing and presentation. IFN-γ is also able to direct antibody production by these cells in such a way as to inhibit IgG1 production and class switching to IgE isotype in plasma cells. (reviewed in:[22-25,53,54])

IFN-γ mediates its biologic effects by interacting with a specific cell surface receptor 90kDa, single chain glycoprotein (IFN-γR) that binds its ligand with high affinity in a species-specific manner. IFN-γR is expressed on nearly all cell types. Whereas this single polypeptide is sufficient to confer ligand binding and processing activity to transfected cells, a second, as yet undefined, component is required to form a functionally active IFN gamma receptor. The IFN-γ receptor (IFN-γR) is the prototypical class II cytokine receptor. Expressed on almost every cell type in the body, the IFN-γR complex comprises two subunit chains, the 90kDa ligand-binding IFN-γR1 chain and IFN-γR2, the accessory or signal transducing chain. IFNγ-R1 binds its ligand with high affinity, a process which then triggers oligomerisation of the IFN-γR1 and IFN-γR2, subunits. Signal transduction events are then initiated, namely the activation of JAK-1 and JAK-2 receptor associated protein tyrosine kinases, phosphorylation of the IFN-γR1 intracellular domain on Tyr440, which is followed by phosphorylation and activation of the latent transcription factor STAT-1α. STAT-1 facilitates the transcriptional regulation of IFN-γ target genes by other DNA-binding factors such as the interferon regulatory factors (IRFs) and the class II transactivator (CIITA).[23,113,114]

Increased levels of IFN-γ production, both in the mucosa and systemically, are a hallmark of Crohn's disease-associated inflammation, but not of ulcerative colitis. As T cells are the major source of IFN-γ, this distinction fits with the former being a T cell driven Th1-like inflammatory disease and the latter not.[76-78,82,83] The increase in mucosal and circulating IFN-γ levels correlates to the number of infiltrating T cells observed in the inflamed mucosa[65,79]and stimulation of the lamina propria CD4 positive (CD4+) T cell population via the CD2/CD28 pathway appears to important in eliciting this IFN-γ production.[80] The gamma delta (γδ) T cell population, particularly V delta 1 T cell receptor positive (Vδ1 TCR+) cells, have also been identified as a major source of IFN-γ production in the lesions of inflamed Crohn's disease intestine[115] and animal models on inflammatory bowel disease in these mice correlated with the development of peripheral and colonic TCR gamma delta+ T cells capable of IFN-gamma production. These results suggest that IFN-gamma may be a common mediator of IBD utilized by pathogenic T cells of distinct phenotype.[116]

Concomitant to the increase in IFN-γ in the mucosa of Crohn's patients, there is an increase in interferon regulatory factor (IRF)-1 expression in lamina propria mononuclear cells (LPMCs). IRF-1 is a transcription factor stimulated by IFN-γ and is responsible for the transcriptional regulation of the expression of several genes implicated in the pathogenesis of inflammatory bowel disease, MHC II molecules, and inducible nitric oxide synthase. IRF-1 also stimulates naive CD4+ T-cells to differentiate into Th1 cells. Increased expression of IRF-1 from patients with Crohn's disease may be relevant to the pathogenesis of Crohn's disease, providing the link between the increase in IFN-γ expression and the phenotypic changes in immune function observed therein.[117] There is a marked induction of MHC class II antigens in the mucosa of Crohn's disease patients, both on the inflamed intestinal epithelial cells and also on the underlying APC population. Antigen processing and presentation function is therefore potentiated during inflammation and may be a factor in Crohn's inflammation. IFN-γ is the most likely candidate for driving this inductive process from evidence obtained from in vitro studies in which an IFN-γ-induced expression of MHC II molecules on human intestinal epithelial cell lines has been observed.[118-120] Further evidence comes from the dose-dependent induction of human leukocyte antigen (HLA)-DR, HLA-DP and HLA-DQ class II molecules observed in biopsy specimens from colonic mucosa, cultured ex vivo with exogenous IFN-γ.[121,122] The expression of B7-2 on lamina propria mononuclear cells (LPMC) isolated from the mucosa of Crohn's disease patients increases proportionately with IFN-γ expression therein. B7-2 is a key costimulatory molecule in the MHC-restricted antigen presentation process and its induction on LPMC in Crohn's inflammation may contribute to the amplification of T cell proliferation and lymphokine production by IFN-γ activated LPMC. B7-2 expression is not observed in ulcerative colitis tissue or control, uninflamed tissue.[77] Nitric oxide (NO) levels are also increased in the inflamed mucosa of Crohn's disease, it's production therein being mediated by the inducible form of the enzyme nitric oxide synthase (iNOS). iNOS is an IRF-1 (and NF-κB) target gene product and the NO it produces may play a role in the pathogenesis of chronic inflammatory bowel disease. iNOS mRNA and protein expression is induced in a panel of human intestinal epithelial cell lines following coculture with IFN-γ added either alone or in combination with IL-1β or TNF-α.[123,124]

5-aminosalicylic acid (5-ASA), a nonsteroidal anti-inflammatory, is commonly used in the treatment and remission maintenance of inflammatory bowel disease. It appears that one of the therapetic mechanisms of actions of 5-ASA is to specifically inhibit the actions of IFN-γ in the intestinal mucosa during the inflammatory process. 5-ASA, 4-ASA, and their N-acetylated metabolites N-acetyl-5ASA and N-acetyl-4ASA inhibit the IFN-γ-induced expression of HLA-DR molecules on HT-29 human intestinal epithelial cells[125]A marked inhibition in binding of IFN-γ to its receptor appears to be the mechanism by which the salicylates achieve this end[126] 5-ASA, but not 4-ASA, also dose-dependently inhibits iNOS expression and NO production by DLD-1 and Caco2 human intestinal epithelial cell lines, suggesting than another mechanism of action is the inhibition of 5-ASA is to block the deleterious effects of NO. Interestingly, the inhibition of iNOS occurs at the transcriptional level, but is independent of IRF-1 and NF-κB.[127]

Interleukin-6 (IL-6)

IL-6 is a typical interleukin in that its biological effects are multi-potent and pleiotropic. IL-6 is an IRF-1 and NF-κB target gene product, produced primarily by B cells, CD4+ T cells, fibroblasts, mast cells, endothelial and epithelial cells. IL-6 promotes cell mediated and humoral immunity by promoting T cell and B cell proliferation and differentiation and also promotes lymphocyte survival by protecting T and B cells from apoptosis. IL-6 is therefore a central factor in determining the life-span of effector lymphocytes through both the haematopoietic and inflammatory processes. IL-6 mediates its biological effects on target cells by binding to the membrane-bound IL-6 receptor (IL-6R), a complex comprising an IL-6 binding α chain and a signal transducing element, gp130. Following ligation of the IL-6Rα chain, gp130 is tyrosine-phosphorylated by JAK tyrosine kinase. Multiple signal-transduction pathways are then initiated by recruitment to the IL-6R complex of a number of signaling molecules, which include STAT, and SHP-2. JAK can also directly activate signaling molecules such as STAT and TEC. These multiple signal-transduction pathways intimately regulate the expression of several genes, which drive the cell growth, differentiation, and survival promoting actions of IL-6. A feature of IL-6 activity is that it can bind to cells lacking surface IL-6 receptor (IL-6R) by means of forming a complex with the soluble form of the IL-6R (sIL-6R), a process known as IL-6 trans signaling. Dysregulated expression of IL-6 and its receptor negatively impacts the survival and function of the lymphocyte compartment and as such, is implicated in a variety of chronic inflammatory conditions [reviewed in refs. 128-131].

Ex vivo organ cultures of inflamed colonic mucosal biopsy tissue from ulcerative colitis[57,59] and Crohn's patients[59] reveals an increase in the levels of IL-6 secreted into culture supernatants following LPS stimulation compared to that seen from healthy, unaffected tissue. This over-expression of IL-6 in inflammatory bowel disease is associated with increased production of other NF-κB-regulated cytokines, namely IL-1β and TNF-α, and increased production of tissue degrading matrix metalloproteinases.[56,85] The levels of IL-6 produced by inflamed ulcerative colitis tissue correlates with the endoscopic grade of inflammation and is significantly higher in patients with intractable disease receiving corticosteroids than in patients with nonintractable disease receiving corticosteroids.[57] Increased basal, as well as poke-weed mitogen (PMA)-induced IL-6 production is observed in peripheral blood mononuclear cell (PBMC) cultures from ulcerative-colitis tissue[65] and also in isolated lamina propria macrophages from Crohn's patients.[56] In situ hybridisation for IL-6 mRNA reveals it's expression in large numbers of immune effector cells throughout the inflamed intestine of both Crohn's and ulcerative colitis patients. However, macroscopically (endoscopic) unaffected areas of mucosal tissue can also contain IL-6 mRNA positive cells. In inflamed mucosa, IL-6 transcripts are predominantly located within macrophage-rich regions of the lamina propria and can be colocalised with endothelial adhesion molecules.[58] That IL-6 mRNA can be detected in macroscopically unaffected as well as inflamed areas, but that IL-6 protein can only be detected in macroscopically involved tissue[57] suggests that IL-6 mRNA expression is a marker of early mucosal inflammation, relative only to histological activity, whereas IL-6 protein production is a marker of established inflammation and is therefore relative to the macroscopic (endoscopic) grade of ulcerative colitis and Crohn's disease activity. Elevated serum IL-6 in Crohn's patients appears to be associated with a high frequency of relapse, plasma levels of IL-6 being significantly higher throughout active and inactive periods of disease activity in patients who present with a relapsing-remitting disease compared to those individuals who maintain a prolonged remission.[132] That neutralising antibodies against the IL-6R suppress established experimental colitis in various animal models of Crohn's disease adds weight to the evidence for a role for IL-6 in inflammatory bowel disease pathology. Neutralising the bioactivity of IL-6 in these models results in an increase in the numbers of apoptotic lamina propria T cells (LPTC),[133] evidence perhaps, that IL-6 contributes to chronic intestinal inflammation by its well-described anti-apoptotic effect on lymphocytes. Mucosal T cells from Crohn's patients exhibit evidence of IL-6 trans signaling via sIL-6R and, similar to the in vivo studies with anti-IL-6 antibodies, neutralisation of sIL-6R with a gp130-Fc fusion protein results in a suppression of colitic activity and an increase in the number of apoptotic LPTC.[133] Thus, the IL-6-sIL-6R trans signaling system appears to mediate the resistance of T cells to apoptosis in Crohn's disease and in doing so, contributes to the perpetuation of chronic intestinal inflammation by blocking the deletion of inflammatory T cells.

Interleukin-12 (IL-12)

IL-12 is a heterodimeric cytokine, produced primarily by APC, of which dendritic cells (DC) are a major source and by Th1 T cells. IL-12 plays a central role in promoting Th1 responses and, hence, cell-mediated immunity. Along with IL-18, IL-12 is a central potentiator of IFN-γ expression. Its activities are mediated through a high-affinity receptor on its target T cells. The receptor is composed of two subunits, designated β1 and β2. Modulation of the expression of IL-12Rβ2 is key in regulating IL-12 responsiveness and also controlling Th1 lineage commitment of target cells, as this sub-unit is the signal-transducing component of the receptor complex. IL-12 signaling results in STAT-4 activation and IFN-γ production. Recent evidence suggests that IL-12 also modulates a number of genes involved in leukocyte trafficking. Thus, IL-12 is not only an important pro-inflammatory cytokine, but also plays a major role in regulating the migration of effector cells. Endogenous IL-12 plays an important role in host defence against infection by a variety of intracellular pathogens but, because of it's Th1 promoting activity, IL-12 appears to be a central component in the genesis of some forms of immunopathology, Crohn's disease being one such condition [reviewed in[19-21,134]].

Immunohistochemical analysis of biopsy sections from the small intestine of Crohn's patients reveals elevated numbers of IL-12-producing macrophages in the lamina propria and muscularis propria. In contrast, these cells are rare or undetectable in patients with ulcerative colitis, other inflammatory bowel conditions, or in healthy tissue. The cytokine profile of cultured CD4+ T cell clones isolated from Crohn's patients shows elevated numbers of IFN-γ secreting cells compared to control T cell populations. The inclusion of a neutralising anti-IL-12 antibody to such cultures impairs the development of the IFN-γ producing CD4+ population therein.[67] IL-12 production by mononuclear cells is increased in actively inflamed areas of Crohn's disease intestine. Quantitative comparative analysis of actively inflamed areas of tissue obtained from ulcerative colitis and Crohn's patients reveals an elevated expression of IL-12, parallel to IFN-γ levels, in Crohn's disease, but no increase in IL-12 or IFN-γ expression in ulcerative colitis tissue. This is further confirmation of the role of Th1 T-helper populations in Crohn's disease and their absence in ulcerative colitis.[70] Messenger RNA for both the p40 and p35 sub-units of IL-12 can be detected in ileal LPMC isolated from Crohn's disease patients.[68,69] These cells secrete biologically significant levels of IL-12 protein when cultured in vitro. The supernatant from these cultures is able to induce IFN-γ production from T cells, an effect which can be blocked in a dose-dependent manner by the inclusion of an anti-IL-12 antibody. In contrast, supernatants from ulcerative colitis or healthy, control LMPC cultures elicit no IFN-γ production in T cell cocultures.[68]

In vitro and ex vivo models have provided insights into the mechanism by which IL-12 elicits it's effects, either alone or in combination with other factors that govern the behaviour of the gut T cell population. IFN-γ messenger RNA can be detected in cultured isolated human LPTC stimulated with IL-12 and also IL-15, but secretion of IFN-Γ protein into culture supernatants only occurs following IL-12 stimulation. IL-12 synergises with either anti-CD2 or anti-CD28 antibodies in inducing IFN-γ synthesis in cultured LPTC in vitro. IL-12 does not induce proliferation of cultured LPTC and does not augment the CD2/CD28 proliferative response on stimulated LPTC. In terms of receptor expression, no transcripts for the IL-12 receptor β1 sub-unit are found in freshly isolated and activated LPTC, whereas β2 sub-unit mRNA is present.[135] Furthermore, IL-12 receptor β2 mRNA expression has been shown to be upregulated in parallel with mRNA for the p40 sub-unit of IL-12 in involved areas of mucosa from Crohn's disease compared to healthy tissue and also that from ulcerative colitis patients. No up-regulation of IL-12Rβ1 is detected in inflamed areas of Crohn's tissue, highlighting the importance of the β2 signal-transducing element of the IL-12R over the β1 sub-unit, in modulating the effects of IL-12 and it's role in Crohn's disease.[136] Activation of T cells in foetal gut explants with anti-CD3 antibody elicits very little IFN-γ production and results in little to no tissue injury. However, the addition of IL-12 with anti-CD3 results in a significant increase in IFN-γ and TNF-α production, associated with a massive increase in tissue injury which is blocked by an inhibitor of the matrix metalloproteinase stromelysin-1. Costimulation of explants with anti-CD3 and IL-18 induces only IFN-γ production but elicits no tissue damage.[137] This suggests that one of the mechanisms by which IL-12 causes the tissue damage to the gut wall in chronic inflammatory bowel disease is activation, directly or indirectly, of matrix metalloproteinases.

CD40 ligand (CD40L) is a type II membrane protein with homology to TNF and is transiently expressed on activated T cells. CD40L ligates CD40, which is expressed most abundantly on DC but also on mononulcear cells and B cells. CD40-CD40L interaction simultaneously activates both the T cells and the APC on which they are expressed and is required for B cell Ig production and for differentiation and maturation of DC and monocytes. CD40-CD40L interaction is essential for the production of IL-12 by the APC and T cells involved and therefore, this interaction may be operational in the pathogenesis of Crohn's disease through it's promotion of IL-12 production. CD40L is expressed on freshly isolated LPTC from Crohn's disease patients and is functional to induce IL-12 production when in coculture with normal monocytes. The inclusion of a neutralising anti-CD40L or anti-CD40 antibody in such cocultures significantly decreases monocyte IL-12 production therein. In vitro activation of lamina propria and peripheral blood T cells from these patients with anti-CD3 antibody elicits increased and prolonged expression of CD40L when compared to cells from control patients. Immunohistochemical analysis indicates that the number of CD40 and CD40L positive cells is significantly increased in inflamed mucosa. The CD40 positive population being determined as B cells and macrophages, with the CD40L positive population being CD4+T cells.[138,139] An elevated expression of CD40 on monocytes from Crohn's disease patients with active disease over levels of expression from control subjects has also been demonstrated and CD40 crosslinking of these PBMC in in vitro cultures elicits a higher level of IL-12 production than is determined from control PBMC.[139]

As IL-12 is strongly implicated as a central factor in the pathology of Crohn's disease, inhibiting its synthesis, secretion or its bioactivity in some way is a target for therapy. Selective inhibition of CD40-CD40L interactions in the mucosal effector cell population may have therapeutic effects in Crohn's disease and selectively blocking the $\beta 2$ sub-unit of IL-12R is perhaps another route of intervention. The major focus of late along these lines however has been the use of neutralising anti-IL-12 antibodies and their effects in murine models of inflammatory bowel disease have been studied in detail. In the hapten reagent-induced, 2,4,6-trinitrobenzene sulphonic acid (TNBS) murine model of colitis, the administration of monoclonal anti-IL-12 antibodies to TNBS-treated mice post-induction of colitis results in an improvement in the histopathological aspects of the disease. Lamina propria CD4+ T cells isolated from anti-IL-12-treated mice do not secrete IFN-γ upon in vitro stimulation.[140] In other, adoptive T cell transfer based models of murine inflammatory bowel disease, anti-IL-12 therapy is also effective in reversing the disease process with lower numbers of IFN-γ-producing T cells being observed in the mucosa of anti-IL-12-treated compared to untreated diseased mice.[141] Confirming the central role for IL-12 in these model immunopathologies, blocking IL-12 rather than IFN-γ appears to be key in reversing the chronic inflammation observed in the disease models.[142] Two possible mechanisms for the actions of IL-12 in the disease pathology have come about from studies incorporating the use of anti-IL-12 antibody therapies in mouse models. Firstly that over-expression of IL-12 breaks tolerance to the resident commensal floral population of the gut, and drives a Th1 mediated response against components of the commensal flora. The addition of anti-IL-12 antibodies restores this tolerance of the T cell population.[143] Secondly, anti-IL-12 therapy appears to protect the local mucosal T cell population from apoptosis.[144]

Interleukin-15 (IL-15)

IL-15 is a cytokine that shares biological activities, but no significant sequence homology, with IL-2. That its actions are similar to IL-2 is due to the usage of common receptor components. IL-15 and IL-2 are members of the four-helix bundle cytokine family and both cytokines transduce signals through the beta (p75) and gamma (p64) chains of the IL-2 receptor (IL-2R) system, but IL-15, like IL-2, binds to its own specific alpha chain, referred to as IL-15Ralpha. IL-15 is produced by macrophages and various other cells in response to environmental stimuli and infectious agents. IL-15 is important for the growth and differentiation of T and B lymphocytes, natural killer cells, macrophages, and monocytes. It induces T cell recruitment to sites of inflammation, T cell proliferation, and cytokine production and rescue from apoptosis. IL-15 may therefore play a pivotal role both in innate and acquired immunity. That is has the potential to influence the major cellular players in the inflammatory process suggests that it also has the potential to play a role in the pathogenesis of various chronic inflammatory disorders, inflammatory bowel disease among them [Reviewed in refs. 145 146,147].

Analysis of IL-15 secretion from ex vivo tissue cultures of Crohn's and ulcerative colitis biopsy samples reveals an increased secretion of IL-15 by tissue from Crohn's patients and ulcerative colitis patients compared to that released by healthy, unaffected tissue.[71,148] Furthermore, the levels of IL-15 secreted by Crohn's tissue appear to be higher than that from ulcerative colitis biopsies.[148] Other studies have shown no detectable IL-15 in the inflamed mucosa of ulcerative colitis tissue, highlighting perhaps IL-15 as another of the cytokines, along with IL-12, IL-18 and perhaps IL-16, that are specific to the pathology of Crohn's disease.[149] Immunohistochemical analysis and in situ hybridisation studies reveal that the macrophage population and also intestinal epithelial cells are the source of the over-expressed IL-15 protein within inflamed mucosa of Crohn's-affected intestine.[71,150] In a large study of cytokine profiles in bacterial enterocolitis (associated with Methicillin-resistant *Staphylococcus aureus* or MRSA) levels of serum IL-15 were significantly elevated in affected patients compared with those of other MRSA infections without enterocolitis, including pneumonia and cholangitis. This increase in serum IL-15 was observed just before clinical manifestation of severe diarrhoea, suggesting that not only is IL-15 associated in the pathogenesis of post-operative enterocolitis, but that its serum levels may be a severity indicator of the disease.[151] In vitro, LPS or IFN-γ stimulation of isolated LPMC from Crohn's biopsy samples elicits IL-15 production, whereas no IL-15 can be detected in supernatants of cultured cells isolated from healthy gut. Moreover, isolated lamina LPTC from Crohn's patients are more responsive to IL-15 as compared with controls, and IL-15 added alone, without an additional primary T cell stimulus, induces IFN-γ and TNF-α production by isolated LPTC from Crohn's patients. In an in vitro coculture system, such LPTC appear to induce CD40-CD40L interaction-dependent TNF-α and IL-12 production by monocytes, an effect that is strongly enhanced by preincubation of the LPTC in IL-15 and results from an IL-15-mediated increase in CD40L expression.[150] In addition to inducing IFN-γ and TNF-α production, recombinant IL-15 induces a dose-dependent proliferative response in in vitro cultures of LPMC isolated from Crohn's-affected mucosa.[71] The overexpression of IL-15 in the inflamed mucosa of Crohn's patients appears therefore to drive local Th1 responses by promoting T cell activation, proliferation and IFN-γ and IL-12 production, the latter via a CD40-CD40L interaction-dependent mechanism. One study suggests that, in contrast to its effects in promoting T cell responses in inflammatory bowel disease, IL-15 appears to inhibit the chemotaxis of the LPMC population. IL-15 down-regulates IL-8 and macrophage chemotactic protein (MCP)-1 production by Caco2 human intestinal epithelial cells as well as by freshly isolated human colonic epithelial cells, an effect which is blocked by the addition of an anti-IL-2 receptor γ-chain antibody.[152]

Interleukin-16 (IL-16)

Interleukin-16 was initially described in 1982 as the first T cell chemoattractant. IL-16 is secreted as a 17kDa peptide, which aggregates into a tetrameric form. This final tetrameric structure is essential for direct interaction with, and cross-linking of IL-16 with its receptor, the CD4 antigen. It is through this interaction with CD4 that IL-16 chemoattracts CD4+ immune cells. In addition to the induction of cell motility of the local CD4+ population, IL-16 also appears to act as a competence growth factor for CD4+ lymphocytes. IL-16 therefore plays a role in both the accumulation and activation of CD4+ cells and could therefore be a key factor in the inflammatory process. Recent in vivo studies have characterized IL-16 as a contributing factor in the regulation of CD4+ cell recruitment and activation at sites of inflammation a number of autoimmune and inflammatory disease states, most recently Crohn's disease. [reviewed in[153-155]] Protein levels for IL-16 are increased in colonic mucosal tissue isolated from patients with Crohn's disease compared to healthy tissue, but IL-16 levels are not increased in samples from ulcerative colitis patients. Additional evidence for a role for IL-16 in chronic intestinal inflammation comes from TNBS mouse model of colitis, in which administration of an anti-IL-16 antibody significantly reduces weight loss, mucosal ulceration, myeloperoxidase activity and increases in mucosal levels of IL-1β and TNF-α in affected animals.[73]

Interleukin-18 (IL-18)

Initially described as interferon gamma inducing factor (IGIF), IL-18 is another key cytokine in the Th1 response. IL-18 is produced by a variety of cells including macrophages, keratinocytes, osteoblasts and intestinal epithelial cells. Its primary action is the induction of IFN-γ production in T cells and natural killer (NK) cells. The importance of IL-18 as a potentiator of IFN-γ even possibly over IL-12 is highlighted by the fact that mice deficient in IL-18 produce diminished amounts of IFN-γ despite the presence of IL-12. IL-18 is a member of the IL-1 family of cytokines and shares common features with the IL-1s, both in structure and function. IL-18 and IL-1β share significant primary amino acid sequences and have a similar tertiary beta-pleated sheet structure. Like IL-1β, IL-18 is synthesised as a biologically inactive precursor molecule lacking a signal peptide. Both the IL-1β and IL-18 precursor forms are cleaved into active, mature molecules by the intracellular cysteine proteinase caspase-1. Caspase-1 inhibitors may therefore be useful as Th1 immunosuppressive agents in limiting the biological activities of IL-18 and also IL-1β. IL-18 induces gene expression and synthesis of TNF, IL-1, FAS ligand (therefore enhancing the FAS ligand-mediated cytotoxicity of CD4+ T and NK cells) and several chemokines. IL-18 mediates its effects by ligating the IL-18 receptor (IL-18R) complex. IL-18R comprises a binding chain termed IL-18Rα (a member of the IL-lR family previously identified as the IL-1R-related protein, IL-1Rrp) and another IL-1R family member, the signaling chain IL-18Rβ. Common to all IL-1/IL-1R family member interactions, this leads to recruitment of IRAK to the IL-18R complex and then activation of TRAF-6, which in turn phosphorylates NF-κB-inducing kinase (NIK), leading to subsequent activation and translocation of NF-κB. IL-18 is an important regulator of innate immunity but it's effects on polarising a Th1 response, similar to IL-12 are what pose a threat in it's potential to participate in the immunopathology of inflammatory bowel disease [Reviewed in [34,156-160]].

Levels of mRNA transcripts for IL-18 are increased in freshly isolated intestinal epithelial cells and LPMCs from Crohn's patients when compared to those seen in cells from ulcerative colitis and uninflamed control patients.[74,111] Serum IL-18 levels are also significantly higher in patients with Crohn's disease than in healthy controls. Intestinal mucosal lymphocytes isolated from Crohn's disease biopsies express functional IL-18 receptors and proliferate readily in response to the addition of recombinant IL-18 when in culture. In contrast, no IL-18 dependent proliferation of control mucosal lymphocytes isolated from healthy tissue occurs. Exogenous IL-18 added to such cultures also up-regulates IL-2 receptor expression in mucosal lymphocytes from Crohn's patients, but not in those from healthy control subjects.[75] In situ hybridisation and immunohistochemical analysis of surgically resected colonic tissue from Crohn's patients reveals that IL-18 mRNA and protein is present in both LPMC (specifically, macrophages and dendritic cells) and the intestinal epithelial cells.[74,75] The IL-18 producing macrophages within the inflamed colonic mucosa of Crohn's-affected intestine infiltrate the lamina propria[75] and more IL-18-positive macrophages are observed therein in Crohn's rather than in ulcerative colitis inflammation or in control tissue and as expected, more cells are localised in inflamed, compared to uninflamed areas.[74] Western blot analysis of mucosal tissue isolated from inflamed and control intestinal tissue shows that an 18. 3-kDa band, consistent with the mature human IL-18 protein, is found predominantly in Crohn's-affected tissue but not in ulcerative colitis samples. Moreover, a second band of 24kDa, consistent with the inactive IL-18 precursor form can be detected both Crohn's disease and ulcerative colitis biopsies and is the sole form detectable by immunoblot analysis of uninflamed control biopsies.[74,111] It appears therefore that regulation of the processing of the precursor form to active IL-18, rather than regulation of de novo synthesis of the cytokine, is key in its role as an inflammatory agent in Crohn's disease. Consistent with this, caspase-1 subunit (p20) levels are increased in biopsy samples from Crohn's patients, whereas in colonic mucosa from uninflamed control tissue, only the p45 precursor of caspase-1 is present.[111] In an ex vivo model of IL-18-induced intestinal inflammation, exogenous IL-18 included in foetal gut explant cultures induces IFN-γ production. So too does the addition of IL-12, but a major difference between these two IFN-γ potentiators is that where the addition of IL-12 also results in massive matrix metalloproteinase-involved tissue damage, the addition of IL-18 does not.[137] Increases in mucosal IL-18 levels, similar to those seen in Crohn's disease, appear to be a feature of both TNBS-induced and NOD/SCID adoptive T cell transfer mouse models of inflammatory bowel disease.[79,161] In one such study, the administration of a high affinity anti-IL-18 antibody to affected mice was shown to block the LPS-induced produced secretion of IFN-γ by approximately 90%.[161] Perhaps, as with anti-IL-12 antibodies, more studies will follow to investigate the potential for anti-IL-18 therapy in Crohn's disease and other conditions where this Th1 cytokine is implicated as a pathological factor.

Chemokines

Chemokines are a large family of small cytokines whose main function is to chemoattract immune effector cells to a number of tissues, including the intestinal epithelium and the underlying mucosal lymphoid aggregates, both in the normal state and during the inflammatory process. Chemokines are divided structurally into two subfamilies, as determined by the arrangement of the first two of four conserved cysteines. These are either adjacent (C-C chemokines) or are separated by one amino acid (C-X-C chemokines). In terms of functional differences between the two sub-families, C-X-C chemokines act preferentially on neutrophils, while the C-C chemokines act on monocytes. Both have additional activities toward basophil and eosinophil granulocytes, T-lymphocytes, natural killer cells and dendritic cells.[162-164] Chemokine receptors belong to the seven-transmembrane-domain type and are coupled to G-proteins.[162,165] Studies in which intestinal epithelial cell lines were screened for the expression and production of these chemotactic agents provided the first evidence that intestinal epithelial cells are a source of chemokines and are therefore able to attract neutrophils and monocytes during mucosal inflammation.[166-169] A number of chemokines and their receptors have since been shown to play a critical role in lymphoid development, mucosal immunity, and inflammation. As such, chemokines have the potential to participate in the pathogenesis of inflammatory bowel disease.[170,171]

Interleukin-8 (IL-8)

Interleukin-8 (IL-8), the prototypical C-X-C or α-chemokine, acts as a motor for the recruitment of neutrophils across the vascular wall and as such, plays a key role in innate host defence mechanisms. The release of IL-8 is triggered by inflammatory signals from a large variety of cells. The diversity of its cellular sources, which include gastrointestinal epithelial cells, indicates the interleukin-like pleiotropy of its functions.[165,172-174] IL-8 transmits its signals through two distinct cell surface receptors, C-X-CReceptor-1 (CXCR-1) and C-X-Creceptor-2 (CXCR-2), that are coexpressed on human neutrophils. Both are 7-transmembrane domain-type proteins functionally coupled to G proteins. Although both receptors bind IL-8 with high affinity, they differ in selectivity for other C-X-C chemokines (a common feature of chemokine receptors is their ligand promiscuity).[172,175,176] Overall, IL-8 acts a key factor in the recruitment and activation of polymorphonuclear neutrophils, cells that are abundant in the intestinal lesions of ulcerative colitis and Crohn's disease. A continued presence of IL-8 in inflamed intestinal mucosa can therefore contribute to the characterisitc tissue damage in inflammatory bowel disease, due to the destructive influence of the matrix-degrading neutrophil granule components that are released upon neutrophil activation following recruitment into the gut wall.

A number of studies employing ex vivo organ culture of biopsy samples as a means of assaying the cytokines secreted from the mucosa of patients with inflammatory bowel disease reveal the presence of high levels of IL-8 in culture supernatants of biopsies, whereas only trace levels are detected in those supernatants from control, uninflamed tissue, tissue from nonspecific colitis or irritable bowel syndrome.[57,60-62] Similarly, IL-8 concentrations in rectal dialysates obtained from patients with active ulcerative colitis are shown to be significantly higher compared to those from control subjects.[177] The production of IL-8 in inflammatory bowel disease also correlates with disease activity, IL-8 protein levels being proportional to macroscopic (endoscopic) grade of inflammation.[57,61,177] Furthermore, the production of IL-8 is significantly higher in patients with intractable disease receiving corticosteroids than in patients with nonintractable disease receiving corticosteroids [57]. Immunohistochemical analysis reveals the major source of increased IL-8 in inflammatory bowel disease to be the infiltrating neutrophils and macrophages[178,179] and in vitro culture of macrophages isolated from patients with ulcerative colitis in the presence of LPS elicits a significant upregulation in IL-8 secretion, whereas no effect is observed in cells isolated from normal colon macrophages.[178] In contrast to the wealth of evidence for the correlation of mucosal and luminally derived IL-8 protein, both as an indicator of intestinal inflammation and as a correlate of disease activity in Crohn's disease and ulcerative colitis, mRNA for IL-8 can generally only be detected in mucosal biopsies from Crohn's but not ulcerative colitis patients, perhaps indicating transient expression of the IL-8 gene or altered mRNA stability in ulcerative colitis versus Crohn's disease.[61,180] The IL-8 protein derived from patients with active ulcerative colitis exhibits chemotactic and activatory effects on isolated primary neutrophils in vitro. This bioactivity has been demonstrated in rectal dialysates from ulcerative colitis patients,[177] colonic mucosal homogenate supernatants[181] and in ex vivo organ culture supernatants[60] in patients with active disease. In each of these studies, the chemotaxis and activation of the neutrophils was blocked by the addition of a neutralising anti-IL-8 antibody.

Monocyte Chemotactic Proteins (MCPs)

The monocyte chemotactic proteins MCP-1 and MCP-3, together with MCP-2 and MCP-4, comprise a subfamily of the C-C or β-chemokines. MCP-1 and MCP-3 are small, inducible chemotactic cytokines, each with their own distinct spectrum of target cells and biological activities.[162,182,183] The signal transduction of MCPs involves at least two, G protein-linked C-C chemokine receptors. MCP-1 binds the C-CReceptor-2 (CCR-2), of which there are two alternatively spliced forms, CCR-2A and CCR-2B, while MCP-3 binds C-CReceptor-3 (CCR3) on target cells.[162,184-187] The most sensitive of target cells to both MCP-1 and MCP-3 are lymphocytes and monocytes.

MCP-1 production in vitro is induced in a variety of cells following stimulation with cytokines (IL-1, TNF-α, IFN-γ), bacterial and viral products or mitogens.[167,168] In vivo, MCP-1 is produced by endothelial cells, LPMCs and epithelial cells and acts to recruit monocytes and macrophages, eosinophils and lymphocytes. Increases in MCP-1 expression at sites of inflammation facilitate increased infiltration therein of these effector cells. MCP-1 production is elevated in inflamed areas of intestinal mucosa in ulcerative colitis and Crohn's disease,[170,188] mRNA and protein for MCP-1 being expressed more abundantly in areas of active Crohn's and ulcerative colitis inflammation compared to the basal levels expressed in unaffected areas or in healthy control tissue. The source of this increased MCP-1 production appears to be the epithelial cells and endothelial cells in the inflamed gut wall[179] as well as smooth muscle cells[189] and the infiltrating monocytes and macrophages therein.[62,189,190]

MCP-3 has the broadest range of target cells and exerts chemotactic and activating influences on monocytes, lymphocytes, eosinophils, natural killer cells, basophils [169,182] and some dendritic cell populations.[182,191] MCP-3 mRNA expression and protein production is observed in control, healthy intestinal mucosa, with both the epithelial [169] and infiltrating monocytic cells[62] therein being the source of MCP-3 protein. MCP-3 protein expression is markedly increased in inflamed intestinal tissue isolated from ulcerative colitis patients, with expression correlating to the extent of epithelial damage.[62,169,192]

Regulated Upon Activation, Normal T cell Expressed and Secreted (RANTES)

RANTES is a C-C chemokine that contributes to the recruitment primarily of memory CD4+ T cells, but also cytotoxic CD8+ T cells, NK cells and also monocytes, macrophages and eosinophils during the inflammatory response.[162,193] Intestinal epithelial cells are a major source of RANTES production, particularly during the inflammatory process.[78,166] Kinetics studies of C-C chemokine secretion from bacterially and agonist challenged human intestinal epithelial cell lines and freshly isolated intestinal epithelial cells show that while some RANTES protein is secreted in the basal state, its production by intestinal epithelial cells is upregulated during inflammation or infection. Interestingly however, this upregulation in RANTES expression occurs much later than the induction of MCP-1 in the same cells.[166] In addition to epithelial cells as a source, mRNA for RANTES is also present in intraepithelial lymphocytes (IEL) and other sub-epithelial lamina propria immune cells[190,194,195] and also activated platelets.[196] RANTES recruits effector cells bearing C-CReceptor-5 (CCR5), the RANTES receptor, to the intestinal mucosa during inflammation. Of any other cell type, CCR-5 is expressed most strongly on lamina propria memory CD4+ T cells; pointing to the central role of RANTES in promoting T cell mediated inflammation in the gut wall.[187,193,197] An abundant expression of CCR-5 on is also however a selective disadvantage as, together with another chemokine receptor that these cells express, C-X chemokine receptor-4 (CXCR4), CCR5 is a coreceptor for the human immunodeficiency virus (HIV)-1. Because lamina propria CD4+ memory T cells can express both of these coreceptors, they are naturally susceptible to infection by both T-(T cell) and M-(macrophage) trophic HIV-1 strains following primary infection by either the mucosal or the parentral route.

In addition to its chemotactic properties, RANTES also enhances certain lymphocyte effector functions, specifically cytotoxic T lymphocyte and NK cell-specific cytolytic responses and also antibody-dependent cell cytotoxicity-specific cytolytic responses.[194] mRNA expression of RANTES is elevated in the mucosa of Crohn's and ulcerative colitis patients compared to controls[190,194 195] and more specifically, is localised to areas of active inflammation, as shown by in situ hybridisation studies where RANTES transcripts were detected most strongly within the granulomas and inflammatory infiltrates of Crohn's affected tissue. RANTES most likely contributes to the selective accumulation of macrophages and memory T helper lymphocytes within these characteristic immune aggregates of Crohn's disease.[78] Even when in remission, Crohn's and ulcerative colitis patients exhibit increased production of RANTES protein. This over-production is associated with the increased platelet activity that is characteristic of inflammatory bowel disease. Thus, in addition to RANTES production by other tissue sources, activated platelets may contribute significantly to the increased production of RANTES in inflammatory bowel disease. Interestingly, Crohn's and ulcerative colitis patients on maintenance doses nonsteroidal anti-inflammatory (5-ASA) medication have lower circulating plasma RANTES levels, suggesting that another benefit of such therapy is in lowering circulating RANTES levels.[196]

Regulatory Cytokines and Growth Factors

Detailed in the introduction to this chapter, the intestinal mucosal environment is critically controlled for inflammation. Adequate and appropriate immune activity towards invading pathogens and other foreign antigens is essential to maintain health. Following successful clearance and the associated inflammatory response however, immune homeostasis must be reestablished. Therefore, regulatory systems exist which mediate the return to the normal state. In addition, the same systems prevent immunogenic activity towards harmless antigens. Regulatory cytokines play a key role in these events and are often found upregulated shortly after the onset of inflammation and consequently. Inadequate or dysregulated expression of regulatory cytokines therefore results in the persistence of an inflammatory condition which in turn drive pathology. The restoration of appropriate levels of the regulatory, and Th2-like cytokines by administration of recombinant protein or gene delivery is therefore a major target for therapeutic intervention in inflammatory bowel disease.

Interleukin-10 (IL-10)

IL-10 (recently reviewed[36-38]) is an important anti-inflammatory mediator that is produced by a variety of cells, but predominantly CD4+ and CD8+ T-cells, macrophages, monocytes, DC, B cells, keratinocytes and epithelial cells. IL-10 acts to control and suppress inflammation, essentially by the down-regulation of pro-inflammatory cytokine production, most likely by its NF-κB blocking activity,[198] and major histocompatability class II antigen (MHCII) expression. IL-10, as well as IL-4 and IL-13, also down-regulates the TNF-α or IL-1β induced production of MCP-1 in vitro by both cultured human intestinal epithelial cell lines and isolated primary intestinal epithelial cells from surgical specimens.[199]

As such, IL-10 can counteract inflammation through its activity on antigen presenting cells, which in turn affect T cell activity. IL-10 can also directly suppress T cell proliferation. Apart from its clear downregulatory role on cellular immunity, IL-10 is a B-cell growth and differentiation factor, involving class switching to IgA and differentiation of B-cells to antibody producing plasma cells. As demonstrated by immunohistochemistry and mRNA in situ hybridisation, intestinal epithelial cells secrete IL-10 in vivo[200] and in in vitro culture systems, this secretion from isolated intestinal epithelial cells is enhanced by stimulation with TNF-α or LPS, or during coculture with lymphocytes or macrophages.[201] Human and murine intestinal epithelial cells also express the IL-10 receptor.[202,203] This suggests that in addition to being both a source of IL-10 and a target for the IL-10 produced by other immune cells, intestinal epithelial cell function is regulated by the autocrine or paracrine actions of IL-10.

Mice that cannot make IL-10 spontaneously develop a chronic intestinal inflammation, which is histologically very similar to inflammatory bowel disease,[204] evidence of the regulatory role that this natural anti-inflammatory factor plays in mucosal immune homeostasis. This is further confirmed by the finding that, following the induction of colitis through the periodic addition of dextran solium sulphate (DSS) to drinking water and the subsequent increase in TNF-α and IL-1β observed in the mucosa of affected BALB C mice as a result, there is a marked increase in IL-10 production. In this model, neutralisation of IL-10 with an anti-IL-10 antibody resulted in exacerbation of the inflammation.[205] Other animal model studies have since demonstrated the efficacy of IL-10 in the treatment of enterocolitis, particularly in the maintenance of tolerance and nonT cell mediated immunity therein.

Freshly isolated human lamina propria T cells secrete significantly higher amounts of IL-10 than peripheral blood T cells upon CD2 or CD3 cross-linking in in vitro culture systems.[206] The polymorphonuclear leukocytes that are, in part, responsible for the release of pro-inflammatory cytokines the inflamed mucosa of patients with intestinal inflammation, are inhibited in their production of these inflammatory mediators in vitro by coculture with IL-10 [63]. The contribution of IL-10 to the maintenance of epithelial barrier integrity is demonstrated in IL-10 deficient mice, which show an increase in ileal and colonic epithelial permeability which precedes the development of mucosal inflammation.[207,208] Intestinal epithelial ells express the nonclassical MHC-class I (MHCI) antigen, CD1d. Ligation of CD1d in vitro induces epithelial IL-10 secretion and attenuates IFN-γ signaling in an IL-10-dependant manner.[209] Evidence for this IL-10-mediated CD1-dependent mechanism being in place in vivo comes from the fact that activation of NK T cells by the CD1d ligand alpha-GalCer, down-regulates intestinal inflammation in DSS-induced murine models of inflammatory bowel disease.[210]

The intestinal epithelium is intercalated with T cells, the so-called intestinal intraepithelial lymphocytes, (IEL) which can be stimulated in vitro to produce IL-10.[211] IEL are mainly of a CD8 positive (CD8+) phenotype and patrol the epithelium, ready to take part in cytolytic responses to luminally acquired antigen as and when required. A considerable proportion of IEL are γδ positive (γδ+) T cells. These regulatory γδ T cells provide another source of IL-10. Outside the gut, γδ T cells mediate IL-10 dependant prevention of death in animals that are subject to acute bacterial infection. This attributes a role to γδ T cells in the control of Th1 responses.[212] Low doses of oral antigen induce antigen-specific regulatory T cells in the gut, which locally release inhibitory cytokines such as TGF-β, IL-4 and IL-10.[44] As low dose oral tolerance is defective in TCR d-/- mice and in animals treated with anti-TCR gd, the intestinal gd T cells population appear to play a significant, and probably IL-10 mediated, immunoregulatory role in low-dose oral tolerance induction.[213] IL-10 does not exclusively exhibit inhibitory effects on T cells but also upregulates the IL-2 receptor and associated IL-2 dependant proliferation[214] and exerts positive influences on clonal expansion of subsets of preactivated T cells[215] Indeed, regulatory tolerogenic T cells must proliferate to IL-2 in order for clonal expansion to occur in the same way as immunogenic T cells must. Groux and collaborators [46] showed that chronic activation of both human and murine CD4+ T cells in vitro in the presence IL-10 gives rise to so-called Tr1 cells, CD4+ T cell clones with low proliferative capacity, producing high levels of IL-10, low levels of IL-2 and no IL-4. Cotransfer of these cells with CD4+CD45Rbhi (naïve phenotype) T cells to SCID mice prevents the onset of colitis.[216] A similar, TGF-β and IL-10 dependent feature resides in CD45Rblo (memory phenotype) T cells, cells capable of preventing the onset of colitis in the SCID CD4+CD45Rbhi transfer colitis model.[48] The down-regulatory effects of Tr1 cells appear not to be restricted to Th1 T cells but also to affect Th2 driven immunity[217] and in fact, transfer of Tr1 cells in mice prior to the induction of immediate hypersensitivity to ovalbumin (OVA) mimics the induction of tolerance towards OVA by repeated nasal or oral administration of antigen. NK cells also appear to exhibit down-regulatory effects on CD4+ T cells in the SCID transfer model of colitis in a perforin dependant manner.[218] As detailed previously, breach of tolerance towards normal intestinal microflora may be the driving force behind inflammatory bowel disease. This appears to be a feature of the TNBS mouse model wherein isolated T cells react to commensal bacterial antigens in vitro. Tolerance in these mice appears to be restored upon systemic treatment with IL-10. This treatment does not however, affect immune reactivity towards heterologous bacterial antigen.[143] While by no means profession APC, IEC may, under certain conditions in which IFN-γ acts to upregulate surface expression of MHCII molecules, present antigen to mucosal T cells[219,220] and in, drive T cell mediated inflammation. IL-10, either alone or in combination with TGF-β, inhibits this IFN-γ-mediated induction of MHCII molecules on IEC.[203,221] Many studies provide evidence for a direct immunomodulatory effect of IL-10 on the T cells that drive immunogenic responses. The immunogenic B subunit of *Staphylococcal* enterotoxin abrogates tolerogenic immune responses to commensal microfolora and self antigen in the intestinal mucosa. IL-10 can counteract this by preventing the enterotoxin B induced activation of T cells and consequent immunogenic activity.[222] Similarly, the phytohaemagglutinin (PHA)-mediated activation and proliferation in vitro of human cytotoxic CD8+ T cells isolated from the epithelium and lamina propria can be blocked by IL-10, regardless of the reported IL-2 receptor-dependent proliferative effects of IL-10 on T cells.[223] IL-10 also acts to regulate the production of prostaglandins, potent pro-inflammatory mediators that are produced by intestinal epithelial cells during the course of inflammation or infection. IL-10, and also IL-4, achieve this end by inhibiting the synthesis of cyclooxygenase (COX)-2, an inducible enzyme required for prostaglandin synthesis.[224] This is reflected in the early up regulation of COX-2 in IL-10 -/- mice.[225] Even the digestion of the extracellular matrix by matrix metalloproteinases, a major component of the tissue destruction and remodelling characteristic in the inflamed gut wall in inflammatory bowel disease, appears to be modulated by IL-10 which down-regulates of the expression of this class of Zn-dependent endopeptidases.[226]

The production of IL-1β and TNF-α by in vitro cultured peripheral and lamina propria monocytes and lymphocytes isolated from patients with ulcerative colitis is down-regulated following topical IL-10 treatment by enema.[227] In this study, equal concentrations of IL-10 can be detected in both normal and ulcerative colitis intestinal mucosal biopsy homogenates. This seems surprising but Autschbach and coworkers[200] also argue against a general deficiency in IL-10 production as a mechanism contributing to the chronic inflammation inflammatory bowel disease. Rather, they suggest that in inflammatory bowel disease, the production of IL-10 is dislocated and is insufficient to down-regulate pro-inflammatory cytokine production by effector immune cells in the lamina propria compartment. In view of its seemingly pivotal role in establishing and maintaining tolerance IL-10 is a good candidate as therapy in the treatment of IBD and has in fact been demonstrated effective in clinical trials.[228] When administered systemically by intravenous injection however, the high levels of exogenous circulating IL-10 result in a number of serious side effects, most related to nonspecific effects on dampening of the Th1 responses. It would seem therefore, that IL-10 is in fact, no better, or even a worse choice in this respect, than conventional synthetic immunosuppressive therapies. As such, the feasibility of local delivery of this natural anti-inflammatory protein was investigated; using recombinant *Lactobacillus lactis* that is engineered to drives the expression and production of murine IL-10. Oral inoculation with this strain leads to a complete abrogation of disease in 50% of mice with DSS-induced colitis and completely prevents the onset of colitis in IL-10-/- mice.[229]

Transforming Growth Factor-β (TGF-β)

TGF-β, a pleiotropic growth factor, belongs to a superfamily of growth factors comprising three mammalian isoforms (TGF-β1, TGF-β2 and TGF-β3), activins/inhibins and bone morphologenic proteins. 80% sequence homology exists among the three TGF-b isoforms, the biological functions of each being very similar. The major difference lies in the potency of the variants, in that TGF-β1 and TGF-β3 are effective at concentrations up to 100-fold lower than TGF-β2. TGF-β has profound inhibitory effects on almost all immune and haemopoietic functions. It is produced by a number of cells, including activated T cells, B cells, DC, macrophages and platelets. It is secreted in an inactive form in which the active component, a 40kDa homodimer of TGF-β, is complexed with a 40kD latency-associated peptide and a 120-160kDa latency binding protein. Active TGF-β is released upon proteolytic cleavage or in acidic conditions. TGF-β mediates its biological effects by binding to one of two high affinity receptors. In accordance with the pleiotropic nature of TGF-β, these receptors are expressed on almost all tissues. The type III TGF-β receptor (TGF-βRIII) is the most highly expressed and is a proteoglycan with a truncated cytoplasmic domain and no signaling motif. Similar to the IL-1RII, TGF-βRIII mediates none of the biological effects of TGF-β but may play a role in facilitating the binding of TGF-β to the type II TGF-β receptor (TGF-βRII). TGF-βRII is a transmembrane .[230-232]

TGF-β inhibits the immune function and proliferation of all T cells, NK cells and macrophages. It downregulates immunogenic responses by antagonising the activity of IL-1, TNF-α, IFN-γ and IL-6, inhibiting the production of IL-12 and stimulating the production of IL-4 and IL-10. TGF-β is therefore a powerful inhibitor of Th 1 responses.[233] The mechanism by which TGF-β can inhibit immune cell proliferation is an arrest of the cell cycle in mid-to-late G1 phase, thereby preventing the synthesis if DNA and progression to S phase.

The pivotal role of TGF-β in immune regulation is demonstrated in TGF-β1-/- mice which display multifocal, mixed inflammatory cell responses and tissue necrosis, resulting in organ failure and death.[234] Excessive inflammatory response with massive infiltration of lymphocytes and macrophages are observed in many organs in these animals. These include the colon, but primarily involve the heart and lungs.[235] SMAD-3 is a key intermediate in the TGF-β signaling pathway and it's absence results in a lethal phenotype similar to the TGF-β1-/- animals. SMAD-3 mutant mice die between 1 and 8 months also as a consequence of immune dysregulation at mucosal surfaces.[236]

IgA antibodies play a major role in antimicrobial defence in the lumen of the intestinal tract by binding the microbes therein and in doing so, preventing breach of the intestinal epithelial barrier by such opportunistic, infectious agents. More recently, IgA has been implicated in a primitive, T cell independent, antibody-mediated mechanism of tolerance towards the indigenous intestinal microflora IgA.[237] IgA, as the mucosal antibody isotype, is unique in that it is adapted not to functioning as a component of the cellular immune response in the lamina propria, but to intraluminal function following secretion by mucosal plasma cells and subsequent transport across the epithelial barrier, facilitated by its secretory component (IgASC). TGF-β1, in its sole immunopotentiatory function, is responsible for directing the class-switch of IgM B cells in lymphoid aggregates in the intestine, to IgA plasma cells.[238] Significantly decreased IgA levels and increased IgG, IgM and IgE isotypes in serum and mucosal secretions occur, especially in the gastrointestinal tract, in TGF-β1-/- mice. This decrease in plasma levels correlates with a decrease in numbers of IgA plasma cells.[239] TGF-β1 also induces an increase in the number of cells expressing surface secretory component (SC) in rat IEC-6 epithelial cell culture.[240] TGF-β1 increases IL-6 secretion by rat intestinal epithelial cells [241] and also synergises with IL-1β in the induction of IL-6 secretion therein.[242] TGF-β induced IL-6 produced by intestinal epithelial cells may represent an important inflammatory stimulus in the intestinal mucosa. However, IL-6 is equally important as TGF-β in the life of an IgA B cell as it drives maturation and proliferation of IgA clones following class switching. This is demonstrated in vitro by an increase of IgA secretion from isolated rat mucosal B cells by rat intestinal epithelial cell-derived IL-6, TGF-β and IL-2.[243] The induction of IL-6 may therefore be a property promoting IgA mediated tolerance. TGF-β also plays a part in driving the class switch of IgM B cells to IgG2b plasma cells.[238]

TGF-β expression is upregulated in intestinal inflammatory conditions. Bacterial products that can initiate an inflammatory response via the induction of IL-8 secretion and consequent chemotaxis of neutrophils also induce the production of TGF-β, in humans colonic epithelial cells.[244] In active inflammatory bowel disease, the number of TGF-β1-positive T cells, neutrophils and mononuclear cells in the lamina propria is increased[245,246] with the highest levels of expression associated with inflammatory cells proximal to the lumen. TGF-β mRNA expression in both Crohn's and ulcerative colitis-associated inflammation correlates with disease activity.[246] However, in contrast to this clear upregulation of TGFβ-1 production in the mucosa, plasma concentrations of TGF-β1 plasma concentrations are not generally significantly altered in Crohn's disease and ulcerative colitis patients when compared to those of control subjects.[247] Epithelial expression of TGF-β1 protein is also increased in biopsies taken from inflamed areas of ileal pouch mucosa in Crohn's disease.[248]

TGF-β1 may play a role in the remodelling and repair of the gut wall in active ulcerative colitis[249] which, if contributory to the reestablishment of epithelial barrier integrity, may be a factor in restoration of tolerance.[45,250] In Crohn's disease, an aberrant localisation, towards the crypt bases and in the lamina propria, of the human serum- and glucocorticoid-regulated protein kinase (h-sgk, a factor upregulated after stress as a consequence of changes in cell volume) expressing cells is observed and correlates with the presence of TGF-β1 protein therein. h-sgk expression in normal and inflamed intestinal mucosa may be regulated by TGF-β1 and may contribute to the pleiotropic actions of TGF-β1 in mucosal macrophages.[251] Further evidence for the upregulation of TGF-β in chronic intestinal inflammation being a response to tissue damage comes from oxazolone-induced mouse models of colitis. If the increase in TGF-β production from mucosal T cells observed in affected animals is blocked by the administration of neutralising anti-TGF-β antibodies, then more severe inflammation results.[252] Furthermore, the induction of low dose oral tolerance towards haptenised colonic protein mediated by antigen-specific CD4+ Th3-type T cells (a discrete but poorly defined population of T cells that appear to produce TGF-β, IL-4 and IL-10[45]), is beneficial in the treatment of TNBS-induced colitis in mouse models, an effect that appears to be TGF-β-dependant.[253] Similarly, intranasal administration of the TGF-β1 gene via a viral vector is effective in treating TNBS colitis, the mechanism seemingly being a TGF-β1-mediated suppression of T helper cell type 1 responses, achieved by the induction of IL-10 expression and IL-12 receptor β2 chain downregulation.[47] The establishment of colitis after the transfer of CD4+CD45Rbhi can be actively inhibited by the cotransfer of a population of $CD4^+\alpha\beta^+$ regulatory T cells that produce immunosuppressive cytokines such as IL-10 and TGF-β1.[46,254]

Taken together, the data obtained from mouse models of inflammatory bowel disease would seem to suggest that TGF-β is a candidate therapeutic agent for the treatment of inflammatory bowel disease. However, an important factor to consider in this context is the potentially harmful effects of unchecked, prolonged tissue-remodelling and repair, driven by TGF-β, in the inflamed gut wall. TGF-β1 induces the production of connective tissue growth factor (CTGF), a component of the repair and remodelling process which contributes to the development of fibrosis and stenosis, both complications in inflammatory bowel disease.[255] The role of TGF-β1 in driving the fibrotic process is demonstrated by the fact that collagen synthesis is inhibited following administration of neutralising anti-TGF-B-1 antibodies in intestinal myofibroblast cultures obtained from a rat model for chronic granulomatous enterocolitis.[256] As such, anti-TFG-β therapy may not be an appropriate route for inflammatory bowel disease therapy.

Interleukin 4, 11 and 13 (IL-4, IL-11 & IL-13)

Interleukin-4 (IL-4) and interleukin-13 (IL-13) are closely related Th2 type cytokines (recently reviewed[32,33]). They display a wide overlap in effects on B lymphocytes, monocytes, dendritic cells and fibroblasts. Both IL-4 and IL-13 genes are situated on adjacent genomic positions and in fact the similarity in their promoters may account for their coordinated expression. The similarity in their properties may also be enforced because both cytokines share the IL-4-α receptor for signaling.[34,35] However, a set number of differences, such as the exclusive expression of IL-4 from Th2 cells or the absence of IL-13 receptors on T-cells, distinguish both cytokines.

The absolute number of IL-4 secreting cells, the major proportion of which are LPL, is reduced in inflamed mucosa of both Crohn's disease and ulcerative colitis patients.[257] Active ulcerative colitis is also associated with decreased colonic IL-13 [72]. In this respect, IL-4 and IL-13 differ from IL-10 and TGF-β, the production of which is increased in inflammatory bowel disease. IL-13 and IL-4 pretreatment of PBMC from both Crohn's disease and ulcerative colitis patients down-regulates PMA-induced IL-1β, TNF-α and IL-6 secretion in cells isolated from biopsies of inactive inflammatory bowel disease. IL-10 displays similar, but more profound effects in that IL-10 oretreatment can downregulate inflammatory cytokine production from PBMC isolated from patients with active disease.[258] There is however, a synergistic effect between IL-10 plus IL-4 and IL-10 plus IL-13 in inhibiting TNF-α and IL-1β secretion from PMA-stimulated monocytes and macrophages.[50] Both IL-13 and IL-4 are potent inhibitors of inducible nitric oxide synthase (iNOS) expression[51] and so act to inhibit the regulated production of the pro-inflammatory mediator, nitric oxide. IL-4 also inhibits the proliferation of IEL.[49]

IL-4, IL-10 and IL-13 down-regulate the production of MCP-1 from stimulated human intestinal epithelial cell lines and freshly isolated intestinal epithelial cells in vitro.[199] Others however, report an enhanced expression of MCP-1 and eotaxin chemokine mRNA expression in intestinal epithelial cells following IL-4 stimulation, although the kinetics of the induction of MCP-1 mRNA were distinct from that observed following pro-inflammatory (e.g., IFN-γ) stimulation. Both IL-4 and IL-13 strongly down-regulate IL-8 secretion from intestinal epithelial cells. The inhibitory effects of IL-4 and IL-13 on chemokine expression in the intestinal mucosa may block the recruitment of immune cells therein during inflammation.[259]

In contrast to IL-10, the role of IL-4 in the mucosa or during intestinal inflammation is much less that of a clear downregulator of immune cascades. In fact, in some Th2-type experimental murine models of inflammatory bowel disease, IL-4 appears to be the driving force of the pathology. Oxazolone-induced murine colitis can be treated with neutralising antibodies to IL-4[252] and IL-4-producing Th2 T cells play a major immunopathological role in the induction of experimental inflammatory bowel disease in T cell receptor(TCR)-α -/- mice.[260] Perhaps then, IL-4 should be regarded as a factor that synergises with IL-10 or TGF-β to achieve its anti-inflammatory effects. Another feature of IL-4 function which argues against a label of a regulatory, anti-inflammatory cytokine is that it promotes class switching of B cells to IgE plasma cells, a phenotype associated with allergy.

IL-4 affects intestinal epithelial cell barrier function and, like IL-10 appears to be important for the maintenance of barrier integrity therein. It selectively enhances horse radish peroxidase (HRP) transport across human T84 colonic epithelial. IL-4 increases neutrophil adhesion to epithelia, but retards neutrophil migration into and across epithelial monolayers.[261]

The use of IL-4 gene transfer by viral vectors as therapy in experimental inflammatory bowel disease[262] seems to elicit only marginal effects although better targeting of the IL-4 to the site of inflammation or the use of IL-4 in combination with conventional immunosuppressive therapies may improve its efficacy. Human trials using local liposome-mediated gene transfer of two anti-inflammatory cytokines, IL-4 and IL-10, in patients with severe inflammatory bowel disease of the rectum are planned, the rationale for the simultaneous use of IL-4 and IL-10 being to shift the immune environment in favour of Th2 immune responses.[263]

IL-11 is a recently described anti-inflammatory, NF-κB inhibiting cytokine that directs the immune response towards a Th2 type reaction (reviewed [39]). IL-11 may act as a suppressor of active electrogenic transport and a potent anti-inflammatory cytokine during active intestinal inflammation in its inhibition of NF-κB activity.[52] Recombinant human IL-11 has been shown to suppress the clinical signs of colitis in HLA-B27 transgenic rats, animals that spontaneously develop chronic inflammation involving the intestine and major joints unless maintained under germ-free conditions,[264] and in TNB-induced[265] and acetic acid-induced experimental colitis in rats.[266] Subcutaneous injection of recombinant IL-11 has been reported as being beneficial in the treatment of psoriasis and reduces disease activity index in some Crohn's disease patients.[267,268] Although by no means vast, the evidence accumulated so far would seem therefore to suggest that IL-11 may be worth pursuing as potential therapeutic agent in the treatment of chronic autoimmune inflammatory conditions, Crohn's disease among them.

Conclusion

Conventional anti-inflammatory and immunosuppressive therapies used for the treatment of inflammatory bowel diseases are, on the whole, effective. Not all patients however, particularly in the case of Crohn's disease, are responsive to such treatments. The side effects of the anti-inflammatories and immunosuppressives used can also often impact quality of life to the same degree as the symptoms of the disease. Thus, the development of improved or novel therapeutics for the treatment of inflammatory bowel disease is a priority. As we learn more about the pathophysiology of Crohn's disease and ulcerative colitis, the central role of cytokines therein becomes increasingly apparent. Targeting these soluble inflammatory factors as a means of therapeutic intervention is therefore an obvious choice, the rationale being to redress the imbalance of inflammatory and regulatory cytokines and to restore immune homeostasis in the gut wall. Blocking the bioactivity of the inflammatory factors by administering neutralising anti-cytokine antibodies, anti-cytokine receptor antibodies or soluble forms of the cytokine receptor has proved to be beneficial in a number of animal models and in some human studies. So too has the administration of recombinant regulatory cytokines. Cytokines are powerful, multi-potent, pleiotropic agents, acting to regulate host immunity as components of a complex communicatory network. Interference with individual component of this process may, while having acute beneficial clinical effects, result in deleterious effects from its down-stream actions on other immune processes. These factors must be taken into consideration when evaluating efficacy and safety data from longer-term animal studies and clinical trials.

In the future, cytokine, chemokine and growth factor-based therapies may provide alternative or complementary option for the treatment of inflammatory bowel disease and may have be a means of treating such conditions safely, effectively and without the complications and side effects of conventional therapies, in all, improving the quality of life even further for those affected with inflammatory bowel disease.

References

1. Andres PG, Friedman LS. Epidemiology and the natural course of inflammatory bowel disease. Gastroenterol Clin North Am 1999; 28:255-281, vii.
2. Delco F, Sonnenberg A. Commonalities in the time trends of Crohn's disease and ulcerative colitis. Am J Gastroenterol 1999; 94:2171-2176.
3. Karlinger K, Gyorke T, Mako E et al. The epidemiology and the pathogenesis of inflammatory bowel disease. Eur J Radiol 2000; 35:154-167.
4. Pera A, Sostegni R, Daperno M et al. Genotype-phenotype relationship in inflammatory bowel disease. Eur J Intern Med 2000; 11:204-209.
5. Andus T, Gross V. Etiology and pathophysiology of inflammatory bowel disease—Environmental factors. Hepatogastroenterology 2000; 47:29-43.

6. Forabosco P, Collins A, Latiano A et al. Combined segregation and linkage analysis of inflammatory bowel disease in the IBD1 region using severity to characterise Crohn's disease and ulcerative colitis. Eur J Hum Genet 2000; 8:846-852.
7. Schreiber S. Genetics of inflammatory bowel disease: A puzzle with contradictions? Gut 2000; 47:746-747.
8. Blum S, Alvarez S, Haller D et al. Intestinal microflora and the interaction with immunocompetent cells. Antonie Van Leeuwenhoek 1999; 76:199-205.
9. Duchmann R, May E, Heike M et al. T cell specificity and cross reactivity towards enterobacteria, bacteroides, bifidobacterium, and antigens from resident intestinal flora in humans [see comments]. Gut 1999; 44:812-818.
10. McKay DM. Intestinal inflammation and the gut microflora. Can J Gastroenterol 1999; 13:509-516.
11. Schultsz C, Van Den Berg FM, Ten Kate FW et al. The intestinal mucus layer from patients with inflammatory bowel disease harbors high numbers of bacteria compared with controls. Gastroenterology 1999; 117:1089-1097.
12. Tiveljung A, Soderholm JD, Olaison G et al. Presence of eubacteria in biopsies from Crohn's disease inflammatory lesions as determined by 16S rRNA gene-based PCR. J Med Microbiol 1999; 48:263-268.
13. Guslandi M. The relationship between gut microflora and intestinal inflammation. Can J Gastroenterol 2000; 14:32.
14. Matsuda H, Fujiyama Y, Andoh A et al. Characterization of antibody responses against rectal mucosa-associated bacterial flora in patients with ulcerative colitis. J Gastroenterol Hepatol 2000; 15:61-68.
15. Dinarello CA. Interleukin-1, interleukin-1 receptors and interleukin-1 receptor antagonist. Int Rev Immunol 1998; 16:457-499.
16. Rosenwasser LJ. Biologic activities of IL-1 and its role in human disease. J Allergy Clin Immunol 1998; 102:344-350.
17. Stylianou E, Saklatvala J. Interleukin-1. Int J Biochem Cell Biol 1998; 30:1075-1079.
18. Bowie A, O'Neill LA. The interleukin-1 receptor/Toll-like receptor superfamily: Signal generators for pro-inflammatory interleukins and microbial products. J Leukoc Biol 2000; 67:508-514.
19. Gately MK, Renzetti LM, Magram J et al. The interleukin-12/interleukin-12-receptor system: Role in normal and pathologic immune responses. Annu Rev Immunol 1998; 16:495-521.
20. Adorini L. Interleukin-12, a key cytokine in Th1-mediated autoimmune diseases. Cell Mol Life Sci 1999; 55:1610-1625.
21. Sinigaglia F, D'Ambrosio D, Panina-Bordignon P et al. Regulation of the IL-12/IL-12R axis: a critical step in T-helper cell differentiation and effector function. Immunol Rev 1999; 170:65-72.
22. Farrar MA, Schreiber RD. The molecular cell biology of interferon-gamma and its receptor. Annu Rev Immunol 1993; 11:571-611.
23. Schreiber RD, Farrar MA. The biology and biochemistry of interferon-gamma and its receptor. Gastroenterol Jpn 1993; 28 Suppl 4:88-94; discussion 95-86.
24. Young HA. Regulation of interferon-gamma gene expression. J Interferon Cytokine Res 1996; 16:563-568.
25. Boehm U, Klamp T, Groot M et al. Cellular responses to interferon-gamma. Annu Rev Immunol 1997; 15:749-795.
26. Moldawer LL. Interleukin-1, TNF alpha and their naturally occurring antagonists in sepsis. Blood Purif 1993; 11:128-133.
27. Riches DW, Chan ED, Winston BW. TNF-alpha-induced regulation and signalling in macrophages. Immunobiology 1996; 195:477-490.
28. Ksontini R, MacKay SL, Moldawer LL. Revisiting the role of tumor necrosis factor alpha and the response to surgical injury and inflammation. Arch Surg 1998; 133:558-567.
29. Kunkel SL, Remick DG, Strieter RM et al. Mechanisms that regulate the production and effects of tumor necrosis factor-alpha. Crit Rev Immunol 1989; 9:93-117.
30. Luster MI, Simeonova PP, Gallucci R et al. Tumor necrosis factor alpha and toxicology. Crit Rev Toxicol 1999; 29:491-511.
31. Chomarat P, Banchereau J. An update on interleukin-4 and its receptor. Eur Cytokine Netw 1997; 8:333-344.
32. Chomarat P, Banchereau J. Interleukin-4 and interleukin-13: Their similarities and discrepancies. Int Rev Immunol 1998; 17:1-52.
33. Romagnani S. Th1/Th2 cells. Inflamm Bowel Dis 1999; 5:285-294.
34. Hart PH, Bonder CS, Balogh J et al. Differential responses of human monocytes and macrophages to IL-4 and IL-13. J Leukoc Biol 1999; 66:575-578.
35. Murata T, Taguchi J, Puri RK et al. Sharing of receptor subunits and signal transduction pathway between the IL-4 and IL-13 receptor system. Int J Hematol 1999; 69:13-20.

36. Stordeur P, Goldman M. Interleukin-10 as a regulatory cytokine induced by cellular stress: Molecular aspects. Int Rev Immunol 1998; 16:501-522.
37. Schreiber S. Interleukin-10 in the intestine [comment]. Gut 1997; 41:274-275.
38. Moore KW, O'Garra A, de Waal Malefyt R et al. Interleukin-10. Annu Rev Immunol 1993; 11:165-190.
39. Schwertschlag US, Trepicchio WL, Dykstra KH et al. Hematopoietic, immunomodulatory and epithelial effects of interleukin- 11. Leukemia 1999; 13:1307-1315.
40. Brombacher F. The role of interleukin-13 in infectious diseases and allergy. Bioessays 2000; 22:646-656.
41. Clark DA, Coker R. Transforming growth factor-beta (TGF-beta). Int J Biochem Cell Biol 1998; 30:293-298.
42. Lawrence DA. Transforming growth factor-beta: A general review. Eur Cytokine Netw 1996; 7:363-374.
43. Wahl SM, Orenstein JM, Chen W. TGF-beta influences the life and death decisions of T lymphocytes. Cytokine Growth Factor Rev 2000; 11:71-79.
44. Weiner HL. Oral tolerance for the treatment of autoimmune diseases. Annu Rev Med 1997; 48:341-351.
45. Chen Y, Inobe J, Marks R et al. Peripheral deletion of antigen-reactive T cells in oral tolerance [published erratum appears in Nature 1995 Sep 21;377(6546):257]. Nature 1995; 376:177-180.
46. Groux H, O'Garra A, Bigler M et al. A CD4+ T-cell subset inhibits antigen-specific T-cell responses and prevents colitis. Nature 1997; 389:737-742.
47. Kitani A, Fuss IJ, Nakamura K et al. Treatment of experimental (Trinitrobenzene sulfonic acid) colitis by intranasal administration of transforming growth factor (TGF)-beta1 plasmid: TGF-beta1-mediated suppression of T helper cell type 1 response occurs by interleukin (IL)-10 induction and IL-12 receptor beta2 chain downregulation. J Exp Med 2000; 192:41-52.
48. Asseman C, Mauze S, Leach MW et al. An essential role for interleukin 10 in the function of regulatory T cells that inhibit intestinal inflammation. J Exp Med 1999; 190:995-1004.
49. Ebert EC, Roberts AI. IL-4 down-regulates the responsiveness of human intraepithelial lymphocytes. Clin Exp Immunol 1996; 105:556-560.
50. Kucharzik T, Lugering N, Adolf M et al. Synergistic effect of immunoregulatory cytokines on peripheral blood monocytes from patients with inflammatory bowel disease. Dig Dis Sci 1997; 42:805-812.
51. Kolios G, Rooney N, Murphy CT et al. Expression of inducible nitric oxide synthase activity in human colon epithelial cells: modulation by T lymphocyte derived cytokines. Gut 1998; 43:56-63.
52. Greenwood-Van Meerveld B, Tyler K, Keith JC, Jr. Recombinant human interleukin-11 modulates ion transport and mucosal inflammation in the small intestine and colon. Lab Invest 2000; 80:1269-1280.
53. Young HA, Hardy KJ. Role of interferon-gamma in immune cell regulation. J Leukoc Biol 1995; 58:373-381.
54. Billiau A. Interferon-gamma: biology and role in pathogenesis. Adv Immunol 1996; 62:61-130.
55. Dionne S, D'Agata ID, Hiscott J et al. Colonic explant production of IL-1and its receptor antagonist is imbalanced in inflammatory bowel disease (IBD). Clin Exp Immunol 1998; 112:435-442.
56. Neurath MF, Fuss I, Schurmann G et al. Cytokine gene transcription by NF-kappa B family members in patients with inflammatory bowel disease. Ann NY Acad Sci 1998; 859:149-159.
57. Ishiguro Y. Mucosal proinflammatory cytokine production correlates with endoscopic activity of ulcerative colitis. J Gastroenterol 1999; 34:66-74.
58. Woywodt A, Ludwig D, Neustock P et al. Mucosal cytokine expression, cellular markers and adhesion molecules in inflammatory bowel disease. Eur J Gastroenterol Hepatol 1999; 11:267-276.
59. Indaram AV, Visvalingam V, Locke M et al. Mucosal cytokine production in radiation-induced proctosigmoiditis compared with inflammatory bowel disease. Am J Gastroenterol 2000; 95:1221-1225.
60. Ina K, Kusugami K, Yamaguchi T et al. Mucosal interleukin-8 is involved in neutrophil migration and binding to extracellular matrix in inflammatory bowel disease. Am J Gastroenterol 1997; 92:1342-1346.
61. Nielsen OH, Rudiger N, Gaustadnes M et al. Intestinal interleukin-8 concentration and gene expression in inflammatory bowel disease. Scand J Gastroenterol 1997; 32:1028-1034.
62. Uguccioni M, Gionchetti P, Robbiani DF et al. Increased expression of IP-10, IL-8, MCP-1, and MCP-3 in ulcerative colitis. Am J Pathol 1999; 155:331-336.
63. Nikolaus S, Bauditz J, Gionchetti P et al. Increased secretion of pro-inflammatory cytokines by circulating polymorphonuclear neutrophils and regulation by interleukin 10 during intestinal inflammation. Gut 1998; 42:470-476.

64. Noguchi M, Hiwatashi N, Liu Z et al. Secretion imbalance between tumour necrosis factor and its inhibitor in inflammatory bowel disease. Gut 1998; 43:203-209.
65. Gotteland M, Lopez M, Munoz C et al. Local and systemic liberation of proinflammatory cytokines in ulcerative colitis. Dig Dis Sci 1999; 44:830-835.
66. Carty E, De Brabander M, Feakins RM et al. Measurement of in vivo rectal mucosal cytokine and eicosanoid production in ulcerative colitis using filter paper. Gut 2000; 46:487-492.
67. Parronchi P, Romagnani P, Annunziato F et al. Type 1 T-helper cell predominance and interleukin-12 expression in the gut of patients with Crohn's disease. Am J Pathol 1997; 150:823-832.
68. Monteleone G, Biancone L, Marasco R et al. Interleukin 12 is expressed and actively released by Crohn's disease intestinal lamina propria mononuclear cells. Gastroenterology 1997; 112:1169-1178.
69. Berrebi D, Besnard M, Fromont-Hankard G et al. Interleukin-12 expression is focally enhanced in the gastric mucosa of pediatric patients with Crohn's disease. Am J Pathol 1998; 152:667-672.
70. Kakazu T, Hara J, Matsumoto T et al. Type 1 T-helper cell predominance in granulomas of Crohn's disease. Am J Gastroenterol 1999; 94:2149-2155.
71. Sakai T, Kusugami K, Nishimura H et al. Interleukin 15 activity in the rectal mucosa of inflammatory bowel disease. Gastroenterology 1998; 114:1237-1243.
72. Vainer B, Nielsen OH, Hendel J et al. Colonic expression and synthesis of interleukin 13 and interleukin 15 in inflammatory bowel disease [In Process Citation]. Cytokine 2000; 12:1531-1536.
73. Keates AC, Castagliuolo I, Cruickshank WW et al. Interleukin 16 is Up-regulated in Crohn's disease and participates in TNBS colitis in mice. Gastroenterology 2000; 119:972-982.
74. Pizarro TT, Michie MH, Bentz M et al. IL-18, a novel immunoregulatory cytokine, is up-regulated in Crohn's disease: expression and localization in intestinal mucosal cells. J Immunol 1999; 162:6829-6835.
75. Kanai T, Watanabe M, Okazawa A et al. Interleukin 18 is a potent proliferative factor for intestinal mucosal lymphocytes in Crohn's disease. Gastroenterology 2000; 119:1514-1523.
76. Breese E, Braegger CP, Corrigan CJ et al. Interleukin-2- and interferon-gamma-secreting T cells in normal and diseased human intestinal mucosa. Immunology 1993; 78:127-131.
77. Noguchi M, Hiwatashi N, Liu Z et al. Enhanced interferon-gamma production and B7-2 expression in isolated intestinal mononuclear cells from patients with Crohn's disease. J Gastroenterol 1995; 30 Suppl 8:52-55.
78. Berrebi D, Banerjee A, Paris R et al. In situ Rantes and interferon-gamma gene expression in pediatric small bowel Crohn's disease. J Pediatr Gastroenterol Nutr 1997; 25:371-376.
79. Camoglio L, Te Velde AA, Tigges AJ et al. Altered expression of interferon-gamma and interleukin-4 in inflammatory bowel disease. Inflamm Bowel Dis 1998; 4:285-290.
80. Fuss IJ, Neurath M, Boirivant M et al. Disparate CD4+ lamina propria (LP) lymphokine secretion profiles in inflammatory bowel disease. Crohn's disease LP cells manifest increased secretion of IFN-gamma, whereas ulcerative colitis LP cells manifest increased secretion of IL-5. J Immunol 1996; 157:1261-1270.
81. McClane SJ, Rombeau JL. Cytokines and inflammatory bowel disease: A review. JPEN J Parenter Enteral Nutr 1999; 23:S20-24.
82. MacDonald TT, Monteleone G, Pender SL. Recent developments in the immunology of inflammatory bowel disease. Scand J Immunol 2000; 51:2-9.
83. Monteleone G, MacDonald TT. Manipulation of cytokines in the management of patients with inflammatory bowel disease. Ann Med 2000; 32:552-560.
84. Murch SH. Local and systemic effects of macrophage cytokines in intestinal inflammation. Nutrition 1998; 14:780-783.
85. Louis E, Ribbens C, Godon A et al. Increased production of matrix metalloproteinase-3 and tissue inhibitor of metalloproteinase-1 by inflamed mucosa in inflammatory bowel disease. Clin Exp Immunol 2000; 120:241-246.
86. Vandenabeele P, Goossens V, Beyaert R et al. Functional requirement of the two TNF receptors for induction of apoptosis in PC60 cells and the role of mitochondria in TNF-induced cytotoxicity. Circ Shock 1994; 44:196-200.
87. Magnusson C, Vaux DL. Signalling by CD95 and TNF receptors: Not only life and death. Immunol Cell Biol 1999; 77:41-46.
88. Gearing AJ, Beckett P, Christodoulou M et al. Processing of tumour necrosis factor-alpha precursor by metalloproteinases. Nature 1994; 370:555-557.
89. Cerretti DP. Characterization of the tumour necrosis factor alpha-converting enzyme, TACE/ADAM17. Biochem Soc Trans 1999; 27:219-223.
90. Killar L, White J, Black R et al. Adamalysins. A family of metzincins including TNF-alpha converting enzyme (TACE). Ann N Y Acad Sci 1999; 878:442-452.
91. Balkwill F. TNF biosynthesis in gut associated immunopathologies. Gut 1999; 45:483.

92. Schreiber S, Nikolaus S, Hampe J et al. Tumour necrosis factor alpha and interleukin 1beta in relapse of Crohn's disease. Lancet 1999; 353:459-461.
93. Lilja I, Gustafson-Svard C, Franzen L et al. Tumor necrosis factor-alpha in ileal mast cells in patients with Crohn's disease. Digestion 2000; 61:68-76.
94. Stokkers PC, Camoglio L, van Deventer SJ. Tumor necrosis factor (TNF) in inflammatory bowel disease: gene polymorphisms, animal models, and potential for anti-TNF therapy. J Inflamm 1995; 47:97-103.
95. van Deventer SJ. Review article: targeting TNF alpha as a key cytokine in the inflammatory processes of Crohn's disease—The mechanisms of action of infliximab. Aliment Pharmacol Ther 1999; 13 Suppl 4:3-8; discussion 38.
96. Papadakis KA, Targan SR. Role of cytokines in the pathogenesis of inflammatory bowel disease. Annu Rev Med 2000; 51:289-298.
97. Mikula CA. Anti-TNF alpha: new therapy for Crohn's disease. Gastroenterol Nurs 1999; 22:245-248.
98. Rutgeerts PJ. Review article: efficacy of infliximab in Crohn's disease—Induction and maintenance of remission. Aliment Pharmacol Ther 1999; 13 Suppl 4:9-15; discussion 38.
99. Present DH. Review article: the efficacy of infliximab in Crohn's disease—Healing of fistulae. Aliment Pharmacol Ther 1999; 13 Suppl 4:23-28; discussion 38.
100. Lofberg R. Treatment of fistulas in Crohn's disease with infliximab. Gut 1999; 45:642-643.
101. Rutgeerts P, D'Haens G, Targan S et al. Efficacy and safety of retreatment with anti-tumor necrosis factor antibody (infliximab) to maintain remission in Crohn's disease. Gastroenterology 1999; 117:761-769.
102. Shanahan F. Anti-TNF therapy for Crohn's disease: A perspective (infliximab is not the drug we have been waiting for). Inflamm Bowel Dis 2000; 6:137-139.
103. Soykan I, Ertan C, Ozden A. Severe anaphylactic reaction to infliximab: Report of a case. Am J Gastroenterol 2000; 95:2395-2396.
104. Hyams JS, Markowitz J, Wyllie R. Use of infliximab in the treatment of Crohn's disease in children and adolescents. J Pediatr 2000; 137:192-196.
105. Dionne S, Hiscott J, D'Agata I et al. Quantitative PCR analysis of TNF-alpha and IL-1 beta mRNA levels in pediatric IBD mucosal biopsies. Dig Dis Sci 1997; 42:1557-1566.
106. Kuboyama S. Increased circulating levels of interleukin-1 receptor antagonist in patients with inflammatory bowel disease. Kurume Med J 1998; 45:33-37.
107. Stokkers PC, van Aken BE, Basoski N et al. Five genetic markers in the interleukin 1 family in relation to inflammatory bowel disease. Gut 1998; 43:33-39.
108. Roussomoustakaki M, Satsangi J, Welsh K et al. Genetic markers may predict disease behavior in patients with ulcerative colitis. Gastroenterology 1997; 112:1845-1853.
109. Nemetz A, Kope A, Molnar T et al. Significant differences in the interleukin-1beta and interleukin-1 receptor antagonist gene polymorphisms in a Hungarian population with inflammatory bowel disease. Scand J Gastroenterol 1999; 34:175-179.
110. Tountas NA, Casini-Raggi V, Yang H et al. Functional and ethnic association of allele 2 of the interleukin-1 receptor antagonist gene in ulcerative colitis. Gastroenterology 1999; 117:806-813.
111. Monteleone G, Trapasso F, Parrello T et al. Bioactive IL-18 expression is up-regulated in Crohn's disease. J Immunol 1999; 163:143-147.
112. McAlindon ME, Hawkey CJ, Mahida YR. Expression of interleukin 1 beta and interleukin 1 beta converting enzyme by intestinal macrophages in health and inflammatory bowel disease. Gut 1998; 42:214-219.
113. Bach EA, Aguet M, Schreiber RD. The IFN gamma receptor: a paradigm for cytokine receptor signaling. Annu Rev Immunol 1997; 15:563-591.
114. Pestka S, Kotenko SV, Muthukumaran G et al. The interferon gamma (IFN-gamma) receptor: A paradigm for the multichain cytokine receptor. Cytokine Growth Factor Rev 1997; 8:189-206.
115. McVay LD, Li B, Biancaniello R et al. Changes in human mucosal gamma delta T cell repertoire and function associated with the disease process in inflammatory bowel disease. Mol Med 1997; 3:183-203.
116. Simpson SJ, Hollander GA, Mizoguchi E et al. Expression of pro-inflammatory cytokines by TCR alpha beta+ and TCR gamma delta+ T cells in an experimental model of colitis. Eur J Immunol 1997; 27:17-25.
117. Clavell M, Correa-Gracian H, Liu Z et al. Detection of interferon regulatory factor-1 in lamina propria mononuclear cells in Crohn's disease. J Pediatr Gastroenterol Nutr 2000; 30:43-47.
118. Colgan SP, Parkos CA, Matthews JB et al. Interferon-gamma induces a cell surface phenotype switch on T84 intestinal epithelial cells. Am J Physiol 1994; 267:C402-410.
119. Donnet-Hughes A, Schiffrin EJ, Huggett AC. Expression of MHC antigens by intestinal epithelial cells. Effect of transforming growth factor-beta 2 (TGF-beta 2). Clin Exp Immunol 1995; 99:240-244.

120. Lopes LM, Hughson E, Anstee Q et al. Vectorial function of major histocompatibility complex class II in a human intestinal cell line. Immunology 1999; 98:16-26.
121. Ishii N, Chiba M, Iizuka M et al. Induction of HLA-DR antigen expression on human colonic epithelium by tumor necrosis factor-alpha and interferon-gamma. Scand J Gastroenterol 1994; 29:903-907.
122. Horie Y, Chiba M, Suzuki T et al. Induction of major histocompatibility complex class II antigens on human colonic epithelium by interferon-gamma, tumor necrosis factor- alpha, and interleukin-2. J Gastroenterol 1998; 33:39-47.
123. Cavicchi M, Whittle BJ. Regulation of induction of nitric oxide synthase and the inhibitory actions of dexamethasone in the human intestinal epithelial cell line, Caco2: Influence of cell differentiation. Br J Pharmacol 1999; 128:705-715.
124. Cavicchi M, Whittle BJ. Potentiation of cytokine induced iNOS expression in the human intestinal epithelial cell line, DLD-1, by cyclic AMP. Gut 1999; 45:367-374.
125. Crotty B, Hoang P, Dalton HR et al. Salicylates used in inflammatory bowel disease and colchicine impair interferon-gamma induced HLA-DR expression. Gut 1992; 33:59-64.
126. Crotty B, Rosenberg WM, Aronson JK et al. Inhibition of binding of interferon-gamma to its receptor by salicylates used in inflammatory bowel disease. Gut 1992; 33:1353-1357.
127. Kennedy M, Wilson L, Szabo C et al. 5-aminosalicylic acid inhibits iNOS transcription in human intestinal epithelial cells. Int J Mol Med 1999; 4:437-443.
128. Hirano T. Interleukin 6 and its receptor: ten years later. Int Rev Immunol 1998; 16:249-284.
129. Heinrich PC, Behrmann I, Muller-Newen G et al. Interleukin-6-type cytokine signalling through the gp130/Jak/STAT pathway. Biochem J 1998; 334:297-314.
130. Hirano T, Ishihara K, Hibi M. Roles of STAT3 in mediating the cell growth, differentiation and survival signals relayed through the IL-6 family of cytokine receptors. Oncogene 2000; 19:2548-2556.
131. Streetz KL, Luedde T, Manns MP et al. Interleukin 6 and liver regeneration. Gut 2000; 47:309-312.
132. Van Kemseke C, Belaiche J, Louis E. Frequently relapsing Crohn's disease is characterized by persistent elevation in interleukin-6 and soluble interleukin-2 receptor serum levels during remission. Int J Colorectal Dis 2000; 15:206-210.
133. Atreya R, Mudter J, Finotto S et al. Blockade of interleukin 6 trans signaling suppresses T cell resistance against apoptosis in chronic intestinal inflammation: Evidence in crohn disease and experimental colitis in vivo. Nat Med 2000; 6:583-588.
134. Okamura H, Kashiwamura S, Tsutsui H et al. Regulation of interferon-gamma production by IL-12 and IL-18. Curr Opin Immunol 1998; 10:259-264.
135. Monteleone G, Parrello T, Luzza F et al. Response of human intestinal lamina propria T lymphocytes to interleukin 12: additive effects of interleukin 15 and 7. Gut 1998; 43:620-628.
136. Parrello T, Monteleone G, Cucchiara S et al. Up-regulation of the IL-12 receptor ss2 chain in Crohn's disease. J Immunol 2000; 165:7234-7239.
137. Monteleone G, MacDonald TT, Wathen NC et al. Enhancing Lamina propria Th1 cell responses with interleukin 12 produces severe tissue injury. Gastroenterology 1999; 117:1069-1077.
138. Liu Z, Colpaert S, D'Haens GR et al. Hyperexpression of CD40 ligand (CD154) in inflammatory bowel disease and its contribution to pathogenic cytokine production. J Immunol 1999; 163:4049-4057.
139. Sawada-Hase N, Kiyohara T, Miyagawa J et al. An increased number of CD40-high monocytes in patients with Crohn's disease. Am J Gastroenterol 2000; 95:1516-1523.
140. Neurath MF, Fuss I, Kelsall BL et al. Antibodies to interleukin 12 abrogate established experimental colitis in mice. J Exp Med 1995; 182:1281-1290.
141. Simpson SJ, Shah S, Comiskey M et al. T cell-mediated pathology in two models of experimental colitis depends predominantly on the interleukin 12/Signal transducer and activator of transcription (Stat)-4 pathway, but is not conditional on interferon gamma expression by T cells. J Exp Med 1998; 187:1225-1234.
142. Davidson NJ, Hudak SA, Lesley RE et al. IL-12, but not IFN-gamma, plays a major role in sustaining the chronic phase of colitis in IL-10-deficient mice. J Immunol 1998; 161:3143-3149.
143. Duchmann R, Schmitt E, Knolle P et al. Tolerance towards resident intestinal flora in mice is abrogated in experimental colitis and restored by treatment with interleukin-10 or antibodies to interleukin-12. Eur J Immunol 1996; 26:934-938.
144. Fuss IJ, Marth T, Neurath MF et al. Anti-interleukin 12 treatment regulates apoptosis of Th1 T cells in experimental colitis in mice. Gastroenterology 1999; 117:1078-1088.
145. Trentin L, Zambello R, Facco M et al. Interleukin-15: A novel cytokine with regulatory properties on normal and neoplastic B lymphocytes. Leuk Lymphoma 1997; 27:35-42.
146. Kirman I, Vainer B, Nielsen OH. Interleukin-15 and its role in chronic inflammatory diseases. Inflamm Res 1998; 47:285-289.

147. Hasan MS, Kallas EG, Thomas EK et al. Effects of interleukin-15 on in vitro human T cell proliferation and activation. J Interferon Cytokine Res 2000; 20:119-123.
148. Vainer B, Nielsen OH, Hendel J et al. Colonic expression and synthesis of interleukin 13 and interleukin 15 in inflammatory bowel disease. Cytokine 2000; 12:1531-1536.
149. Inoue S, Matsumoto T, Iida M et al. Characterization of cytokine expression in the rectal mucosa of ulcerative colitis: correlation with disease activity. Am J Gastroenterol 1999; 94:2441-2446.
150. Liu Z, Geboes K, Colpaert S et al. IL-15 is highly expressed in inflammatory bowel disease and regulates local T cell-dependent cytokine production. J Immunol 2000; 164:3608-3615.
151. Mayumi T, Takezawa J, Takahashi H et al. IL-15 is elevated in the patients of postoperative enterocolitis. Cytokine 1999; 11:888-893.
152. Lugering N, Kucharzik T, Maaser C et al. Interleukin-15 strongly inhibits interleukin-8 and monocyte chemoattractant protein-1 production in human colonic epithelial cells. Immunology 1999; 98:504-509.
153. Cruikshank WW, Kornfeld H, Center DM. Signaling and functional properties of interleukin-16. Int Rev Immunol 1998; 16:523-540.
154. Center DM, Kornfeld H, Ryan TC et al. Interleukin 16: Implications for CD4 functions and HIV-1 progression. Immunol Today 2000; 21:273-280.
155. Cruikshank WW, Kornfeld H, Center DM. Interleukin-16. J Leukoc Biol 2000; 67:757-766.
156. Dinarello CA, Novick D, Puren AJ et al. Overview of interleukin-18: More than an interferon-gamma inducing factor. J Leukoc Biol 1998; 63:658-664.
157. Kohno K, Kurimoto M. Interleukin 18, a cytokine which resembles IL-1 structurally and IL-12 functionally but exerts its effect independently of both. Clin Immunol Immunopathol 1998; 86:11-15.
158. Dinarello CA. IL-18: A TH1-inducing, proinflammatory cytokine and new member of the IL-1 family. J Allergy Clin Immunol 1999; 103:11-24.
159. Golab J. Interleukin 18—Interferon gamma inducing factor—A novel player in tumour immunotherapy? Cytokine 2000; 12:332-338.
160. McInnes IB, Gracie JA, Leung BP et al. Interleukin 18: A pleiotropic participant in chronic inflammation. Immunol Today 2000; 21:312-315.
161. Holmes S, Abrahamson JA, Al-Mahdi N et al. Characterization of the in vitro and in vivo activity of monoclonal antibodies to human IL-18. Hybridoma 2000; 19:363-367.
162. Sozzani S, Locati M, Allavena P et al. Chemokines: a superfamily of chemotactic cytokines. Int J Clin Lab Res 1996; 26:69-82.
163. Feng L. Role of chemokines in inflammation and immunoregulation. Immunol Res 2000; 21:203-210.
164. Rossi D, Zlotnik A. The biology of chemokines and their receptors. Annu Rev Immunol 2000; 18:217-242.
165. Baggiolini M, Loetscher P, Moser B. Interleukin-8 and the chemokine family. Int J Immunopharmacol 1995; 17:103-108.
166. Yang SK, Eckmann L, Panja A et al. Differential and regulated expression of C-X-C, C-C, and C-chemokines by human colon epithelial cells. Gastroenterology 1997; 113:1214-1223.
167. Warhurst AC, Hopkins SJ, Warhurst G. Interferon gamma induces differential upregulation of alpha and beta chemokine secretion in colonic epithelial cell lines. Gut 1998; 42:208-213.
168. Kolios G, Wright KL, Jordan NJ et al. C-X-C and C-C chemokine expression and secretion by the human colonic epithelial cell line, HT-29: differential effect of T lymphocyte- derived cytokines. Eur J Immunol 1999; 29:530-536.
169. Wedemeyer J, Lorentz A, Goke M et al. Enhanced production of monocyte chemotactic protein 3 in inflammatory bowel disease mucosa. Gut 1999; 44:629-635.
170. van Deventer SJ. Review article: Chemokine production by intestinal epithelial cells: A therapeutic target in inflammatory bowel disease? Aliment Pharmacol Ther 1997; 11 Suppl 3:116-120; discussion 120-111.
171. Papadakis KA, Targan SR. The role of chemokines and chemokine receptors in mucosal inflammation. Inflamm Bowel Dis 2000; 6:303-313.
172. Hoch RC, Schraufstatter IU, Cochrane CG. In vivo, in vitro, and molecular aspects of interleukin-8 and the interleukin-8 receptors. J Lab Clin Med 1996; 128:134-145.
173. Atta ur R, Harvey K, Siddiqui RA. Interleukin-8: An autocrine inflammatory mediator. Curr Pharm Des 1999; 5:241-253.
174. Zeilhofer HU, Schorr W. Role of interleukin-8 in neutrophil signaling. Curr Opin Hematol 2000; 7:178-182.
175. Murphy PM. Neutrophil receptors for interleukin-8 and related CXC chemokines. Semin Hematol 1997; 34:311-318.

176. Williams EJ, Haque S, Banks C et al. Distribution of the interleukin-8 receptors, CXCR1 and CXCR2, in inflamed gut tissue. J Pathol 2000; 192:533-539.
177. Keshavarzian A, Fusunyan RD, Jacyno M et al. Increased interleukin-8 (IL-8) in rectal dialysate from patients with ulcerative colitis: Evidence for a biological role for IL-8 in inflammation of the colon. Am J Gastroenterol 1999; 94:704-712.
178. Grimm MC, Elsbury SK, Pavli P et al. Interleukin 8: Cells of origin in inflammatory bowel disease. Gut 1996; 38:90-98.
179. MacDermott RP. Chemokines in the inflammatory bowel diseases. J Clin Immunol 1999; 19:266-272.
180. Arai F, Takahashi T, Furukawa K et al. Mucosal expression of interleukin-6 and interleukin-8 messenger RNA in ulcerative colitis and in Crohn's disease. Dig Dis Sci 1998; 43:2071-2079.
181. Cole AT, Pilkington BJ, McLaughlan J et al. Mucosal factors inducing neutrophil movement in ulcerative colitis: the role of interleukin 8 and leukotriene B4. Gut 1996; 39:248-254.
182. Proost P, Wuyts A, Van Damme J. Human monocyte chemotactic proteins-2 and -3: structural and functional comparison with MCP-1. J Leukoc Biol 1996; 59:67-74.
183. Van Coillie E, Van Damme J, Opdenakker G. The MCP/eotaxin subfamily of CC chemokines. Cytokine Growth Factor Rev 1999; 10:61-86.
184. Charo IF. CCR2: from cloning to the creation of knockout mice. Chem Immunol 1999; 72:30-41.
185. Sanders SK, Crean SM, Boxer PA et al. Functional differences between monocyte chemotactic protein-1 receptor A and monocyte chemotactic protein-1 receptor B expressed in a jurkat T cell. J Immunol 2000; 165:4877-4883.
186. Loetscher P, Pellegrino A, Gong JH et al. The Ligands of CXC Chemokine Receptor 3, I-TAC, Mig, and IP10, Are Natural Antagonists for CCR3. J Biol Chem 2001; 276:2986-2991.
187. Fernandez EJ, Wilken J, Thompson DA et al. Comparison of the structure of vMIP-II with eotaxin-1, RANTES, and MCP- 3 suggests a unique mechanism for CCR3 activation(,). Biochemistry 2000; 39:12837-12844.
188. MacDermott RP, Sanderson IR, Reinecker HC. The central role of chemokines (chemotactic cytokines) in the immunopathogenesis of ulcerative colitis and Crohn's disease. Inflamm Bowel Dis 1998; 4:54-67.
189. Grimm MC, Elsbury SK, Pavli P et al. Enhanced expression and production of monocyte chemoattractant protein-1 in inflammatory bowel disease mucosa. J Leukoc Biol 1996; 59:804-812.
190. Mazzucchelli L, Hauser C, Zgraggen K et al. Differential in situ expression of the genes encoding the chemokines MCP-1 and RANTES in human inflammatory bowel disease. J Pathol 1996; 178:201-206.
191. Dieu-Nosjean MC, Vicari A, Lebecque S et al. Regulation of dendritic cell trafficking: a process that involves the participation of selective chemokines. J Leukoc Biol 1999; 66:252-262.
192. Helwig U, Lammers KM, Gionchetti P et al. MCP-3 in inflammatory bowel disease. Gut 2000; 47:155.
193. Agace WW, Roberts AI, Wu L et al. Human intestinal lamina propria and intraepithelial lymphocytes express receptors specific for chemokines induced by inflammation. Eur J Immunol 2000; 30:819-826.
194. Taub DD, Ortaldo JR, Turcovski-Corrales SM et al. Beta chemokines costimulate lymphocyte cytolysis, proliferation, and lymphokine production. J Leukoc Biol 1996; 59:81-89.
195. Olsson J, Poles M, Spetz AL et al. Human immunodeficiency virus type 1 infection is associated with significant mucosal inflammation characterized by increased expression of CCR5, CXCR4, and beta-chemokines. J Infect Dis 2000; 182:1625-1635.
196. Fagerstam JP, Whiss PA, Strom M et al. Expression of platelet P-selectin and detection of soluble P-selectin, NPY and RANTES in patients with inflammatory bowel disease. Inflamm Res 2000; 49:466-472.
197. Lapenta C, Boirivant M, Marini M et al. Human intestinal lamina propria lymphocytes are naturally permissive to HIV-1 infection. Eur J Immunol 1999; 29:1202-1208.
198. Schottelius AJ, Mayo MW, Sartor RB et al. Interleukin-10 signaling blocks inhibitor of kappaB kinase activity and nuclear factor kappaB DNA binding. J Biol Chem 1999; 274:31868-31874.
199. Kucharzik T, Lugering N, Pauels HG et al. IL-4, IL-10 and IL-13 down-regulate monocyte-chemoattracting protein-1 (MCP-1) production in activated intestinal epithelial cells. Clin Exp Immunol 1998; 111:152-157.
200. Autschbach F, Braunstein J, Helmke B et al. In situ expression of interleukin-10 in noninflamed human gut and in inflammatory bowel disease. Am J Pathol 1998; 153:121-130.
201. Napolitano LM, Buzdon MM, Shi HJ et al. Intestinal epithelial cell regulation of macrophage and lymphocyte interleukin 10 expression. Arch Surg 1997; 132:1271-1276.

202. Bourreille A, Segain JP, Raingeard de la Bletiere D et al. Lack of interleukin 10 regulation of antigen presentation-associated molecules expressed on colonic epithelial cells. Eur J Clin Invest 1999; 29:48-55.
203. Denning TL, Campbell NA, Song F et al. Expression of IL-10 receptors on epithelial cells from the murine small and large intestine. Int Immunol 2000; 12:133-139.
204. Kuhn R, Lohler J, Rennick D et al. Interleukin-10-deficient mice develop chronic enterocolitis [see comments]. Cell 1993; 75:263-274.
205. Tomoyose M, Mitsuyama K, Ishida H et al. Role of interleukin-10 in a murine model of dextran sulfate sodium- induced colitis. Scand J Gastroenterol 1998; 33:435-440.
206. Braunstein J, Qiao L, Autschbach F et al. T cells of the human intestinal lamina propria are high producers of interleukin-10 [see comments]. Gut 1997; 41:215-220.
207. Madsen KL, Malfair D, Gray D et al. Interleukin-10 gene-deficient mice develop a primary intestinal permeability defect in response to enteric microflora. Inflamm Bowel Dis 1999; 5:262-270.
208. Lu J, Philpott DJ, Saunders PR et al. Epithelial ion transport and barrier abnormalities evoked by superantigen-activated immune cells are inhibited by interleukin-10 but not interleukin-4. J Pharmacol Exp Ther 1998; 287:128-136.
209. Colgan SP, Hershberg RM, Furuta GT et al. Ligation of intestinal epithelial CD1d induces bioactive IL-10: Critical role of the cytoplasmic tail in autocrine signaling. Proc Natl Acad Sci USA 1999; 96:13938-13943.
210. Saubermann LJ, Beck P, De Jong YP et al. Activation of natural killer T cells by alpha-galactosylceramide in the presence of CD1d provides protection against colitis in mice. Gastroenterology 2000; 119:119-128.
211. Lundqvist C, Melgar S, Yeung MM et al. Intraepithelial lymphocytes in human gut have lytic potential and a cytokine profile that suggest T helper 1 and cytotoxic functions. J Immunol 1996; 157:1926-1934.
212. Hsieh B, Schrenzel MD, Mulvania T et al. In vivo cytokine production in murine listeriosis. Evidence for immunoregulation by gamma delta+ T cells. J Immunol 1996; 156:232-237.
213. Fujihashi K, Dohi T, Kweon MN et al. gammadelta T cells regulate mucosally induced tolerance in a dose- dependent fashion. Int Immunol 1999; 11:1907-1916.
214. Cohen SB, Katsikis PD, Feldmann M et al. IL-10 enhances expression of the IL-2 receptor alpha chain on T cells. Immunology 1994; 83:329-332.
215. Pawelec G, Pohla H, Schlotz E et al. Interleukin 10 is a human T cell growth factor in vitro. Cytokine 1995; 7:355-363.
216. Asseman C, Powrie F. Interleukin 10 is a growth factor for a population of regulatory T cells. Gut 1998; 42:157-158.
217. Cottrez F, Hurst SD, Coffman RL et al. T Regulatory Cells 1 Inhibit a Th2-Specific Response In Vivo. J Immunol 2000; 165:4848-4853.
218. Fort MM, Leach MW, Rennick DM. A role for NK cells as regulators of CD4+ T cells in a transfer model of colitis. J Immunol 1998; 161:3256-3261.
219. Brandeis JM, Sayegh MH, Gallon L et al. Rat intestinal epithelial cells present major histocompatibility complex allopeptides to primed T cells. Gastroenterology 1994; 107:1537-1542.
220. Bland PW, Warren LG. Antigen presentation by epithelial cells of the rat small intestine. II. Selective induction of suppressor T cells. Immunology 1986; 58:9-14.
221. Zimmerman MJ, Radford-Smith GR, Jewell DP. The effect of interleukin-10 and transforming growth factor beta-1 on HLA-DR expression in colonic epithelial cells. Mediators Inflamm 1998; 7:7-11.
222. Ito K, Takaishi H, Jin Y et al. Staphylococcal enterotoxin B stimulates expansion of autoreactive T cells that induce apoptosis in intestinal epithelial cells: regulation of autoreactive responses by IL-10. J Immunol 2000; 164:2994-3001.
223. Ebert EC. IL-10 enhances IL-2-induced proliferation and cytotoxicity by human intestinal lymphocytes. Clin Exp Immunol 2000; 119:426-432.
224. Niiro H, Otsuka T, Tanabe T et al. Inhibition by interleukin-10 of inducible cyclooxygenase expression in lipopolysaccharide-stimulated monocytes: Its underlying mechanism in comparison with interleukin-4. Blood 1995; 85:3736-3745.
225. Shattuck-Brandt RL, Varilek GW, Radhika A et al. Cyclooxygenase 2 expression is increased in the stroma of colon carcinomas from IL-10(-/-) mice [see comments]. Gastroenterology 2000; 118:337-345.
226. Pender SL, Breese EJ, Gunther U et al. Suppression of T cell-mediated injury in human gut by interleukin 10: role of matrix metalloproteinases. Gastroenterology 1998; 115:573-583.
227. Schreiber S, Heinig T, Thiele HG et al. Immunoregulatory role of interleukin 10 in patients with inflammatory bowel disease. Gastroenterology 1995; 108:1434-1444.

228. van Deventer SJ, Elson CO, Fedorak RN. Multiple doses of intravenous interleukin 10 in steroid-refractory Crohn's disease. Crohn's disease Study Group. Gastroenterology 1997; 113:383-389.
229. Steidler L, Hans W, Schotte L et al. Treatment of murine colitis by *Lactococcus lactis* secreting interleukin-10 [see comments]. Science 2000; 289:1352-1355.
230. Massague J, Wotton D. Transcriptional control by the TGF-beta/Smad signaling system. Embo J 2000; 19:1745-1754.
231. Miyazono K. TGF-beta signaling by Smad proteins. Cytokine Growth Factor Rev 2000; 11:15-22.
232. Verschueren K, Huylebroeck D. Remarkable versatility of Smad proteins in the nucleus of transforming growth factor-beta activated cells. Cytokine Growth Factor Rev 1999; 10:187-199.
233. Strober W, Kelsall B, Fuss I et al. Reciprocal IFN-gamma and TGF-beta responses regulate the occurrence of mucosal inflammation. Immunol Today 1997; 18:61-64.
234. Shull MM, Ormsby I, Kier AB et al. Targeted disruption of the mouse transforming growth factor-beta 1 gene results in multifocal inflammatory disease. Nature 1992; 359:693-699.
235. Kulkarni AB, Huh CG, Becker D et al. Transforming growth factor beta 1 null mutation in mice causes excessive inflammatory response and early death. Proc Natl Acad Sci USA 1993; 90:770-774.
236. Yang X, Letterio JJ, Lechleider RJ et al. Targeted disruption of SMAD3 results in impaired mucosal immunity and diminished T cell responsiveness to TGF-beta. Embo J 1999; 18:1280-1291.
237. Macpherson AJ, Gatto D, Sainsbury E et al. A primitive T cell-independent mechanism of intestinal mucosal IgA responses to commensal bacteria. Science 2000; 288:2222-2226.
238. Zan H, Cerutti A, Dramitinos P et al. CD40 engagement triggers switching to IgA1 and IgA2 in human B cells through induction of endogenous TGF-beta: evidence for TGF-beta but not IL-10-dependent direct S mu—>S alpha and sequential S mu—>S gamma, S gamma—>S alpha DNA recombination. J Immunol 1998; 161:5217-5225.
239. van Ginkel FW, Wahl SM, Kearney JF et al. Partial IgA-deficiency with increased Th2-type cytokines in TGF-beta 1 knockout mice. J Immunol 1999; 163:1951-1957.
240. McGee DW, Aicher WK, Eldridge JH et al. Transforming growth factor-beta enhances secretory component and major histocompatibility complex class I antigen expression on rat IEC-6 intestinal epithelial cells. Cytokine 1991; 3:543-550.
241. McGee DW, Beagley KW, Aicher WK et al. Transforming growth factor-beta enhances interleukin-6 secretion by intestinal epithelial cells. Immunology 1992; 77:7-12.
242. McGee DW, Beagley KW, Aicher WK et al. Transforming growth factor-beta and IL-1 beta act in synergy to enhance IL-6 secretion by the intestinal epithelial cell line, IEC-6. J Immunol 1993; 151:970-978.
243. Goodrich ME, McGee DW. Preferential enhancement of B cell IgA secretion by intestinal epithelial cell-derived cytokines and interleukin-2. Immunol Invest 1999; 28:67-75.
244. Sanfilippo L, Li CK, Seth R et al. Bacteroides fragilis enterotoxin induces the expression of IL-8 and transforming growth factor-beta (TGF-beta) by human colonic epithelial cells. Clin Exp Immunol 2000; 119:456-463.
245. Xian CJ, Xu X, Mardell CE et al. Site-specific changes in transforming growth factor-alpha and -beta1 expression in colonic mucosa of adolescents with inflammatory bowel disease. Scand J Gastroenterol 1999; 34:591-600.
246. Babyatsky MW, Rossiter G, Podolsky DK. Expression of transforming growth factors alpha and beta in colonic mucosa in inflammatory bowel disease. Gastroenterology 1996; 110:975-984.
247. Sturm A, Schulte C, Schatton R et al. Transforming growth factor-beta and hepatocyte growth factor plasma levels in patients with inflammatory bowel disease. Eur J Gastroenterol Hepatol 2000; 12:445-450.
248. Campbell AP, Smithson J, Lewis C et al. Altered expression of TGF alpha and TGF beta 1 in the mucosa of the functioning pelvic ileal pouch. J Pathol 1996; 180:407-414.
249. Sambuelli A, Diez RA, Sugai E et al. Serum transforming growth factor-beta1 levels increase in response to successful anti-inflammatory therapy in ulcerative colitis. Aliment Pharmacol Ther 2000; 14:1443-1449.
250. Lee HO, Miller SD, Hurst SD et al. Interferon gamma induction during oral tolerance reduces T-cell migration to sites of inflammation [see comments]. Gastroenterology 2000; 119:129-138.
251. Waldegger S, Klingel K, Barth P et al. h-sgk serine-threonine protein kinase gene as transcriptional target of transforming growth factor beta in human intestine. Gastroenterology 1999; 116:1081-1088.
252. Boirivant M, Fuss IJ, Chu A et al. Oxazolone colitis: A murine model of T helper cell type 2 colitis treatable with antibodies to interleukin 4. J Exp Med 1998; 188:1929-1939.
253. Neurath MF, Fuss I, Kelsall BL et al. Experimental granulomatous colitis in mice is abrogated by induction of TGF-beta-mediated oral tolerance. J Exp Med 1996; 183:2605-2616.
254. Powrie F, Leach MW. Genetic and spontaneous models of inflammatory bowel disease in rodents: Evidence for abnormalities in mucosal immune regulation. Ther Immunol 1995; 2:115-123.

255. Dammeier J, Brauchle M, Falk W et al. Connective tissue growth factor: A novel regulator of mucosal repair and fibrosis in inflammatory bowel disease? Int J Biochem Cell Biol 1998; 30:909-922.
256. van Tol EA, Holt L, Li FL et al. Bacterial cell wall polymers promote intestinal fibrosis by direct stimulation of myofibroblasts. Am J Physiol 1999; 277:G245-255.
257. Karttunnen R, Breese EJ, Walker-Smith JA et al. Decreased mucosal interleukin-4 (IL-4) production in gut inflammation. J Clin Pathol 1994; 47:1015-1018.
258. Kucharzik T, Lugering N, Weigelt H et al. Immunoregulatory properties of IL-13 in patients with inflammatory bowel disease; comparison with IL-4 and IL-10. Clin Exp Immunol 1996; 104:483-490.
259. Lugering N, Kucharzik T, Kraft M et al. Interleukin (IL)-13 and IL-4 are potent inhibitors of IL-8 secretion by human intestinal epithelial cells. Dig Dis Sci 1999; 44:649-655.
260. Iijima H, Takahashi I, Kishi D et al. Alteration of interleukin 4 production results in the inhibition of T helper type 2 cell-dominated inflammatory bowel disease in T cell receptor alpha chain-deficient mice. J Exp Med 1999; 190:607-615.
261. Colgan SP, Resnick MB, Parkos CA et al. IL-4 directly modulates function of a model human intestinal epithelium. J Immunol 1994; 153:2122-2129.
262. Hogaboam CM, Vallance BA, Kumar A et al. Therapeutic effects of interleukin-4 gene transfer in experimental inflammatory bowel disease. J Clin Invest 1997; 100:2766-2776.
263. Rogy MA, Beinhauer BG, Reinisch W et al. Transfer of interleukin-4 and interleukin-10 in patients with severe inflammatory bowel disease of the rectum. Hum Gene Ther 2000; 11:1731-1741.
264. Peterson RL, Wang L, Albert L et al. Molecular effects of recombinant human interleukin-11 in the HLA-B27 rat model of inflammatory bowel disease. Lab Invest 1998; 78:1503-1512.
265. Qiu BS, Pfeiffer CJ, Keith JC, Jr. Protection by recombinant human interleukin-11 against experimental TNB- induced colitis in rats. Dig Dis Sci 1996; 41:1625-1630.
266. Keith JC, Jr., Albert L, Sonis ST et al. IL-11, a pleiotropic cytokine: Exciting new effects of IL-11 on gastrointestinal mucosal biology. Stem cells 1994; 12:79-89; discussion 89-90.
267. Sands BE, Bank S, Sninsky CA et al. Preliminary evaluation of safety and activity of recombinant human interleukin 11 in patients with active Crohn's disease. Gastroenterology 1999; 117:58-64.
268. Bank S, Sninsky C, Robinson M et al. Safety and activity evaluation of rhIL-11 in subjects with active Crohn's disease. Gastroenterology 1997; 112:A927.

Index

B

C

D

E

F

G

H

I

J

K

L

M

N

O

P

R

S

T

U

V

X

Z

GPSR Compliance

The European Union's (EU) General Product Safety Regulation (GPSR) is a set of rules that requires consumer products to be safe and our obligations to ensure this.

If you have any concerns about our products, you can contact us on ProductSafety@springernature.com

In case Publisher is established outside the EU, the EU authorized representative is:

Springer Nature Customer Service Center GmbH
Europaplatz 3
69115 Heidelberg, Germany

Batch number: 10755805

Printed by Printforce, the Netherlands